Natural Gas

Natural Gas

A Basic Handbook

Second Edition

James G. Speight, Ph.D., D.Sc.
CD&W Inc., Laramie, Wyoming, USA

Gulf Professional Publishing
An imprint of Elsevier

Gulf Professional Publishing is an imprint of Elsevier
50 Hampshire Street, 5th Floor, Cambridge, MA 02139, United States
The Boulevard, Langford Lane, Kidlington, Oxford, OX5 1GB, United Kingdom

British Library Cataloguing-in-Publication Data
A catalogue record for this book is available from the British Library

Library of Congress Cataloging-in-Publication Data
A catalog record for this book is available from the Library of Congress

ISBN: 978-0-12-809570-6

For Information on all Gulf Professional Publishing publications
visit our website at https://www.elsevier.com/books-and-journals

Working together
to grow libraries in
developing countries

www.elsevier.com • www.bookaid.org

Publisher: Joe Hayton
Acquisition Editor: Katie Hammon
Editorial Project Manager: Peter Jardim/Joanna Collett
Production Project Manager: Sruthi Satheesh
Cover Designer: Matthew Limbert

Typeset by MPS Limited, Chennai, India

Contents

Preface to the second edition

Natural gas (also called marsh gas and swamp gas in older texts) is a gaseous fossil fuel that is found in oil fields and natural gas field. While it is commonly grouped in with other fossil fuels and sources of energy, there are many characteristics of natural gas that make it unique. The term natural gas is often extended to gases and liquids from the recently developed shale formations as well as gas produced from biological sources (biogas). However, for the purposes of this book, the petroliferous natural gas is placed under the category of conventional gas, while petroliferous gas from tight formations and the nonpetroliferous gases (such as biogas and landfill gas) are placed under the term nonconventional gas.

The last four decades of the 20th century have seen not only perturbations of energy supply systems but also changes in attitudes of governments and voters alike toward environmental issues. Thus environmental issues will be with us as long as we cut down trees, manufacture consumer goods, and burn fossil fuels for energy. And it is this latter issue that is the subject of this text.

The supplies of crude petroleum in North America continue to dwindle and there must be a definite movement to developing crude oil from tight formations and new sources of energy. In the general context of the present text, coal and natural gas are continuing sources of energy but renewable energy sources are needed. However, until renewable energy sources fulfill the goal of replacing the fossil energy sources, the environment will continue to be affected the resulting effects on the environment are major issues that need to be continually addressed.

The continued use of coal and natural gas as combustible fuel is a reality. And so is the generation of gaseous products that are not indigenous to the atmosphere, at least in any great quantities, and the effect of these products on the surrounding flora and fauna. On this issue, it is necessary to consider the gradual rise in temperature of the Earth's atmosphere that has become evident during the past four decades. This so-called *greenhouse effect* has many proponents and many opponents and is the subject of much debate. Whether or not the issue will be resolved in the next decade remains open to question and speculation.

It is a fact that the emissions of carbon dioxide into the atmosphere are known to come from the combustion of fossil fuels particularly the heavier fuels such as coal. It is believed that there is a strong need to move to the higher hydrogen/carbon fossil fuels to combat carbon dioxide production. Thus there is a need not only to promote the use of natural gas as a fuel for combustion, which produces less carbon dioxide per unit of fuel, but also to understand the chemistry and engineering of the combustion of fossil fuels and the means by which the gaseous pollutants are produced. Following on from this, it may then be possible to design suitable new methods by which their emission to the atmosphere is not only reduced but also completely mitigated. It is the purpose of this book to outline the current methods and known technologies that will aid in the development of processes by which this might be accomplished.

Gas processing, although generally understandable using chemical and/or physical principles, still requires an attempt to alleviate some of the confusion that arises from uncertainties in the terminology. This book corrects any such uncertainties.

The success of the first edition in European countries has prompted the publication of this edition. However, the text is almost completely rewritten and updated. Without assigning blame to any particular industry, there is a focus on the fossil fuel industries and many gas-generating industries that was not evident in the first edition. Descriptions are more detailed on the basis that the book will serve as an educational and problem-solving text.

As for the first edition, Part I of this book deals with the origins of process gases and also contains chapters dealing with recovery, properties and composition. Part II deals with the chemistry and engineering aspects of the methods, and principles involved, by which the gas streams produced during industrial operations might be cleaned from their noxious constituents.

Thus although gas processing employs different process types, there is always an overlap between the various concepts. Therefore, when necessary, cross-referencing is employed so that the reader will not miss any particular aspect of the processing operations.

The sections relating to testing procedures contain references to the relevant standard test method. In some cases, reference is given to older methods as well as to the current methods. Even though some of the older test methods are no longer in use, in so far as they have been replaced by newer methods, it is considered useful to refer to these methods as they have played an important role in the evolution of the newer methods. Indeed, some of the older methods are still preferred by many laboratories, hence their inclusion here.

Thus this book also presents some of the methods that are generally applied to the analysis of natural gas and other gaseous products. There are, of course, many analytical methods the can be applied to the analysis of natural gas and other fuel gases, but they vary with sample condition and composition. More specifically, this book cites the more common methods used to define chemical and physical properties of the sample. Moreover, any of the methods described herein might also be applied to the analysis of sample for environmental purposes.

The data derived from any one, or more, of the standard test methods cited in this text give an indication of the characteristics of the natural gas (or condensate) and products as well as options for gas processing as well as for the prediction of product properties. Other properties may also be required for more detailed evaluation of the natural gas and for comparison between gaseous feedstocks and product yields/properties even though they may not play any role in dictating which gas processing operations are necessary for production of the products.

However, proceeding from the raw evaluation data to full-scale production is not the preferred step. Further evaluation of gas processability is usually through the use of a pilot-scale operation followed by scale-up to a demonstration-size plant. It will then be possible to develop accurate and realistic relationships using the data obtained from the actual plant operations. After that, feedstock mapping can play an important role to assist in the various *tweaks* that are needed to maintain a healthy process to produce saleable products with the necessary properties.

Indeed, the use of physical properties for feedstock evaluation and product slate has continued in refineries and in process research laboratories to the present time and will continue for some time. It is, of course, a matter of choosing the relevant and meaningful properties to meet the nature of the task.

Finally, attempts have been made to render each chapter a stand-alone segment of the book. While every effort is made to ensure adequate cross-referencing, sufficient information is included in each chapter to give the reader the necessary background. Also, to be chemically correct, it is must be recognized that hydrocarbon molecules (hydrocarbon oils) contain carbon atoms and hydrogen atoms *only*.

In summary, this edition will also provide an easy-to-use reference source to compare the scientific and technological aspects of gas-processing operations and the means by which the environment might be protected.

Dr. James G. Speight,
Laramie, Wyoming
May 2018.

Part I

Origin and Properties

History and use

1.1 Introduction

Natural gas (also called marsh gas and swamp gas in older texts) is a gaseous fossil fuel that is found in oil fields and natural gas field. While it is commonly grouped in with other fossil fuels and sources of energy, there are many characteristics of natural gas that make it unique. The term natural gas is often extended to gases and liquids from the recently developed shale formations (Kundert and Mullen, 2009; Aguilera and Radetzki, 2014; Khosrokhavar et al., 2014; Speight, 2017b) as well as gas (biogas) produced from biological sources (John and Singh, 2011; Ramroop Singh, 2011; Singh and Sastry, 2011). However, for the purposes of this book, the petroliferous natural gas is placed under the category of conventional gas, while petroliferous gas from tight formations and the nonpetroliferous gases (such as biogas and landfill gas) are placed under the term *nonconventional gas* (sometime called *unconventional gas*) (Chapter 3: Unconventional gas).

Although the terminology and definitions involved with the natural gas technology are quite succinct, there may be those readers that find the terminology and definitions somewhat confusing. *Terminology* is the means by which various subjects are named so that reference can be made in conversations and in writings and so that the meaning is passed on. *Definitions* are the means by which scientists and engineers communicate the nature of a material to each other either through the spoken or through the written word. Thus, the terminology and definitions applied to natural gas (and, for that matter, to other gaseous products and fuels) are extremely important and have a profound influence on the manner by which the technical community and the public perceive that gaseous fuel. For the purposes of this book, natural gas and those products that are isolated from natural gas during recovery [such as natural gas liquids (NGLs), gas condensate, and natural gasoline] are a necessary part of this text. Thus:

Conventional gas
    Associated gas
    Nonassociated gas
    Gas condensate
Unconventional gas
    Gas hydrates
    Biogas
    Coalbed methane
    Coal gas
    Flue gas
    Gas in geopressurized zones
    Gas in tight formations

Natural Gas. DOI: https://doi.org/10.1016/B978-0-12-809570-6.00001-1

Landfill gas
Manufactured gas
Refinery gas
Shale gas
Synthesis gas
A more meaningful categorization of these gases would be as fuel gases with a third category that includes the gases produced in manufacturing processes. Thus:
Conventional natural gas
   Associated gas
   Nonassociated gas
   Gas condensate
Unconventional gas
   Gas hydrates
   Coalbed methane
   Gas in geopressurized zones
   Gas in tight formations
   Shale gas
Manufactured gas
   Biogas
   Coal gas
   Flue gas
   Landfill gas
   Refinery gas
   Synthesis gas
These categorizations are based on the source of the gas or the method of production of the gas which also has some relationship to the composition of the gas. Nevertheless, whatever, the source or origin, natural gas and other fuel gases are vital components of the energy supply of the world and form a necessary supply chain for energy production:
reservoir gas → produced gas − wellhead gas → transported gas → stored gas → sales gas

Specifically, the term *natural gas* is the generic term that is applied to the mixture of and gaseous hydrocarbon derivatives and low-boiling liquid hydrocarbon derivatives (typically up to and including hydrocarbon derivatives such as *n*-octane, $CH_3(CH_2)_6CH_3$, boiling point $125.1-126.1°C$, $257.1-258.9°F$) (Tables 1.1 and 1.2) that is commonly associated with petroliferous (petroleum-producing, petroleum-containing) geologic formations (Mokhatab et al., 2006; Speight, 2014a).

From a chemical standpoint, natural gas is a mixture of hydrocarbon compounds and nonhydrocarbon compounds with crude oil being much more complex than natural gas (Mokhatab et al., 2006; Speight, 2012, 2014a). The fuels that are derived from this natural product supply more than one quarter of the total world energy supply. The more efficient use of natural gas is of paramount importance and the technology involved in processing both feedstocks will supply the industrialized nations of the world for (at least) the next five decades until suitable alternative forms of energy (such as biogas and other nonhydrocarbon fuels) are readily available (Boyle, 1996; Ramage, 1997; Rasi et al., 2007, 2011; Speight, 2011a,b,c, 2008). Any gas sold, however, to an industrial or domestic consumer must meet designated specification that is designed according to the use of the gas.

Table 1.1 Constituents of natural gas

Constituent	Formula	% v/v
Methane	CH_4	>85
Ethane	C_2H_6	3−8
Propane	C_3H_8	1−5
n-Butane	C_4H_{10}	1−2
iso-Butane	C_4H_{10}	<0.3
n-Pentane	C_5H_{12}	1−5
iso-Pentane	C_5H_{12}	<0.4
Hexane, heptane, octane[a]	C_nH_{2n+2}	<2
Carbon dioxide	CO_2	1−2
Hydrogen sulfide	H_2S	1−2
Oxygen	O_2	<0.1
Nitrogen	N_2	1−5
Helium	He	<0.5

[a]Hexane (C_6H_{14}) and higher molecular weight hydrocarbon derivatives up to octane as well as benzene (C_6H_6) and toluene ($C_6H_5CH_3$).

Table 1.2 Differentiation of the constituents of natural gas

Hydrocarbon constituents	
Dry gas or natural gas	Methane (CH_4)
	Ethane (C_2H_6)
Liquefied petroleum gas	Propane (C_3H_8)
	n-Butane (C_4H_{10})
	iso-Butane (C_4H_{10})
Natural gas liquids	Pentane isomers (C_5H_{15})
	Hexane isomers (C_6H_{14})
	Heptane isomers (C_7H_{16})
	Octane isomers (C_8H_{18})
	Condensate ($\geq C_5H_{12}$)
	Natural gasoline ($\geq C_5H_{12}$)
	Naphtha ($\geq C_5H_{12}$)
Nonhydrocarbon constituents	
	Carbon dioxide (CO_2)
	Hydrogen sulfide (H_2S)
	Water (H_2O)
	Nitrogen (N_2)
	Carbonyl sulfide (COS)

Typically, in field operations, the composition of natural gas (which affects the specific gravity), especially of associated gas, can vary significantly as the product flowing out of the well can change with variability of the production conditions as

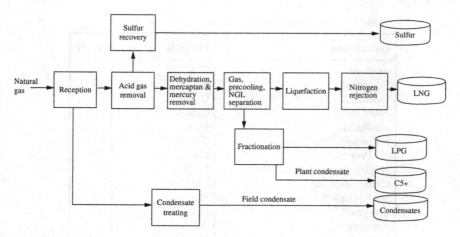

Figure 1.1 Typical gas processing sequence.

well as the change of pressure as gas is removed from the reservoir (Burruss and Ryder, 2003, 2014). Constituents of the gas that were in the liquid phase under the pressure of the reservoir can revert to the gas phase as the reservoir pressure is reduced by gas removal.

As a result, it should not be a surprise that at each stage of natural gas production, wellhead treating, transportation, and processing, analysis of the gas to determine the composition and properties of the gas by standard test methods is an essential part of the chemistry and technology of natural gas. Use of analytical methods offers (Speight, 2018) vital information about the behavior of natural gas during recovery, wellhead processing, transportation, gas processing, and use (Fig. 1.1). The data produced from the test methods are the criteria by means of which the suitability of the gas for use and the potential for interference with the environment.

1.2 History

Natural gas is a naturally occurring gaseous fossil fuel that is found in gas-bearing formations, oil-bearing formations—coalbed methane is often referred to (incorrectly) as natural gas or as coal gas due to lack of standardization of the terminology (Levine, 1993; Speight, 2013b, 2014a). For clarification, natural gas is not the same as town gas, although the history of natural gas cleaning prior to sales has its beginnings in town gas cleaning (Speight, 2013a).

Town gas is a generic term referring to manufactured gas produced for sale to consumers and municipalities. The terms coal gas, manufactured gas, producer gas, and syngas [synthetic natural gas (SNG)] are also used for gas produced from coal. Depending on the processes used for its creation, town gas is a mixture of hydrogen, carbon monoxide, methane, and volatile hydrocarbons with small amounts of carbon dioxide and nitrogen as impurities.

Thus, town gas is manufactured from coal and the terms coal gas, manufactured gas, producer gas, and synthesis gas (syngas), and SNG are also in regular use for gases produced from coal (Chapter 3: Unconventional gas) (Speight, 2013b). Also, by way of definition and clarification, town gas is a flammable gaseous fuel made by the destructive distillation of coal. It contains a variety of calorific gases including hydrogen, carbon monoxide, methane, and other volatile hydrocarbon derivatives, together with small quantities of noncalorific gases such as carbon dioxide and nitrogen. Town gas, although not currently used to any great extent in the United States, is still generated and used in some countries and is used in a similar way to natural gas in the United States. This is a historical technology and is not usually economically and environmentally competitive with modern sources of natural gas.

Most town gas-generating plants located in the eastern United States in the late 19th century and early 20th century were ovens that heated bituminous coal in airtight chambers to produce coke through the carbonization process. The gas driven off from the coal was collected and distributed through networks of pipes to residences and other buildings where it was supplied to industrial and domestic users—natural gas did not come into widespread use until the last half of the 20th century. The coal tar collected in the bottoms of the gashouse ovens was often used for roofing and other waterproofing purposes, and when mixed with sand and gravel (aggregate) was used for paving streets (road asphalt). The coal tar asphalt has been replaced by asphalt produced from crude oil (Speight, 2014a, 2015, 2018; ASTM, 2017). Thus, prior to the development of resources, virtually all fuel and lighting gas was manufactured from coal and the history of natural gas has its roots in town gas production and use (Speight, 2013b). Thus, with the onset of industrial expansion after World War II, natural gas has become one of the most important raw materials consumed by modern industries to provide raw materials for the ubiquitous plastics and other products as well as feedstocks for the energy and transportation industries.

1.2.1 Timeline

Natural gas has been known for many centuries, but initial use for the gas was more for religious purposes rather than as a fuel. For example, gas wells were an important aspect of religious life in ancient Persia because of the importance of fire in their religion. In classical times these wells were often flared and must have been, to say the least, awe inspiring (Forbes, 1964).

These types of gas leaks became prominent in the religions of India, Greece, and Persia where the inhabitants of the region were unable to explain the origin of the fires and regarded the origin of the flames as divine, or supernatural, or both. As a result, the energy value of natural gas was not recognized until approximately 900 BC in China and the Chinese drilled the first known natural gas well in 211 BC. Crude pipelines (probably state-of-the-art pipelines at the time) were constructed from bamboo stems to transport the gas, where it was used to boil sea water, removing the salt as a residue product, after which the water was condensed and, therefore, drinkable (Abbott, 2016).

The uses of natural gas did not necessarily parallel its discovery. In fact, the discovery of natural gas dates from ancient times in the Middle East. During recorded historical time, there was little or no understanding of what natural gas was, it posed somewhat of a mystery to man. Sometimes, such things as lightning strikes would ignite natural gas that was escaping from under the Earth's crust. This would create a fire coming from the Earth, burning the natural gas as it seeped out from underground. These fires puzzled most early civilizations and were the root of much myth and superstition. One of the most famous of these types of flames was found in ancient Greece, on Mount Parnassus approximately 1000 BC. A goat herdsman came across what looked like a *burning spring*, a flame rising from a fissure in the rock. The Greeks, believing it to be of divine origin, built a temple on the flame. This temple housed a priestess who was known as the Oracle of Delphi, giving out prophecies she claimed were inspired by the flame.

In Europe, natural gas was unknown until it was discovered in Great Britain in 1659 and Britain was the first country to commercialize the use of natural gas. In 1785, natural gas produced from coal was used to light houses, as well as streetlights. Manufactured natural gas of this type (as opposed to naturally occurring gas) was first brought to the United States in 1816, when it was used to light the streets of Baltimore, Maryland. This manufactured gas was much less efficient, and less environment-friendly, than modern natural gas that comes from underground.

Naturally occurring natural gas was discovered and identified in America as early as 1626, when French explorers discovered natives igniting gases that were seeping into and around Lake Erie (Table 1.3).

In 1821 in Fredonia, United States, residents observed gas bubbles rising to the surface from a creek. William Hart, considered as America's *father of natural gas*, dug there the first natural gas well in North America (Speight, 2014a). In 1859, Colonel Edwin Drake, a former railroad conductor (the origin of the title "Colonel" is unknown but seemed to impress the townspeople), dug the first well. Drake found

Table 1.3 Abbreviated timeline for the use of natural gas

1620	French missionaries recorded that Indians ignited gases near Lake Erie
1785	Natural gas is introduced for home lighting and street lighting
1803	Gas lighting system patented in London by Frederick Winsor
1812	First gas company founded in London
1815	Metering for households, invented in 1815 by Samuel Clegg
1816	First US gas company (using manufactured gas) founded in Baltimore
1817	First natural gas from the wellhead used in Fredonia, NY for house lighting
1840	Fifty or more US cities were burning public utility gas
1850	Thomas Edison postulated replacing gas lighting by electric lighting
1859	Carl Auer von Welsbach in Germany developed a practical gas mantle
1885	Depleted reservoirs are used for the first time to store gas

crude oil and natural gas at 69 ft below the surface of the Earth. More recently, natural gas was discovered because of prospecting for crude oil. However, the gas was often an unwelcome by-product because, as any gas containing in the reservoirs were tapped during the drilling process, the drillers were forced to discontinue the drilling operations that allow the gas to vent freely into the air. Currently, and particularly after the crude oil shortages of the 1970s, natural gas has become an important source of energy in the world.

Throughout the 19th century, natural gas was used almost exclusively as source of light and its use remained localized because of lack of transport structures, making difficult to transport large quantities of natural gas through long distances. There was an important change in 1890 with the invention of leak proof pipeline coupling but transportation of natural gas to long distance customers did not become practical until the 1920s as a result of technological advances in pipelines. Moreover, it was only after World War II that the use of natural gas grew rapidly because of the development of pipeline networks and storage systems.

1.2.2 Formation

Although covered in more detail elsewhere (Chapter 2: Origin and production), it is pertinent at this point, to give a brief coverage of the formation of natural gas to place it in context.

Just as crude oil is a product of decomposed organic matter (often referred to as organic debris or detritus), natural gas is also a product of the decomposition of organic matter. The organic matter is the remains of ancient flora and fauna that was deposited over the past 550 million years. This organic debris is mixed with mud, silt, and sand on the sea floor, gradually becoming buried over time. Sealed off in an oxygen-free (anaerobic) environment and exposed to increasing amounts of pressure and an unknown amount of heat, the organic matter underwent a decomposition process in which hydrocarbons (and nonhydrocarbons) were the products that converted it into hydrocarbons. The lowest boiling of these hydrocarbons exist in the gaseous state under normal conditions and become known collectively as natural gas. In the purest hydrocarbon form, natural gas is a colorless, odorless gas composed primarily of methane. These hydrocarbons are highly flammable compounds.

Thus, natural gas, like crude oil and often occurring in conjunction with crude oil, has been generated over geological time from deep-lying source rock, sometimes called the *kitchen*, which contains organic debris. Nevertheless, the actual chemical paths involved in the maturation of the organic debris are largely unknown and, therefore, subject to speculation.

Thus, it has been speculated, but not absolutely accepted (Speight, 2014a), that the deeper and hotter the source rock, the more likelihood of gas being produced. However, there is considerable discussion about the heat to which the organic precursors have been subjected. Cracking temperatures ($\geq 300°C$, $\geq 572°F$) are not by any means certain as having played a role in natural gas formation. Maturation of the organic debris through temperature effects occurred over geological time (millennia) and shortening the time to laboratory time and increasing the temperature to above and beyond the cracking temperature (at which the chemistry changes) does not offer conclusive proof of natural gas (and crude oil) formation involving high temperatures (Speight, 2014a). Once the natural gas has formed, its fate depends on two critical characteristics of the surrounding rock: (1) porosity and (2) permeability.

The term *porosity* refers to the amount of empty space contained within the grains of a rock. Highly porous rocks, such as a sandstone formation, typically has a porosity on the order of 5%—25% v/v (percent volume of the rock), which gives the formation a substantial amount of space for the storage of fluids—the term *reservoir fluids* includes natural gas, crude oil, and water. On the other hand, the term *permeability* is a measure of the degree to which the pore spaces in a rock are interconnected and, therefore amenable to fluid flow. A highly permeable rock will permit gas and liquids to flow easily through the rock, while a low-permeability rock will not allow fluids to pass through. This latter term is characteristic of shale formations and tight formations (Chapter 3: Unconventional gas).

After natural gas forms, it will tend to rise toward the surface through pore spaces in the rock because of its low density compared to the surrounding rock. Thus, at some point during or after the maturation process, the gas and crude oil migrated from the source rock either upward or sideways or in both directions (subject to the structure of the accompanying and overlying geological formations) through the underground sediments through fissures and faults until the gas enters a geological formation (*reservoir*) that retains or *traps* the gas through the presence of impermeable basement rock and cap rock. Has this not occurred, there is the distinct likelihood that most of the natural gas would percolate through the surface formations and escape into the atmosphere.

It is rare that the source rock and the reservoir were one and the same and the reservoir may be many miles from the source rock. Thus, a natural gas field may have a series of layers of crude oil/gas and gas reservoirs in the subsurface. In some instances, the natural gas and crude oil parted company leading to the occurrence of reservoirs containing only gas (nonassociated gas).

The techniques to discover gas are essentially those used to discover crude oil (Speight, 2014a). When using seismic techniques, gas slows down the velocity of the seismic waves to produce a characteristic and stronger reflection. Over time, as more knowledge of a hydrocarbon province is obtained, better recognition of the characteristics and amplitude of the seismic reflection from gas can lead to a greater chance of success.

Thus, potential natural gas reservoirs can be located with seismic testing methods similar to those used for petroleum exploration (Chapter 2: Origin and production) (Speight, 2014a). In such tests, gas prospectors use seismic trucks or more advanced three-dimensional tools that involve setting off a series of small charges near the surface of the Earth to generate seismic waves thousands of feet below ground in underlying rock formations. By measuring the travel times of these waves through the Earth at acoustic receivers (*geophones*), geophysicists can construct a pictorial representation of the subsurface structure and identify potential gas deposits. However, to verify whether the rock formation actually contains economically recoverable quantities of natural gas or other hydrocarbons, an exploratory well must be drilled. Once the viability of a site is determined, vertical wells are drilled to penetrate the overlying impermeable cap rock and reach the reservoir. Natural buoyancy and reservoir pressure then brings the gas to the surface, where it can be processed and sent to the consumers.

1.3 Conventional gas

Natural gas resources, like crude oil resources, are typically divided into two categories: (1) conventional gas and (2) unconventional gas (Mokhatab et al., 2006; Islam, 2014; Speight, 2014a; AAPG, 2015). For the purposes of this text, the term unconventional gas resources also include coalbed methane and natural gas from shale formations and from tight formations as well as biogas and landfill gas (Brosseau, 1994; Briggs, 1988; Rice, 1993; John and Singh, 2011; Ramroop Singh, 2011; Singh and Sastry, 2011; Speight, 2011a, 2017b). Conventional gas is typically found in reservoirs with a permeability greater than 1 milliDarcy (>1 mD) and can be extracted by means of traditional recovery methods. In contrast, unconventional gas is found in reservoirs with relatively low permeability (<1 mD) and hence cannot be extracted by conventional methods (Speight, 2016a,b).

1.3.1 Associated gas

Associated or *dissolved* natural gas occurs either as free gas in a petroleum reservoir or as gas in solution in the petroleum. Gas that occurs as a solution in the petroleum is *dissolved* gas, whereas the gas that exists in contact with the petroleum (*gas cap*) is *associated* gas (Fig. 1.2).

Crude oil cannot be produced without producing some associated gas, which consists of low-boiling hydrocarbon constituents that are emitted from solution in

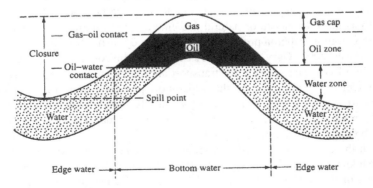

Figure 1.2 An anticlinal reservoir containing oil and associated gas.

the crude oil as the pressure is reduced on the way to, and on, the surface. Designs for well completion and reservoir management protocols are used to minimize the production of associated gas to retain the maximum energy in the reservoir and thus increase ultimate recovery of the crude oil (Parkash, 2003; Gary et al., 2007; Hsu and Robinson, 2017; Speight, 2014a, 2017a).

Crude oil in the reservoir with minimal or no dissolved associated gas (*dead crude oil* or *dead oil*) is rare and is often difficult to produce as there is little reservoir energy to drive the oil into the production well and to the surface. Thus, *associated* or *dissolved* natural gas occurs either as free gas or as gas in solution in the petroleum. Gas that occurs as a solution in the petroleum is *dissolved* gas whereas the gas that exists in contact with the petroleum is *associated* gas—the *gas cap* is an example of associated gas (Parkash, 2003; Mokhatab et al., 2006; Gary et al., 2007; Hsu and Robinson, 2017; Speight, 2014a, 2017a).

After the production fluids are brought to the surface, the gas is treated to separate out the higher molecular weight NGLs which are treated in a liquid petroleum gas (LPG) processing plant (refining plant) to provide propane and butane, either separately or as a mixture of the two. By definition, NGLs include ethane, propane, butanes, and pentanes and higher molecular weight hydrocarbon derivatives (C_{6+}). While NGLs are gaseous at underground pressure, these constituents condense at atmospheric pressure and turn into liquids. The composition of natural gas can vary by geographic region, the geological age of the deposit, the depth of the gas, and many other factors. Natural gas that contains a lot of NGLs and condensates is referred to as wet gas, while gas that is primarily methane, with little to no liquids in it when extracted, is referred to as dry gas.

These liquids are hydrocarbon derivatives that are removed (condensed) as a liquid from a hydrocarbon stream (natural gas) that is typically in a vapor phase (i.e., natural gas). They are kept in a liquid state for storage, shipping, and consumption. There has been a movement to classify NGLs are components of natural gas that are separated from the gas state in the form of liquids on the basis of vapor pressure, for example: (1) low vapor pressure—condensate and (2) high vapor pressure—liquefied natural gas and/or LPG. However, the boundaries drawn in this manner are arbitrary and caution is advised when using such a classification scheme.

Mixtures of these higher molecular weight hydrocarbon derivatives are often referred to as gas condensate or natural gasoline and the mixture has the characteristics of low-boiling naphtha produced in a refinery by distillation and cracking processes (Parkash, 2003; Gary et al., 2007; Hsu and Robinson, 2017; Speight, 2014a, 2017a). The LPG stored is ready for transport and the nonvolatile residue (i.e., nonvolatile under the conditions of the separation process), after the propane and butane are removed, is gas condensate (or, simply, condensate), which is mixed with the crude oil or exported as a separate product (low-boiling naphtha) (Mokhatab et al., 2006; Speight, 2014a).

Thus, in the case of associated gas, crude oil may be assisted up the wellbore by gas lift (Mokhatab et al., 2006; Speight, 2014a) in which the gas is compressed into the annulus of the well and then injected by means of a gas lift valve near the bottom of the well into the crude oil column in the tubing. At the top of the well the crude oil and gas mixture passes into a separation plant (consisting of high-pressure and low-pressure separators) in which the gas pressure is reduced considerably in two stages. The crude oil and water exits the bottom of the lower pressure separator, from where it is pumped to tanks for separation of the crude oil and water. The gas produced in the separators is recompressed and the gas that comes out of solution with the produced crude oil (surplus gas) is then treated to separate out the NGLs that are treated in a gas plant to provide propane and butane or a mixture of the two (LPG). The higher boiling residue, after the propane and butane are removed, is condensate, which is mixed with the crude oil or exported as a separate product. At each stage of this process (often referred to under the collective term *wellhead processing*), the composition of the gaseous and liquid products should be monitored to determine separator efficiency as well as for safety reasons (Colborn et al., 2011).

The gas itself is then *dry* and, after compression, is suitable to be injected into the natural gas system where it substitutes for natural gas from the nonassociated gas reservoir. Pretreated associated gas from other fields can also enter the system at this stage. Another use for the gas is as fuel for the gas turbines on site. This gas is treated in a fuel gas plant to ensure it is clean and at the correct pressure. The startup fuel gas supply will be from the main gas system, but facilities exist to collect and treat low-pressure gas from the various other plants as a more economical fuel source.

Other components such as carbon dioxide (CO_2), hydrogen sulfide (H_2S), mercaptans (thiols; RSH), as well as trace amounts of other constituents may also be present. Thus, there is no single composition of components which might be termed *typical natural gas* because of the variation in composition of the gas from different reservoirs, even from different wells from the same reservoir. Methane and ethane constitute the bulk of the combustible components; carbon dioxide (CO_2) and nitrogen (N_2) are the major noncombustible (inert) components.

1.3.2 Nonassociated gas

In addition to the natural gas found in petroleum reservoirs, there are also those reservoirs in which natural gas is the sole occupant and is referred to as *nonassociated gas*. As with associated gas, the principal constituent of nonassociated gas is methane—higher molecular weight hydrocarbon derivatives may also be present but in lower quantities than found in associated gas. Carbon dioxide is also a common constituent of nonassociated natural gas and trace amounts of rare gases, such as helium, may also occur, and certain natural gas reservoirs are a source of these rare gases.

Thus, *nonassociated ga*s (sometimes called *gas well gas*) is produced from geological formations that typically do not contain much, if any, crude oil, or higher boiling hydrocarbon derivatives (*gas liquids*) than methane. The nonassociated gas recovery system is somewhat simpler than the associated gas recovery system. The gas flows up the well under its own energy, through the wellhead control valves and along the flow line to the treatment plant.

The nonassociated gas recovery system is somewhat simpler than the associated gas recovery system. The gas flows up the well under its own energy, through the wellhead control valves and along the flow line to the treatment plant. Treatment requires the temperature of the gas to be reduced to a point dependent upon the pressure in the pipeline so that all liquids that would exist at pipeline temperature and pressure condense and are removed.

Processing of nonassociated gas is somewhat less complicated than processing associated gas. Typically, nonassociated gas flows up the production well under the reservoir energy and then through the wellhead control valves and along the flow line to the wellhead processing plant. At this stage, the first processing option is to reduce the temperature of the gas to a point dependent upon the pressure in the pipeline so that the higher molecular weight constituents which would exist as liquids at the temperature and pressure of the pipeline condense to a liquid phase and are removed in a separator. The temperature is reduced by expanding the gas through a Joule–Thomson valve, although other methods of removal do also exist (Mokhatab et al., 2006; Speight, 2014a). Briefly, the Joule–Thomson effect (also known as the Joule–Kelvin effect, the Kelvin–Joule effect, or the Joule–Thomson expansion) relates to the temperature change of a gas or liquid when it is forced through a valve while kept insulated so that no heat is exchanged with the environment.

Water in the gas stream must also be removed to mitigate the potential for the formation of gas hydrates (Gornitz and Fung, 1994; Collett, 2002; Buffett and Archer, 2004; Collett et al., 2009; Demirbaş, 2010a,b,c; Boswell and Collett, 2011; Chong et al., 2016) which would block flow lines and have the potential for explosive dissociation. One method for water removal from the gas stream, involves the injection of ethylene glycol ($HOCH_2CH_2OH$, also referred to as *glycol*) which combines with the water and is later recovered in a glycol plant (Mokhatab et al., 2006; Speight, 2014a). The treated gas then passes from the top of the treatment vessel and into the pipeline. The water is treated in a glycol plant to recover the glycol and the fraction of the natural gas stream that has been isolated as NGLs is sent, as additional feedstock, to the LPG plant. Alternatively, the lower boiling constituents of the NGLs may be used as feedstock for the production of petrochemicals (Parkash, 2003; Gary et al., 2007; Hsu and Robinson, 2017; Speight, 2014a, 2017a).

Finally, one other aspect of gas processing that requires attention (Chapter 3: Unconventional gas) and is worthy of mention here, is the removal of sulfur from natural gas. The potential of sulfur-containing constituents, such as hydrogen sulfide (H_2S) and mercaptans (RSH) to corrode shipping equipment (such as pipelines) is high—especially in the presence of water (Speight, 2014b). Once the hydrogen sulfide has been removed by a suitable wellhead treatment process—it is environmentally undesirable to flare the hydrogen sulfide, so where there are significant quantities in the gas stream, it is converted into elemental sulfur and used for the manufacture of sulfuric acid and other products (Chapter 3: Unconventional gas). The sulfur can be transported over long distances by being pumped as a liquid at a temperature on the order of 120°C (248°F) through an insulated pipeline, which is maintained at this temperature by a counter flow of hot pressurized water.

Finally, it would be remissed if other types of gases were not included here since these gases are becoming a blend stock for natural gas in many gas processing operations.

1.3.3   Gas condensate

Gas condensate (sometimes referred to as *condensate*) is a mixture of low-boiling hydrocarbon liquids obtained by condensation of the vapors of these hydrocarbon constituents either in the well or as the gas stream emits from the well. Gas condensate is predominately pentane (C_5H_{12}) with varying amounts of higher-boiling hydrocarbon derivatives (up to C_8H_{18}) but relatively little methane or ethane; propane (C_3H_8), butane (C_4H_{10}) may be present in condensate by dissolution in the liquids. Depending upon the source of the condensate, benzene (C_6H_6), toluene ($C_6H_5CH_3$), xylene isomers ($CH_3C_6H_4CH_3$), and ethyl benzene ($C_6H_5C_2H_5$) may also be present (Mokhatab et al., 2006; Speight, 2014a).

The terms *condensate* and *distillate* are often used interchangeably to describe the liquid produced in tanks, but each term stands for a different material. Along with large volumes of gas, some wells produce a water-white or light straw-colored liquid that resembles low-boiling naphtha (Mokhatab et al., 2006; Speight, 2014a). The liquid has been called *distillate* because it resembles the products obtained from crude oil in refineries by distilling the volatile components from crude oil.

Lease condensate, so-called because it is produced at the lease level from oil or gas wells, is the most common type of gas condensate and is typically a clear or translucent liquid. The API gravity of lease condensate ranges between 45 and 75°API but, on the other hand, lease condensate with a lower API gravity can be black or near black color and, like crude oil, has higher concentrations of higher molecular weight constituents. This condensate is generally recovered at atmospheric temperatures and pressures from wellhead gas production and can be produced along with large volumes of natural gas and lease condensates with higher API gravity contains more NGLs, which include ethane, propane, and butane, but not many higher molecular weight hydrocarbon derivatives.

1.3.4   Other definitions

In addition to the definitions presented earlier, there are several other definitions that have been applied to natural gas from conventional formations that can also be applied to gas from any source.

Rich gas has a high heating value and a high hydrocarbon dew point. However, the terms *rich gas* and *lean gas*, as used in the gas processing industry, are not precise indicators of gas quality but only indicate the relative amount of NGLs in the gas stream. Thus, *lean* gas is gas in which methane is the predominant major constituent with other hydrocarbon constituents in the low minority, while *wet* gas contains considerable amounts of the higher molecular weight hydrocarbon derivatives than dry gas (Table 1.4). When referring to NGLs in the natural gas stream, the

Table 1.4 Composition of dry gas, wet gas, and gas condensate

Component	Dry gas	Wet gas	Condensate
Carbon dioxide, CO_2	0.10	1.41	2.37
Nitrogen, N_2	2.07	0.25	0.31
Methane, CH_4	86.12	92.46	73.19
Ethane, C_2H_6	5.91	3.18	7.80
Propane, C_3H_8	3.58	1.01	3.55
n-Butane, $n\text{-}C_4H_{10}$		0.24	1.45
iso-Butane, $i\text{-}C_4H_{10}$	1.72	0.28	0.71
n-Pentane, $n\text{-}C_5H_{12}$	–	0.08	0.68
iso-Pentane, $i\text{-}C_5H_{12}$	0.50	0.13	0.64
Hexane isomers, C_6H_{14}	–	0.14	1.09
Heptane isomers-plus[a], $\geq C_7H_{16}$	–	0.82	8.21

[a]Indicates higher molecular weight hydrocarbons.

term *gallons per thousand cubic feet* of gas is used as a measure of hydrocarbon richness.

Sour gas contains hydrogen sulfide and the equally odorous mercaptans, whereas *sweet* gas contains very little, if any, hydrogen sulfide or mercaptan. *Residue gas* is natural gas from which the higher molecular weight hydrocarbon derivatives have been extracted and *casinghead gas* is derived from crude oil but is separated at the separation facility at the wellhead.

The term residue (as in *residue gas*) is used in relation to gas as a direct opposite as it is applied to crude oil in a refinery. In the refinery, the residue is the distillation residue of crude oil from which the lower molecular weight constituents have been removed. In natural gas technology, residue gas is natural gas from which the higher molecular weight constituents have been removed during gas processing operations (Chapter 3: Unconventional gas and Chapter 7: Process classification) to leave methane (the lower boiling constituent) as residue gas.

Other terms applied to natural gas typically apply to the method by which the gas occurs in the reservoir. By way of explanation, natural gas is generated by any combination of (1) primary thermogenic degradation of organic matter, (2) secondary thermogenic decomposition of petroleum, and (3) biogenic degradation of organic matter. Gas generated by thermogenic and biogenic pathways may both exist in the same shale reservoir.

After production, the gas is stored in the reservoir formation in three different ways: (1) by adsorption, which refers to *adsorbed gas* that is physically attached (adsorption) or chemically attached (chemisorption) to organic matter or to clay minerals, (2) nonadsorbed gas, which refers to *free gas* (also referred to as *nonassociated gas*) that occurs within the pore spaces in the reservoir rock or in spaces created by the rock cracking (fractures or microfractures), and (3) by solution, also referred to as *associated gas*, which is gas that exists in solution in liquids such as petroleum, heavy oil, and (in the current context) in the gas condensate that occurs in some tight reservoirs with the gas (Speight, 2014a).

The amount of adsorbed gas component (which is, typically, methane) usually increases with an increase in organic matter or surface area of organic matter and/or clay. On the beneficial side, a higher free-gas (nonassociated) content in unconventional tight reservoirs generally results in higher initial rates of production because the free gas resides in fractures and pores and, when production is commenced moves easier through the fractures (induced channels) relative to any adsorbed gas. However, the high, initial flow rate will decline rapidly to a low steady rate as the nonassociated gas is produced leaving the adsorbed gas to move to the well as it is slowly released from the shale.

1.4 Use

Natural gas is a versatile, clean-burning, and efficient fuel that is used in a wide variety of applications as well as for the production of a variety of chemicals, especially when the natural gas is used as the starting point for the production of synthesis gas (a mixture of hydrogen and carbon monoxide) and thence to the various chemicals (Table 1.5).

After the discovery by the Chinese more than 2000 years ago that the energy in natural gas could be harnessed and used as a heat source, the use of natural gas has grown (Mokhatab et al., 2006; Speight, 2014a). In the late 19th century and in the early 20th century, natural gas played a subsidiary role to coal gas insofar as coal gas was used for street lighting and for building lighting and provided what was known as gaslight (Mokhatab et al., 2006; Speight, 2013b). However, as the 20th century progressed and moved into the 21st century, the discovery of large reserves

Table 1.5 Examples of routes to chemicals from natural gas via synthesis gas

Starting material	Intermediate	Product
Methane	Synthesis gas	Oxo synthesis
		Alcohols
		Aldehydes
		Fischer–Tropsch synthesis
		Naphtha
		Diesel
		Kerosene
		Lubricants
		Waxes
		Carbonylation
		Formic acid
		Methanol
		Acetic acid
		Dimethyl ether
		Formaldehyde

of natural gas in various countries as well as improved distribution of gas has made possible a wide variety of uses in homes, businesses, factories, and power plants and natural gas is becoming a global energy source (Nersesian, 2010; Hafner and Tagliapietra, 2013).

However, the uses of natural gas did not necessarily parallel its discovery and during recorded historical time, there was little or no understanding of what natural gas was, it posed somewhat of a mystery to man. Prior to the development of natural gas supplies and transmission in the United States during 1940s and 1950s, virtually all fuel and lighting gas was manufactured, and the by-product coal tar was an important feedstock for the chemical industry. The development of manufactured gas paralleled that of the industrial revolution and urbanization.

However, in the current contact, it is perhaps at least as awe-inspiring if not predictable considering the current attention to the environment—considering the history of the use of other fossil fuels such as coal and crude oil during the 20th century—that the use of natural gas is superseding the use of crude oil and coal in many countries. During that time, natural gas was generally flared as a product of limited use until the depletion of crude oil reserves in the late 20th century caused a back-and-forth concern about the future lack of energy-producing fuels (Speight, 2011a, 2014a; Speight and Islam, 2016).

Once the transportation of natural gas was possible over considerable distances, the increased use of natural gas led to innovations from the discovery of new uses for natural gas which included the use of natural gas by industrial consumers. In fact, the fastest growing use of natural gas is for the generation of electric power and, to a large extent, has been a replacement fuel many formerly coal-fired power plants and oil-fired power plants. Natural gas power plants usually generate electricity in gas turbines (which are derived from jet engines), directly using the hot exhaust gases from the combustion process.

As the use of natural gas has increased and diversified, the need for knowledge of the composition of the gas has also increased (Mokhatab et al., 2006; Speight, 2018). Natural gas has many applications: for domestic use, for industrial use, and for transportation. In addition, natural gas is also a raw material for many common products such as paints, fertilizer, plastics, antifreeze, dyes, photographic film, medicines, and explosives. Along with these newer uses, there has been an increased need not only for the compositional analysis of natural gas but also for analytical data that provide other information about the behavior of natural gas.

Natural gas-fired power plants are currently among the cheapest power plants to construct which is a reversal of previous trends where operating costs were generally higher than those of coal-fired power plants because of the relatively high cost of natural gas. In addition, natural gas-fired plants have greater operational flexibility than coal-fired power plants because they can be fired up and turned down rapidly. Because of this, many natural gas plants in the United States were originally used to provide additional capacity (peak capacity) at times when electricity demand was especially high, such as the summer months when air conditioning is widely used. During much of the year, these natural gas peak plants were idle, while coal-fired power plants typically provided base-load power. However, since

2008, natural gas prices in the United States have fallen significantly, and natural gas is now increasingly used as base-load power as well as intermediate-load power source in many cities. Natural gas can also be used to produce both heat and electricity simultaneously [cogeneration or combined heat and power (CHP)]. Cogeneration systems are highly efficient, able to put 75%–80% of the energy in gas to use. Trigeneration systems, which provide electricity, heating, and cooling, can reach even higher efficiencies using natural gas.

Natural gas also has a broad range of other uses in industry, not only as a source of both heat and power and as source of valuable hydrogen that is necessary for crude oil refining as well as for producing plastics and chemicals. Most hydrogen gas (H_2) production, for example, comes from reacting high temperature water vapor (steam) with methane—steam-methane reforming reaction followed by the water gas shift reaction:

$$CH_4 + H_2O \rightarrow CO + 3H_2 \text{ (steam-methane reforming reaction)}$$

$$CO + H_2O \rightarrow CO_2 + H_2 \text{ (Water-gas shift reaction)}$$

Natural gas-fired plants have greater operational flexibility than coal plants because they can be fired up and turned down rapidly. Because of this, many natural gas plants were originally used to provide peaking capacity at times when electricity demand was especially high, such as the summer months when air conditioning is widely used. During much of the year, the natural gas peak-plants were in low use or idle, while coal-fired power plants typically provided base load power. However, (1) with the current (and projected prolonged) plentiful supplies of natural gas, (2) lower natural gas prices, and (3) the projected environmental benefits of natural gas use vis-à-vis coal, natural gas is now increasingly used as a base and intermediate load power source in many places.

The integrated gasification combined cycle plant can be used as an example of the benefits of gas-fired power generation. The natural gas is combusted in a gas turbine unit that is connected to a generator after which the hot exhaust gases are then passed through a heat exchanger to generate steam for a steam turbine. By using this approach, a natural gas combined cycle power plant can reach efficiencies at least on the order of 50%, compared to a lower efficiency (30%–35%) for a similar megawatt size coal-fired power plant.

Furthermore, the hydrogen produced from natural gas can itself be used as a fuel. The most efficient way to convert hydrogen into electricity is by using a fuel cell, which combines hydrogen with oxygen to produce electricity, water, and heat. Although the process of reforming natural gas to produce hydrogen still has associated carbon dioxide emissions, the amount released for each unit of electricity generated is much lower than for a combustion turbine.

As part of the industrial use of natural gas, there is the need for analysis before the products (in this context, the gaseous products) are used by industrial and domestic consumers. Detection of even the slightest amounts of impurities can be an indication of process inefficiency and whether or not the gas is suitable for the designated use. In fact, one of the most important tasks in gas technology, especially in the context of petroleum-related natural gas is the need for reliable values of the volumetric and thermodynamic properties for pure low-boiling hydrocarbon derivatives and their mixtures. These properties are important in the design and operation of much of the processing equipment (Poling et al., 2001).

For example, reservoir engineers and process engineers use pressure-volume-temperature relationships and phase behavior of reservoir fluids (1) to estimate the amount of oil or gas in a reservoir, (2) to develop a recovery process for a crude oil or gas field, (3) to determine an optimum operating condition in a gas—liquid separator unit, (4) to determine the need for a wellhead processing system to protect a pipeline from corrosion, and (5) to design suitable gas processing options. However, the most advanced design approaches or the most sophisticated simulation experiments cannot guarantee the optimum design or operation of a unit (or protection of a pipeline) if the physical properties are not known. For these reasons accurate knowledge of the properties of the gas is an extremely increasingly important aspect of gas technology.

Natural gas can also be used to produce both heat and electricity simultaneously (CHP). Cogeneration systems are highly efficient, able to put 75%–80% of the energy in gas to use. *Trigeneration* systems, which provide electricity, heating, and cooling, can reach even higher efficiencies.

Natural gas sees a broad range of other industrial uses, as a source of both heat and power and as an input for producing plastics and chemicals. For example, most of the hydrogen gas production comes from reacting high temperature water vapor (steam) with methane. The resulting hydrogen has wide use in crude oil refineries in order to produce marketable products from heavy crude oil, extra heavy cried oil and tar sand bitumen (Speight, 2014a, 2017a) as well as to produce ammonia for fertilizer. Although the process of reforming natural gas to hydrogen still has associated carbon dioxide emissions, the amount released for each unit of electricity generated is much lower than for a combustion turbine.

Hydrogen produced from natural gas can itself be used as a fuel—the most efficient way to convert hydrogen into electricity is by using a fuel cell, which combines hydrogen with oxygen to produce electricity, water, and heat:

$$2H_2 + O_2 \rightarrow 2H_2O + Heat$$

Compressed natural gas (CNG) has been used as a transportation fuel (Chapter 5: Recovery, storage, and transportation), mostly in public transit. The natural gas, which is compressed at over 3000 psi to 1% of the volume that the gas would occupy at normal atmospheric pressure, can be burned in an internal combustion engine that has been appropriately modified. Approximately 0.1% v/v of the natural gas consumed in the United States has been used to power vehicles, representing the energy content of more than 5 million barrels of oil (US EIA, 2012).

Compared to gasoline, vehicles powered by CNG emit less carbon monoxide, nitrogen oxides (NO_x), and particulate matter. However, a disadvantage of CNG is the low energy density compared with the higher energy density of liquid fuels. A gallon of CNG has approximately one quarter of the energy in a gallon of

Table 1.6 Natural gas liquids (NGLs), uses, products, and consumers

NGL	Chemical formula	Uses	Other uses
Ethane	C_2H_6	Ethylene production Power generation	Plastics Antifreeze Detergents
Propane	C_3H_8	Heating fuel Transportation petrochemical feedstock	Plastics

Butanes: *n*-butane and *iso*-butane	C_4H_{10}	Petrochemical
		  feedstock
		Refinery feedstock
		Blend stock for
		  gasoline
			Plastics
			Synthetic
			  rubber

Condensate	C_5H_{12}
	Higher boiling
	  hydrocarbons
		Petrochemical
		  feedstock
		Additive to gasoline
		Diluent for heavy
		  crude oil
			Solvents

gasoline and, therefore, vehicles powered by CNG require larger fuel tanks (compared to vehicles powered by liquid fuels).

Thus, a more suitable use for natural gas in the transportation sector may be as a resource to generate electricity for plug-in vehicles or hydrogen for fuel cell vehicles, which can reduce emissions savings on the order of 40% (or more).

NGLs which are, by definition, hydrocarbon derivatives also have use other than fuel components (Table 1.6) because of the hydrocarbon constituents. Thus, there are many uses for NGLs that span almost all sections of the industrial chemicals economy. NGLs are used as feedstocks for petrochemical plants, burned for space heat and cooking, and blended into vehicle fuel.

The chemical composition of NGLs from different sources is similar, yet their applications vary widely. Ethane occupies the largest share of the filed production of NGLs and is used almost exclusively to produce ethylene, which is then converted into plastic products. By contrast, the majority of the propane, by contrast, is burned for heating, although a substantial amount is used as petrochemical feedstock. A blend of propane and butane, sometimes referred to as *autogas* (*auto-gas*), is a popular fuel in some parts of Europe, Turkey, and Australia. Natural gasoline (pentanes plus) can be blended into various kinds of fuel for combustion engines and is useful in energy recovery from wells and tar sand (oil sand) deposits.

A challenge with the use of NGLs is that they are (1) expensive to handle, (2) store, and (3) transport compared to refined products. NGLs are highly flammable and require high pressure and/or low temperature to be maintained in the liquid state for shipping and handling. The flammability of these liquids necessitates the use of special trucks, ships, and storage tanks.

References

AAPG, 2015. Unconventional energy resources: 2015 Review. Nat. Resour. Res. 24 (4), 443–508. American Association of Petroleum Geologists, Tulsa Oklahoma.

Abbott, M., 2016. The Economics of the Gas Supply Industry. Routledge Publishers, Taylor & Francis Group, New York.

Aguilera, R.F., Radetzki, M., 2014. The shale revolution: global gas and oil markets under transformation. Miner. Econ. 26 (3), 75–84.

ASTM, 2017. Annual Book of Standards. ASTM International, West Conshohocken, PA.

Boswell, R., Collett, T.S., 2011. Current perspectives on gas hydrate resources. Energy Environ. Sci. 4, 1206–1215.

Boyle, G. (Ed.), 1996. Renewable Energy: Power for a Sustainable Future. Oxford University Press, Oxford.

Briggs, J., February 1988. Municipal Landfill Gas Condensate. Report No. EPA/600/S2-87/090. Environmental Protection Agency Hazardous Waste Engineering Research Laboratory, Cincinatti, OH.

Brosseau, J., 1994. Trace gas compound emissions from municipal landfill sanitary sites. Atmos. Environ. 28 (2), 285–293.

Buffett, B., Archer, D., 2004. Global inventory of methane clathrate: sensitivity to changes in the deep ocean. Earth Planet. Sci. Lett. 227 (3–4), 185.

Burruss, R.C., Ryder, R.T., 2003. Composition of Crude Oil and Natural Gas Produced from 14 Wells in the Lower Silurian "Clinton" Sandstone and Medina Group, Northeastern Ohio and Northwestern Pennsylvania. Open-File Report 03-409, United States Geological Survey, Reston, VA.

Burruss, R.C., Ryder, R.T., 2014. Composition of natural gas and crude oil produced from 10 wells in the lower Silurian "Clinton" sandstone, Trumbull County, Ohio. In: Ruppert, L. F., Ryder, R.T. (Eds.), Coal and Petroleum Resources in the Appalachian Basin; Distribution, Geologic Framework, and Geochemical Character. United States Geological Survey, Reston, VA, Professional Paper 1708.

Chong, Z.R., Yang, S.H.B., Babu, P., Linga, P., Li, X.S., 2016. Review of natural gas hydrates as an energy resource: prospects and challenges. Appl. Energy 162, 1633–1652.

Colborn, T., Kwiatkowski, C., Schultz, K., Bachran, M., 2011. Natural gas operations from a public health perspective. Hum. Ecol. Risk Assess. 17, 1039–1056.

Collett, T.S., 2002. Energy resource potential of natural gas hydrates. Am. Assoc. Pet. Geol. Bull. 86, 1971–1992.

Collett, T.S., Johnson, A.H., Knapp, C.C., Boswell, R., 2009. Natural gas hydrates: a review. In: Collett, T.S., Johnson, A.H., Knapp, C.C., Boswell, R. (Eds.), Natural Gas Hydrates – Energy Resource Potential and Associated Geologic Hazards. American Association of Petroleum Geologists, Tulsa, OK, pp. 146–219. , AAPG Memoir No. 89.

Demirbaş, A., 2010a. Methane from gas hydrates in the Black Sea. Energy Sources Part A 32, 165–171.

Demirbaş, A., 2010b. Methane hydrates as potential energy resource: Part 1—Importance, resource and recovery facilities. Energy Convers. Manage. 51, 1547–1561.

Demirbaş, A., 2010c. Methane hydrates as potential energy resource: Part 2—Methane production processes from gas hydrates. Energy Convers. Manage. 51, 1562–1571.

Forbes, R.J., 1964. Studies in Ancient Technology. E. J. Brill, Leiden.

Gary, J.G., Handwerk, G.E., Kaiser, M.J., 2007. Petroleum Refining: Technology and Economics, fifth ed. CRC Press, Taylor & Francis Group, Boca Raton, FL.

Gornitz, V., Fung, I., 1994. Potential distribution of methane hydrate in the world's oceans. Glob. Biogeochem. Cycles 8, 335–347.

Hafner, M., Tagliapietra, S., 2013. The Globalization of Natural Gas Markets: New Challenges and Opportunities for Europe. Claeys & Casteels Law Publishers, Deventer, International Specialized Book Services, Portland Oregon.

Hsu, C.S., Robinson, P.R. (Eds.), 2017. Handbook of Petroleum Technology. Springer International Publishing AG, Cham.

Islam, M.R., 2014. Unconventional Gas Reservoirs. Elsevier, Amsterdam.

John, E., Singh, K., 2011. Production and properties of fuels from domestic and industrial waste. In: Speight, J.G. (Ed.), The Biofuels Handbook. Royal Society of Chemistry, London, pp. 333–376.

Khosrokhavar, R., Griffiths, S., Wolf, K.-H., 2014. Shale gas formations and their potential for carbon storage: opportunities and outlook. Environ. Processes 1 (4), 595–611.

Kundert, D., Mullen, M., April 14–16, 2009. Proper evaluation of shale gas reservoirs leads to a more effective hydraulic-fracture stimulation. Paper No. SPE 123586. In: Proceedings. SPE Rocky Mountain Petroleum Technology Conference, Denver, Colorado.

Levine, J.R., 1993. Coalification: the evolution of coal as a source rock and reservoir rock for oil and gas, Stud. Geol., 38. Am. Assoc. Petrol. Geol., pp. 39–77.

Mokhatab, S., Poe, W.A., Speight, J.G., 2006. Handbook of Natural Gas Transmission and Processing. Elsevier, Amsterdam.

Nersesian, R.L., 2010. Energy for the 21st Century, second ed. M.E. Sharpe, Armonk, NY.

Parkash, S., 2003. Refining Processes Handbook. Gulf Professional Publishing, Elsevier, Amsterdam.

Poling, B.E., Prausnitz, J.M., O'Connell, J.P., 2001. The Properties of Gases and Liquids, fifth ed. McGraw-Hill, New York.

Ramage, J., 1997. Energy: A Guidebook. Oxford University Press, Oxford.

Ramroop Singh, N., 2011. Biofuels. In: Speight, J.G. (Ed.), The Biofuels Handbook. Royal Society of Chemistry, London, pp. 160–198.

Rasi, S., Veijanen, A., Rintala, J., 2007. Trace compounds of biogas from different biogas production plants. Energy 32, 1375–1380.

Rasi, S., Lantela, J., Rintala, J., 2011. Trace compounds affecting biogas energy utilization— a review. Energy Convers. Manage. 52 (12), 3369–3375.

Rice, D.D., 1993. Composition and origins of coalbed gas, Stud. Geol., 38. Am. Assoc. Petrol. Geol., pp. 159–184.

Singh, K., Sastry, M.K.S., 2011. Production of fuels from landfills. In: Speight, J.G. (Ed.), The Biofuels Handbook. Royal Society of Chemistry, London, pp. 408–453.

Speight, J.G., 2008. Synthetic Fuels Handbook: Properties, Processes, and Performance. McGraw-Hill, New York.

Speight, J.G., 2011a. The Refinery of the Future. Gulf Professional Publishing, Elsevier, Oxford.

Speight, J.G., 2011b. An Introduction to Petroleum Technology, Economics, and Politics. Scrivener Publishing, Salem, MA.

Speight, J.G. (Ed.), 2011c. The Biofuels Handbook. Royal Society of Chemistry, London.

Speight, J.G., 2012. Crude Oil Assay Database. Knovel, New York. Available at: http://www.knovel.com/web/portal/browse/display?_EXT_KNOVEL_DISPLAY_bookid = 5485 &VerticalID = 0.

Speight, J.G., 2013a. Shale Gas Production Processes. Gulf Professional Publishing, Elsevier, Oxford.

Speight, J.G., 2013b. The Chemistry and Technology of Coal, third ed. CRC Press, Taylor & Francis Group, Boca Raton, FL.

Speight, J.G., 2014a. The Chemistry and Technology of Petroleum, fifth ed. CRC Press, Taylor & Francis Group, Boca Raton, FL.

Speight, J.G., 2014b. Oil and Gas Corrosion Prevention. Gulf Professional Publishing, Elsevier, Oxford.

Speight, J.G., 2015. Handbook of Petroleum Product Analysis, second ed. John Wiley & Sons Inc, Hoboken, NJ.

Speight, J.G., 2016a. Introduction to Enhanced Recovery Methods for Heavy Oil and Tar Sands, second ed. Gulf Publishing Company, Taylor & Francis Group, Waltham MA.

Speight, J.G., 2016b. Handbook of Hydraulic Fracturing. John Wiley & Sons Inc, Hoboken, NJ.

Speight, J.G., 2017a. Handbook of Petroleum Refining. CRC Press, Taylor & Francis Group, Boca Raton, FL.

Speight, J.G., 2017b. Deep Shale Oil and Gas. Gulf Professional Publishing, Elsevier, Oxford.

Speight, J.G., 2018. Handbook of Natural Gas Analysis. John Wiley & Sons Inc, Hoboken, NJ.

Speight, J.G., Islam, M.R., 2016. Peak Energy—Myth or Reality. Scrivener Publishing, Beverly, MA.

US EIA, 2012. Natural Gas Consumption by End Use. United States Energy Information Administration. Washington, DC. https://www.eia.gov/dnav/ng/ng_cons_sum_dcu_nus_a.htm.

Origin and production

2

2.1 Introduction

Natural gas occurs deep beneath the Earth's surface and consists predominantly of methane (CH_4), but also contains small amounts of hydrocarbon gas liquids and nonhydrocarbon gases (Speight, 2014a, 2017a; Faramawy et al., 2016). While natural gas from different sources may seemingly have similar behavior, the composition of the gas is not necessarily the same and should be analyzed with this in mind (Esteves et al., 2016). In fact, changes in composition can occur from one step in the production train to the next depending upon the composition of the gas and the processes used in the train.

reservoir gas → produced gas → wellhead gas → transported gas → stored gas → sales gas

Since the different gas supplies enter the gas system at different locations, the exact composition at any site will vary among the different regions and over time. As an example, the heating value will depend on the composition of the gas. The anticipation of differences in the composition must also consider the differences in the reservoir gas due to differences in the maturation processes from one locale to another (Speight, 2014a; Faramawy et al., 2016). Thus, in terms of natural gas definition, there are three basic approaches to defining the gas: (1) a qualitative description of the natural gas by origin, type, and constituents, (2) classification by characteristics based upon testing procedures, and (3) classification as a result of the concentration of specific constituents in preference to other constituents. In addition, the recent increases in domestic natural gas supplies have been made possible by two technologies (1) horizontal drilling and (2) hydraulic fracturing, which allow energy companies to develop natural gas supplies once considered to be inaccessible (Speight, 2016).

In order to understand natural gas as it exists in the reservoir or is produced at the wellhead, understanding the origin of the gas can offer pointers to the natural gas scientist or natural gas engineer as to the properties and behavior of the gas and also the means by which the gas originated. The key is to recognize that natural gas varies in composition not only because of the reservoir in which the gas exists but also because of the relative placement of wells within a reservoir.

The following sections describe natural gas from its origin in the Earth to production thereby confirming that the variations in gas composition (which influence properties and behavior) that, in turn, influence the processes selected for gas processing (gas cleaning) (Chapter 4: Composition and Properties, Chapter 7: Process Classification, Chapter 8: Gas Cleaning Processes).

Natural Gas. DOI: https://doi.org/10.1016/B978-0-12-809570-6.00002-3

2.2 Origin

There are many different theories as to the origin of the various fossil fuels, specifically (in the current context) natural gas and crude oil (Speight, 2014a; Faramawy et al., 2016). The most generally accepted theory is that natural gas and crude oil are formed when organic matter or *organic debris* (such as the remains of a plants or animals) is compressed under the Earth, at very high pressure for periods of geologic time that are measured in terms of millions of years (millennia) (Table 2.1). At the start of the oil- and gas-producing periods, the remains of plants and animals decayed and built up in thick layers and, over time as sediment and mud and other debris piled on top of the organic matter, metamorphosis of the organic debris occurred. As a result, the sediment, mud, and other inorganic debris were changed to rock, causing pressure to be put on the organic matter. The increasing pressure compressed the organic matter and, combined with other subterranean effects, decomposed the individual constituents into natural gas and crude oil (Table 2.1).

The deep-lying source rock that contains the organic precursors is often referred to as the *kitchen* (Speight, 2014a). In theory, the deeper and hotter the kitchen, the more likelihood of gas being produced from the precursor material. The increase of heat with depth is due to the presence of a *geothermal gradient* that is generally on the order of $25-30°C/km$ ($0.008-0.009°C$ per foot of depth or $8-9°C$ per 1000 ft of depth; $0.015-0.016°F$ per foot of depth or $15-16°F$ per 1000 ft of depth), that is, approximately 1 degree for every 100 ft below the surface of the Earth. However, this does not mean that nonassociated gas will always be found at greater depths than crude oil. Natural gas and crude oil have migrated from the *kitchen*, in a sideways direction and/or in an upward direction, until the natural gas and crude oil are trapped in a reservoir that forms part of a subsurface formation. Thus, a field

Table 2.1 Timeline for the original of natural gas

Approximate time period	Events
400–300 million years ago	Tiny sea plants and animals died and were buried on the ocean floor; over time they were covered with layers of silt and sand
300–100 million years ago	The organic debris started change by simple chemical reactions
100–50 million years ago	The organic debris was buried deeper and deeper; pressure increased and (possibly) temperature increased (but, as noted earlier, the level of the temperature is largely unknown and, at best, very speculative)
50–1 million years ago	The organic debris reacted, under the prevalent conditions underground, to produce methane and other hydrocarbon products that eventually because natural gas that migrated to reservoirs where it was trapped in the reservoir rock

may have a series of layers of natural gas—crude oil and (nonassociated) natural gas reservoirs at levels on the order of (or more than) 3000 ft below the surface.

The methane (natural gas) produced in this manner is referred to as *thermogenic* methane (thermogenic natural gas) because it is presumed that the pressure effects and the increased depth of the precursors includes the influence of heat to convert the organic matter to natural gas and crude oil. However, the actual temperature is not known and, like the remainder of the thermal theory is, at best, speculative.

For the organic chemist, it is well known that laboratory studies at specified temperatures on the order of 300—350°C (570—650°F) will convert organic materials to methane and other hydrocarbons gases. Thus the temperatures used in the laboratory to demonstrate that high temperatures used in the laboratory to substitute for the lack of geological time in the laboratory is not only speculative thinking but is erroneous thinking. Many laboratory experimentalists often ignore the fact that higher temperatures can (and do) change the chemistry of the process and the findings may not be completely true. The fact remains that the high temperature involved in the thermogenic origin of natural gas and other fossil fuels is not conclusively proven.

Methane (natural gas) can also be formed through the transformation of organic matter by microorganisms. This type of methane is referred to as *biogenic* methane (*biogenic* natural gas). In this process, the methane-producing microorganisms (*methanogens*) chemically break down organic matter to produce methane. The *methanogen* microorganisms are commonly found in oxygen-deficient areas near the surface of the Earth. Thus, the formation of methane in this manner usually takes place close to the surface of the Earth, and the methane produced is usually lost into the atmosphere, which can be seen as an environmental problem. In certain circumstances, however, this methane can be trapped underground, recoverable as natural gas. An example of biogenic methane is landfill gas. Waste-containing landfills produce a relatively large amount of natural gas, from the decomposition of the waste materials that they contain. New and evolving technologies are allowing this gas to be collected to add to the supply of natural gas, especially the production of gas from biomass sources (John and Singh, 2011; Ramroop Singh, 2011; Singh and Sastry, 2011).

A third way in which methane (and natural gas) may be formed is through *abiogenic* processes. Extremely deep under the Earth's crust, there exist hydrogen-rich gases and as these gases rise toward the surface of the Earth, they may interact through the catalytic activity of minerals in the absence of oxygen (Speight, 2013a, 2017b). This interaction may result in a reaction to form products that are found in the atmosphere (including nitrogen, oxygen, carbon dioxide, argon, and water). If these gases are under very high pressure as they move toward the surface of the Earth, they are likely to form methane (*abiogenic methane*), similar to thermogenic methane. Whatever be the mode of formation of methane, the standard test methods remain the same (ASTM, 2017). Methane is methane! However, there may be some minor differences in applying some of the test methods and interpreting the data when methane is not the only constituents of the gas to be tested is not only methane, but other constituents are present with the methane.

Table 2.2 General properties of unrefined natural gas (left-hand number) and refined natural gas (right-hand number)

Relative molar mass	20—16
Carbon content (% w/w)	73—75
Hydrogen content (% w/w)	27—25
Oxygen content (weight % w/w)	0.4—0
Hydrogen-to-hydrogen atomic ratio	3.5—4.0
Density relative to air @15°C	1.5—0.6
Boiling temperature (°C/1 atm)	−162
Autoignition temperature (°C)	540—560
Octane number	120—130
Methane number	69—99
Vapor flammability limits (% v/v)	5—15
Flammability limits	0.7—2.1
Lower heating/calorific value (Btu/lb)	900
Methane concentration (% v/v)	100—80
Ethane concentration (% v/v)	5—0
Nitrogen concentration (% v/v)	15—0
Carbon dioxide concentration (% v/v)	5—0
Sulfur concentration (ppm, w/w)	5—0

Because of the manner in which it is formed, and the diverse nature of the precursors, natural gas contains constituents other than methane which affect the composition and the properties of the gas (Table 2.2). While the major constituent of natural gas is indeed methane, there are components such as saturated hydrocarbon derivatives (C_nH_{2n+2}), low-boiling aromatic hydrocarbon derivatives (derivatives of benzene), nitrogen, carbon dioxide (CO_2), hydrogen sulfide (H_2S), and mercaptan derivatives (RSH, also called thiol derivative), as well as trace amounts of other constituents such as the always-valuable helium and argon. In cases where the natural gas does not contain hydrogen sulfide, there may also be a relative lack of carbon dioxide, which is not true for the gases produced from coal or from any type of biomass (Speight, 2011, 2013b).

2.3 Exploration

Historically, in the modern fossil fuel era, natural gas was discovered as a consequence of the exploration for crude oil. The methods used to discover natural gas reservoirs are essentially those used to crude oil reservoirs. In addition, in the early days of the natural gas industry, the only way of locating natural gas reservoirs was to search for surface evidence of the underground formations (capable of being reservoirs) such as seepages of natural gas or crude oil from the underground formations as was evidenced in various countries of the Middle East (Chapter 1: History and Use).

However, the only way of being certain that a natural gas reservoir exists is to drill an exploratory well. This consists of actually digging into the Earth's crust to allow geologists to study the composition of the underground rock layers in detail (Burnett, 1995). In addition to looking for natural gas reservoirs by drilling an exploratory well, geologists also examine the drill cuttings and fluids to gain a better understanding of the geologic features of the area. However, drilling an exploratory well is an expensive, time-consuming effort. Therefore, exploratory wells are only drilled in areas where other data has indicated a high probability of gas-containing formations.

The search for natural gas now begins with geologists locating the types of rock that are usually found near to natural gas and crude oil reservoirs. The methods used include seismic surveys that are used to estimate the correct places to drill wells. Seismic surveys use echoes from a vibration source at the Earth's surface (usually a vibrating pad under a truck built for this purpose) to generate information about the subterranean formation. If required, dynamite may be used to provide the necessary underground vibrations.

Technological innovation in the exploration and production sector is necessary if the industry is to continually increase the production of natural gas to meet rising demand. New technologies serve to make the exploration and production of natural gas more efficient, safe, and environment-friendly. Despite the fact that natural gas deposits are continually being found deeper in the ground, in remote, inhospitable areas that provide a challenging environment in which to produce natural gas, the exploration and production industry has not only kept up its production pace, but in fact has improved the general nature of its operations.

However, even though technology advances have allowed for phenomenal increases in the success rate of locating natural gas reservoirs, the process of exploring for natural gas reservoirs is still filled with uncertainty, as well as trial-and-error. The search for a reservoir that actually contains gas (there can be a reservoir rock formation that is empty) and which is often thousands of feet below remains complex.

2.3.1 Geological survey

The exploration for natural gas typically begins with a *geological examination* (*geological survey*) of the surface structure of the Earth and determination of the areas where there is a high probability that a natural gas reservoir exists. In fact, as early as the mid-1800s, it was realized that anticlinal slopes had a particularly increased chance of containing natural gas. These anticlinal slopes are areas where the Earth has folded up on itself, forming the dome shape that is characteristic of a great number of reservoirs.

By surveying and mapping the surface and subsurface characteristics of a certain area, the geologist can extrapolate which areas are most likely to contain a natural gas reservoir. The geologist has many tools at his disposal to do so, from the outcroppings of rocks on the surface or in valleys and gorges, to the geologic information attained from the rock cuttings and samples obtained from the digging of irrigation ditches, water wells, and other oil and gas wells. This information is combined to allow the geologist to make inferences as to the fluid content, porosity, permeability, age, and formation sequence of the rocks underneath the surface of a particular area.

Once the geologist has located an underground formation in which natural gas and/or crude oil formation can exist, further tests can be performed to gain more detailed data about the potential reservoir area. These tests (commonly performed by a *geophysical team*) allow for the more accurate mapping of underground formations.

2.3.2 Seismic survey

Perhaps the biggest breakthrough in the exploration for natural gas exploration came through the use of *seismology*.

Seismology is the study of the movement of energy, in the form of seismic waves, through the Earth's crust and interacts differently with various types of underground formations. The *seismograph*, an instrument used to detect and record earthquakes, is able to pick up and record the vibrations of the Earth that occur during an earthquake. When seismology is applied to the search for natural gas, seismic waves, emitted from a source, are sent into the Earth and the seismic waves interact differently with the underground formation (underground layers), each with its own properties. The waves are reflected back toward the source by each formation. It is this reflection that allows for the use of seismology in discovering the character and properties of underground formations leading to conclusions about the potential for one or more of the formation to contain natural gas.

In the early days of seismic exploration, seismic waves were created using dynamite. These carefully planned, small explosions created the requisite seismic waves, which were then picked up by the geophones, generating data to be interpreted by geophysicists, geologists, and reservoir engineers.

In practice, using seismology for exploring onshore areas involves artificially creating seismic waves, the reflection of which are then picked up by sensitive pieces of equipment (*geophones*) embedded in the ground. The data gathered up by these geophones are then transmitted to a seismic recording truck, which records the data for further interpretation. The source of seismic waves (usually a preset underground explosion) creates vibrations which reflect off of the different layers of the Earth, to be picked up by geophones on the surface and relayed to a seismic recording truck to be interpreted and logged.

Furthermore, due to environmental concerns and improved technology, it is often no longer feasible to use explosive charges to generate the needed seismic waves. Instead, most seismic crews use nonexplosive seismic technology to generate the required data. This nonexplosive technology usually involves the use of a large heavy wheeled or tracked vehicle carrying special equipment designed to create a large impact or series of vibrations. These impacts or vibrations create seismic waves similar to those created by dynamite. In the seismic truck shown, the large piston in the middle is used to create vibrations on the surface of the Earth, sending seismic waves that are used to generate useful data.

The same sort of process is used in *offshore seismic exploration*. When exploring for natural gas that may exist thousands of feet below the seabed floor, which may itself be thousands of feet below sea level, a slightly different method of seismic exploration is used. Instead of trucks and geophones, a ship is used to pick up the seismic data. Instead of geophones, offshore exploration uses *hydrophones*, which are designed to pick up seismic waves underwater. These hydrophones are towed behind the ship in various configurations depending on the needs of the geophysicist. Instead of using dynamite or impacts on the seabed floor, the seismic ship uses a large air gun, which releases bursts of compressed air under the water, creating seismic waves that can travel through the Earth's crust and generate the seismic reflections that are necessary.

The development of seismic imaging in three dimensions greatly changed the nature of natural gas exploration. This technology uses traditional seismic imaging techniques, combined with powerful computers and processors, to create a three-dimensional (3-D) model of the subsurface layers. Four-dimensional (4-D) seismology expands on this, by adding time as a dimension, allowing exploration teams to observe how subsurface characteristics change over time. Exploration teams can now identify natural gas prospects more easily, place wells more effectively, reduce the number of dry holes drilled, reduce drilling costs, and cut exploration time.

In the last two decades, it has become relatively easy to use computers to assemble seismic data that is collected from the field. This allows for the processing of much larger amounts of data, increasing the reliability and informational content of the seismic model. There are three main types of computer assisted exploration models: two-dimensional (2-D), 3-D, and most recently 4-D. These imaging techniques, while relying mainly on seismic data acquired in the field, are becoming more and more sophisticated. Computer technology has advanced so far that it is now possible to incorporate the data obtained from different types of tests, such as logging, production information, and gravimetric testing which can all be combined to create a "visualization" of the underground formation. By this means, geologists and geophysicists are able to combine all of their sources of data to compile one clear, complete image of subsurface geology. An example of this is the use of an interactive computer by a geologist to enable visualization of the seismic data thereby allowing exploration of the subsurface layers.

2-D seismic imaging refers to using the data collected from seismic exploration activities to develop a cross-sectional picture of the underground formations. The geophysicist interprets the seismic data obtained from the field, taking the vibration recordings of the seismograph and using them to develop a conceptual model of the composition and thickness of the various layers of rock underground. This process is normally used to map underground formations, and to make estimates based on the geologic structures to determine where it is likely that deposits may exist.

There also exists a technique using basic seismic data known as *direct detection* in which white bands (*bright spots*) that often appeared on seismic recording strips indicated hydrocarbon reservoirs. The nature or porous rock containing natural gas could often result in reflecting stronger seismic reflections than normal, water filled rock. Therefore, in these circumstances, the actual natural gas reservoir could be detected directly from the seismic data. However, this concept is not universally applicable since many of the *bright spots* do not contain hydrocarbon derivatives, and many deposits of hydrocarbon derivatives are not indicated by *bright spots* on the seismic data.

One of the biggest recent innovations in computer-aided exploration techniques was the development of *3-D seismic imaging*. 3-D imaging utilizes seismic field data to generate a 3-D picture of underground formations and geologic features. This, in essence, allows the geophysicist and geologist to see a clear picture of the composition of the Earth's crust in a particular area. Thus, in the exploration for natural gas, as an actual image could be used to estimate the probability of formations existing in a particular area, and the characteristics of that potential formation. This technology has been extremely successful in raising the success rate of exploration efforts. In fact, using 3-D seismic imaging has been estimated to increase the likelihood of successful reservoir location by 50%.

Although this technology is very useful, it is also very costly. The generation of 3-D images requires data to be collected from several thousand locations, as opposed to 2-D imaging, which only requires several hundred data points. As such, 3-D imaging is a much more involved and prolonged process. Therefore, it is usually used in conjunction with other exploration techniques. For example, a geophysicist may use traditional 2-D modeling and examination of geologic features to determine if there is a probability of the presence of natural gas. Once these basic techniques are used, 3-D seismic imaging may be used only in those areas that have a high probability of containing reservoirs. In addition to broadly locating natural gas and crude oil reservoirs, 3-D seismic imaging allows for the more accurate placement of wells to be drilled. This increases the productivity of successful wells, allowing for more natural gas and crude oil to be extracted from the ground.

One of the latest breakthroughs in seismic exploration, and the modeling of underground rock formations, has been the introduction of *4-D seismic imaging*. This type of imaging is an extension of 3-D imaging technology. However, instead of achieving a simple, static image of the underground, in 4-D imaging the changes in structures and properties of underground formations are observed over time (the fourth dimension). Hence, the technique is also referred to as *4-D time lapse imaging*.

Various seismic readings of a particular area are taken at different times, and this sequence of data is fed into a powerful computer. The different images are amalgamated, to create a sort of "movie" of what is going on under the ground. Through studying how seismic images change over time, geologists can gain a better understanding of many properties of the rock, including underground fluid flow, viscosity, temperature, and saturation. The 4-D seismic images can also be used by geologists to evaluate the properties of a reservoir, including the rate of depletion of natural gas once production has begun.

2.3.3 Magnetometers

In addition to using seismology to gather data concerning the composition of the Earth's crust, the magnetic properties of underground formations can be measured to generate geological and geophysical data. This is accomplished through the use of a *magnetometer*, which is a device that can measure the small differences in the Earth's magnetic field.

A magnetometer is a scientific instrument used to measure the strength and/or direction of the magnetic field in the vicinity of the instrument. The Earth's magnetism varies from place to place and differences in the magnetic field of the Earth (the magnetosphere) can be caused either by (1) the differing nature of subterranean rock formations and (2) the interaction between charged particles from the Sun and the magnetosphere. It is the former that is of interest to natural gas and crude oil exploration, since the differing mineralogy of underground formations can have a different effect on the gravitational field of the Earth (Ballard, 2007).

Early magnetometers were large and bulky and only able to survey a small area at a time. Modern magnetometers are much more subtle in their operations and accurately measure any minute differences between the formations with very sensitive equipment and allow geophysicists to estimate the structure of the underground formations and whether or not the formations have the potential for containing natural gas.

2.3.4 Logging

Well logging is a method used for recording rock and fluid properties to find gas and oil containing zones in subterranean formations.

Drilling of an exploratory or developing well is the first contact that a geologist has with the actual contents of the subsurface geology. Logging, in its many forms, consists of using this opportunity to gain a fuller understanding of what actually lies beneath the surface. In addition to providing information specific to that particular well, vast archives of historical logs exist for geologists interested in the geologic features of a given, or similar, area. The actual logging procedure consists of lowering a *logging tool* on the end of a wireline into a natural gas well or crude oil well to measure the rock and fluid properties of the formation. An interpretation of these measurements is then made to locate and quantify potential depth zones containing natural gas and crude oil.

Logging tools developed over the years measure the electrical, acoustic, radioactive, electromagnetic, and other properties of the rocks and their contained fluids. Logging is performed at various intervals during the drilling of the well and when the total depth is drilled, which could range in depths from 1000 to 25,000 ft or more. The data is recorded to a printed record (*well log*) that can, through modern computers, be transmitted digitally to other locations. Two of the most prolific and often performed tests include *standard logging* and *electric logging*.

Standard logging consists of examining and recording the physical aspects of a well. For example, the drill cuttings (rock that is displaced by the drilling of the well) are all examined and recorded, allowing geologists to physically examine the subsurface rock. Also, core samples are taken, which consists of lifting a sample of underground rock intact to the surface, allowing the various layers of rock, and their thickness, to be examined. These cuttings and cores are often examined using powerful microscopes, which can magnify the rock and allows the geologist to examine the porosity and fluid content of the subsurface rock, and to gain a better understanding of the Earth in which the well is being drilled.

Electric logging consists of lowering a device used to measure the electric resistance of the rock layers in the downhole portion of the well. This is achieved by running an electric current through the rock formation and measuring the electrical resistance that the current encounters along its way. This gives geologists an idea of the fluid content and characteristics. A newer version of electric logging (*induction electric logging*) provides much the same types of readings but is more easily performed and provides data that is more easily interpreted.

In recent years, a new technique, *logging while drilling* (LWD, also called *measurement while drilling*, MWD), has been introduced, which provides similar information about the well. Instead of sensors being lowered into the well at the end of wireline cable, the sensors are integrated into the frill string and the measurements are made while the well is being drilled. While wireline well logging occurs after the drill string is removed from the well, LWD measures geological parameters while the well is being drilled. However, because there is no high bandwidth telemetry path available (i.e., no wires to the surface), the data is either recorded downhole and retrieved when the drill string is removed from the hole, or the measurement data is transmitted to the surface via pressure pulses in the mud fluid column of the well.

The MWD methods allow for the collection of data from the bottom of a well as it is being drilled. Drilling teams can access up to the second information on the exact nature of the rock formations being encountered by the drill bit. This improves drilling efficiency and accuracy in the drilling process, allows better formation evaluation as the drill bit encounters the underground formation, and reduces the chance of formation damage and blowouts.

2.4 Reservoirs

A reservoir is a subsurface rock structure, with sufficient size and closure that contains a 3-D network of interconnected void (pore) space and is overlain by a fine-grained water-saturated rock. Natural gas is typically located in reservoirs that are porous rocks (often sandstone) surrounded by impermeable materials and water. It is the exploration methods that have been used to define the structure and characteristics of the reservoir. Some reservoirs occur at depths up to 2 miles under the surface, while others may be as much as 5 miles under the surface. In addition, there are also subsea reservoirs, that is, reservoirs that are located under the floor of the ocean (Max, 2000).

Table 2.3 Information required for safe development of a natural gas resource

Stage of development	Information needed	Comment
Origin	Strata mineralogy	Identify potential for gas adsorption
Exploration	Gas quality	Identify potential for gas processing
	Gas reserves	Identify economic aspects of resource development
Reservoir	Reservoir mineralogy	Tendency of gas constituents to adsorb or react with mineral
Production	Gas quality	Identify need for wellhead processing prior to transportation
Transportation	Gas composition	Identify potential for corrosion during transportation

Whatever the location of the reservoir, the gas must be extracted by the application of methods that assure maximum recovery of the gas. More recently, substantial reserves of natural gas have been discovered in tight formations, such as shale formation where the permeability is extremely low. Furthermore, the methods selected to recover the gas will vary with the reservoir and even from wells drilled across a gas reservoir due to differences such as (1) reservoir thickness, (2) reservoir extent, (3) reservoir pressure, (4) reservoir depth, (5) mineralogy of the reservoir, (6) gas composition, and (7) water content, all of which are required in order for the project to proceed with the potential for success (Table 2.3) (Speight, 2018).

The subsurface porous permeable rock body or formation that has the capability to store and transmit fluids (natural gas, crude oil, and water) and has been created by the sequential steps of deposition, conversion, migration, and entrapment. Thus, porosity and permeability are the key features of the pore structure of the reservoir and are important factors in the production of natural gas and crude oil from the reservoir.

Natural gas is derived from aquatic plants and animals that lived and died hundreds of millions of years ago. Their remains mixed with mud and sand in layered deposits that, over the millennia, were geologically transformed into sedimentary rock. Gradually the organic matter decomposed and eventually formed natural gas and/or crude oil which migrated from the original source beds to more porous and permeable rocks, such as *sandstone* and *siltstone*, where it finally became trapped in the rock formation that was to become the reservoir. A series of reservoirs within a common rock structure or a series of reservoirs in separate but neighboring formations is commonly referred to as an *oil field* or, in the context of this book, a *gas field*. A group of natural gas fields or crude oil fields is often found in a single geologic environment known as a *sedimentary basin* or *province*. When a hydrocarbon reservoir is identified, it is important to also identify the types of fluids that are present, along with their main physicochemical characteristics (Fig. 2.1). Generally, that information is obtained by performing a pressure-volume-temperature analysis on a fluid sample taken from the reservoir.

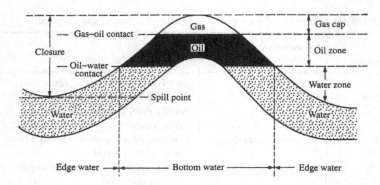

Figure 2.1 Examples of some of the parameters necessary to identify reservoir structure using an anticlinal reservoir containing oil and associated gas as the example (see also Table 2.2 and Table 2.5).

Obtaining preliminary values of properties such as: molar percentage of heptane and heavier components (% mole of C_{7+}), molecular weight of the original fluid (MW), maximum retrograde condensation (MRC), and dew point pressure (P_d). Most of these properties are essential for exploitation of a gas condensate reservoir and early availability of the data allows the initiation of reservoir studies that will ensure an efficient exploitation and maximize the final recovery of the liquids present in the reservoir. The only parameter needed to use these correlations is the value of the gas condensate ratio of the fluid during the early stage of production.

2.4.1 Natural gas reservoirs

A reservoir generally is a geological formation that is made up of layers of porous, sedimentary rock, such as sandstone, in which the gas can collect. However, for retention of the gas each trap must have an impermeable base rock and an impermeable cap rock to prevent further movement of the gas. Such formations, known as *reservoirs* or *traps* (i.e., naturally occurring storage areas) vary in size and can retain varying amounts of gas. There are a number of different types of these formations, but the most common is, characteristically, a folded rock formation such as an anticline as occurs in many natural gas and crude oil reservoirs (Fig. 2.1). On the other hand, a reservoir may be formed by a geological fault that occurs when the normal sedimentary layers sort of split vertically, so that impermeable rock shifts down to trap natural gas in the more permeable limestone or sandstone layers. Essentially, the geological formation which layers impermeable rock over more porous, oil and gas rich sediment has the potential to form a reservoir.

Just as processes play a role in determining the composition and type of gaseous products, reservoirs (especially reservoir mineralogy) also play a major role in determining the composition and, hence, the behavior, of natural gas. Different minerals adsorb gases at different rates leading to a requirement for determining the mineralogical composition of the rock within the reservoir (Ballard, 2007). The data related to mineralogical composition can lead to estimates of the gaseous constituents not produced through a well and the mean by which these constituents are held within the reservoir. In addition, the pressure within the reservoir may also be a major influence of the properties of the gas as it appears at the wellhead. For example, in reservoir pressure varies depending on the depth and, because of the pressure, some of the typically gaseous constituents (at STP) of natural gas may be in liquid form and therefore appear at the surface as liquids (gas condensate) (Chapter 9: Gas Condensate).

The formations that contain natural gas are varied in mineralogy and require considerable knowledge and expertise to produce the gas efficiently and economically. Thus, as the development of a natural gas commences, the structure and mineralogy of the reservoir (Fig. 2.1) increases in importance, especially when the highly adsorbing clay minerals are present. Thus the four most important items to address during a mineralogical analysis of the reservoir rock are: (1) the specific elements that are in the reservoir rock, (2) the characteristics of these elements, such as the atomic weight and atomic number, (3) the concentrations of the elements, % w/w, ppm w/w, and (4) the minerals or mineral phases with which the elements are associated. These data are especially important for in situ analysis of the reservoir and the gas where knowledge of the phase relationships in the reservoir (involving determination of the dew point of the gas) is a first-order control on the development of the resource. Additionally, the geological data should describe any lateral and vertical variations of reservoir characteristics (Chopra et al., 1990; Damsleth, 1994), and how those characteristics will affect the production method and the composition of the produced gas.

In essence, the reservoir is a subsurface trap where the basement rock and the cap rock are in place to trap the gas and the crude oil and prevent any further migration of the natural gas and crude oil. Typically, whether it is a natural gas reservoir (unassociated gas) or a natural gas—crude oil reservoir (associated natural gas), the reservoir rock is a porous medium made up of pores conventionally envisaged as irregularly shaped holes in rock some $1-100$ μm in length and diameter connected to maybe six other pores—there can be some 10^6 pores in 1 cc of rock and possibly over 10^{22} pores in a typical reservoir. However, the variables that is typically sought to describe a reservoir are those which influence the amount, the position, the accessibility, and the flow of fluids through the reservoir.

The mineralogy of a rock determines its physical and chemical properties. In a reservoir rock, physical properties such as density, sonic velocity, compressibility, and wetting properties contribute to essential reservoir properties such as porosity, permeability, and fluid saturations. Also, the chemical properties of a reservoir rock are especially important in drilling and production operations in order to predict how the formation will react when exposed to foreign materials such as drilling mud and production chemicals. The knowledge of mineralogy also provides information about the deposition and diagenesis of reservoir rocks, which further helps in understanding flow characteristics in a reservoir.

Furthermore, the identification of the lithology (i.e., identification of the rock types) is fundamental to all reservoir characterization because the physical and chemical properties of the rock that holds hydrocarbon derivatives and/or water affect the response of every tool used to measure formation properties. Understanding reservoir lithology is the foundation from which all other petrophysical calculations are made. In fact, the lithology of a reservoir impacts the petrophysical calculations in numerous ways. The depositional environment and sediments being deposited will define the grain size, its sorting, and its distribution within the reservoir interval. In most sandstone reservoirs, the depositional environment controls the porosity—permeability relationship—two essential factors that determine gas flow within and from the reservoir to the production well.

Natural gas reservoirs, like crude oil reservoirs, exist in many forms such as the dome (syncline—anticline) structure (Fig. 2.1), with water below, or a dome of gas with a crude oil rim and water below the oil. When the water is in direct contact with the gas, pressure effects may dictate that a considerable portion of the natural gas (20% v/v or more) is dissolved in the crude oil as well as in the water. As gas is produced (or recovered) from the reservoir, the reservoir pressure declines allowing the dissolved gas to enter the gas phase. In addition, and because of the variability of reservoir structure, gas does not always flow equally to wells placed throughout the length, breadth, and depth of the reservoir and at equal pressure. Recovery wells must be distributed throughout the reservoir to recover as much of the gas as efficiently as possible.

For carbonate formations, the rock formations typically consist of interbedded sequences of carbonates, dolomites, anhydrite, salt, and shale layers. The keys to reservoir development within the carbonate layers are the original grain size and how it has been altered by chemical diagenetic processes. As these chemical reactions take place, the pore-size distribution and porosity level will change (e.g., by dolomitization). Carbonate reservoir porosity is also greatly enhanced by weathering, dissolution, and fracturing (Speight, 2016). The texture of clay minerals and their composition are requisite for understanding their influence on reservoir characteristics and the properties of the natural gas and crude oil (Archer, 1985; Hurst and Archer, 1986; Speight, 2014a; Saha et al., 2017).

2.4.2 Natural gas reservoirs

In addition to the natural gas found in reservoirs that contain crude oil, there are also reservoirs in which natural gas may be the sole occupant. Again, the principal constituent of natural gas is methane, but other hydrocarbon derivatives, such as ethane, propane, and butane, may also be present, but these constituents are typically in lower proportions than found in natural gas from crude oil reservoirs. Carbon dioxide is also a common constituent of natural gas. Trace amounts of rare gases, such as helium, may also occur, and certain natural gas reservoirs are a source of these rare gases.

Since natural gas has a low density, once formed it will rise toward the surface of the Earth through loose, shale type rock and other material. Most of this methane will simply rise to the surface and dissipate into the air. However, a great deal of this methane will rise up into geological formations that trap the gas under the ground. These formations are made up of layers of porous, sedimentary rock with a denser, impermeable layer of rock on top. This impermeable rock traps the natural gas under the ground. If these formations are large enough, they can trap a great deal of natural gas underground, in what is known as a reservoir.

There are a number of different types of these formations, but the most common is created when the impermeable sedimentary rock appears as *dome*-shaped formation that traps all of the natural gas (Speight, 2014a). There are a number of ways that this sort of dome may be formed. For instance, faults are a common location for oil and natural gas deposits to exist. A fault occurs when the normal sedimentary layers sort of split vertically, so that impermeable rock shifts down to trap natural gas in the more permeable limestone or sandstone layers. Essentially, the geological formation which layers impermeable rock over more porous, oil and gas rich sediment has the potential to form a reservoir.

Once a potential natural gas deposit has been located by a team of exploration geologists and geophysicists, it is up to a team of drilling experts to actually dig down to where the natural gas is thought to exist. Although the process of digging deep into the Earth's crust to find deposits of natural gas that may or may not actually exist seems daunting, a number of innovations and techniques have been developed that increase the efficiency of drilling for natural gas, with a concurrent decrease in cost. The advance of technology has also contributed greatly to the increased efficiency and success rate for drilling natural gas wells. However, the search for natural gas always carries some drawbacks and there is always the inherent risk that no natural gas will be found.

However, on the positive side, if a new well, once drilled, does in fact come in contact with a natural gas reservoir, it is developed to allow for the extraction of this natural gas, and is termed a *development well* or *productive well*. At this point, with the well drilled and hydrocarbon derivatives present, the well may be completed to facilitate its production of natural gas. However, if the exploration team was incorrect in its estimation of the existence of marketable quantity of natural gas at a well site, the well is termed a *dry well*, and production does not proceed. Then, in order to successfully bring natural gas to the surface, a hole must be drilled through the impermeable rock to release the natural gas that is under pressure. Thus natural gas that is trapped under in a subsurface formation can be recovered by drilling through the impermeable rock. Gas in these reservoirs is typically under pressure, allowing it to escape from the reservoir on its own. The decision of whether or not to drill a well into a reservoir depends on several factors, not the least of which is the economic characteristics of the potential natural gas reservoir.

The exact placement of the drill site depends on a variety of factors, including the nature of the potential formation to be drilled, the characteristics of the subsurface geology, and the depth and size of the target deposit. After the geophysical team identifies the optimal location for a well, it is necessary for the drilling company to ensure that they complete all the necessary steps to ensure that they can legally drill in that area. This usually involves securing permits for the drilling operations, establishment of a legal arrangement to allow the natural gas company to extract and sell the resources under a given area of land, and a design for gathering lines that will connect the well to the pipeline. There are a variety of potential owners of the land and mineral rights of a given area.

In order to combat the presence of sulfur compounds in natural gas (many pipeline owners specify maximum sulfur content), scrubbers are installed, usually at or near the wellhead. The scrubbers serve primarily to remove sand and other large-particle impurities such as hydrogen sulfide. Heaters also ensure that the temperature of the gas does not drop too low. With natural gas that contains even low quantities of water, natural gas hydrates tend to form when the temperature decreases. These hydrates are solid or semisolid compounds, resembling ice-like crystals (Chapter 3: Unconventional Gas) (Berecz and Balla-Achs, 1983; Gudmundsson et al., 1998; Sloan, 1997, 2000). When gas hydrates accumulate, the passage of natural gas through valves and gathering systems can be seriously impeded. To reduce the occurrence of hydrates, small natural gas-fired heating units are typically installed along the gathering pipe wherever it is likely that hydrates may form.

A reservoir containing *wet* gas (Table 2.4) with a large amount of valuable *natural gas liquids* (any hydrocarbon derivatives other than methane such as ethane, propane, and butane) and even light crude oil (>30° API) and condensate has to be treated carefully. When the reservoir pressure drops below the critical point for the mixture, the liquids may condense into a liquid phase out and remain in the reservoir. Thus it is necessary to implement a *cycling* process in which the wet gas is produced to the surface and the natural gas liquids are condensed as a separate stream and the gas is compressed and injected back into the reservoir to maintain the pressure.

There are many reservoirs that produce gas condensate, and each reservoir will produce gas condensate with its own unique composition. However, in general, gas condensate has a specific gravity on the order of ranging from 0.5 to 0.8 and is composed of higher molecular weight hydrocarbon up to an including dodecane (C_{12}). Propane and butane, normally gases at standard temperature and pressure may also occur in gas condensate as gas soluble in the liquid hydrocarbon

Table 2.4 Composition of dry gas, wet gas, and condensate

Component	Dry gas	Wet gas	Condensate
CO_2	0.10	1.41	2.37
N_2	2.07	0.25	0.31
C_1	86.12	92.46	73.19
C_2	5.91	3.18	7.80
C_3	3.58	1.01	3.55
i-C_4	1.72	0.28	0.71
n-C_4	–	0.24	1.45
i-C_5	0.50	0.13	0.64
n-C_5	–	0.08	0.68
C_6	–	0.14	1.09
C_7+	–	0.82	8.21

derivatives. The C_8 to C_{12} hydrocarbon derivatives include the higher boiling hydrocarbon compounds such as cyclohexane derivatives and aromatic derivative such as benzene, toluene ($C_6H_5CH_3$), xylene isomers (*ortho-*, *meta-*, and *para-* $CH_3C_6H_4CH_3$), and ethyl benzene ($C_6H_5C_2H_5$). In addition, the gas condensate may contain additional impurities such as hydrogen sulfide, thiol derivatives (also called mercaptans, RSH), carbon dioxide, cyclohexane (C_6H_{12}), and low molecular weight aromatics such as benzene (C_6H_6), toluene (C_6H5CH_3), ethylbenzene ($C_6H_5CH_2CH_3$), and xylene derivatives ($H_3CC_6H_4CH_3$) (Mokhatab et al., 2006; Speight, 2014a).

The composition of the gas condensate can have a major influence on the production of gas and condensate at the wellhead. For example, when condensation occurs in the reservoir, the phenomenon known as condensate blockage can halt flow of the liquids to the wellbore. Hydraulic fracturing is the most common mitigating technology in siliciclastic reservoirs (reservoirs composed of clastic rocks), and acidizing is used in reservoirs composed of carbonate mineral rocks (generally referred to as carbonate reservoirs) (Speight, 2016). Briefly, clastic rocks are composed of fragments, or clasts, of preexisting minerals and rock. A clast is a fragment of geological detritus, chunks and smaller grains of rock broken off other rocks by physical weathering. The geological term *clastic* is used with reference to sedimentary rocks as well as to particles in sediment transport whether in suspension or as bed load, and in sedimentary deposits.

The most common reservoir rocks are sandstone (SiO_2), limestone ($CaCO_3$), and dolomite [a mixed calcium−magnesium carbonate mineral ($CaCO_3 \cdot MgCO_3$)]. The four basic elements of a reservoir system include: (1) the source rock, which is the rock containing the organic material that is converted into natural gas and/or crude oil; (2) a migratory pathway, which is a route for the partially formed or completely formed natural gas and crude oil to migrate from the source rock to the reservoir; (3) the reservoir rock, which is a rock of suitable porosity that can store and suitable permeability that can allow the fluid to move through the reservoir to a production well, and (4) a seal, which is an impermeable cap rock that prevents the upward escape of the natural gas and crude oil to the surface of the Earth.

2.4.3 Crude oil reservoirs

In order to process and transport associated dissolved natural gas, it must be separated from the oil in which it is dissolved. This separation of natural gas from crude oil is most often done using equipment installed at or near the wellhead. The actual process used for water-crude oil-natural gas separation starts at the wellhead and the equipment used can vary considerably. Although dry pipeline quality natural gas is virtually identical across different geographic areas, raw natural gas from different regions will vary in composition (Table 2.2) and therefore separation requirements may emphasize or deemphasize any one of several optional separation processes. Typically, it is expected that a separator vessel will cause the water-oil-gas to separate into the three individual phases. The most basic type of separator is known as a conventional separator and consists of a simple closed tank, where the force of gravity serves to separate the heavier liquids like oil, and the lighter gases, like natural gas.

However, in many instances, natural gas is dissolved in oil underground primarily due to the formation pressure and may give rise to what is known as foamy oil. In some cases, when natural gas and crude oil is produced, it is possible (but subject to pressure and temperature effects in the reservoir) that the natural gas will readily separate from the crude oil. In these cases, separation of oil and gas is relatively easy, and the two hydrocarbon derivatives are sent in separate ways for further processing. In many instances, however, specialized equipment is necessary to separate oil and natural gas. An example of this type of equipment is the low-temperature separator and this type of separator is most often used for wells producing high-pressure natural gas as well as light crude oil or gas condensate. These separators use pressure differentials to cool the wet natural gas and separate the oil and condensate.

In the separation process, the natural gas enters the separator, being cooled slightly by a heat exchanger. The gas then travels through a high-pressure liquid *knockout pot*, which serves to remove any liquids into a low-temperature separator. The gas then flows into this low-temperature separator through a choke mechanism, which expands the gas as it enters the separator. This rapid expansion of the gas allows for the lowering of the temperature in the separator. After liquid removal, the dry gas then travels back through the heat exchanger and is warmed by the incoming wet gas. By varying the pressure of the gas in various sections of the separator, it is possible to vary the temperature, which causes the oil and some water to be condensed out of the wet gas stream. This basic pressure–temperature (PT) relationship can work in reverse as well, to extract gas from a liquid oil stream. In addition, the potential for corrosion of the equipment and pipelines due to the presence of water and acid gases (hydrogen sulfide, H_2S, and carbon dioxide, CO_2) is high (Speight, 2014b). Hence, the need to know the composition of the natural gas mixture that is produced at the wellhead.

2.4.4 Gas condensate reservoirs

Natural-gas condensate is a low-density mixture of hydrocarbon liquids that are present as gaseous components in the raw natural gas produced from many natural gas fields. Some gas species within the raw natural gas will condense to a liquid state if the temperature is reduced to below the hydrocarbon dew-point temperature at a definitive pressure.

A gas-condensate reservoir (also called a *dew point reservoir*) is a reservoir in which condensation causes a liquid to leave the gas phase. The condensed liquid remains immobile at low concentrations. Thus, the gas produced at the surface will have a lower liquid content, and the producing gas–oil ratio therefore rises. This process of MRC continues until a point of maximum liquid volume is reached. The term *retrograde* is used because generally vaporization, rather than condensation, occurs during isothermal expansion. After the dew point is reached, because the composition of the produced fluid changes, the composition of the remaining reservoir fluid also changes.

Typically, a gas condensate reservoir will have a reservoir temperature located between the critical point and the cricondentherm on the reservoir fluid PT diagram. This is one way of identifying a gas condensate reservoir—any other definition—such as condensate–gas ratio or molecular weight of the C^{7+} fraction or the API gravity of the C_{7+} fraction may leave gaps in knowledge of the behavior of the reservoir and the condensate (Thomas et al., 2009).

Drip gas, so named because it can be drawn off the bottom of small chambers (called *drip chambers* or *drips*) sometimes installed in pipelines from gas wells, is another name for natural-gas condensate, a naturally occurring form of gasoline obtained as a byproduct of natural gas extraction. Drip gas is defined in the United States Code of Federal Regulations as consisting of butane, pentane, and hexane derivatives. Within set ranges of distillation, drip gas may be extracted and used as a cleaner and solvent as well as a lantern and stove fuel. Accordingly, each type of condensate (including drip gas, natural gasoline, and casinghead gas) requires a careful compositional analysis for an estimation of the potential methods of preliminary purification at the wellhead facilities prior to transportation through a pipeline to a gas processing plant or a refinery (Speight, 2014a, 2017a).

Because gas condensate is typically liquid in ambient conditions and also has very low viscosity (Chapter 9: Gas Condensate), it is often used as a diluent for highly viscous heavy crude oil that cannot otherwise be efficiently transported by means of a pipeline. In particular, condensate (or low-boiling naphtha from a refinery) is frequently mixed with bitumen from tar sand (called oil sand in Canada) to create the blend known as *Dilbit*. However, caution is required when condensate having an unidentified composition is blended with heavy oil, extra heavy oil, and/or tar sand bitumen, since the potential for incompatibility of the blended material may become a reality.

This is especially true if the condensate is composed predominantly of *n*-alkane hydrocarbon derivatives of the type (pentane, C_5H_{12}, and heptane, C_7H_{16}, as well as other low-boiling liquid alkane derivatives). These hydrocarbon derivatives are routinely used in laboratory deasphalting and in commercial deasphalting units in which the asphaltene fraction is produced as a solid insoluble as a solid product from the heavy oil or bitumen feedstock (Speight, 2014a, 2015a).

2.5 Reservoir fluids

Reservoir fluids are the fluids (including gases and solids) that exist in a reservoir and each fluid requires application of one or more of the relevant standard test methods (Table 2.5) (ASTM D2017) to estimate behavior of the fluid both in and out of the reservoir. The fluid type must be determined very early in the life of a reservoir (often before sampling or initial production) because fluid type is the critical factor in many of the decisions that must be made about producing the fluid from the reservoir.

Table 2.5 Data required for the identification of reservoir fluids, including natural gas and gas condensate

Formation volume factor	The ratio of a phase volume (water, oil, gas, or gas plus oil) at reservoir conditions, relative to the volume of a surface phase (water, oil, or gas) at standard conditions resulting when the reservoir material is brought to the surface. Denoted mathematically as B_w (bbl/STB), B_0 (bbl/STB), B_g (ft^3/SCF), and B_t (bbl/STB)
Solution gas−oil ratio	The amount of surface gas that can be dissolved in a stock tank oil when brought to a specific pressure and temperature. Denoted mathematically as R_s (SCF/STB)
Solution oil−gas ratio	The amount of surface condensate that can be vaporized in a surface gas at a specific pressure and temperature; sometimes referred to as liquid content. Denoted mathematically as r_s (STB/MMSCF)
Liquid specific gravity	The ratio of density of any liquid measured at standard conditions (usually 14.7 psia and 60°F) to the density of pure water at the same standard conditions. Denoted mathematically as γ_o (where water = 1)
API gravity	A common measure of the specific gravity of crude oil, defined by $\gamma_{API} = (141.5/\gamma_o) - 131.5$, with units in °API
Gas specific gravity	The ratio of density of any gas at standard conditions (14.7 psia and 60°F) to the density of air at standard conditions; based on the ideal gas law ($pV = nRT$), gas gravity is also equal to the gas molecular weight divided by air molecular weight ($M_{air} = 28.97$). Denoted mathematically as γ_g (where air = 1)
Bubble point pressure	At a given temperature, this condition occurs when an oil releases an infinitesimal bubble of gas from solution when pressure drops below the bubble point
Retrograde pressure	At a given temperature, this condition occurs when a gas condenses a drop of oil from solution when pressure drops below the dew point; also called "retrograde dew point pressure"
Saturation pressure	An oil at its bubble point pressure or a gas at its dew point pressure
Critical point	The pressure and temperature of a reservoir fluid where the bubble point pressure curve meets the retrograde dew point pressure curve; a unique state where all properties of the bubble point oil are identical to the dew point gas
Composition	Quantifies the amount of each component in a reservoir mixture, usually reported in mole fraction. Typical components in crude oil reservoir mixtures include the nonhydrocarbons N_2, CO_2, and H_2S and the hydrocarbons C_1 C_2, C_3, i-C_4, n-C_4, i-C_5, n-C_5, C_6, and C_{7+} (heptanes-plus)
Saturated condition	A condition where an oil and gas are in thermodynamic equilibrium, that is, the chemical force exerted by each component in the oil phase is equal to the chemical force exerted by the same component in the gas phase, thereby eliminating mass transfer of components from one phase to the other.
Undersaturated condition	A condition when crude oil or natural gas is in a single phase but not at its saturation point (bubble point or dew point), that is, the mixture is at a pressure greater than its saturation pressure

Reservoirs contain complex mixtures of fluids and the behavior of these fluids is strongly dependent on chemical properties and physical properties. Heavy crude oil is a hydrocarbonaceous fluid (even a semisolid) that is also a multicomponent mixture composed of nonhydrocarbon derivatives and a variety of hydrocarbon derivatives—especially of the alkane series of hydrocarbon derivatives. Typical hydrocarbon derivatives encountered in heavy oil are the higher boiling hydrocarbon derivatives—the amount depending on the original source materials and the maturation pathways; volatile hydrocarbon derivatives boiling lower that C_{12} are not present in a ready abundance. Thus the properties of reservoir fluids can play a key role in the design and optimization of injection/production strategies and surface facilities for efficient reservoir management. Inaccurate fluid characterization often leads to high uncertainties in in-place-volume estimates and recovery predictions, and hence affects asset value. Indeed, evaluating reservoir fluids the application of standard test methods to property determination is an important aspect of prospect evaluation, development planning, and reservoir management. Although many different types of fluid exist, the composition of the reservoir fluids provides important information that can influence the recovery process.

To produce a wellhead product from a reservoir there has to be flow of the fluids to the wellbores through the heterogeneous porous media of the reservoir. Fluid movements within the reservoir are governed by the local fluid potential gradients (Whitson and Belery, 1994) and reservoir effective permeability, the injection and production points and the fluid viscosities. However, fluid flow is also governed by the character of the fluids in the reservoir.

The term *fluid* as used by reservoir engineers can mean the contents of the reservoir that are in the gaseous state or in the liquid state and the fluids exist either as (1) a solid phase, of which wax is an example, (2) a liquid phase, of which crude oil is the example, and (3) a gaseous phase, of which natural gas is the example. Gas and liquids coexist in a given reservoir. More specifically in the current context, a reservoir typically contains three main *fluids*: (1) natural gas, (2) crude oil, and (3) water with minor constituents being acid gases (carbon dioxide and hydrogen sulfide). These components will vary greatly in combination and proportion within each reservoir and, in the case of heavy crude oil, the amount of gas will be substantially less than would be found in a conventional oil reservoir. Reservoir fluids vary greatly in composition and chemical properties. Typically, heavy oil has an API gravity between 10 and 20 degrees and is more viscous than conventional crude oil and has the commonality of being, or ability to be, in the liquid state and, therefore, has mobility in the reservoir. As a result of this mobility in the reservoir, heavy crude oil can be recovered from a reservoir by the use of conventional (including enhanced) oil recovery techniques (Speight, 2009, 2014a).

In reservoirs that contain natural gas and crude oil, it is often assumed (with reasonable justification) that natural gas, being the least dense, is found at the top of the reservoir, with the crude oil beneath the natural gas, typically followed by water at the base of the reservoir. However, rather than discrete zones of natural gas, crude oil, and water, it is more likely that a reservoir will consist of several boundary zones, such as (1) an oil-in-water zone, (2) a water-in-oil zone, (3) a gas-in-oil zone, (4) a gas-in-water zone, as well as various zones of mixed composition. It is safe to assume that in the reservoir, zones are not always ordered according to density. Thus the distribution of the fluids in a reservoir rock is not always dependent on the density of the reservoir fluids as well as on the properties of the rock.

However, if the pores of the reservoir rock are of uniform size and evenly distributed, there is: (1) an upper zone where the pores are filled mainly by gas (the gas cap), (2) a middle zone in which the pores are occupied principally by oil with gas in solution, and (3) a lower zone with its pores filled by water. A certain amount of water (approximately 10%−30%) occurs along with the oil in the middle zone. There is a transition zone from the pores occupied entirely by water to pores occupied mainly by oil in the reservoir rock, and the thickness of this zone depends on the densities and interfacial tension of the oil and water as well as on the sizes of the pores. Similarly, there is some water in the pores in the upper gas zone that has at its base a transition zone from pores occupied largely by gas to pores filled mainly by oil.

The water found in the oil and gas zones (*interstitial water*, the compositional analysis of which can often provide information about the mineralogy of the reservoir from the presence of water-soluble minerals) and usually occurs as (1) collars around grain contacts, (2) a filling of pores with unusually small throats connecting with adjacent pores, or, to a much smaller extent, (3) as wetting films on the surface of the mineral grains when the rock is preferentially wet by water. The 3-D reservoir network allows continuity to exist for the hydrocarbon derivatives by means of connections on every side of the sand grains. The so-called gas−oil and oil−water contacts are generally regarded as being horizontal but have been known to exist as a very gentle incline. On occasion, part of an accumulation of the oil or gas has its lower boundary marked, not by the water-bearing zone of the reservoir rock but by an adjacent sealing rock that has characteristics similar to those of the cap rock. When the pressure and temperature conditions are suitable in relation to the proportions and the nature of the gas and oil, there may be no gas cap but only oil, with dissolved gas overlying the water.

The water (brine) produced with crude oil from the reservoir is brought to the surface with the crude oil. Because the water has been in contact with the oil, it contains some of the chemical characteristics of the formation and the oil itself. Oil and gas wells produce more water than oil (7 bbl/bbl-of-crude oil in some fields). The composition (salt content) of coproduced water determines the need for anti-scaling additives. There are strict regulations to limit disposal and beneficial use options as well as environmental impacts that pertain to oil field waters.

Reservoir temperatures may vary up to 90°C (194°F) or even higher, while surface conditions are around 20°C (68°F). Pressure can vary from its atmospheric value (or lower in the case of vacuum distillation) to a number in the hundred million Pascals (Pa). Within such an ample range of conditions, hydrocarbon fluids undergo severe transformations and exist as a single phase (gas, liquid, or solid) or coexist in several forms (liquid plus gas, solid plus liquid, vapor plus solid, or even in liquid-plus-liquid combinations). Understanding the methods by which the hydrocarbon fluids interact with and react to their thermodynamic surroundings is essential to understanding the PT diagram or the PT envelope. Each envelope represents a thermodynamic boundary separating the two-phase conditions (inside the envelope) from the single-phase region (outside). The correct identification of the type of hydrocarbon fluid is critical for the proper design and development of the correct production strategy for the field under consideration.

Typically, fluids from a crude oil reservoir are brought to the surface as a mixture of natural gas, crude oil, and water, which is then sent to a surface production facility before they can be disposed or sold to an industrial consumer (e.g., a refinery). A surface production facility is the system in charge of the separation of the well stream fluids into its three single phase components—oil, gas, and water—and of their transport and processing into marketable products and/or their disposal in an environmentally acceptable manner. Once separated, the oil, natural gas, and water follow different paths. Water is typically reinjected for reservoir pressure maintenance operations. The oil usually goes through a process of dehydration, which removes basic sediments and hydrocarbon fluids are assumed to comprise two components—stock tank oil and surface gas.

2.6 Production

Production is the process of extracting the hydrocarbon derivatives followed by processing (at the wellhead or in a gas processing facility) and separating the mixture of liquid hydrocarbon derivatives, gas, water, and solids, removing the constituents that are nonsaleable, and selling the liquid hydrocarbon derivatives and gas (Fig. 2.2). Production sites often handle crude oil from more than one well and, after any necessary wellhead treatment, is sent to a refinery for processing whereas natural gas may receive more assiduous processing to remove impurities either in the field (at the wellhead) or at a nearby natural gas processing plant.

Offshore drilling operations used to be some of the most risky and dangerous undertakings but improved offshore drilling rigs, dynamic positioning devices, and

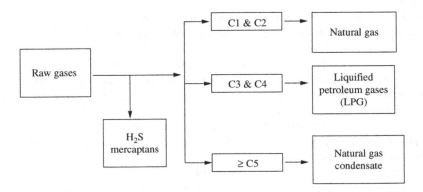

Figure 2.2 Schematic representation of natural gas processing.

modern navigation systems allow efficient offshore drilling in waters more than 10,000 ft deep. For offshore drilling, the wells are deep beneath the surface of the ocean, and artificial platforms are constructed on the surface. The first offshore rig was built and used in 1869, but it was not until 1974 that drilling was done far out in the ocean in deep water—namely in the Gulf of Mexico. The original rigs were designed to work solely in very shallow water, but the modern rigs have a similar four-legged design to the earliest models but are able to drill in very deep water. In either case (i.e., onshore or offshore), once the gas reservoir is found and penetrated, gas flows up through the well to the surface of the ground and into large pipelines. Some of the gases that are produced along with methane, such as ethane, propane, and butane (*gas liquids*) are separated and cleaned at a gas processing plant. The gas liquids, once removed, are employed individually or collectively for various uses.

Deep-sea rigs have specific components that allow them to function efficiently and the two most important features are (1) the subsea drilling template and (2) the blowout preventer. The subsea drilling template connects the drilling site to the platform at the surface of the water, and the blowout preventer is in place to prevent oil or gas from leaking into the water. Also, modern offshore (deep-sea) rigs fall into two main types of deep-sea rigs: (1) movable rigs and (2) unmovable rigs. As the name indicates, moveable rigs can move from location to location and drill in multiple places, while unmovable rigs remain in one place only. There are also various other types of rigs, including drill ships and drilling barges (Speight, 2015b).

During the many years when natural gas was produced as a byproduct of crude oil productions its value was largely ignored. In many cases large volumes of natural gas were flared with no effort being made to conserve this valuable material. Only a slight improvement was realized when many of the first sales contracts were written for ridiculously low prices. Until recently, large volumes of natural gas were sold for less than the cost of compression to the sales pressures. Such contracts completely ignored the intrinsic value of this high-quality fuel. This poor situation was made worse when these distress prices were recognized by a federal regulatory body as proper price levels for natural gas moving in interstate commerce.

In fact, natural gas reservoirs, like crude oil reservoirs, exist in many forms such as the dome (syncline—anticline) structure, with water below, or a dome of gas with a crude oil interface and water below the crude oil. When the water is in direct contact with the gas, pressure effects may dictate that a considerable portion of the gas (20% v/v or more) is dissolved in the crude oil as well as in the water. As the gas is produced the reservoir pressure will decline allowing the gas to enter the gas phase. In addition, and because of the variability of reservoir structure, gas does not always flow equally throughout the length, breadth, and depth of the reservoir and at equal pressure, recovery wells must be distributed throughout the reservoir to recover as much of the gas as possible.

As the gas pressure in the reservoir declines, the reservoir energy declines and the natural gas requires stimulation for continued production. Furthermore, reduction in the gas pressure may allow compaction of the reservoir rock by the weight of rock above eventually resulting in subsidence of the surface above the reservoir. This can be gradual process or a sudden catastrophic process depending on the structures of the above geological features.

A reservoir containing *wet* gas with a large amount of valuable natural gas liquids (ethane, propane, and higher molecular weight hydrocarbon derivatives) and even light crude oil and condensate has to be treated carefully. When the reservoir pressure drops below the critical point for the mixture, the liquids may condense out and remain in the reservoir. Thus it is necessary to implement a *cycling* process in which the wet gas is produced to the surface and the natural gas liquids are condensed as a separate stream and the gas is compressed and injected back into the reservoir to maintain the pressure.

The production of natural gas from a reservoir is inherently more efficient than the production of crude oil from a reservoir. In volumetric reservoirs, at moderate depths with high formation permeabilities, recoveries of 90% v/v of the gas originally in place are common. However, in many cases, high recoveries are not attained. In a reservoir with a low permeability, the producing rates may fall below the economic limit, while large portions of the original gas are still in the reservoir. The minimum economic gas production rate is raised by the production of liquids or by any other difficulties which tend to increase operating problems and expenses. In reservoirs with strong water drives and unequal water invasion, large amounts of gas may be lost in the residual gas saturation or by premature watering out of the producing wells. In such cases, the recovery of the gas originally in place may be as low as 50% of the volume in place.

Thus the suitability of a natural gas reservoir for the production of gas is impacted by evaluation and implementation of sound reservoir management and investment decisions during different stages of field life (delineation, primary depletion, waterflooding, infill primary depletion, waterflooding, infill drilling, and flooding). Reservoir characterization has a big impact on field development decisions and process and facilities design. The two most challenging issues in reservoir characterization are: (1) estimation and description of key reservoir properties (porosity, permeability, transmissivity, geometry, saturations and permeability, transmissivity, geometry, saturations and pressures, and rock−fluid interactions) at locations not pressures, and rock−fluid interactions at locations not sampled by well data, and (2) representation of these properties as effective properties for reservoir simulation properties and prediction of reservoir performance.

Nevertheless, the necessary properties are evident when the reservoir is first prepared for gas production through the following categories: (1) gas wells, (2) well completion, (3) the wellhead, which must be suitable for preliminary treatment of the gas when necessary, (4) well treatment, and (5) natural gas production. Each of these necessary issues are presented in the following sections.

2.6.1 Gas wells

The well development is an essential part of the production process and is instituted immediately after exploration has located a reservoir (or *field*) that can economically produce natural gas. This involves the construction of one or more wells from the beginning (often referred to as *spudding*) to either abandonment if no hydrocarbon derivatives are found, or to well completion if hydrocarbon derivatives are found in sufficient quantities. The well development program depends upon the quality of the natural gas and the reservoir pressure.

Gas wells are technically similar to crude oil wells with casing, tubing, and a wellhead and controls at the top (Speight 2007). Conventional wells use casings *telescoped* inside each other and given pressure resistance by cement. However, with *expanded tubulars* each tubular casing is expanded against the previous one by pumping a tool called a *mandrel* down the casing. Thus the well can either be thinner and cheaper for a given size of final tubing, or alternatively, using the conventional 20-in. diameter external casing, there is space to install a wider and thus higher capacity tubing.

When gas is contained in deep, relatively tight, low-permeability reservoirs, it is often economic to drill large-diameter wells with horizontal sections through the reservoirs to collect more gas. In addition, gas production can be increased by fracturing the reservoir rock using over-pressure and by pumping in ceramic beads that is maintaining the integrity of the fractures. It is also attractive to install *smart wells* that have measurement gauges permanently installed downhole. This eliminates the need for lowering gauges down the wells during production. With the high gas flow rates, sand can often be pulled into the well with the gas and therefore wearing parts are hardened against sand erosion.

2.6.2 Well completion

Once a natural gas or oil well is drilled, and it has been verified that commercially viable quantities of natural gas are present for extraction, the well must be completed to allow for the flow of crude oil or natural gas out of the formation and up to the surface. This process includes strengthening the well hole with casing, evaluating the pressure and temperature of the formation, and then installing the proper equipment to ensure an efficient flow of natural gas out of the well.

Since oil is commonly associated with natural gas deposits, a certain amount of natural gas may be obtained from wells that were drilled primarily for oil production. In some cases, this *associated natural gas* is used to help in the production of oil, by providing pressure in the formation for the oils extraction. The associated natural gas may also exist in large enough quantities to allow its extraction along with the oil.

Condensate wells are wells that contain natural gas, as well as a liquid condensate that is predominantly hydrocarbon derivatives. Thus the gas condensate is actually a mixture of hydrocarbon derivatives that are liquid at ambient temperature and pressure mixture that is often separated from the natural gas either at the wellhead, or during the processing of the natural gas. Depending on the type of well that is being drilled, completion may differ slightly. It is important to remember that natural gas, being lighter than air and given the appropriate conditions, will naturally rise to the surface of a well. Because of this, in many natural gas and condensate wells, lifting equipment and well treatment are not necessary.

Completing a well consists of a number of steps; installing the well casing, completing the well, installing the wellhead, and installing lifting equipment or treating the formation should that be required. The *well casing* consists of a series of metal tubes installed in the freshly drilled hole and installation of the casing is an important part of the drilling and completion process. Casing serves to strengthen the sides of the well hole, ensure that no oil or natural gas seeps out of the well hole as it is brought to the surface, and to keep other fluids or gases from seeping into the formation through the well. In addition to strengthening the production well, the casing also provides a conduit to allow hydrocarbon derivatives to be extracted without intermingling with other fluids and formations found underground. It is also instrumental in preventing blowouts, allowing the formation to be sealed from the top should dangerous pressure levels be reached. The types of casing used depend on the subsurface characteristics of the well, including the diameter of the well (which is dependent on the size of the drill bit used) and the pressures and temperatures experienced throughout the well. In most wells, the diameter of the well hole decreases the deeper it is drilled, leading to a type of conical shape that must be considered when installing casing. There are five different types of well casing, which include: (1) conductor casing, (2) surface casing, (3) intermediate casing, (4) liner string, and (5) production casing.

Conductor casing is installed first, usually prior to the arrival of the drilling rig. The hole for conductor casing is often drilled with a small auger drill, mounted on the back of a truck. Conductor casing, which is usually no more than 20–50 ft long, is installed to prevent the top of the well from caving in and to help in the process of circulating the drilling fluid up from the bottom of the well. Onshore, this casing is usually 16–20 in. in diameter, while offshore casing usually measures 30–42 in. The conductor casing is cemented into place before drilling begins.

Surface casing is the next type of casing to be installed and can be anywhere from a few hundred to 2000 ft long and is smaller in diameter than the conductor casing. When installed, the surface casing fits inside the top of the conductor casing. The primary purpose of surface casing is to protect freshwater deposits near the surface of the well from being contaminated by leaking hydrocarbon derivatives or salt water from deeper underground. It also serves as a conduit for drilling mud returning to the surface and helps protect the drill hole from being damaged during drilling. Surface casing, like conductor casing, is also cemented into place. Regulations often dictate the thickness of the cement to be used, to ensure that there is little possibility of freshwater contamination.

Intermediate casing is usually the longest section of casing found in a well. The primary purpose of intermediate casing is to minimize the hazards that come along with subsurface formations that may affect the well. These include abnormal underground pressure zones, underground shale formations, and formations that might have otherwise contaminated the well, such as underground salt water deposits. In many instances, even though there may be no evidence of an unusual underground formation, intermediate casing is run as insurance against the possibility of such a formation affecting the well. These intermediate casing areas may also be cemented into place for added protection. *Liner strings* are sometimes used instead of intermediate casing and are commonly run from the bottom of another type of casing to the open well area. However, liner strings are usually just attached to the previous casing with *hangers*, instead of being cemented into place. This type of casing is thus less permanent than intermediate casing.

Production casing (alternatively called the *oil string* or *long string*) is installed last and is the deepest section of casing in a well. This casing that provides a conduit from the surface of the well to the crude oil producing formation. The size of the production casing depends on a number of considerations, including the lifting equipment to be used, the number of completions required, and the possibility of deepening the well at a later time. For example, if it is expected that the well will be deepened at a later date, the production casing must be wide enough to allow the passage of a drill bit when required.

Once the casing has been set, and in most cases cemented into place, proper lifting equipment is installed to bring the hydrocarbon derivatives from the formation to the surface. Once the casing is installed, tubing is inserted inside the casing, from the opening well at the top, to the formation at the bottom. The hydrocarbon derivatives that are extracted run up this tubing to the surface. This tubing may also be attached to pumping systems for more efficient extraction, should that be necessary.

Well completion commonly refers to the process of finishing a well so that it is ready to produce oil or natural gas. In essence, completion consists of deciding on the characteristics of the intake portion of the well in the targeted hydrocarbon formation. There are a number of types of completions, including: (1) open hole completion, (2) conventional perforated completion, (3) sand exclusion completion, (4) permanent completion, (5) multiple zone completion, and (6) drain hole completion. The use of any type of completion depends on the characteristics and location of the hydrocarbon formation to be mined.

Open hole completion is the most basic type and is only used in very competent formations, which are unlikely to cave in. An open hole completion consists of simply running the casing directly down into the formation, leaving the end of the piping open, without any other protective filter. Very often, this type of completion is used on formations that have been treated with hydraulic of acid fracturing. *Conventional perforated completion* consists of production casing being run through the formation. The sides of this casing are perforated, with tiny holes along the sides facing the formation, which allows for the flow of hydrocarbon derivatives into the well hole, but still provides a suitable amount of support and protection for the well hole. The process of actually perforating the casing involves the use of specialized equipment designed to make tiny holes through the casing, cementing, and any other barrier between the formation and the open well. In the past, *bullet perforators* were used, which were essentially small guns lowered into the well. The guns, when fired from the surface, sent off small bullets that penetrated the casing and cement. Currently, *jet perforating* is preferred and consists of small, electrically ignited charges, lowered into the well that, when ignited, poke tiny holes through to the formation, in the same manner as bullet perforating.

Sand exclusion completion is designed for production in an area that contains a large amount of loose sand. These completions are designed to allow for the flow of natural gas and oil into the well, but at the same time prevent sand from entering the well. Sand inside the well hole can cause many complications, including erosion of casing and other equipment. The most common method of keeping sand out of the well hole are screening or filtering systems.

Permanent completion is the type of completion in which the completion, and wellhead, is assembled and installed only once. Installing the casing, cementing, perforating, and other completion work is done with small diameter tools to ensure the permanent nature of the completion. Completing a well in this manner can lead to significant cost savings compared to other types.

Multiple zone completion is the practice of completing a well such that hydrocarbon derivatives from two or more formations may be produced simultaneously, without mixing with each other. For example, a well may be drilled that passes through a number of formations on its way deeper underground, or alternately, it may be efficient in a horizontal well to add multiple completions to drain the formation most effectively. Although it is common to separate multiple completions so that the fluids from the different formations do not intermingle, the complexity of achieving complete separation is often a barrier. In some instances, the different formations being drilled are close enough in nature to allow fluids to intermingle in the well hole. When it is necessary to separate different completions, hard rubber packing instruments are used to maintain separation.

Drain hole completion is a form of horizontal or slant drilling and consists of drilling out horizontally into the formation from a vertical well, essentially providing a "drain" for the hydrocarbon derivatives to run down into the well. In certain formations, drilling a drain hole completion may allow for more efficient and balanced extraction of the natural gas. This type of completion is more commonly associated with oil wells than with natural gas wells.

2.6.3 Wellhead

The wellhead consists of an assemblage of necessary equipment that is mounted at the opening of the well to regulate and monitor the extraction of hydrocarbon derivatives from the underground formation. The equipment also prevents leaking of natural gas or crude oil out of the well and prevents blowouts due to high pressure formations. Generally, the wellhead consists of three components: (1) the casing head, (2) the tubing head, and (3) the *Christmas tree*.

The *casing head* consists of heavy fittings that provide a seal between the casing and the surface. The casing head also serves to support the entire length of casing that is run all the way down the well. This piece of equipment typically contains a gripping mechanism that ensures a tight seal between the head and the casing itself. The tubing head is much like the casing head and provides a seal between the tubing, which is run inside the casing, and the surface. Like the casing head, the tubing head is designed to support the entire length of the casing, as well as provide connections at the surface, which allow the flow of fluids out of the well to be controlled.

The *Christmas tree* (so named, because of its many branches that make it appear somewhat like a Christmas tree) is the piece of equipment that fits atop the casing and tubing heads and contains tubes and valves that serve to control the flow of hydrocarbon derivatives and other fluids out of the well. The *Christmas tree* is the most visible part of a producing well and allows for the surface monitoring and regulation of the production of hydrocarbon derivatives from a producing well.

2.6.4 Well treatment

Well treatment is another method of ensuring the efficient flow of hydrocarbon derivatives out of a formation. Essentially, this type of well stimulation consists of injecting acid, water, or gases into the well to open up the formation and allow the crude oil to flow through the formation more easily. Acidizing a well consists of injecting acid (usually hydrochloric acid) into the well. In limestone or carbonate formations, the acid dissolves portions of the rock in the formation, opening up existing spaces to allow for the flow of crude oil. Fracturing consists of injecting a fluid into the well, the pressure of which "cracks" or opens up fractures already present in the formation. In addition to the fluid being injected, *propping agents* are also used and consist of sand, glass beads, epoxy, or silica sand, and serve to prop open the newly widened fissures in the gas-bearing or oil-bearing formation (Speight, 2016).

Hydraulic fracturing involves the injection of water into the formation, while carbon dioxide fracturing uses gaseous carbon dioxide. Fracturing, acidizing, and lifting equipment may all be used on the same well to increase permeability. For example, carbon dioxide-sand fracturing involves using a mixture of sand and liquid carbon dioxide to fracture formations, creating and enlarging cracks through which oil and natural gas may flow more freely. The carbon dioxide then vaporizes, leaving only sand in the formation, holding the newly enlarged cracks open. Because there are no other substances used in this type of fracturing, there are no *leftovers* from the fracturing process that must be removed. This type of fracturing effectively opens the formation and allows for increased recovery of natural gas and (1) does not damage the deposit, (2) generates no below ground wastes, and (3) protects groundwater resources.

Because it is a low-density gas under pressure, the completion of natural gas wells usually requires little more than the installation of casing, tubing, and the wellhead. Natural gas, unlike crude oil, is much easier to extract from an underground formation. However, as deeper and less conventional natural gas wells are drilled, it is becoming more common to use stimulation techniques on gas wells.

2.6.5 Gas production

Once the well is completed, it may begin to produce natural gas. In some instances, the hydrocarbon derivatives that exist in pressurized formations will naturally rise up through the well to the surface. This is most commonly the case with natural gas. Since natural gas is lighter than air, once a conduit to the surface is opened, the pressurized gas will rise to the surface with little or no interference. This is most common for formations containing natural gas alone, or with a light condensate. In these scenarios, once the *Christmas tree* is installed, the natural gas will flow to the surface on its own.

In order to more fully understand the nature of the well, a potential test is typically run in the early days of production. This test allows well engineers to determine the maximum amount of natural gas that the well can produce in a 24-hour period. From this, and other knowledge of the formation, the engineer may make an estimation of the *most efficient recovery rate* (the rate at which the greatest amount of natural gas may be extracted without harming the formation itself).

As the gas pressure in the reservoir declines, the reservoir energy (i.e., reservoir pressure) declines, and the gas requires stimulation for continued production. Furthermore, reduction in the gas pressure may allow compaction of the reservoir rock by the weight of rock above eventually resulting in subsidence of the surface above the reservoir. This can be gradual process or a sudden catastrophic process depending on the structures of the geological formation above the reservoir.

Thus another important aspect of producing wells is the *decline rate*. When a well is first drilled, the formation is under pressure and produces natural gas at a very high rate. As more natural gas is extracted from the formation, the production rate of the well decreases (the *decline rate*). Certain techniques, including lifting equipment and well stimulation, can increase the production rate of a well. In some natural gas wells, and oil wells that have associated natural gas, it is more difficult to ensure an efficient flow of hydrocarbon derivatives up the well. The underground formation may be very *tight* (i.e., low permeability such as is found in shale formations and other tight formations) making the movement of natural gas—crude oil through the formation and up the well a very slow and inefficient process (Chapter 3: Unconventional Gas). In these cases, lifting equipment or well treatment is required.

Lifting equipment consists of a variety of specialized equipment used to help lift natural gas or crude oil out of a formation. The most common lifting method is known as *rod pumping* in which the pumping mechanism is powered by a surface pump that moves a cable and rod up and down in the well, providing the lifting pressure required to bring the oil to the surface. The most common type of cable rod lifting equipment is the *horse head (balanced conventional beam, sucker rod)* pump. These pumps are recognizable by the distinctive shape of the cable feeding fixture, which resembles a horse's head.

References

ASTM, 2017. Annual Book of Standards. ASTM International, West Conshohocken, PA.

Archer, J.S., 1985. Reservoir volumetrics and recovery factors. In: Dawe, R., Wilson, D. (Eds.), Developments in Crude Oil Engineering — 1. Elsevier Applied Science Publishers, New York.

Ballard, B.D., 2007. Quantitative mineralogy of reservoir rocks using Fourier transform infrared spectroscopy. Paper No. SPE-113023-STU. Proceedings of SPE Annual Technical Conference and Exhibition, Anaheim, California, 11—14 November, Society of Crude oil Engineers, Richardson, TX.

Berecz, E., Balla-Achs, M., 1983. Gas Hydrates. Elsevier, Amsterdam.

Burnett, A., 1995. A quantitative X-ray diffraction technique for analyzing sedimentary rocks and soils. J. Test. Eval. 23 (2), 111—118.

Chopra, A.K., Severson, C.D., Carhart, S.R., 1990. Evaluation of geostatistical techniques for reservoir characterization. Paper No. SPE-20734-MS. Proceedings of SPE Annual Technical Conference and Exhibition, 23—26 September, New Orleans, Louisiana, Society of Crude oil Engineers, Richardson, TX.

Damsleth, E., 1994. Mixed reservoir characterization methods. Paper No. SPE-27969-MS. Proceedings. University of Tulsa Centennial Crude oil Engineering Symposium, Tulsa, Oklahoma, 29—31 August, Society of Crude oil Engineers, Richardson, TX.

Esteves, I.A.A.C., Sousa, G.M.R.P.L., Silva, R.J.S., Ribeiro, R.P.P.L., Eusébio, M.F.J., Mota, J.P.B., 2016. A sensitive method approach for chromatographic analysis of gas streams in separation processes based on columns packed with an adsorbent material. Adv. Mater. Sci. Eng. 2016, Article ID 3216267. Hindawi Publishing Corporation: <https://doi.org/10.1155/2016/3216267> (accessed 01.12.17.).

Faramawy, S., Zaki, T., Sakr, A.A.-E., 2016. Natural gas origin, composition, and processing: a review. J. Nat. Gas Sci. Eng. 34, 34—54.

Gudmundsson, J.S., Andersson, V., Levik, O.I., Parlaktuna, M., 1998. Hydrate concept for capturing associated gas. Proceedings of SPE European Crude Oil Conference, 20—22 October, The Hague, Netherlands.

Hurst, A., Archer, J.S., 1986. Sandstone reservoir description: an overview of the role of geology and mineralogy. Clay Miner. 21, 791—809.

John, E., Singh, K., 2011. Production and properties of fuels from domestic and industrial waste. In: Speight, J.G. (Ed.), The Biofuels Handbook. Royal Society of Chemistry, London, pp. 333—376.

Max, M.D. (Ed.), 2000. Natural Gas in Oceanic and Permafrost Environments. Kluwer Academic Publishers, Dordrecht.

Mokhatab, S., Poe, W.A., Speight, J.G., 2006. Handbook of Natural Gas Transmission and Processing. Elsevier, Amsterdam.

Ramroop Singh, N., 2011. Biofuels. In: Speight, J.G. (Ed.), The Biofuels Handbook. Royal Society of Chemistry, London, pp. 160—198.

Saha, R., Uppaluri, R.V.S., Tiwari, P., 2017. Effect of mineralogy on the adsorption characteristics of surfactant-reservoir rock system. Colloids Surf. A 531, 121—132.

Singh, K., Sastry, M.K.S., 2011. Production of fuels from landfills. In: Speight, J.G. (Ed.), The Biofuels Handbook. Royal Society of Chemistry, London, pp. 408—453.

Sloan, E.D., 1997. Clathrates of Hydrates of Natural Gas. Marcel Dekker Inc, New York.

Sloan, E.D., 2000. Clathrates hydrates: the other common water phase. Ind. Eng. Chem. Res. 39, 31123—33129.

Speight, J.G., 2007. Natural Gas: A Basic Handbook. GPC Books, Gulf Publishing Company, Houston, TX.

Speight, J.G., 2009. Enhanced Recovery Methods for Heavy Oil and Tar Sands. Gulf Publishing Company, Houston, TX.

Speight, J.G. (Ed.), 2011. The Biofuels Handbook. Royal Society of Chemistry, London.

Speight, J.G., 2013a. Shale Gas Production Processes. Gulf Professional Publishing, Elsevier, Oxford.

Speight, J.G., 2013b. The Chemistry and Technology of Coal, third ed. CRC Press, Taylor and Francis Group, Boca Raton, FL.

Speight, J.G., 2014a. The Chemistry and Technology of Crude oil, fifth ed. CRC Press, Taylor & Francis Group, Boca Raton, FL.

Speight, J.G., 2014b. Oil and Gas Corrosion Prevention. Gulf Professional Publishing, Elsevier, Oxford.

Speight, J.G., 2015a. Handbook of Crude oil Product Analysis, second ed. John Wiley & Sons Inc, Hoboken, NJ.

Speight, J.G., 2015b. Handbook of Offshore Oil and Gas Operations. Gulf Professional Publishing, Elsevier, Oxford.

Speight, J.G., 2016. Handbook of Hydraulic Fracturing. John Wiley & Sons Inc, Hoboken, NJ.

Speight, J.G., 2017a. Handbook of Crude oil Refining. CRC Press, Taylor & Francis Group, Boca Raton, FL.

Speight, J.G., 2017b. Deep Shale Oil and Gas. Gulf Professional Publishing, Elsevier, Oxford.

Speight, J.G., 2018. Handbook of Natural Gas Analysis. John Wiley & Sons Inc, Hoboken, NJ.

Thomas, F.B., Bennion, D.B., Andersen, G., 2009. Gas condensate reservoir performance. J. Can. Crude Oil Technol. 48 (7), 18–24.

Whitson, C.H., Belery, P., 1994. Compositional gradients in crude oil reservoirs. Paper No. SPE28000. Proceedings of Centennial Crude oil Engineering Symposium. University of Tulsa, Tulsa, Oklahoma, 29–31 August, Society of Crude oil Engineers, Richardson, TX.

Unconventional gas

3

3.1 Introduction

In addition to conventional natural gas (Chapter 1: History and use and Chapter 2: Origin and production), there are several types of unconventional gas resources that are currently produced and these are (alphabetically rather than in order of preference): (1) methane hydrates—natural gas that occurs at low temperature and high pressure regions such as the sea bed and is made up of a lattice of frozen water, which forms a *cage* around the methane; (2) biogas, which is a gas produced from various types of biomass; (3) coalbed methane—natural gas that occurs in conjunction with coal seams, coal gas, which is a gas produced by the thermal decomposition or gasification of coal; (4) flue gas, which is a gas from various industrial source that is sent up a flue for dispersal; (5) gas in geopressurized zones—natural underground formations that are under unusually high pressure for their depth; (6) gas in tight formations, which is a gas located in reservoirs in which the permeability is zero or, at best very low; (7) landfill gas, which is a gas produced by the decomposition of landfill materials; (8) manufactured gas, which is a fuel−gas mixture made from other solid, liquid, or gaseous materials, such as coal, coke, oil, or natural gas—examples are retort coal gas, coke oven gas, water gas, carbureted water gas, producer gas, oil gas, reformed natural gas, and reformed propane or liquefied petroleum gas (LPG); (9) refinery gas, also called petroleum gas, which is a gas that emanates from the top of a refinery distillation column or from any other refinery process; (10) shale gas, which is a gas that is recovered from shale formation; and (11) synthesis gas, also known as *syngas*, which is a mixture of carbon monoxide (CO) and hydrogen (H_2) and is produced from a wide range of carbonaceous feedstocks and is used as a fuel gas as well as to produce a wide range of chemicals (Mokhatab et al., 2006; Speight, 2013b, 2014a).

Typically, natural gas produced from shale reservoirs and other tight reservoirs has been classed under the general title *unconventional gas*. The production process requires stimulation by horizontal drilling coupled with hydraulic fracturing because of the pack of permeability in the gas-bearing formation. The boundary between conventional gas and unconventional gas resources is not well defined, because they result from a continuum of geologic conditions. Coal seam gas, more frequently called coal bed methane (CBM), is frequently referred to as unconventional gas. Tight shale gas and gas hydrates are also placed into the category of *unconventional gas*.

Shale is, in this context, a low-permeability reservoir rock which prohibits natural movement of the gas to a well. Moreover, maximization or optimization of reservoir producibility can only be achieved by a thorough understanding of the

Natural Gas. DOI: https://doi.org/10.1016/B978-0-12-809570-6.00003-5

occurrence and properties of the shale gas resources as well as the producibility of the gas from the reservoir.

The boundary between conventional gas and unconventional gas resources is not well defined, because they result from a continuum of geologic conditions. Coal seam gas, more frequently called coalbed methane (coal bed methane), is frequently referred to as unconventional gas. Gas hydrates are of particular importance because they represent a high resource that will be tapped commercially in the not-too-distant future. Tight shale gas, although in commercial production is also placed into the category of *unconventional gas* because of the nature of the reservoir and the extreme methods (fracturing) needed for production of the gas.

3.2 Gas hydrates

The concept of natural gas production from *methane hydrate* (also called *gas hydrate, methane clathrate, natural gas hydrate (NGH), methane ice, hydro-methane, methane ice, fire ice*) is relatively new but does offer the potential to recover hitherto unknown reserves of methane that can be expected to extend the availability of natural gas (Giavarini et al., 2003, 2005; Giavarini and Maccioni, 2004; Makogon et al., 2007; Makogon, 2010; Wang and Economides, 2012; Yang and Qin, 2012). In terms of gas availability from this resource, 1 L of solid methane hydrate can contain up to 168 L of methane gas).

The methane in gas hydrates is predominantly generated by bacterial degradation of organic matter in low oxygen environments. Organic matter in the uppermost few centimeters of sediments is first attacked by aerobic bacteria, generating carbon dioxide, which escapes from the sediments into the water column. In this region of aerobic bacterial activity sulfate derivatives ($-SO_4$) are reduced to sulfide derivatives ($-S$). If the sedimentation rate is low (<1 cm per 1000 years), the organic carbon content is low ($<1\%$), and oxygen is abundant, and the aerobic bacteria use up all the organic matter in the sediments. However, when sedimentation rate is high, and the organic carbon content of the sediment is high, the pore waters in the sediments are anoxic at depths of less than 1 ft or so and methane is produced by anaerobic bacteria.

The two major conditions that promote hydrate formation are thus: (1) high gas pressure and low gas temperature and (2) the gas at or below its water dew point with free water present (Sloan, 1998b; Collett et al., 2009). The hydrates are believed to form by migration of gas from depth along geological faults, followed by precipitation, or crystallization, on contact of the rising gas stream with cold sea-water. At high pressures methane hydrates remain stable at temperatures up to 18°C (64°F) and the typical methane hydrate contains one molecule of methane for every six molecules of water that forms the ice cage. However, the methane (hydro-carbon)—water ratio is dependent on the number of methane molecules that fit into the cage structure of the water lattice.

Table 3.1 Miscellaneous properties of ice and methane hydrate

Property	Ice	Methane hydrate
Dielectric constant at 273 K	94	58
Isothermal Young's modulus at 268 K	9.5	8.4
Poisson's ratio	0.33	0.33
Bulk modulus (272 K)	8.8	5.6
Shear modulus (272 K)	3.9	2.4
Bulk density (g/cm^3)	0.916	0.912
Thermal conductivity at 263 K (W/m-K)	2.23	0.49

Chemically, gas hydrates are nonstoichiometric compounds formed by a lattice of hydrogen bonded molecules (host) which engage low molecular weight gases or volatile liquids (guest) with specific properties that differentiate them from ice (Table 3.1) (Bishnoi and Clarke, 2006). No actual chemical bond exists between guest and host molecules. Hydrate formation is favored by low temperature and high pressure (Makogon, 1997; Sloan, 1998a; Lorenson and Collett, 2000; Carrol, 2003; Seo et al., 2009). Most methane hydrate deposits also contain small amounts of other hydrocarbon hydrates; these include ethane hydrate and propane hydrate. In fact, gas hydrates of current interest are composed of water and the following molecules: methane, ethane, propane, isobutane, normal butane, nitrogen, carbon dioxide, and hydrogen sulfide. However, other nonpolar components such as argon (Ar) and ethyl cyclohexane ($C_6H_{11} \cdot C_2H_5$) can also form hydrates. Typically, gas hydrates form at temperatures on the order of 0°C (32°F) and elevated pressures (Sloan, 1998a).

In the hydrate structure, methane is trapped within a cage-like crystal structure composed of water molecules in a structure that resembles packed snow or ice (Lorenson and Collett, 2000). Under the appropriate pressure, gas hydrates can exist at temperatures significantly above the freezing point of water, but the stability of the hydrate derivatives depends on pressure and gas composition and is also sensitive to temperature changes (Stern et al., 2000; Stoll and Bryan, 1979; Collett, 2001; Belosludov et al., 2007; Collett, 2010). For example, methane plus water at 600 psia forms hydrate at 5°C (41°F), while at the same pressure, methane with 1% v/v propane forms a gas hydrate at 9.4°C (49°F). Hydrate stability can also be influenced by other factors, such as salinity.

Methane hydrates are restricted to the shallow lithosphere (i.e., at depths less than 6000 ft below the surface). The necessary conditions for the formation of hydrates are found only either in polar continental sedimentary rocks where surface temperatures are less than 0°C (32°F) or in oceanic sediment at water depths greater than 1000 ft where the bottom water temperature is on the order of 2°C (35°F).

Caution is advised when drawing generalities about the formation and the stability of gas hydrates. Methane hydrates are also formed during natural gas production operations, when liquid water is condensed in the presence of methane at high pressure. Higher molecular weight hydrocarbons such as ethane and propane can also form hydrates, although larger molecules (butane hydrocarbons and pentane hydro-carbons) cannot fit into the water cage structure and, therefore, tend to destabilize the formation of hydrates (Belosludov et al., 2007). However, for this text, the emphasis is focused on methane hydrates.

3.2.1 Occurrence

Estimates of the global inventory of methane hydrates have exceeded 10^{19} g of carbon, which is comparable with the estimates of potentially recoverable coal, oil, and natural gas. The size of this type of resource has motivated speculations about a release of methane in response to climate change. Increases in temperature or decreases in pressure (through changes in sea level) tend to dissociate the hydrate, thereby releasing methane into the near-surface environment (Li et al., 2016). In fact, such releases of methane from clathrate have been invoked to explain abrupt increases in the atmospheric concentration of methane during the last glacial cycle.

On a unit volume basis, gas hydrates contain a high amount of gas. For example, 1 cubic yard of hydrate disassociates at ambient temperature and pressure to form approximately 160 cubic yards ($4320\,\text{ft}^3$) of natural gas plus 0.8 cubic yards ($21.6\,\text{ft}^3$) of water. The natural gas component of gas hydrates is typically dominated by methane, but other natural gas components (e.g., ethane, propane, and carbon dioxide) can also be incorporated into a hydrate. The origin of the methane in a hydrate can be either thermogenic gas or biogenic gas. Bacterial gas formed during early digenesis of organic matter can become part of a gas hydrate in continental shelf sediment. Similarly, thermogenic gas leaking to the surface from a deep thermogenic gas accumulation can form a gas hydrate in the same continental shelf sediment (Collett et al., 2009).

Generally, methane hydrates are common constituents of the shallow marine geosphere, and they occur both in deep sedimentary structures, and form outcrops on the ocean floor. Methane hydrates are believed to form by migration of gas from depth along geological faults, followed by precipitation, or crystallization, on contact of the rising gas stream with cold seawater. When drilling in natural gas-bearing and in crude oil-bearing formations submerged in deep water, the reservoir gas may flow into the well bore and form gas hydrates owing to the low temperatures and high pressures found during deep water drilling. The gas hydrates may then flow upward with drilling mud or other discharged fluids. When the hydrates rise, the pressure in the annulus decreases and the hydrates dissociate into gas and water. The rapid gas expansion ejects fluid from the well, reducing the pressure further, which leads to more hydrate dissociation (sometimes explosive dissociation) and further fluid ejection.

As might be anticipated, the gas that is evolved from gas hydrates under laboratory conditions varies over a considerable range and is variously composed of the following hydrocarbon gases−liquids: methane (CH_4), ethane (C_2H6), and the sum of the C_{3+} hydrocarbon derivatives composed of propane (C_3H_8), n-butane ($n\text{-}C_4H_{10}$), and isobutane ($i\text{-}C_4H_{10}$) (Lorenson and Collett, 2000). Other hydrocarbon gases are present in some sediments include: n-pentane (C_5H_{12}), iso-pentane (C_5H_{12}), neo-pentane (neo-C_5H_{12}), cyclopentane (cyclo-C_5H_{12}), n-hexane

(n-C_6H_{14}), iso-hexane (i-C_6H_{14}), neo-hexane (neo-C_6H_{14}), n-heptane (n-C_7H_{16}), iso-heptane (i-C_7H_{16}), and methylcyclohexane ($cyclo$-$C_6H_{11} \cdot CH_3$) (Lorenson and Collett, 2000). Hydrogen sulfide (H_2S) has been detected in gas hydrates but, because hydrogen sulfide is water-soluble, the gas hydrate dissociation measurements have the potential to be contaminated with some sediment and pore water, it is also possible that in some cases the hydrogen sulfide may not have existed within the gas hydrate structure (Lorenson and Collett, 2000).

The size of the oceanic methane hydrate reservoir is not well defined and estimates of its size have varied considerably over a wide range. However, improvements in understanding the nature of the gas hydrate resource have revealed that hydrates only form in a narrow range of depths (such as in the area of continental shelves) and typically are found at low concentrations (0.9−1.5% v/v) at sites where they do occur. Recent estimates constrained by direct sampling suggest the global inventory lies on the order of 2074 trillion cubic feet (2075×10^{12} tcf) of gas or 2.074 quadrillion cubic feet (2.074×10^{15} qcf) of gas (US DOE, 2011).

3.2.2 Structure and morphology

Methane hydrate ($CH_4 \cdot 5.75H_2O$ or $4CH_4 \cdot 23H_2O$) is a solid clathrate compound in which a large amount of methane is trapped within a crystal structure of water, forming a solid similar to ice. In fact, the physical properties of bulk gas hydrates are remarkably close to those of pure ice. As a result, the properties of sediments containing hydrate in the pore space are similar to sediments containing nonhydrate (normal) ice. The morphology of the gas hydrates has large effects on sedimentary physical properties, from seismic velocity on a large scale to borehole electrical resistivity on a smaller scale and, therefore, gas hydrate morphology impacts the amount of gas hydrate saturation estimated from geophysical data (Gabitto and Tsouris, 2010; Du Frane et al., 2011, 2015). More generally, at the molecular level, methane hydrate consists of gas molecules surrounded by cages of water molecules (Bishnoi and Clarke, 2006). Each water cage encloses a space of a particular size, and only a gas molecule small enough to fit within this site can be hosted in that specific hydrate structure. The typical methane clathrate hydrate composition is usually represented as $(CH_4)_4(H_2O)_{23}$ or 1 mol of methane for every 5.75 mol of water, corresponding to 13.4% w/w, but the actual composition is dependent on the number of methane molecules that can be accommodated by the various cage structures of the water lattice. Structurally, methane forms a structural hydrate with two dodecahedral water cages (i.e., 12 water molecules) and six tetradecahedral water cages (i.e., 14 water molecules) per unit cell. The morphology of gas hydrates determines the basic physical properties of the sediment−hydrate matrix (Holland et al., 2008; Gabitto and Tsouris, 2010; Holland and Schultheiss, 2014; Collett et al., 2015). In fact, knowledge of gas hydrate morphology and bulk properties will provide the necessary data to determine the mechanism of dissociation (Llamedo et al., 2004).

The composition of the hydrate phase seems to be affected by the consumed amount of natural gas, which results in a variation of heating value of retrieved gas from mixed hydrate derivatives as a function of the formation temperature (Seo et al., 2009). This indicates that the fractionation of the hydrate phase with higher molecular weight hydrocarbon molecules is enhanced in silica gel pores. In addition, when higher molecular weight hydrocarbon derivatives are depleted in the vapor phase during the formation of mixed hydrate, different hydrate structures can coexist together.

Once formed, hydrates can block pipeline and processing equipment. They are generally then removed by reducing the pressure, heating them, or dissolving them by chemical means (methanol is commonly used). Care must be taken to ensure that the removal of the hydrates is carefully controlled, because of the potential for the hydrate to undergo a phase transition from the solid hydrate to release water and gaseous methane at a high rate when the pressure is reduced. The rapid release of methane gas in a closed system can result in a rapid increase in pressure.

It is preferable to prevent hydrates from forming or blocking equipment. In terms of inhibiting hydrate formation, the most common thermodynamic inhibitors are methanol, monoethylene glycol (MEG), and diethylene glycol (DEG). MEG is preferred over DEG for applications where the temperature is expected to be $-10°C$ (14°F) or lower due to high viscosity at low temperatures. The vapor pressure of triethylene glycol is too low to be suited as an inhibitor injected into a gas stream. In recent years, development of other forms of hydrate inhibitors have been developed, such as kinetic hydrate inhibitors (which slow the rate of hydrate formation) and antiagglomerates, which do not prevent hydrates forming, but do prevent them sticking together to block equipment.

When drilling in crude oil-bearing and gas-bearing formations submerged in deep water, the reservoir gas may flow into the well bore and form gas hydrates owing to the low temperatures and high pressures found during deep water drilling. The gas hydrates may then flow upward with drilling mud or other discharged fluids. When the hydrates rise, the pressure in the annulus decreases and the hydrates dissociate into gas and water. The rapid gas expansion ejects fluid from the well, reducing the pressure further, which leads to more hydrate dissociation and further fluid ejection. The resulting violent expulsion of fluid from the annulus is one potential cause or contributor to a blowout. This behavior is also of value to the analyst who has to handle these hydrates for analytical purposes.

The physical properties of hydrates are dependent upon the morphology of the hydrate. In terms of hydrate in sediment, which is a common means of detecting the presence of the hydrates, all of the physical parameters depend strongly on the hydrate saturation in pore space. This fact significantly reduces the impact of low-hydrate saturations on the measured physical parameters, an effect that is particularly pronounced at the hydrate saturations characteristic of many natural systems (Gabitto and Tsouris, 2010; Du Frane et al., 2011, 2015).

3.2.3 Gas production

Gas production from gas hydrates can be achieved by methods such as: (1) thermal stimulation, (2) depressurization, or (3) chemical inhibition.

The gas content of gas hydrates typically involves the controlled decomposition of the hydrate in a *dissociation chamber* and allowed to decompose at room temperature. The gas pressure of the chamber will increase and reach a stable value over a period of time after which the hydrate gas is ready for analysis. The volume of water that remains in the dissociation chamber is then measured and stored in a glass ampoule for determination of the chemical composition and to ensure that no gases remain dissolved. Gas analysis will show that the gas released from the hydrate is predominantly methane with trace-to small- (typically <5% v/v) of ethane, propane, isobutane, normal butane, nitrogen, carbon dioxide, and hydrogen sulfide (Gabitto and Tsouris, 2010).

To this point, the heat of dissociation (ΔH_d) is the enthalpy change to dissociate the hydrate phase to a vapor and aqueous liquid, with values given at temperatures just above the ice point. Also, the heat of dissociation is a function of the number of crystal hydrogen bonds (generally assumed to be the same as the hydration number) (Liu et al., 2010). However, the value of the heat of dissociation is relatively constant for molecules which occupy the same cavity, within a wide range of component sizes.

The enthalpy of dissociation may be determined via the univariant slopes of phase equilibrium lines (ln P vs $1/T$) using the Clausius−Clapeyron relationship (Sloan, 2006):

$$\left[(\Delta H_d) = - zR_d(\ln P)/d(1/T) \right]$$

Gas hydrates may formally be referred to as chemical compounds because they have a fixed composition at a certain pressure and temperature. However, hydrates are the compounds of a molecular type. They form because of the Van der Waals attraction forces between the molecules. Covalent bonding is absent in the gas hydrates because during their formation there is no pairing of valence electrons and no spatial redistribution of electron cloud density.

NGHs are metastable minerals, where the formation and dissociation depend on the pressure and temperature, composition of gas, salinity of the reservoir water, and the characteristics of the porous medium in which they were formed. Hydrate crystals in reservoir rocks can be dispersed in the pore space without the destruction of pores; however, in some cases, the rock is affected. Hydrates can be in the form of small nodules (from 5 to 12 cm in size), in the form of small lenses, or in the form of layers that can be several meters thick.

The composition of NGHs is determined by the composition of the gas and water, and the pressure and temperature which existed at the time of their formation. Over geologic time, there will be changes in the thermodynamic conditions and the vertical and lateral migration of gas and water; therefore, the composition of hydrate can change both due to the absorption of free gas and the recrystallization of already-formed hydrate. Based on the cores taken while drilling in gas hydrate deposits, the hydrate usually consists of methane with small admixtures of heavier components. However, in a number of cases the hydrate contains a significant volume of higher molecular weight gases.

The recovery of methane gas from NGHs can be achieved by dissociating the solid gas hydrate structure into gas and water. The conventional transportation method then can be used to transport this dissociated gas. The methods that have been proposed for the extraction of natural gas from NGHs include: (1) depressurization, (2) thermal stimulation, (3) chemical inhibitor injections, and (4) carbon dioxide replacement (Castaldi et al., 2007; Lee et al., 2013; Makogon, 2010). The method for reducing pressure (the depressurization method) for obtaining gas from NGHs is reported to be more effective and economical than the thermal stimulation method (Demirbaş, 2010).

The depressurization method operates by lowering the pressure inside the well with embedded and adjacent zones of gas hydrates, which results in dissociation of methane gas (Fan et al., 2017; Wang et al., 2017). The target is to lower the pressure below the dissociation pressure, which results in decomposition of hydrates into gas and water. Ahmadi et al. (2007) studied the production of natural gas from dissociation of NGHs in a confined and pressurized reservoir utilizing the depressurizing method. The depressurization method proved to be a successful method with both well pressure and reservoir temperature being the sensitive parameters for gas production and overall well output (Ahmadi et al., 2007).

In thermal stimulation method, heat is applied to NGH system at constant pressure to increase the temperature above the dissociation temperature, to decompose the hydrates to produce methane gas. Different heat sources can be used to apply direct heat such as steam injection, hot water or other liquids, or indirect heat by electric current or sonication. The methane gas, mixed with hot water returned to the surface, is then separated from hot water to be used as a pure gas (Goel et al., 2001).

The third method for dissociation of NGHs is chemical inhibition, in which a chemical inhibitor is injected adjacent to the hydrate to displace the NGH equilibrium conditions beyond the thermodynamic conditions of hydrate stability zone. Chemical inhibitor, e.g., methanol is used and injected into the methane hydrate-bearing layers that results in separation of methane gas from the solid hydrate structure.

The choice of method for methane extraction from NGHs not only depends on the geological locations and thermodynamic conditions of NGH, but also on amounts of hydrate deposits, capital and maintenance cost, environmental impact, and simplicity of the selected method. The thermal stimulation and chemical inhibition methods are relatively more expensive. In comparison, depressurization method for production of methane gas from NGHs is considered to be most economical and efficient method. Methane gas is an attractive fuel for its least greenhouse effect, because it produces less carbon dioxide during its combustion as compared to all other petroleum-derivate fuels. The real positive contribution of methane gas fuel for minimization of environmental pollution must also be considered in more details together with its extraction from NGHs and potential as an unconventional future energy resource (Davies, 2001; Demirbaş, 2006).

Thus it is necessary to use much of the energy contained in the gas hydrate deposits for heating the rock layers near the gas hydrate deposits. Preliminary estimates show that the coefficient of extraction of the gas hydrate can be as high as 50%−70%. However, from total world potential resource it has been estimated that the coefficient of extraction should average from 17% to 20%.

For offshore conditions, with the depths of water ranging from 0.7 to 2.5 km, effective production of gas from gas hydrate deposits in the majority of the cases may occur when hydrate saturation of porous media exceeds 30%−40%. However, each geologic region will have to be studied in detail to establish the minimal hydrate saturation that is required. To change a gas hydrate deposit to natural gas, it is necessary to (1) decrease reservoir pressure to lower than equilibrium one, (2) increase the temperature to higher than the equilibrium temperature, (3) inject active reagents, which facilitate the decomposition of hydrate, and (4) use some new technology. The easiest method is to lower the reservoir pressure in the gas hydrate deposit—this method is only feasible when free gas is found below the gas hydrate deposit.

Several properties of gas hydrates are unique. For example, 1 m^3 of water may accommodate up 207 m^3 of methane to form 1.26 m^3 of solid hydrate, whereas without gas, 1 m^3 of water freezes to form 1.09 m^3 of ice. One volume of methane hydrate at a pressure of 3800 psi and temperature of 0°C (32°F) contains 164 volumes of gas. In a hydrate, 80% v/v is occupied by water and 20% v/v by gas. Thus, 164 m^3 of gas are contained in a volume of 0.2 m^3 of hydrate.

The dissociation of methane hydrate by increasing the temperature in a constant volume will be accompanied by a substantial increase in pressure. For methane hydrate formed at a pressure of 3800 psi and a temperature of 0°C (32°F), it is possible to obtain a pressure increase of up to 23,500 psi. Hydrate density depends on its composition, pressure, and temperature. Depending on the composition of gas, pressure, and temperature, the density of the hydrates range be on the order of 0.9 g/cm^3 (Table 3.2)—typically falling in the range 0.8−1.2 g/cm^3—and, thus, methane hydrate will float to the surface of the sea or of a lake unless it is bound in place by being formed in or anchored to sediment.

In the thermal stimulation method of decomposition and gas release, the temperature of the formation is raised by the injection of heated fluid or potentially direct heating of the formation. Thermal stimulation is energy intensive and will lead to relatively slow, conduction limited dissociation of gas hydrates unless warmer pore fluids become mobilized and increase the volume of the formation exposed to

Table 3.2 Density and molar volume of gas hydrates

Gas	Formula	Molecular weight	Density (g/cm^3)	Mole volume (cm^3/mol)
CH_4	$CH_4 \cdot 6H_2O$	124	0.910	136.264
CO_2	$CO_2 \cdot 6H_2O$	152	1.117	136.078
C_2H_6	$C_2H_6 \cdot 7H_2O$	308	0.959	162.669
C_3H_8	$C_3H_8 \cdot 17H_2O$	350	0.866	404.157
$i\text{-}C_4H_{10}$	$i\text{-}C_4H_{10} \cdot 17H_2O$	354	0.91	403.996

higher temperatures. The endothermic nature of gas hydrate dissociation also presents a challenge to thermal stimulation—the cooling associated with dissociation (and, in some cases, gas expansion) will partially offset artificial warming of the formation, meaning that more heat must be introduced to drive continued dissociation and prevent formation of new gas hydrate. In terrestrial settings thermal stimulation must be carefully controlled to minimize permafrost thawing, which might lead to unintended environmental consequences and alter the permeability seal for the underlying gas hydrate deposits.

On the other hand, the depressurization method of decomposition and gas release does not require large energy expenditure and can be used to drive dissociation of a significant volume of gas hydrate relatively rapidly (Lorenson and Collett, 2000). On the other hand, the chemical inhibition exploits the fact that gas hydrate stability is inhibited in the presence of certain organic (e.g., glycol) or ionic (seawater or brine) compounds. Seawater or other inhibitors might be needed during some stages of production of gas from methane hydrate deposits but would not be the primary means of dissociating gas hydrate nor used for an extended period or on a large scale.

3.2.4 Other properties

The properties of methane hydrate (and other gas hydrates) typically focus on the bulk properties of the hydrate derivatives, particularly on the dissociation of the hydrate derivatives into water and gas. However, there are other properties that are also of importance.

The presence of methane gas hydrates can alter the thermal properties of host sediments. The high thermal diffusivity of gas hydrate accelerates the heating rate in the surrounding sediment and, in addition, the decomposition of gas hydrates can destabilize the host sediment and lead to well failure. Also, warm pipelines also must be routed to avoid regions of outcropping or surficial gas hydrates by providing a numerical basis for predicting the thermal response of hydrate-bearing sediment to changes induced naturally or through human action.

Thus, the physical properties of gas hydrate derivatives trapped in sediments are extremely valuable for (1) detecting the presence of these compounds, (2) estimating the amount of gas hydrates trapped in the sediments, and (3) developing processes to exploit this resource. Unfortunately, little is known about the physical properties of NGH deposits in nature, making their detection by remote geophysical surveys difficult. However, in some cases, such as in the case of marine sediments, the presence of gas hydrates can dramatically alter some of the typical physical properties of the sediment, which can be detected by field measurements and by downhole logs (Collett, 2001, 2010). In these cases, properties such as (1) electrical resistivity, (2) electrical conductivity, (3) specific heat, (4) thermal conductivity, and (5) thermal diffusivity assume an increased importance. In many cases, it is only by using knowledge of such properties that gas hydrate derivatives can be detected in sediments.

3.2.4.1 Electrical conductivity and resistivity

An electrical conductor is a material that allows the flow of an electric current in one or more directions. The electrical current is generated by the flow of negatively charged electrons, positively charged holes, and positive or negative ions in some cases. Thus, electrical conductivity (also called *specific conductance*, σ) is the reciprocal of electrical resistivity, and is a measure of the ability of a material to conduct an electric current:

$$\sigma = \frac{1}{\rho}$$

Electrical resistivity (also called *resistivity, specific electrical resistance*, or *volume resistivity*) is a measure of the ability of a material to oppose the flow of an electric current. A low resistivity indicates a material that readily allows the flow of an electric current.

A gas hydrate has a low electrical conductivity, which provides a suitable target for marine controlled source electromagnetic (CSEM) surveys. CSEM sounding measures the amplitude and phase of electromagnetic energy propagating through the seafloor at one or more frequencies, and these data can be inverted to obtain the spatial distribution of conductivity. The combined use of both seismic and electromagnetic methods can help distinguish between gas (low velocity and high resistivity) and gas hydrate (high velocity and high resistivity) to map both the upper and lower boundaries of gas hydrate deposits.

CSEM studies have demonstrated the sensitivity of this method in assessing general gas hydrate concentration, saturation, and distribution patterns (Schwalenberg et al., 2005; Evans, 2007; Weitemeyer et al., 2006, 2011). Quantifying the estimates of hydrate volume, however, requires knowledge of the electrical conductivity of gas hydrates in combination with petrophysical mixing relations established from theory and experiment (Collett and Ladd, 2000; Ellis et al., 2008).

Hydrates normally exclude the salt in the pore fluid from which it forms, and thus they have high electric resistivity just as ice and sediments containing hydrates have a higher resistivity compared to sediments without gas hydrates (Judge, 1982). The unconsolidated sediments in the upper several hundred meters of the marine sediment section (50% v/v porosity) normally have a very low resistivity of about 1 ohm-m. If the saturation in the pore space is on the order of 15−20% v/v saturation (7−10% v/v of the sediment), the resistivity increases by about a factor of 2.

3.2.4.2 Specific heat

The specific heat of a substance or material is the amount of heat per unit mass required to raise the temperature of the substance or material by 1°C. The relationship between heat and temperature change is usually expressed in the following form where c is the specific heat:

$$Q = cm\Delta T$$

Table 3.3 Temperature and pressure dependence of selected physical properties

Temperature dependence (pressure held at 31.5 MPa, 4570 psi)	Temperature range
λ (W/m K)$=-(2.28 \pm 0.05)\ 10^{-4}\ T(^{\circ}C) + (0.62 \pm 0.02)$ [a]κ (m^2/s)$=(5.04 \pm 0.02)\ 10^{-5}/T(K) + (1.25 \pm 0.05)\ 10^{-7}$	-20°C to 17°C $1-17^{\circ}$C $(274-290$ K$)$
c_p (J/kg K)$=(6.1 \pm 0.3)\ T(^{\circ}C) + (2160 \pm 100)$	$1-17^{\circ}$C
Pressure dependence (temperature held at 14.4°C, 58°F)	
λ (W/m K)$=(2.54 \pm 0.06)\ 10^{-4}\ P(MPa) + (0.61 \pm 0.02)$ κ (m^2/s)$=(2.87 \pm 0.08)\ 10^{-10}\ P(MPa) + (3.1 \pm 0.2)\ 10^{-7}$ c_p (J/kg K)$=(3.30 \pm 0.06)\ P(MPa) + (2140 \pm 100)$	

[a]The T^{-1} dependence of the κ fit requires input temperatures in Kelvin.

In this equation, Q is the heat added, c is the specific heat, m is the mass, and ΔT is the change in temperature. However, the relationship does not apply if a phase change occurs because the heat added or removed during a phase change does not change the temperature.

Hydrate breakdown is an endothermic process, absorbing heat while the surrounding sediment cools. Because the specific heat, c_p, of methane hydrates is approximately equal to one half that of water, hydrate-bearing sediment stores less heat which can then be made available to help fuel dissociation (Waite et al., 2006, 2007). When estimating the efficiency of hydrate dissociation, neglecting the reduced contribution of methane hydrates to the host sediment's specific heat results in an overestimate of the dissociation rate and, hence, the methane production rate. There is a temperature and pressure dependence (Table 3.3) (USGS, 2007).

3.2.4.3 Thermal conductivity

Thermal conductivity is a measure of the ability of a substance or material to conduct heat. Heat transfer occurs at a lower rate in materials of low thermal conductivity than in materials of high thermal conductivity. The thermal conductivity of the substance or material may depend on temperature. The reciprocal of thermal conductivity is the thermal resistivity.

By way of a reference point, the thermal conductivity, λ, of water is approximately equal to that of methane hydrate, so the thermal conductivity beneath permafrost or in marine settings is essentially independent of methane hydrate content (Ross et al., 1978; Stoll and Bryan, 1979; Cook and Leaist, 1983; Ashworth et al., 1985; Tse and White, 1988). Methane hydrate is generally found in abundance below, rather than in, permafrost, but the presence of methane hydrates in ice-dominated permafrost can measurably increase the geothermal gradient because the thermal conductivity of ice is approximately four times that of methane hydrate.

In contrast with most well-defined crystalline structures, in which the thermal conductivity falls with increasing temperature following a $T-1$ dependence (for $T>100$ K), the thermal conductivity of clathrate hydrates increases slightly with increasing temperature (Tse and White, 1988). The thermal conductivity of clathrate hydrates is five times lower than that of ice near the melting point, and even lower (by a factor >20) at lower temperatures. The temperature dependence of thermal conductivity in clathrate hydrates is characteristic of an amorphous material (Tse and White, 1988).

In sediment-bearing hydrates, the thermal conductivity reflects the competing effects of the thermal conductivity of the phases involved, their volume fraction, and their spatial distribution. The electrical conductivity is controlled by the availability and mobility of ions. A gradual reduction in conductivity is measured during hydrate formation even though ion exclusion keeps available ions within the unfrozen water. Mechanical properties are strongly influenced by both soil properties and the hydrate loci.

3.2.4.4   Thermal diffusivity

The thermal diffusivity of a substance or material is the thermal conductivity divided by the density and specific heat capacity at constant pressure and is a measure the rate of transfer of heat of a substance or material from the hot side to the cold side. The thermal diffusivity is a material-specific property for characterizing unsteady heat conduction and describes how quickly a material reacts to a change in temperature. In order to predict cooling processes or to simulate temperature fields, the thermal diffusivity must be known.

In the case of gas hydrates, because the thermal diffusivity, κ, of water (Waite et al., 2006, 2007) is approximately half that of methane hydrate, hydrate-bearing sediment can change temperature more rapidly than water-bearing sediment. This characteristic presents a potential geohazard for crude oil production from deepwater production sites overlain by hydrate-bearing layers. If high-temperature hydrocarbons in the wellbore dissociate hydrate in the surrounding sediment, the strength of the sediment decreases, potentially causing well failure or localized submarine landslides.

3.3   Other types of gases

In addition to gas hydrate derivatives, there are several types of unconventional gas resources that arise from different sources and/or are currently produced by methods other than those used for conventional gas production and require processing before sale to the consumer: (1) biogas, (2) coalbed methane, (3) coal gas, (4) flue gas, (5) gas in geopressurized zones, (6) gas in tight formations, (7) landfill gas, (8) refinery gas, and (9) shale gas (Mokhatab et al., 2006; Speight, 2011—biofuel book, 2013b, 2014a).

The boundary between conventional gas and unconventional gas resources is not well defined, because they result from a continuum of geologic conditions. Typically, natural gas produced from shale reservoirs and other tight reservoirs has been categorized under the general title *unconventional gas*, because the production process requires stimulation by horizontal drilling coupled with hydraulic fracturing because of the pack of permeability in the gas-bearing formation (Speight, 2016). Low permeability of the reservoir rocks prohibits natural movement of the gas to a well. Moreover, maximization or optimization of reservoir producibility can only be achieved by a thorough understanding of the occurrence and properties of the resources as well as the producibility of the gas from the reservoir.

In addition and with the onset of renewable energy programs, it would be remiss not to mention prominent gases produced from biomass and waste materials, viz. biogas and landfill gas. Both types of gas contain methane and carbon dioxide as well as various other constituents and are often amenable to gas processing methods that are applied to natural gas.

3.3.1 Biogas

Biogas (often called *biogenic gas* and sometimes incorrectly known as *swamp gas*) typically refers to a biofuel gas produced by (1) anaerobic digestion with anaerobic organisms, which digest material inside a closed system or (2) fermentation of bio-degradable organic matter including manure, sewage sludge, municipal solid waste, biodegradable waste, or any other biodegradable feedstock, under anaerobic conditions (Table 3.4). Examples of biomass are: (1) wood and wood processing wastes, (2) agricultural crops and waste materials, (3) food, yard, and wood waste in garbage, and (4) animal manure and human sewage, which are all potential sources of biogas (biogenic gas). The process of biogas production (typically an anaerobic process) is a multistep biological process, where the originally complex and big-sized organic solid wastes are progressively transformed in simpler and smaller sized organic compounds by different bacteria strains up to have a final energetically worthwhile gaseous product and a semisolid material (digestate) that is rich in nutrients and thus suitable for its utilization in farming.

Table 3.4 The composition of gas from various carbonaceous fuels

Composition	Coal gas (%)	Coke oven gas (%)	Biogas (%)	Digester gas (%)	Landfill gas (%)	Natural Gas (%)
Hydrogen (H_2)	14.0	51.9	18.0		0.1	–
Carbon monoxide (CO)	27.0	2.0	24.0		0.1	–
Carbon dioxide (CO_2)	4.5	5.5	6.0	30.0	47.0	–
Oxygen (O_2)	0.6	0.3	0.4	0.7	0.8	–
Methane (CH_4)	3.0	32.0	3.0	64.0	47.0	90.0
Nitrogen (N_2)	50.9	4.8	48.6	2.0	3.7	5.0
Ethane (C_2H_6)	–	–	–			5.0

Table 3.5 Examples of biogas composition

Constituents	Source-1	Source-2	Source-3
Methane, CH_4 (% v/v)	50−60	60−75	60−75
Carbon dioxide, CO_2 (% v/v)	38−34	33−19	33−19
Nitrogen, N_2 (% v/v)	5−0	1−0	1−0
Oxygen, O_2 (% v/v)	1−0	<0.5	<0.5
Water, H_2O (% v/v)	6 (~40°C)	6 (~40°C)	6 (~40°C)
Hydrogen sulfide H_2S (mg/m^3)	100−900	1000−4000	3000−10,000
Ammonia, NH_3 (mg/m^3)	−	−	50−100

Source-1: Household waste.
Source-2: Wastewater treatment plant sludge.
Source-3: Agricultural waste.

Biogas production (typically an anaerobic process) is a multistep process in which originally complex organic (liquid or solid) wastes are progressively transformed into low molecular weight products by different bacteria strains (Esposito et al., 2012). Biogas can also be produced by pyrolysis of biomass (freshly harvested or as a biomass waste). Thus the name biogas gathers a large variety of gases under the name resulting from specific treatment processes, starting from various organic waste such as livestock manure, food waste, and sewage are all potential sources of biogenic gas, or biogas, which is usually considered a form of renewable energy and is often categorized according to the source (Table 3.5). In spite of the potential differences in composition, biogas can be processed (upgraded) to the standards required for natural gas though choice of the relevant processing sequence is depending upon the composition of the gas (Chapter 7: Process classification and Chapter 8: Gas cleaning processes).

During the combustion biomass, various kinds of impurities are generated and some of them occur in the flue gas and most of the contaminants in the flue gas are related to the composition of the biomass. If the combustion is incomplete (i.e., carried out in a deficiency of oxygen), soot, unburned matter, toxic dioxin derivatives may also occur in the flue gas.

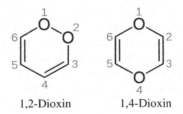

1,2-Dioxin   1,4-Dioxin

In addition, metals (Cui et al., 2013), such as lead (Pb), also occur in the ash and may even evaporate during combustion and react, condense, and/or sublime during cooling in the boiler. Upstream of the gas cleaning installation, normally at a temperature $<200°C$ ($<390°F$), all metals will occur as solid particles, except mercury which evaporates during combustion and reacts in the boiler but remains mainly in its gaseous form. The impurities in the biogas are harmful if they are emitted to the atmosphere and gas cleaning units must be installed to eliminate or at least reduce this problem. The degree of cleaning depends on federal, regional and local regulations, but regional and local authorities, organizations, and individuals have often an opinion on an actual plant due to its size and location.

More generally, contaminants aside, in terms of composition, biogas is primarily a mixture of methane (CH_4) and inert carbonic gas (CO_2), but variations in the composition of the source material lead to variations in the composition of the gas (Table 3.5) (Speight, 2011). Water (H_2O), hydrogen sulfide (H_2S), and particulates are removed if present at high levels or if the gas is to be completely cleaned. Carbon dioxide is less frequently removed, but it must also be separated to achieve pipeline quality gas. If the gas is to be used without extensive cleaning, it is sometimes cofired with natural gas to improve combustion. Biogas cleaned up to pipeline quality is called renewable natural gas.

Finally, natural gas is classified as fossil fuel (Speight, 2014a), whereas biomethane is defined as a nonfossil fuel (Speight, 2011) and is further characterized or described as a green source of energy. Noteworthy at this point is that methane, whatever the source (fossil fuel or nonfossil fuel) and when released into the atmosphere, is approximately 20 times more potent as a greenhouse gas than carbon dioxide. Organic matter from which biomethane is produced would release the carbon dioxide into the atmosphere if simply left to decompose naturally, while other gases that are produced during the decomposition process such as, e.g., nitrogen oxide(s) would make an additional contribution to the greenhouse effect.

3.3.2 Coalbed methane

Just as natural gas is often located in the same reservoir as with crude oil, a gas (predominantly methane) can also be found trapped within coal seams where it is often referred to as coalbed methane (or *coal bed methane*, CBM, sometimes referred to as *coalmine methane*, CMM). The gas occurs in the pores and cracks in the coal seam and is held there by underground water pressure. To extract the gas, a well is drilled into the coal seam and the water is pumped out (*dewatering*), which allows the gas to be released from the coal and brought to the surface.

However, the occurrence of methane in coal seams is not a new discovery and methane (also called *firedamp*) was known to coal miners for at least 150 years (or more) before it was *rediscovered* and developed as coalbed methane (Speight, 2013b). To the purist, CMM is the fraction of coalbed methane that is released during the mining operation (referred to in the older literature as *firedamp* by miners because of its explosive nature). In practice, the terms coalbed methane and coalmine methane may usually refer to different sources of gas—both forms of gas, whatever the name, are equally dangerous to the miners.

Coalbed methane is relatively pure compared to conventional natural gas, containing only very small proportions of higher molecular weight hydrocarbon derivatives such as ethane and butane and other gases (such as hydrogen sulfide and carbon dioxide). Because coal is a solid, very high carbon content mineral, there are usually no liquid hydrocarbon derivatives contained in the produced gas. The coal bed (coal seam) must first be dewatered to allow the trapped gas to flow through the formation to produce the gas. Consequently, coalbed methane usually has a lower heating value, and elevated levels of carbon dioxide, oxygen, and water that must be treated to an acceptable level, given the potential to be corrosive.

The gas from coal seams can be extracted by using technologies that are similar to those used to produce conventional gas, such as using well-bores. However, complexity arises from the fact that the coal seams are generally of low permeability and tend to have a lower flow rate (or permeability) than conventional gas systems, gas is only sourced from close to the well and as such a higher density of wells is required to develop a coalbed methane resource as an unconventional resource (such as tight gas) than a conventional gas resource. Technologies such as horizontal and multilateral drilling with hydraulic fracturing are sometimes used to create longer, more open channels that enhance well productivity but not all coal seam gas wells require application of this technique.

Unlike natural gas from conventional reservoirs, coalbed methane contains very little higher molecular weight hydrocarbon derivatives such as propane or butane or condensate. It often contains up to a few percent carbon dioxide. Some coal seams contain little methane, with the predominant coal seam gas being carbon dioxide. In fact, coalbed methane has been suggested with sufficient justification that the materials comprising a coalbed fall broadly into the following two categories: (1) volatile low-molecular-weight materials that can be liberated from the coal by pressure reduction, mild heating, or solvent extraction and (2) materials that will remain in the solid state after the separation of volatile components.

Typically, with some exceptions, coalbed gas is typically in excess of 90% v/v methane and, subject to gas composition data, may be suitable for introduction into a commercial pipeline with little or no treatment (Mokhatab et al., 2006; Speight, 2007; 2013a). Methane within coalbeds is not structurally trapped by overlying geologic strata, as in the geologic environments typical of conventional gas deposits (Speight, 2013a, 2014a). Only a small amount (on the order of 5–10% v/v) of the coalbed methane is present as free gas within the joints and cleats of coalbeds. Most of the coalbed methane is contained within the coal itself (adsorbed to the sides of the small pores in the coal).

3.3.3 Coal gas

Coal gas is any gaseous product that is produced by gasification of coal or by carbonization of coal (Speight, 2013b).

Coal carbonization is used for processing of coal to produce coke using metallurgical grade coal (Speight, 2013a). The process involves heating coal in the absence of air to produce coke and is a multistep complex process and variety of solid liquids and gaseous products are produced, which contain many valuable products. The various products from coal carbonization in addition to coke are (1) coke oven gas, (2) coal tar, (3) low-boiling oil, also called light oil, and (4) aqueous solution of ammonia and ammonium salts. With the development of the steel industry there was a continuous development in coke oven plants during the latter half of 19th century to improve the process conditions, recovery of chemicals and this continued during the 20th century to adapt to environmental pollution control strategies and energy consumption measures.

The carbonization can be carried out at various temperatures (Table 3.6), although low temperature or high temperature is preferred. Low-temperature carbonization is used to produce liquid fuels, while high-temperature carbonization is used to produce gaseous products (Speight, 2013a). Low-temperature carbonization (approximately 450−750°C, 840−1380°F) is used to produce liquid fuels (with smaller amounts of gaseous products), while the high-temperature carbonization process (approximately 900°C, 1650°F) is used to produce gaseous products. The high-temperature carbonization process gaseous product is less while liquid products are large and the production of tar is relatively low because of the cracking of the secondary (liquid products and tar) products (Speight, 2013a).

Most of these products have been replaced by natural gas (the city and town gas works is an industry of the past) but still find use in some parts of the world. The products of coal gasification are varied insofar as the gas composition varies with the system employed. It is emphasized that the gas product must be first freed from any pollutants such as particulate matter and sulfur compounds before further use, particularly when the intended use is a water gas shift or methanation.

Gases of high calorific value are obtained by low-temperature or medium-temperature carbonization of coal. The gases obtained by the carbonization of any given coal change in a progressive manner with increasing temperature (Table 3.6). The composition of coal gas also changes during the course of carbonization at a given temperature and secondary reactions of the volatile products are important in determining gas composition (Speight, 2013a; ASTM, 2017).

Table 3.6 Effect of carbonization temperature on the composition of coal gas

Gas composition and yields (% v/v)	Temperature of carbonization					
	500°C	600°C	700°C	800°C	900°C	1000°C
CO_2	5.7	5.0	4.4	4.0	3.2	2.5
Unsaturates	3.2	4.0	5.2	5.1	4.8	4.5
CO	5.8	6.4	7.5	8.5	9.5	11.0
H_2	20.0	29.0	40.0	47.0	50.0	51.0
CH_4	49.5	47.0	36.0	31.0	29.5	29.0
C_2H_6	14.0	5.3	4.5	3.0	1.0	0.5
Yield (m^3/tonne)	62.3	102	176	238	278	312
CV (MJ/m^3)	39.0	29.0	26.5	24.4	22.3	22.3
Yield (MJ/tonne)	2118	2960	4660	5810	6200	6960

Low heat-content gas (low-Btu gas) is produced during the gasification of when the oxygen is not separated from the air and, thus, the gas product invariably has a low heat-content (ca. $150-300$ Btu/ft^3). Low heat-content gas is also the usual product of in situ gasification of coal which is used essentially as a technique for obtaining energy from coal without the necessity of mining the coal. The process is a technique for utilization of coal which cannot be mined by other techniques.

The nitrogen content of low heat-content gas ranges from somewhat less than 33% v/v to slightly more than 50% v/v and cannot be removed by any reasonable means; the presence of nitrogen at these levels renders the product gas to be low heat-content. The nitrogen also strongly limits the applicability of the gas to chemical synthesis. Two other noncombustible components (water, H_2O, and carbon dioxide, CO_2) lower the heating value of the gas further. Water can be removed by condensation and carbon dioxide by relatively straightforward chemical means.

The two major combustible components are hydrogen and carbon monoxide; the H_2/CO ratio varies from approximately 2:3 to approximately 3:2 but methane may also make an appreciable contribution to the heat content of the gas. Of the minor components hydrogen sulfide is the most significant and the amount produced is, in fact, proportional to the sulfur content of the feed coal. Any hydrogen sulfide present must be removed by one, or more, of several procedures. Low heat-content gas is of interest to industry as a fuel gas or even, on occasion, as a raw material from which ammonia, methanol, and other compounds may be synthesized.

Medium heat-content gas (medium-Btu gas) has a heating value in the range $300-550$ Btu/ft^3 and the composition is much like that of low heat-content gas, except that there is virtually no nitrogen. The primary combustible gases in medium heat-content gas are hydrogen and carbon monoxide. Medium heat-content gas is considerably more versatile than low heat-content gas; like low heat-content gas, medium heat content gas may be used directly as a fuel to raise steam or used through a combined power cycle to drive a gas turbine, with the hot exhaust gases employed to raise steam.

Medium heat-content gas is especially amenable to the production of (1) methane, (2) higher molecular weight hydrocarbon derivatives by the Fischer−Tropsch synthesis, (3) methanol, and (4) a variety of synthetic chemicals (Chadeesingh, 2011; Speight, 2013a). The reactions used to produce medium heat-content gas are the same as those employed for low heat-content gas synthesis, the major difference being the application of a nitrogen barrier (such as the use of pure oxygen) to keep diluent nitrogen out of the system.

In medium heat-content gas, the hydrogen−carbon monoxide ratio varies from 2:3 to ca. 3:1 and the increased heating value correlates with higher methane and hydrogen contents as well as with lower carbon dioxide contents. In fact, the nature of the gasification process used to produce the medium heat-content gas has an effect on the ease of subsequent processing. For example, the carbon dioxide-acceptor product is available for use in methane production because it has (1) the desired hydrogen−carbon dioxide ratio just exceeding 3:1, (2) an initially high methane content, and (3) relatively low carbon dioxide content and low water content.

High heat-content gas (high-Btu gas) is almost pure methane and often referred to as *synthetic natural gas* or *substitute natural gas* (SNG). However, to qualify as SNG, a product must contain at least 95% methane; the energy content in the order of 980–1080 Btu/ft^3.

The commonly accepted approach to the synthesis of high heat content gas is the catalytic reaction of hydrogen and carbon monoxide.

$$3H_2 + CO \rightarrow CH_4 + H_2O$$

To avoid catalyst poisoning, the feed gases for this reaction must be quite pure and, therefore, impurities in the product are rare. The water produced by the reaction is removed by condensation and recirculated as very pure water through the gasification system. The hydrogen is usually present in slight excess to ensure that the toxic carbon monoxide is reacted.

The carbon monoxide-hydrogen reaction is not the most efficient way to produce methane because of the exothermicity of the reaction. Also, the methanation catalyst is subject to poisoning by sulfur compounds and the decomposition of metals can destroy the catalyst. Hydrogasification may be employed to minimize the need for methanation.

$$C_{coal} + 2H_2 \rightarrow CH_4$$

The product of this reaction is not pure methane and additional methanation is required after hydrogen sulfide and other impurities are removed.

3.3.4   Flue gas

Flue gas (sometimes called *exhaust gas* or *stack gas*) is the gas that emanates from combustion plants and which contains the reaction products of fuel and combustion air and residual substances such as particulate matter (dust), sulfur oxides, nitrogen oxides, and carbon monoxide (Table 3.7). When burning coal and/or waste materials, hydrogen chloride and hydrogen fluoride may be present in the flue gas as well as hydrocarbon derivatives and heavy metal derivatives. In many countries, as part of a national environmental protection program, exhaust gases must comply with strict governmental regulations regarding the limit values of pollutants such as dust, sulfur and nitrogen oxides and carbon monoxide. To meet these limit values combustion plants are equipped with flue gas cleaning systems such as gas scrubbers and dust filters.

As with other gases, the composition of flue gas depends on the type of fuel and the combustion conditions, e.g., the air ratio value. Many flue gas components are air pollutants and must therefore, due to governmental regulations, be eliminated or minimized by special cleaning procedures before the gas is released to the atmosphere. For example, flue gas produced by the combustion of fossil fuels, such as coal and crude oil can contain a significant amount of sulfur derivatives. In fact, when carbonaceous fuels, such as fossil fuels, are burned, approximately 90% + w/w of the sulfur in the feedstock is converted to sulfur dioxide (SO_2), which occurs under

Table 3.7 Constituents of flue gas

Constituents	Comment
Nitrogen	The main constituent (79 vol. %) of air; fed to the combustion as part of the combustion air but is not involved directly in the combustion process. Minor quantities of this combustion air related nitrogen are, together with the nitrogen released from the fuel, responsible for the formation of the nitrogen oxides
Carbon dioxide	A colorless and odorless gas with a slightly sour taste; produced during all combustion processes including respiration
Water vapor	Hydrogen contained in the fuel will react with oxygen and form water (H_2O); together with the water content of the fuel and the combustion air, exists either as flue gas humidity (at higher temperatures) or as condensate (at lower temperatures)
Oxygen	The oxygen that has not been consumed by the combustion process remains as part of the flue gas and is a measure for the efficiency of the combustion
Carbon monoxide	A colorless, odorless, toxic gas; formed predominantly during incomplete combustion of carbonaceous fuels
Oxides of nitrogen	Formed in combustion processes nitrogen of the fuel and, at high temperatures, also of the combustion air reacts to a certain amount with oxygen of the combustion air and forms first nitric oxide (fuel-NO and thermal-NO)
Sulfur dioxide	A colorless, toxic gas with a pungent smell. It is formed through oxidation of sulfur that is present in the fuel; with water or condensate, forms sulfurous acid and sulfuric acid
Hydrogen sulfide	A toxic odorous gas; a component of crude oil and natural gas and is therefore present in refineries and natural gas plants; also generated during some other industrial processes
Hydrocarbons	An extensive group of chemical compounds that are composed of hydrogen and carbon; occur in crude oil, natural gas, and coal; formed through incomplete combustion processes
Hydrocyanic acid	A very toxic liquid with a boiling point of only 25.6°C; it may exist in flue gases of incineration plants
Ammonia	Relevant in flue gases in connection with denitrification plants
Hydrogen halides	Occur in flue gas from the combustion of coal and/or waste material; the formation of the hydrogen halides hydrogen chloride and hydrogen fluoride may result which form aggressive acids in humid atmospheres
Solids (dust, soot)	Solid pollutants in flue gases originate from the incombustible components of solid or liquid fuels; include oxides of silica, aluminum, and calcium in case of coal

normal conditions of temperature in the furnace and the oxygen fed to the combustor. However, when there is an excess of oxygen present, the sulfur dioxide is oxidized to sulfur trioxide (SO_3) and at higher temperatures (approximately 800°C, 1470°F), the formation of sulfur trioxide is favored.

As with other gases, the composition of flue gas depends upon the composition and properties of the fuel being combusted but, nevertheless, the composition will usually consist predominantly of nitrogen (usually on the order of 60% v/v and higher derived from the combustion process using air as the oxidant, carbon dioxide, and water vapor), as well as any excess oxygen (also derived from the air). Flue gas can also contain a small percentage of a number of pollutants, such as (1) particulate matter, e.g., soot, (2) carbon monoxide, (3) nitrogen oxides, NO_x, and (4) sulfur oxides, SO_x. The potential for the presence of hydrocarbon derivatives is low unless the combustion process used a minimal of air and then also uses a hydrocarbonaceous feedstock. The particulate matter is composed of very small particles of solid materials and very small liquid droplets which give flue gases their smoky appearance. The nitrogen oxides are derived from the nitrogen in the ambient air as well as from any nitrogen-containing compounds in the fuel. The sulfur dioxide is derived from any sulfur-containing compounds in the fuel.

At power plants, flue gas is often treated with a series of chemical processes and scrubbers, which remove pollutants such as sulfur dioxide and sulfur trioxide (Speight, 2013a). Flue gas desulfurization units capture the sulfur dioxide (and the sulfur trioxide, if present) and electrostatic precipitators or fabric filters remove particulate matter produced by burning fossil fuels, particularly coal. Nitrogen oxides are treated either by modifications to the combustion process to prevent their formation, or by high temperature or catalytic reaction with ammonia (NH_3) or urea (H_2NCONH_2). In either case, the aim is to produce nitrogen gas, rather than nitrogen oxides. In the United States, there is a rapid deployment of technologies to remove mercury from flue gas (Scala et al., 2013). This is typically accomplished by absorption on sorbents or by capture in inert solids as part of the flue gas desulfurization process. Such scrubbing can lead to meaningful recovery of sulfur for further industrial use (Mokhatab et al., 2006; Speight, 2014a).

Examples of common flue gas cleaning processes are: (1) a wet scrubbing process, which uses a slurry of alkaline sorbent, usually limestone or lime, or seawater to scrub the gases, (2) a spray-dry scrubbing process, which uses similar sorbent slurries as described in the first category, (3) a wet sulfuric acid process, which allows the recovery of sulfur in the form of commercial quality sulfuric acid, (4) a process commonly referred to as a SNOX flue gas desulfurization process, which removes sulfur dioxide, nitrogen oxides, and particulate matter from flue gases, and (5) a dry sorbent injection process, which introduces powdered hydrated lime or other sorbent material into the exhaust ducts to eliminate sulfur dioxide and sulfur trioxide from the process emissions (Chapter 7: Process classification and Chapter 8: Gas cleaning processes).

In terms of flue gas (and biogas) cleaning, the separation of the acid gaseous pollutants, the separation of hydrogen chloride, sulfur dioxide, and hydrogen fluorides (and any vestiges thereof) is essential. These gases form the greater part of the polluting constituents affected by absorption, preferentially by means of lime products (CaO and Ca(OH)$_2$) and also by sodium-based products (NaOH, NaHCO$_3$, and Na$_2$CO$_3$). The available gas cleaning processes for the absorption of the acid gaseous pollutants can be classified into the following three groups: (1) dry sorption, (2) spray absorption/drying, and (3) wet scrubbing.

The separation of the fly ash and the metals occurring in the form of particulate matter at the boiler outlet takes place likewise via filtration, so that this process step can be integrated into the absorption process for the acid gaseous pollutants. The separation of dioxins, furans, and those metals, in particular mercury, present in gaseous form at the boiler outlet generally takes place by adsorption on activated carbon, zeolites, open hearth furnace coke, or bentonite. For adsorption, either static or a moving bed adsorber unit or a filter layer adsorber unit may be employed, which offers the possibility of integration of the adsorption process with the filtering out of fly ash and reaction products from the four basic gas cleaning concepts. Removal of nitrogen oxides (NO_x) from the flue gases can be achieved in conjunction with the above pollution control systems by the use of selective noncatalytic reduction.

3.3.5 Gas in geopressurized zones

The term *geopressure* refers to a reservoir fluid (including gas) pressure that significantly exceeds hydrostatic pressure (which is on the order of 0.4−0.5 psi per foot of depth) and may even approach overburden pressure (on the order of 1.0 psi per foot of depth). Thus geopressurized zones are natural underground formations that are under unusually high pressure for their depth. The geopressurized zones are formed by layers of clay that are deposited and compacted very quickly on top of more porous, absorbent material such as sand or silt. Water and natural gas that are present in this clay are squeezed out by the rapid compression of the clay and enter the more porous sand or silt deposits. Geopressured reservoirs frequently are associated with substantial faulting and complex stratigraphy, which can make correlation, structural interpretation, and volumetric mapping subject to considerable uncertainty.

The pressures in sand-shale sequences generally are attributed to undercompaction of thick sequences of marine shales and reservoirs in such depositional sequences tend to be geologically complex. This can cause considerable uncertainty in reserves estimates at all stages of development and production of gas from the reservoir. Geologic complexity contributes to the uncertainty of estimating the in-place reserves and lack of a thorough understanding of the producing mechanisms contributes to uncertainty in estimates of reserves that are based on pressure/production performance.

In addition, geopressurized zones are typically located at great depths, usually 10,000−25,000 ft below the surface of the Earth. The combination of all these factors makes the extraction of natural gas in geopressurized zones quite complicated. However, of all the unconventional sources of natural gas, geopressurized zones are estimated to hold the greatest amount of gas.

The amount of natural gas in these geopressurized zones is uncertain although unproven estimates indicate that 5000−49,000 tcf (5000−49,000 × 10^{12} ft^3) of natural gas may exist in these areas. Like gas hydrates, the gas in the geopressurized zones offers an opportunity for future supplies of natural gas. However, the combination of the above factors makes the extraction of natural gas or crude oil located in geopressurized zones quite complicated (Speight, 2017b).

3.3.6 Gas in tight formations

The term *tight formation* refers to a formation consisting of extraordinarily impermeable, hard rock (Speight, 2013a). Tight formations are relatively low permeability, nonshale, sedimentary formations that can contain oil and gas. A *tight reservoir*
(tight sands) is a low-permeability sandstone reservoir that produces primarily dry
natural gas. A tight gas reservoir is one that cannot be produced at economic flow
rates or recover economic volumes of gas unless the well is stimulated by a large
hydraulic fracture treatment and/or produced using horizontal wellbores. This definition also applies to coalbed methane and tight carbonate reservoirs—shale gas
reservoirs are also included by some observers (but not in this text).

Typically, tight formations which formed under marine conditions contain less
clay and are more brittle, and thus more suitable for hydraulic fracturing than formations formed in fresh water which may contain more clay. The formations
become more brittle with an increase in quartz content (SiO_2) and carbonate content
(such as calcium carbonate, $CaCO_3$, or dolomite, $CaCO_3 \cdot MgCO_3$).

By way of explanation and comparison, in a conventional sandstone reservoir
the pores are interconnected so that natural gas and crude oil can flow easily
through the reservoir and to the production well. Conventional gas typically is
found in reservoirs with permeability greater than 1 milliDarcy (mD) and can be
extracted via traditional techniques (Fig. 3.1). However, in tight sandstone formations, the pores are smaller and are poorly connected (if at all) by very narrow
capillaries, which results in low permeability and immobility of the natural gas.
Such sediments typically have an effective permeability of less than 1 mD. In contrast, unconventional gas is found in reservoirs with relatively low permeability
(<1 mD) (Fig. 3.1) and hence cannot be extracted via conventional methods.

The tight gas is contained in lenticular or blanket reservoirs that are relatively
impermeable, which occurs downdip from water-saturated rocks and cuts across
lithologic boundaries. They often contain a large amount of in-place gas but exhibit
low recovery rates. Gas can be economically recovered from the better quality continuous tight reservoirs by creating downhole fractures with explosives or hydraulic
pumping. The nearly vertical fractures provide a pressure sink and channel for the
gas, creating a larger collecting area so that the gas recovery is at a faster rate.
Sometimes massive hydraulic fracturing is required, using half million gallons of

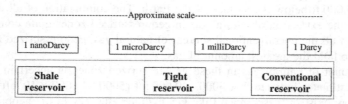

Figure 3.1 Representation of the differences in permeability of shale reservoirs, tight
reservoirs, and conventional reservoirs.

gelled fluid and a million pounds of sand to keep the fractures open after the fluid has been drained away.

Natural gas from formations designated as tight formations (i.e., formations having zero to extremely low permeability) offers additional energy-producing resources. This gas can be recovered from such formations that typically function as both the reservoir rock and the source rock. The tight formations that yield gas and oil are organic-rich shale formations that were previously regarded only as source rocks and seals for gas accumulating in the strata near sandstone and carbonate reservoirs of traditional onshore gas development. In terms of chemical makeup, shale gas is typically a dry gas composed primarily of methane, but some formations do produce wet gas, while crude oil from tight formations is typically more volatile than many crude oils from conventional reservoirs.

The challenge in treating such gases is the low (or differing) hydrogen sulfide/carbon dioxide ratio and the need to meet pipeline specifications. By way of recall, a specification is the collected data that give adequate control of natural gas (or condensate) behavior in a gas processing plant or refinery or for sales of the gas. More accurately, the specifications are derived from the set of tests and data limits applicable to the natural gas or to a finished product in order to ensure that every batch is of satisfactory and consistent quality at release for sales. The specifications should include all critical parameters in which variations would be likely to affect the safety and in-service use of the product.

3.3.7 Landfill gas

Landfill gas, which is often included under the umbrella definition of biogas, is also produced from the decay of organic wastes (such as municipal solid waste that contains organic materials) but these wastes may then be biomass-type materials (Lohila et al., 2007; Staley and Barlaz, 2009; Speight, 2011). Landfill sites offer another underutilized source of biogas. When municipal waste is buried in a landfill, bacteria break down the organic material contained in garbage such as newspapers, cardboard, and food waste, producing gases such as carbon dioxide and methane. Rather than allowing these gases to go into the atmosphere, where they contribute to global warming, landfill gas facilities can capture them, separate the methane, and combust it to generate electricity, heat, or both.

Landfill gas is produced by wet organic waste decomposing under anaerobic conditions in a biogas. In fact, landfill gas is a product of three processes: (1) evaporation of volatile organic compounds such as low-boiling solvents, (2) chemical reactions between waste components, and (3) microbial action, especially methanogenesis. The first two processes depend strongly on the nature of the waste—the most dominant process in most landfills is the third process whereby anaerobic bacteria decomposes organic waste to produce biogas, which consists of methane and carbon dioxide together with traces of other compounds. Despite the heterogeneity of waste, the evolution of gases follows well-defined kinetic pattern in which the formation of methane and carbon dioxide commences approximately 6 months after depositing the landfill material. The evolution of landfill gases reaches a maximum at approximately 20 years, then declines over the course of several decades.

Thus landfill gas is another source of gas generated by microorganisms from the decomposition of the biodegradable organic fraction of municipal solid waste. This generally occurs under semicontrolled conditions in a landfill; its constituents depend on the composition and age of the waste. Large variations may exist in the composition of landfill gas due to differences in sources of municipal solid waste and operating conditions at the landfill. The three main gas constituents (methane, CH_4, carbon dioxide, CO_2, and hydrogen sulfide, H_2S) are used to characterize landfill gas.

In the process, the waste is covered and mechanically compressed by the weight of the material that is deposited above. This material prevents oxygen exposure thus allowing anaerobic microbes to thrive. Biogas builds up and is slowly released into the atmosphere if the site has not been engineered to capture the gas. Landfill gas released in an uncontrolled way can be hazardous since it can become explosive when it escapes from the landfill and mixes with oxygen.

As should be expected, the amount of methane that is produced varies significantly based on composition of the waste (Staley and Barlaz, 2009). The efficiency of gas collection at landfills directly impacts the amount of energy that can be recovered—closed landfills (those no longer accepting waste) collect gas more efficiently than open landfills (those that are still accepting waste). The gas is a complex mix of different gases created by the action of microorganisms within a landfill. Typically, landfill gas is composed of 45−60% v/v methane, 40−60% v/v carbon dioxide, 0−1.0% v/v hydrogen sulfide, 0−0.2% v/v hydrogen (H_2), trace amounts of nitrogen (N_2), low molecular weight hydrocarbon derivatives (dry volume basis), and water vapor (saturated). The specific gravity of landfill gas is approximately 1.02−1.06. Trace amounts of other volatile organic compounds comprise the remainder (typically, 1−2% v/v or less) and these trace gases include a large array of species, such as low-molecular weight hydrocarbon derivatives. Other minor components include hydrogen sulfide, nitrogen oxides, sulfur dioxide, nonmethane volatile organic compounds, polycyclic aromatic hydrocarbon derivatives, polychlorinated dibenzodioxin derivatives, and polychlorinated dibenzofuran derivatives (Brosseau, 1994; Rasi et al., 2007). All of the aforementioned agents are harmful to human health at high doses.

Landfill gas collection is typically accomplished through the installation of wells installed vertically and/or horizontally in the waste mass. Design heuristics for vertical wells call for about one well per acre of landfill surface, whereas horizontal wells are normally spaced about 50−200 ft apart on center. Efficient gas collection can be accomplished at both open and closed landfills, but closed landfills have systems that are more efficient, owing to greater deployment of collection infrastructure since active filling is not occurring. On average, closed landfills have gas collection systems that capture approximately 84% v/v of produced gas, compared to approximately 67% v/v for open landfills.

Landfill gas can also be extracted through horizontal trenches instead of vertical wells. Both systems are effective at collecting. Landfill gas is extracted and piped to a main collection header, where it is sent to be treated or flared. The main collection header can be connected to the leachate collection system to collect condensate forming in the pipes. A blower is needed to pull the gas from the collection wells to the collection header and further downstream.

The gas produced within a landfill site can be collected for various uses, such as direct utilization on site in a boiler or any type of combustion system, providing heat. Electricity can also be generated on site through the use of microturbines, steam turbines, or fuel cells (Sullivan, 2010). The landfill gas can also be sold off site and sent into natural gas pipelines. This approach requires the gas to be processed into pipeline quality, e.g., by removing various contaminants and components. As should be expected, the amount of methane that is produced varies significantly based on composition of the waste (Staley and Barlaz, 2009). The efficiency of gas collection at landfills directly impacts the amount of energy that can be recovered—closed landfills (those no longer accepting waste) collect gas more efficiently than open landfills (those that are still accepting waste).

However, landfill gas cannot be distributed through utility natural gas pipelines unless it is cleaned up to less than 3% carbon dioxide and a few parts per million of hydrogen sulfide, because carbon dioxide and hydrogen sulfide corrode the pipelines (Speight, 2014b). Thus landfill gas must be treated to remove impurities, condensate, and particulates. Hence, the need for analysis to determine the composition of the gas. However, the treatment system depends on the end use: (1) minimal treatment is needed for the direct use of gas in boiler, furnaces, or kilns and (2) using the gas in electricity generation typically requires more in-depth treatment.

Treatment systems are divided into primary and secondary treatment processing. Primary processing systems remove moisture and particulates. Gas cooling and compression are common in primary processing. Secondary treatment systems employ multiple cleanup processes, physical and chemical, depending on the specifications of the end use. Two constituents that may need to be removed are siloxane derivatives and sulfur-containing compounds, which are damaging to equipment and significantly increase maintenance cost. Adsorption and absorption are the most common technologies used in secondary treatment processing. Also, landfill gas can be converted to high-Btu gas by reducing the amount of carbon dioxide, nitrogen, and oxygen in the gas.

The high-Btu gas can be piped into existing natural gas pipelines or in the form of compressed natural gas or liquid natural gas. Compressed natural gas and liquid natural gas can be used on site to power hauling trucks or equipment or sold commercially. Three commonly used methods to extract the carbon dioxide from the gas are membrane separation, molecular sieve, and amine scrubbing (Chapter 7: Process classification and Chapter 8: Gas cleaning processes). Oxygen and nitrogen are controlled by the design and operation of the landfill since the primary cause for oxygen or nitrogen in the gas is intrusion from outside into the landfill because of a difference in pressure.

Landfill gas condensate is a liquid that is produced in landfill gas collection systems and is removed as the gas is withdrawn from landfills. Production of condensate may be through natural or artificial cooling of the gas or through physical processes such as volume expansion. The condensate is composed principally of water and organic compounds. Often the organic compounds are not soluble in water and the condensate separates into a watery (aqueous) phase and a floating organic (hydrocarbon) phase which may constitute up to 5% v/v of the liquid.

A large number of acid and base/neutral compounds are typically present in the aqueous phase but, as expected, are dependent upon the compounds types in the landfill. The organic phase can consist of hydrocarbon derivatives, xylene isomers, chloroethane derivatives, chloroethylene derivatives, benzene, toluene, other priority pollutants, and trace moisture.

3.3.8 Manufactured gas

Manufactured gas is a fuel—gas mixture made from other solid, liquid, or gaseous materials, such as coal, coke, or crude oil and should not be confused with natural gas. The principal types of manufactured gas are retort coal gas, coke oven gas, water gas, carbureted water gas, producer gas, oil gas, reformed natural gas, and reformed propane or LPG. Several processes for making SNG from coal have been developed. Most of the manufactured gas, as it fits into the context of this book, is produced by any one of three processes which, at the present time, find limited use but are still active processes on some areas and these are: (1) coal carbonization process, (2) the carbureted water gas process, and (3) oil gas process.

The *coal carbonization process* was the primary commercial mode of manufacturing gas from ca. 1816 to 1875. After 1875, newer processes and technologies gradually replaced coal carbonization. Coal gas was produced through the distillation of bituminous coal in heated, anaerobic vessels called retorts. In this process, coal was broken down into its volatile components through the action of heat in an anaerobic environment. During the retorting phase, approximately 40% w/w of the coal is converted into volatile gases and liquids, while the remainder of the coal is converted into solids, primarily coke.

From the retort, some of the gaseous products are condensed (to liquids) and others remain in the gaseous state. The liquids (also called *liquors*) consist of water and coal tar. The products remaining in the gaseous phase are cooled to produce additional coal tar after which the gas is cooled further to remove any other nongaseous impurities, such as ammonia and sulfur compounds. These are removed by washing the gas in water and by running the gas through beds of moist lime or moist iron oxides. After this final purification process, the coal gas is sent to storage.

The *carbureted water gas process* consists of enriching a form or coal gas, known as water gas (blue gas), to increase the heat content. Thus by injecting oil into a vessel containing heated water gas, the oil and vapor combined, form a gaseous fuel with a thermal content of approximately 300—350 Btu/ft^3. Typically, a carbureted water gas plant consists of a brick-lined, cylindrical, steel vessel, the generator, the carburetor (carbureted), and a superheater. As the gas exits the generator, it is passed into the carburetor where the oil is introduced into the vapor.

The *oil gas process* is very similar to carbureted water gas process but consists of steam-cracking the oil in a steam environment to produce the raw gas rather than by distilling coal. From the generator, the gas is passed to a vaporizer, where it is enriched with additional injections of oil and then routed through a superheater. After exiting the superheater, the gas is scrubbed and processed for distribution in much the same way as was carbureted water gas. Many of the same waste products associated with the production of coal gas, notably tars containing polynuclear aromatic hydrocarbons (or polyaromatic hydrocarbons) were also generated during gas manufacture.

3.3.9 Refinery gas

The terms *refinery gas* or *petroleum gas* are often used to identify LPG or even gas that emanates as light ends (gases and volatile liquids) from the atmospheric distillation unit or from any one of several other refinery processes.

For the purpose of this text, refinery gas not only describes LPG but also natural gas and refinery gas (Mokhatab et al., 2006; Gary et al., 2007; Speight, 2014a, 2017a; Hsu and Robinson, 2017). In this chapter, each gas is, in turn, referenced by its name rather than the generic term *petroleum gas*. However, the composition of each gas varies and recognition of this is essential before the relevant testing protocols are selected and applied. Thus refinery gas (fuel gas) is the noncondensable gas that is obtained during distillation of crude oil or treatment (cracking, thermal decomposition) of petroleum (Table 3.8) (Speight, 2014a).

Refinery gas is produced in considerable quantities during the different refining processes and is used as fuel for the refinery itself and as an important feedstock for the production of petrochemicals. It consists mainly of hydrogen (H_2), methane

Table 3.8 General summary of product types and distillation range

Product	Lower carbon limit	Upper carbon limit	Lower boiling point (°C)	Upper boiling point (°C)	Lower boiling point (°F)	Upper boiling point (°F)
Refinery gas	C_1	C_4	−161	−1	−259	31
Liquefied petroleum gas	C_3	C_4	−42	−1	−44	31
Naphtha	C_5	C_{17}	36	302	97	575
Gasoline	C_4	C_{12}	−1	216	31	421
Kerosene/diesel fuel	C_8	C_{18}	126	258	302	575
Aviation turbine fuel	C_8	C_{16}	126	287	302	548
Fuel oil	C_{12}	$>C_{20}$	216	421	>343	>649
Lubricating oil	$>C_{20}$		>343		>649	
Wax	C_{17}	$>C_{20}$	302	>343	575	>649
Asphalt	$>C_{20}$		>343		>649	
Coke	$>C_{50}^a$		$>1000^a$		$>1832^a$	

[a]Carbon number and boiling point difficult to assess; inserted for illustrative purposes only.

Table 3.9 Origin of petroleum-related gases

Gas	Origin
Natural gas	Occurs naturally with or without crude oil
	A gaseous combination of hydrocarbons
	Predominantly C_1 through C_4 hydrocarbons
	May also contain gas condensate or natural gasoline
Refinery gas	A combination of gases produced by distillation
process gas	Products from the cracking of crude oil
	Consists of C_2 to C_4 hydrocarbons including olefin gases
	Boiling range of approximately $-51°C$ to $-1°C$ ($-60°F$ to $30°F$)
Tail gas	A combination of hydrocarbons from the distillation of products from catalytically cracked feedstocks
	Predominantly of C_1 to C_4 hydrocarbons

(CH_4), ethane (C_2H_6), propane (C_3H_8), butane (C_4H_{10}), and olefins ($RCH{=}CHR^1$, where R and R^1 can be hydrogen or a methyl group) and may also include off-gases from petrochemical processes (Table 3.9). Olefins such as ethylene ($CH_2{=}CH_2$, boiling point: $-104°C$, $-155°F$), propene (propylene, $CH_3CH{=}CH_2$, boiling point: $-47°C$, $-53°F$), butene (butene-1, $CH_3CH_2CH{=}CH_2$, boiling point: $-5°C$, 23°F) *iso*-butylene (($CH_3)_2C{=}CH_2$, $-6°C$, 21°F), *cis*- and *trans*-butene-2 ($CH_3CH{=}CHCH_3$, boiling point: ca. 1°C, 30°F), and butadiene ($CH_2{=}CHCH{=}CH_2$, boiling point: $-4°C$, 24°F) as well as higher boiling olefins are produced by various refining processes.

Still gas is broad terminology for low-boiling hydrocarbon mixtures and is the lowest boiling fraction isolated from a distillation (*still*) unit in the refinery (Speight, 2014a, 2017a)). If the distillation unit is separating light hydrocarbon fractions, the still gas will be almost entirely methane with only traces of ethane (CH_3CH_3) and ethylene ($CH_2{=}CH_2$). If the distillation unit is handling higher boiling fractions, the still gas might also contain propane ($CH_3CH_2CH_3$), butane ($CH_3CH_2CH_2CH_3$), and their respective isomers. *Fuel gas* and still gas are terms that are often used interchangeably but the term *fuel gas* is intended to denote the product's destination to be used as a fuel for boilers, furnaces, or heaters.

A group of refining operations that contributes to gas production are the thermal cracking and catalytic cracking processes. The thermal cracking processes (such as the coking processes) produce a variety of gases, some of which may contain olefin derivatives ($>C{=}C<$). In the visbreaking process, fuel oil is passed through externally fired tubes and undergoes liquid phase cracking reactions, which result in the formation of lower boiling fuel oil components. Substantial quantities of both gas and carbon are also formed in coking (both fluid coking and delayed coking) in addition to the middle distillate and naphtha. When coking a residual fuel oil or heavy gas oil, the feedstock is preheated and contacted with hot carbon (coke), which causes extensive cracking of the feedstock constituents of higher molecular weight to produce lower molecular weight products ranging from methane, LPGs and naphtha, to gas oil and heating oil. Products from coking processes tend to be unsaturated and olefin-type components predominate in the tail gases from coking processes.

The various catalytic cracking processes, in which higher boiling gas oil fractions are converted into gaseous products, various naphtha fractions, fuel oil, and coke by contacting the feedstock with the hot catalyst. Thus both catalytic and thermal cracking processes, the latter being now largely used to produce chemical raw materials, result in the formation of unsaturated hydrocarbon derivatives, particularly ethylene (CH_2=CH_2), but also propylene (propene, CH_3CH=CH_2), isobutylene (isobutene, $(CH_3)_2C$=CH_2) and the n-butenes (CH_3CH_2CH=CH_2, and CH_3CH=$CHCH_3$) in addition to hydrogen (H_2), methane (CH_4) and smaller quantities of ethane (CH_3CH_3), propane ($CH_3CH_2CH_3$), and butane isomers ($CH_3CH_2CH_2CH_3$, $(CH_3)_3CH$). Diolefins such as butadiene (CH_2=$CHCH$=CH_2) are also present.

In a series of reforming processes, distillation fractions which include paraffin derivatives and naphthene derivatives (cyclic nonaromatic) are treated in the presence of hydrogen and a catalyst to produce lower molecular weight products or are isomerized to more highly branched hydrocarbon derivatives. Also, the catalytic reforming process not only results in the formation of a liquid product of higher octane number but also produce substantial quantities of gaseous products. The composition of these gases varies in accordance with process severity and the properties of the feedstock. The gaseous products are not only rich in hydrogen but also contain hydrocarbon derivatives from methane to butane derivatives, with a preponderance of propane ($CH_3CH_2CH_3$), n-butane ($CH_3CH_2CH_2CH_3$), and isobutane (($CH_3)_3CH$). Since all catalytic reforming processes require substantial recycling of a hydrogen stream, it is normal to separate reformer gas into a propane ($CH_3CH_2CH_3$) and/or a butane ($CH_3CH_2CH_2CH_3$/$(CH_3)_3CH$) stream, which becomes part of the refinery LPG production, and a lower boiling gaseous fraction, part of which is recycled.

A further source of refinery gas is produced by the hydrocracking process which is a high-pressure pyrolysis process carried out in the presence of fresh and recycled hydrogen. The feedstock is again heavy gas oil or residual fuel oil, and the process is mainly directed at the production of additional middle distillates and gasoline. Since hydrogen is to be recycled, the gases produced in this process again must be separated into lighter and heavier streams; any surplus recycle gas and the LPG from the hydrocracking process are both saturated.

Both hydrocracker and catalytic reformer tail gases are commonly used in catalytic desulfurization processes (Speight, 2014a, 2017a). In the latter, feedstocks ranging from light to vacuum gas oils are passed at pressures on the order of 500–1000 psi with hydrogen over a hydrofining catalyst. This results mainly in the conversion of organic sulfur compounds to hydrogen sulfide:

$$[S]_{feedstock} + H_2 \rightarrow H_2S + \text{Hydrocarbon derivatives}$$

The process also has the potential to produce lower boiling hydrocarbon derivatives by hydrocracking.

Refinery gas, usually in more than one stream, is typically treated for hydrogen sulfide removal and gas sales are usually on a thermal content (calorific value and heating value) basis with some adjustment for variation in the calorific value and hydrocarbon type (Speight, 2014a).

3.3.10 Shale gas

Shale is a sedimentary rock characterized by low permeability mainly compositing of mud, silts, and clay minerals but, however, this composition varies with burial depth and tectonic stresses. Shale reservoirs have a permeability which is substantially lower than the permeability of other tight reservoirs. Natural gas and light tight oil (sometimes erroneously referred to as shale oil—by way of definition, shale oil is the liquid product produced by the decomposition of the kerogen component of oil shale) are confined in the pore spaces of these impermeable shale formations. On the other hand, oil shale is a kerogen-rich petroleum source rock that was not buried under the correct maturation conditions to experience the temperatures required to generate oil and gas (Speight, 2014a).

Current interest, in the context of this book, is focused on the gas which is present in the shale pores, which is the well-known shale gas. Gas-rich shale formations are organic-rich shale formations that were previously regarded only as source rocks and seals for gas accumulating in the strata near sandstone and carbonate reservoirs of traditional onshore gas.

The gaseous noncondensable products released upon oil shale thermal processing (retorting) should be called retorting gases and not shale gases. The natural gas is associated with shale formations and such gas is commonly referred to as *shale gas*—to define the origin of the gas rather than the character and properties (Speight, 2013b). Thus shale gas is natural gas produced from shale formations that typically function as both the reservoir and source rocks for the natural gas (Speight, 2013b).

The gas in a shale formation is present as a free gas in the pore spaces or is adsorbed by clay minerals and organic matters (Ross et al., 1978). The free gas will be produced upon well completion, but the production of adsorbed gas depends on the pressure drop needed for desorption. It is, therefore, essential to know the relative amounts of free and adsorbed gas in shales.

Thus, shale gas is a natural gas produced from shale formations (Table 3.10) that typically function as both the reservoir and source rocks for the natural gas. In terms of chemical makeup, shale gas is typically a dry gas composed primarily of methane (60–95% v/v), but some formations do produce wet gas. The Antrim and New Albany plays have typically produced water and gas. Gas shale formations are organic-rich shale formations that were previously regarded only as source rocks and seals for gas accumulating in the strata near sandstone and carbonate reservoirs of traditional onshore gas. In the United States, unconventional gas accumulations account for about 2 tcf of gas production per year, some 10% of total gas output. In the rest of the world, however, gas is predominantly recovered from conventional accumulations.

Table 3.10 Composition of raw shale gas

Component	Marcellus	Appalachian	Haynesville	Eagle Ford
Methane	97.1	79.1	96.3	74.6
Ethane	2.4	17.7	1.1	13.8
Propane	0.1	0.6	0.21	5.4
C_{4+}	<0.02	<0.04	0.2	4.5
C_{6+}	<0.01	0.00	0.06	0.5
Carbon Dioxide	0.04	0.07	1.8	1.5
Nitrogen	0.3	2.5	0.4	0.2
Properties				
HHV (Btu/ft^3)	1031	1133	1009	1307
Dew point (°F)	−96.8	−41.3	9.7	119.6
Wobbe number	1367	1397	1320	1490

Gas treatment may begin at the wellhead—condensates and free water usually are separated at the wellhead using mechanical separators, he observes. Gas, condensate, and water are separated in the field separator and are directed to separate storage tanks and the gas flows to a gathering system. After the free water has been removed, the gas is still saturated with water vapor, and depending on the temperature and pressure of the gas stream, may need to be dehydrated or treated with methanol to prevent hydrates as the temperature drops. But this may not be always the case in actual practice.

3.3.11   Synthesis gas

Synthesis gas (also known as *syngas*) is a mixture of carbon monoxide (CO) and hydrogen (H_2) that is used as a fuel gas but is produced from a wide range of carbonaceous feedstocks and is used to produce a wide range of chemicals. The production of synthesis gas, i.e., mixtures of carbon monoxide and hydrogen has been known for several centuries and can be produced by gasification of carbonaceous fuels. However, it is only with the commercialization of the Fischer—Tropsch reaction that the importance of synthesis gas has been realized.

Synthesis gas can be produced from any one of several carbonaceous feedstocks (such as a crude oil residuum, heavy oil, tar sand bitumen, and biomass) by gasification (partially oxidizing) the feedstock (Speight, 2011, 2013a, 2014a,b):

$$[2CH]_{feedstock} + O_2 \rightarrow 2CO + H_2$$

The initial partial oxidation step consists of the reaction of the feedstock with a quantity of oxygen insufficient to burn it completely, making a mixture consisting of carbon monoxide, carbon dioxide, hydrogen, and steam.

Success in partially oxidizing heavy feedstocks such as heavy crude oil, extra heavy crude oil, and tar sand bitumen feedstocks depends mainly on the properties of the feedstock and the burner design. The ratio of hydrogen to carbon monoxide in the product gas is a function of reaction temperature and stoichiometry and can be adjusted, if desired, by varying the ratio of the steam to the feedstock.

3.4 Olefins and diolefins

The earlier sections have focused on natural gas and refinery gases, which are primarily produced in crude oil refineries as the low-boiling fractions of distillation and cracking processes or in gas plants that separate natural gas and natural gas liquids. Substances in both categories have high vapor pressure and moderate to high water solubility. The gas mixtures are composed primarily of paraffin derivatives and olefin derivatives, mostly containing one to six carbon atoms (C_1 to C_6 or, in some cases, to C_8). Some of the mixtures may contain varying amounts of other components, including hydrogen, nitrogen, and carbon dioxide. The refinery gas streams also contain olefin constituents that are produced by various cracking processes and some streams also contain varying amounts of other chemicals including ammonia, hydrogen, nitrogen, hydrogen sulfide, mercaptans, carbon monoxide, carbon dioxide, 1,3-butadiene, and/or benzene.

Olefin derivatives are not typical constituents of natural gas but do occur in refinery gases, which can be complex mixtures of hydrocarbon gases and nonhydrocarbon gas (Table 3.9) (Speight, 2014a, 2017a). Some gases may also contain inorganic compounds, such as hydrogen, nitrogen, hydrogen sulfide, carbon monoxide, and carbon dioxide. Many low molecular weight olefins (such as ethylene and propylene) and diolefins (such as butadiene) which are produced in the refinery are isolated for petrochemical use (Speight, 2014a). The individual products are: (1) ethylene, (2) propylene, and (3) butadiene.

Ethylene (C_2H_4) is a normally gaseous olefinic compound having a boiling point of approximately $-104°C$ ($-155°F$) which may be handled as a liquid at very high pressures and low temperatures. Ethylene is made normally by cracking an ethane or naphtha feedstock in a high-temperature furnace and subsequent isolation from other components by distillation. The major uses of ethylene are in the production of ethylene oxide, ethylene dichloride, and the polyethylene polymers. Other uses include the coloring of fruit, rubber products, ethyl alcohol, and medicine (anesthetic).

Propylene concentrates are mixtures of propylene and other hydrocarbons, principally propane and trace quantities of ethylene, butylenes, and butanes. Propylene concentrates may vary in propylene content from 70% mol up to over 95% mol and may be handled as a liquid at normal temperatures and moderate pressures. Propylene concentrates are isolated from the furnace products mentioned in the preceding paragraph on ethylene. Higher purity propylene streams are further purified by distillation and extractive techniques. Propylene concentrates are used in the production of propylene oxide, isopropyl alcohol, polypropylene, and the synthesis of isoprene. As is the case for ethylene, moisture in propylene is critical.

Butylene concentrates are mixtures of butene-1, *cis-* and *trans*-butene-2, and, sometimes, isobutene (2-methyl propylene) (C_4H_8).

Butene-1

cis-Butene-2

trans-Butene-2

iso-Butene (2-methylpropene, 2-methyl propylene)

These products are stored as liquids at ambient temperatures and moderate pressures. Various impurities such as butane, butadiene, and the C_5 hydrocarbons are generally found in butylene concentrates. The majority of the butylene concentrates are used as a feedstock for either: (1) an alkylation plant, where isobutane and butylenes are reacted in the presence of either sulfuric acid or hydrofluoric acid to form a mixture of C_7 to C_9 paraffins used in gasoline, or (2) butylene dehydrogenation reactors for butadiene production.

Butadiene (C_4H_6, CH_2=$CHCH$=CH_2) is a normally gaseous hydrocarbon having a boiling point of $-4.38°C$ ($24.1°F$), which may be handled as a liquid at moderate pressure. Ambient temperatures are generally used for long-term storage due to the easy formation of butadiene dimer (4-vinyl cyclohexene-l). Butadiene is produced by two major methods: the catalytic dehydrogenation of butane or butylenes or both, and as a by-product from the production of ethylene. In either case, the butadiene must be isolated from other components by extractive distillation techniques and subsequent purification to polymerization-grade specifications by fractional distillation. The largest end use of butadiene is as a monomer for production of GR-S synthetic rubber. Butadiene is also chlorinated to form 2-chloro butadiene (chloroprene) (CH_2=$CHCCl$=CH_2) that is a feedstock used to produce neoprene (a polychloroprene rubber).

The major quality criteria for butadiene are the various impurities that may affect the polymerization reactions for which butadiene is used. The gas chromatographic examination of butadiene (ASTM, 2017) can be employed to determine the gross purity as well as C_3, C_4, and C impurities. Most of these hydrocarbons are innocuous to polymerization reactions, but, some, such as butadiene-1,2 and pentadiene-1,4, are capable of polymer cross-linking.

References

ASTM, 2017. Annual Book of Standards. ASTM International, West Conshohocken, PA.

Ahmadi, G., Ji, C., Smith, D.H., 2007. Production of natural gas from methane hydrate by a constant downhole pressure well. Energy Convers. Manage. 48, 2053–2068.

Ashworth, T., Johnson, L.R., Lai, L.P., 1985. Thermal conductivity of pure ice and tetrahydrofuran clathrate hydrates. High Temp. - High Pressures 17 (4), 413–419.

Belosludov, V.R., Subbotin, O.S., Krupskii, D.S., Belosludov, R.V., Kawazoe, Y., Kudoh, J., 2007. Physical and chemical properties of gas hydrates: theoretical aspects of energy storage application. Mater. Trans. 48 (4), 704–710.

Bishnoi, P.R., Clarke, M.A., 2006. Natural gas hydrates. Encyclopedia of Chemical Processing. Taylor & Francis Publishers, Philadelphia, PA.

Brosseau, J., 1994. Trace gas compound emissions from municipal landfill sanitary sites. Atmos.-Environ. 28 (2), 285–293.

Carrol, J.J., 2003. Natural Gas Hydrates. Gulf Professional Publishing, Burlington, VT.

Castaldi, M.J., Zhou, Y., Yegulalp, T.M., 2007. Down-hole combustion method for gas production from methane hydrates. J. Pet. Sci. Eng. 56, 176–185.

Chadeesingh, R., 2011. The Fischer-Tropsch process. In: Speight, J.G. (Ed.), The Biofuels Handbook. The Royal Society of Chemistry, London, pp. 476–517. , Part 3, Chapter 5.

Collett, T.S., 2001. Natural-gas hydrates; resource of the twenty-first century?. J. Am. Assoc. Pet. Geol. 74, 85–108.

Collett, T.S., 2010. Physical Properties of Gas Hydrates: A Review. Journal of Thermodynamics Volume 2010, Article ID 271291; doi:10.1155/2010/271291; https://www.hindawi.com/journals/jther/2010/271291/ (accessed 01.11.17.).

Collett, T.S., Ladd, J.W., 2000. Detection of gas hydrate with downhole logs and assessment of gas hydrate concentrations (saturations) and gas volumes on the blake ridge with electrical resistivity log data. In: Proceedings. Ocean Drilling Program Sci. Results, vol. 164, pp. 179–191.

Collett, T.S., Johnson, A.H., Knapp, C.C., Boswell, R., 2009. Natural gas hydrates: a review. In: Collett, T.S., Johnson, A.H., Knapp, C.C., Boswell, R. (Eds.), Natural Gas Hydrates – Energy Resource Potential and Associated Geologic Hazards. American Association of Petroleum Geologists, Tulsa, OK, pp. 146–219. , AAPG Memoir No. 89.

Collett, T.S., Bahk, J.J., Baker, R., Boswell, R., Divins, D., Frye, M., et al., 2015. Methane hydrates in nature – current knowledge and challenges. J. Chem. Eng. Data 60 (2), 319–329.

Cook, J.G., Leaist, D.G., 1983. An exploratory study of the thermal conductivity of methane hydrate. Geophys. Res. Lett. 10 (5), 397–399.

Cui, H., Turn, S.Q., Keffer, V., Evans, D., Foley, M., 2013. Study on the fate of metal elements from biomass in a bench-scale fluidized bed gasifier. Fuel 108, 1–12.

Davies, P., 2001. The new challenge of natural gas. In: Proceedings. OPEC and the Global Energy Balance: Towards A Sustainable Future. Vienna, September 28.

Demirbaş, A., 2006. The importance of natural gas in the world. Energy Sources Part B 1, 413–420.

Demirbaş, A., 2010. Methane hydrates as a potential energy resource: Part 2 – Methane production processes from gas hydrates. Energy Convers. Manage. 51, 1562–1571.

Du Frane, W.L., Stern, L.A., Weitemeyer, K.A., Constable, S., Pinkston, J.C., Roberts, J.J., 2011. Electrical properties of polycrystalline methane hydrate. Geophys. Res. Lett. 38, L09313. Available from: https://doi.org/10.1029/2011GL047243.

Du Frane, W.L., Stern, L.A., Constable, S., Weitemeyer, K.A., Smith, M.M., Roberts, J.J., 2015. Electrical properties of methane hydrate + sediment mixtures. J. Geophys. Res. 120, 4773–4783.

Ellis, M.H., Minshull, T.A., Sinha, M.C., Best, A.I. 2008. Joint seismic/electrical effective medium modelling of hydrate-bearing marine sediments and an application to the Vancouver Island Margin. In: Proceedings. 6th International Conference on Gas Hydrates, 5586, Vancouver, Canada.

Esposito, G., Frunzo, L., Liotta, F., Panico, A., Pirozzi, F., 2012. Bio-methane potential tests to measure the biogas production from the digestion and co-digestion of complex organic substrates. Open Environ. Eng. J. 5, 1–8.

Evans, R.L., 2007. Using CSEM techniques to map the shallow section of the seafloor: from the coastline to the edges of the continental slope. Geophysics 72 (2), WA105–WA116.

Fan, Z., Sun, C., Kuang, Y., Wang, B., Zhao, J., Song, Y., 2017. MRI analysis for methane hydrate dissociation by depressurization and the concomitant ice generation. Energy Procedia 105, 4763–4768.

Gabitto, J., Tsouris, C., 2010. Physical properties of gas hydrates: a review. J. Thermodyn. Volume 2010, Article ID 271291. https://www.hindawi.com/journals/jther/2010/271291/citations/.

Gary, J.G., Handwerk, G.E., Kaiser, M.J., 2007. Petroleum Refining: Technology and Economics, fifth ed. CRC Press, Taylor & Francis Group, Boca Raton, FL.

Giavarini, C., Maccioni, F., 2004. Self-preservation at low pressure of methane hydrates with various gas contents. Ind. Eng. Chem. Res. 43, 6616–6621.

Giavarini, C., Maccioni, F., Santarelli, M.L., 2003. Formation kinetics of propane hydrate. Ind. Eng. Chem. Res. 42, 1517–1521.

Giavarini, C., Maccioni, F., Santarelli, M.L., 2005. Characterization of gas hydrates by modulated differential scanning calorimetry. Pet. Sci. Technol. 23, 327–335.

Goel, N., Wiggins, M., Shah, S., 2001. Analytical modeling of gas recovery from in situ hydrates dissociation. J. Pet. Sci. Eng. 29, 115–127.

Holland, M., Schultheiss, P., 2014. Comparison of methane mass balance and x-ray computed tomographic methods for calculation of gas hydrate content of pressure cores. Mar. Pet. Geol. 58 (A), 168–177.

Holland, M., Schultheiss, P., Roberts, J., Druce, M., July 2008. Observed gas hydrate morphologies in marine sediments. In: Proceedings. 6th International Conference on Gas Hydrates (ICGH 08) Vancouver, British Columbia, Canada.

Hsu, C.S., Robinson, P.R. (Eds.), 2017. Handbook of Petroleum Technology. Springer International Publishing AG, Cham.

Judge, A., 1982. Natural gas hydrate in Canada. In: Proceedings. 4th Canadian Permafrost Conference, pp. 320–328.

Lee, S., Lee, Y., Lee, J., Lee, H., Seo, Y., 2013. Experimental verification of methane-carbon dioxide replacement in natural gas hydrates using a differential scanning calorimeter. Environ. Sci. Technol. 47, 13184–13190.

Li, S., Zheng, R., Xu, X., Hou, J., 2016. Natural gas hydrate dissociation by hot brine injection. Pet. Sci. Technol. 34, 422–428.

Liu, C.L., Ye, Y.G., Meng, Q.C., 2010. Determination of hydration number of methane hydrates using micro-laser Raman spectroscopy. Guang Pu Xue Yu Guang Pu Fen Xi 30 (4), 963–966 (in Chinese) https://www.ncbi.nlm.nih.gov/pubmed/20545140.

Llamedo, M., Anderson, R., Tohidi, B., 2004. Thermodynamic prediction of clathrate hydrate dissociation conditions in mesoporous media. Am. Mineral. 89 (8–9), 1264–1270.

Lohila, A., Laurila, T., Tuovinen, J.-P., Aurela, M., Hatakka, J., Thum, T., et al., 2007. Micrometeorological measurements of methane and carbon dioxide fluxes at a municipal landfill. Environ. Sci. Technol. 41 (8), 2717–2722.

Lorenson, T.D., Collett, T.S., 2000. Gas content and composition of gas hydrate from sediments of the Southeastern North American Continental Margin. In: Paull, C.K., Matsumoto, R., Wallace, P.J., Dillon, W.P., (Eds.), Proceedings of the Ocean Drilling Program, Scientific Results, vol. 164, pp. 37–46.

Makogon, Y.F., 1997. Hydrates of Hydrocarbons. PennWell Books, Tulsa, OK.

Makogon, Y.F., 2010. Natural gas hydrates – a promising source of energy. J. Nat. Gas Sci. Eng. 2 (1), 49–59.

Makogon, Y.F., Holditch, S.A., Makogon, T.Y., 2007. Natural gas hydrates – a potential energy source for the 21st century. J. Pet. Sci. Eng. 56 (1–3), 14–31.

Mokhatab, S., Poe, W.A., Speight, J.G., 2006. Handbook of Natural Gas Transmission and Processing. Elsevier, Amsterdam.

Rasi, S., Veijanen, A., Rintala, J., 2007. Trace compounds of biogas from different biogas production plants. Energy 32, 1375–1380.

Ross, R.G., Anderson, P., Backstrom, G., 1978. Effects of H and D order on the thermal conductivity of ice phases. J. Chem. Phys. 68 (9), 3967–3972.

Scala, F., Anacleria, C., Cimino, S., 2013. Characterization of a regenerable sorbent for high temperature elemental mercury capture from flue gas. Fuel 108, 13–18.

Schwalenberg, K., Willoughby, E., Mir, R., Edwards, R.N., 2005. Marine gas hydrate electromagnetic signatures in cascadia and their correlation with seismic blank zones. First Break 23, 57–63.

Seo, Y., Kang, S.P., Jang, W., 2009. Structure and composition analysis of natural gas hydrates: 13C NMR spectroscopic and gas uptake measurements of mixed gas hydrates. J. Phys. Chem. 113 (35), 9641–9649.

Sloan Jr., E.D., 1998a. Gas hydrates: review of physical/chemical properties. Energy Fuels 12 (2), 191–196.

Sloan Jr., E.D., 1998b. Clathrate Hydrates of Natural Gases, second ed. Marcel Dekker Inc, New York.

Sloan Jr., E.D., 2006. Clathrate Hydrates of Natural Gases, third ed. Marcel Dekker Inc, New York.

Speight, J.G. (Ed.), 2011. The Biofuels Handbook. Royal Society of Chemistry, London.

Speight, J.G., 2013a. The Chemistry and Technology of Coal, third ed. CRC Press, Taylor & Francis Group, Boca Raton, FL.

Speight, J.G., 2013b. Shale Gas Production Processes. Gulf Professional Publishing, Elsevier, Oxford.

Speight, J.G., 2014a. The Chemistry and Technology of Petroleum, fifth ed. CRC Press, Taylor & Francis Group, Boca Raton, FL.

Speight, J.G., 2014b. Gasification of Unconventional Feedstocks. Gulf Professional Publishing, Elsevier, Oxford.

Speight, J.G., 2016. Hydrogen in refineries. In: Stolten, D., Emonts, B. (Eds.), Hydrogen Science and Engineering: Materials, Processes, Systems, and Technology. Wiley-VCH Verlag GmbH & Co, Weinheim, pp. 3–18. (Chapter 1).

Speight, J.G., 2017a. Handbook of Petroleum Refining. CRC Press, Taylor & Francis Group, Boca Raton, FL.

Speight, J.G., 2017b. Deep Shale Oil and Gas. Gulf Professional Publishing, Elsevier, Oxford.

Staley, B., Barlaz, M.A., 2009. Composition of municipal solid waste in the United States and implications for carbon sequestration and methane yield. J. Environ. Eng. 135 (10), 901–909.

Stern, L., Kirby, S., Durham, W., Circone, S., Waite, W.F., 2000. Laboratory synthesis of pure methane hydrate suitable for measurement of physical properties and decomposition behavior. In: Max, M.D. (Ed.), Proceedings. Natural Gas Hydrate in Oceanic and Permafrost Environments. Kluwer Academic Publishers, Dordrecht, pp. 323–348.

Stoll, R.G., Bryan, G.M., 1979. Physical properties of sediments containing gas hydrates. J. Geophys. Res. 84 (B4), 1629–1634.

Sullivan, P., 2010. The Importance of Landfill Gas Capture and Utilization in the U.S. Earth Engineering Center. Columbia University, New York, http://www.scsengineers.com/wp-content/uploads/2015/03/
Sullivan_Importance_of_LFG_Capture_and_Utilization_in_the_US.pdf   (accessed 11.11.17.).

Tse, J.S., White, M.A., 1988. Origin of glassy crystalline behavior in the thermal properties of clathrate hydrates: a thermal conductivity study of tetrahydrofuran hydrate. J. Phys. Chem. 92 (17), 5006–5011.

US DOE, 2011. Energy Resource Potential of Methane Hydrate: An Introduction to the Science and Energy Potential of a Unique Resource. NETL, The Energy Lab. Unites States Department of Energy, Washington, DC. Available from: https://www.netl.doe.gov/File%20Library/Research/Oil-Gas/methane%20hydrates/MH-Primer2011.pdf.

USGS, July 2007. Thermal Properties of Methane Gas Hydrates. Fact Sheet 2007-3041. United States Geological Survey, Reston, VA. <https://pubs.usgs.gov/fs/2007/3041/>.

Waite, W.F., Gilbert, L.Y., Winters, W.J., Mason, D.H., 2006. Estimating Thermal Diffusivity and Specific Heat from Needle Probe Thermal Conductivity Data: Review of Scientific Instruments, vol. 77, Paper No. 044904, https://doi.org/10.1063/1.2194481.

Waite, W.F., Stern, L.A., Kirby, S.H., Winters, W.J., Mason, D.H., 2007. Simultaneous determination of thermal conductivity, thermal diffusivity and specific heat in methane hydrate. Geophys. J. Int. 169, 767–774.

Wang, B., Dong, H., Fan, Z., Zhao, J., Song, Y., 2017. Gas production from methane hydrate deposits induced by depressurization in conjunction with thermal stimulation. Energy Procedia 105, 4713–4717.

Wang, X., Economides, M.J., 2012. Natural gas hydrates as an energy source – Revisited 2012. In: Proceedings. SPE International Petroleum Technology Conference 2012, vol. 1. Society of Petroleum Engineers, Richardson, TX, pp. 176–186.

Weitemeyer, K.A., Constable, S.C., Key, K.W., Behrens, J.P., 2006. First results from a marine controlled-source electromagnetic survey to detect gas hydrates offshore Oregon. Geophys. Res. Lett. 33, L03304.

Weitemeyer, K.A., Constable, S.C., Tréhu, A.M., 2011. A marine electromagnetic survey to detect gas hydrate at hydrate ridge, Oregon. Geophys. J. Int . Available from: https://doi.org/10.1111/j.1365-246X.2011.05105.x.

Yang, X., Qin, M., 2012. Natural gas hydrate as potential energy resources in the future. Adv. Mater. Res. 462, 221−224.

Further reading

Parkash, S., 2003. Refining Processes Handbook. Gulf Professional Publishing, Elsevier, Amsterdam.

Speight, J.G., 2018. Handbook of Natural Gas Analysis. John Wiley & Sons Inc, Hoboken, NJ.

Composition and properties

4.1 Introduction

The gaseous mixtures considered in this volume are mixtures of various constituents that may or may not vary over narrow limits. Typically, these gases fall into the general category of fuel gases and each gas is any one of several fuels that, at standard conditions of temperature and pressure, are gaseous. Before sale of the gas to the consumer actions, it is essential to give consideration of the variability of the composition of gas streams (Table 4.1) and the properties of the individual constituents and their effects on gas behavior, even when considering the hydrocarbon constituents only (Chapter 3: Unconventional gas). If not, the properties of the gas may be unstable and the ability of the gas to be used for the desired purpose will be seriously affected.

Thus there is no single composition of components which might be termed *typical* natural gas. Methane and ethane often constitute the bulk of the combustible components; carbon dioxide (CO_2) and nitrogen (N_2) are the major noncombustible (inert) components. Thus sour gas is a natural gas that occurs mixed with higher levels of sulfur compounds (such as hydrogen sulfide, H_2S, and mercaptans or thiols, RSH) and which constitutes a corrosive gas. The sour gas requires additional processing for purification (Mokhatab et al., 2006; Speight, 2014). Olefins are also present in the gas streams from various refinery processes and are not included in liquefied petroleum gas but are removed for use in petrochemical operations (Crawford et al., 1993).

The analysis of any gas streams, compared to liquid streams, is relatively simple because bulk characterization of a single phase is implicit. However, when gas condensate is present, the analysis is more complicated (ASTM, 2017; Speight, 2018). In the case of gas condensate, besides bulk analysis, there may be interest on surface-composition (often quite distinct to that of the bulk phase). Compositional analysis in which the components of the mixture are identified, may be achieved by (1) physical means, which is measurement of physical properties; (2) pure chemical means, which is measurement of chemical properties; or more commonly (3) by physicochemical means. Gas analysis is even more may be dangerous and difficult if the composition is a complete unknown. However, when some main constituents are known to occur, the analysis gains accuracy (and may be easier) if the known component is removed; this is particularly important in the case of water vapor, which may condense on the instruments, or constituents when the molecular behavior may complicate spectral analyses.

Crude oil-related gases (including natural gas) and refinery gases (process gases) as well as product gases produced from petroleum refining, upgrading, or natural gas processing facilities are a category of saturated and unsaturated gaseous

Natural Gas. DOI: https://doi.org/10.1016/B978-0-12-809570-6.00004-7

Table 4.1 General properties of unrefined natural gas (*left-hand data*) and refined natural gas (*right-hand data*)

Relative molar mass	20−16
Carbon content (% w/w)	73−75
Hydrogen content (% w/w)	27−25
Oxygen content (% w/w)	0.4−0
Hydrogen-to-hydrogen atomic ratio	3.5−4.0
Density relative to air @15°C	1.5−0.6
Boiling temperature (°C/1 atm)	−162
Autoignition temperature (°C)	540−560
Octane number	120−130
Methane number	69−99
Vapor flammability limits (% v/v)	5−15
Flammability limits	0.7−2.1
Lower heating/calorific value (Btu)	900
Methane concentration (% v/v)	100−80
Ethane concentration (% v/v)	5−0
Nitrogen concentration (% v/v)	15−0
Carbon dioxide concentration (% v/v)	5−0
Sulfur concentration (ppm, w/w)	5−0

Source: http://www.visionengineer.com/env/alt_ng_prop.php.

hydrocarbons, predominantly in the C_1 to C_6 carbon number range. Some gases may also contain inorganic compounds, such as hydrogen, nitrogen, hydrogen sulfide, carbon monoxide, and carbon dioxide. As such, petroleum and refinery gases (unless produced as a salable product that must meet specifications prior to sale) are often unknown or variable composition and toxic (API, 2009). The site-restricted petroleum and refinery gases (i.e., those not produced for sale) often serve as fuels consumed on-site, as intermediates for purification and recovery of various gaseous products, or as feedstocks for isomerization and alkylation processes within a facility.

As with petroleum, natural gas from different wells varies widely in composition and analyses (Mokhatab et al., 2006; Speight, 2014), and the proportion of nonhydrocarbon constituents can vary over a very wide range. The nonhydrocarbon constituents of natural gas can be classified as two types of materials: (1) diluents, such as nitrogen, carbon dioxide, and water vapors and (2) contaminants, such as hydrogen sulfide and/or other sulfur compounds. Thus a particular natural gas field could require production, processing, and handling protocols different from those used for gas from another field.

The diluents are noncombustible gases that reduce the heating value of the gas and are on occasion used as *fillers* when it is necessary to reduce the heat content of the gas. On the other hand, the contaminants are detrimental to production and transportation equipment in addition to being obnoxious pollutants. Thus the primary reason for gas refining is to remove the unwanted constituents of natural gas and to separate the gas into its various constituents. The processes are analogous to the distillation unit in a refinery where the feedstock is separated into its various constituent fractions before further processing to products. The major diluents or contaminants of natural gas are: (1) acid gas, which is predominantly hydrogen sulfide although carbon dioxide does occur to a lesser extent; (2) water, which includes all entrained free water or water in condensed forms; (3) liquids in the gas, such as higher boiling hydrocarbons as well as pump lubricating oil, scrubber oil, and, on occasion, methanol; and (4) any solid matter that may be present, such as fine silica (sand) and scaling from the pipe.

Like any other refinery product, natural gas (or, for that matter, any fuel gas) must be processed to prepare it for final use and to ascertain the extent of contaminants that could cause environmental damage. Furthermore, gas processing is a complex industrial process designed to clean raw (dirty, contaminated) gas by separating impurities and various nonmethane and fluids to produce what is known as *pipeline quality* dry natural gas. However, there are many variables in treating natural gas and nonconventional refinery gas that are dependent upon the properties of the gas streams.

These properties are either temperature-independent or values of some basic properties at a fixed temperature. However, the composition of natural gas is not universally constant, as it is normally drawn from several production fields. Variations in the composition of gas delivered by pipelines can be caused by: (1) variations in the proportion of the contribution from various sources at a given supply location and (2) time variations within a given supply source. In addition, the composition of a gas stream from a source or at a location can also vary over time which can cause difficulties in resolving the data from the application of standard test methods (Klimstra, 1978; Liss and Thrasher, 1992).

4.2 Types of gases

The gaseous products that occur in a refinery comprise mixtures that vary from natural; gas to gases produced during refining (*refinery gas, process gas*). The constituents of each type of gas may be similar (except for the olefin-type gases produced during thermal processes), but the variations of the amounts of these constituents cover wide ranges. Each type of gas may be analyzed by similar methods although the presence of high-boiling hydrocarbons and nonhydrocarbon species such as carbon dioxide and hydrogen sulfide may require slight modifications to the various analytical test methods (Table 4.2) (ASTM, 2017; Speight, 2018).

However, at the onset, it is necessary that the analyst determine the type of gas to be analyzed. Mixtures of the various constituents named earlier are commonly encountered in material testing, and the composition varies depending upon the source and intended use of the material. Other nonhydrocarbon constituents of these mixtures are important analytes since they may be useful products or may be undesirable as a source of processing problems. Some of these components are helium,

Table 4.2 Selection of standard test methods commonly applied to the determination of gas quality

ASTM D792. 2017. Standard Test Methods for Density and Specific Gravity (Relative Density) of Plastics by Displacement

ASTM D1434. Standard Test Method for Determining Gas Permeability Characteristics of Plastic Film and Sheeting

ASTM D1505. 2017. Standard Test Method for Density of Plastics by the Density-Gradient Technique, D1505. West Conshohocken, PA

American Society for Testing and Materials ASTM D1945. 2017. Standard Test Method for Analysis of Natural Gas by Gas Chromatography

ASTM D3588. 2017. Standard Practice for Calculating Heat Value, Compressibility Factor, and Relative Density of Gaseous Fuels

ASTM D1826. 2017. Standard Test Method for Calorific (Heating) Value of Gases in Natural Gas Range by Continuous Recording Calorimeter

ASTM D1070. 2017. Standard Test Methods for Relative Density of Gaseous Fuels

ASTM D4084. 2017. Standard Test Method for Analysis of Hydrogen Sulfide in Gaseous Fuels — Lead Acetate Reaction Rate Method

ASTM 5199. 2017. Standard Test Methods for Measuring the Nominal Thickness of Geosynthetics

ASTM D5454. 2017. Standard Test Method for Water Vapor Content in Gaseous Fuels Using Electronic Moisture Analyzer

hydrogen, argon, oxygen, nitrogen, carbon monoxide, carbon dioxide, sulfur, and nitrogen containing compounds, as well as higher molecular weight hydrocarbons. Thus the desired testing of these hydrocarbon mixtures involves (1) identification of the type of gas for component speciation and quantitation and (2) the influence of the composition not only on the bulk physical or chemical properties but, more important, the influence on the composition on the performance of the test method.

Raw natural gas varies greatly in composition (Table 4.3) and the constituents can be several of a group of saturated hydrocarbons from methane to butane (Table 4.4) and nonhydrocarbons. The treatment required to prepare natural gas for distribution as an industrial or household fuel is specified in terms of the use and environmental regulations. Briefly, natural gas contains hydrocarbons and nonhydrocarbon gases. Hydrocarbon gases are methane (CH_4), ethane (C_2H_6), propane (C_3H_8), butanes (C_4H_{10}), pentanes (C_5H_{12}), hexane (C_6H_{14}), heptane (C_7H_{16}), and sometimes trace amounts of octane (C_8H_{18}), and higher molecular weight hydrocarbons. Some aromatics (BTX—benzene (C_6H_6), toluene ($C_6H_5CH_3$), and the xylenes ($CH_3C_6H_4CH_3$)) can also be present, raising safety issues due to their toxicity. The nonhydrocarbon gas portion of the natural gas contains nitrogen (N_2), carbon dioxide (CO_2), helium (He), hydrogen sulfide (H_2S), water vapor (H_2O), and other sulfur compounds (such as carbonyl sulfide (COS) and mercaptans (e.g., methyl mercaptan, CH_3SH)) and trace amounts of other gases.

Carbon dioxide and hydrogen sulfide are commonly referred to as *acid gases* since they form corrosive compounds in the presence of water. Nitrogen, helium, and carbon dioxide are also referred to as *diluents* since none of these burn, and

Table 4.3 Composition of associated natural gas from a petroleum well

Category	Component	Amount (%)
Paraffinic	Methane (CH_4)	70−98
	Ethane (C_2H_6)	1−10
	Propane (C_3H_8)	Trace−5
	Butane (C_4H_{10})	Trace−2
	Pentane (C_5H_{12})	Trace−1
	Hexane (C_6H_{14})	Trace−0.5
Cyclic	Cyclopropane (C_3H_6)	Traces
	Cyclohexane (C_6H_{12})	Traces
Aromatic	Benzene (C_6H_6), others	Traces
Nonhydrocarbon	Nitrogen (N_2)	Trace−15
	Carbon dioxide (CO_2)	Trace−1
	Hydrogen sulfide (H_2S)	Trace occasionally
	Helium (He)	Trace−5
	Other sulfur and nitrogen compounds	Trace occasionally
	Water (H_2O)	Trace−5

Table 4.4 Possible constituents of natural gas and refinery process gas streams

Gas	Molecular weight	Boiling point 1 atm °C (°F)	Density at 60°F (15.6°C), 1 atm	
			g/L	Relative to air = 1
Methane	16.043	− 161.5 (−258.7)	0.6786	0.5547
Ethylene	28.054	− 103.7 (−154.7)	1.1949	0.9768
Ethane	30.068	− 88.6 (−127.5)	1.2795	1.0460
Propylene	42.081	− 47.7 (−53.9)	1.8052	1.4757
Propane	44.097	− 42.1 (−43.8)	1.8917	1.5464
1,2-Butadiene	54.088	10.9 (51.6)	2.3451	1.9172
1,3-Butadiene	54.088	− 4.4 (24.1)	2.3491	1.9203
1-Butene	56.108	− 6.3 (20.7)	2.4442	1.9981
cis-2-Butene	56.108	3.7 (38.7)	2.4543	2.0063
trans-2-Butene	56.108	0.9 (33.6)	2.4543	2.0063
iso-Butene	56.104	− 6.9 (19.6)	2.4442	1.9981
n-Butane	58.124	− 0.5 (31.1)	2.5320	2.0698
iso-Butane	58.124	− 11.7 (10.9)	2.5268	2.0656

thus they have no heating value. Mercury can also be present either as a metal in vapor phase or as an organometallic compound in liquid fractions. Concentration levels are generally very small, but even at very small concentration levels, mercury can be detrimental due its toxicity and its corrosive properties (reaction with aluminum alloys).

A natural gas stream traditionally has high proportions of natural gas liquids (NGLs) and is referred to as rich gas. NGLs are constituents such as ethane, propane, butane, and pentanes and higher molecular weight hydrocarbon constituents. The higher molecular weight constituents (i.e., the C_{5+} product) are commonly referred to as gas condensate or natural gasoline (sometimes, on occasion, erroneously called *casinghead gas*).

When referring to NGLs in the gas stream, the term gallon per thousand cubic feet is used as a measure of high-molecular weight hydrocarbon content. On the other hand, the composition of nonassociated gas (sometimes called well gas) is deficient in NGLs. The gas is produced from geological formations that typically do not contain much, if any, hydrocarbon liquids.

Many sources of natural and petroleum gases contain sulfur compounds that are odorous, corrosive, and poisonous to catalysts used in gaseous fuel processing. In fact, sulfur odorants are added (in the ppm range, i.e., 1−4 ppm v/v) to natural gas and liquefied petroleum gases for safety purposes. Some odorants are unstable and react to form compounds having lower odor thresholds. Quantitative analysis of these odorized gases ensures that odorant injection equipment is performing to specification.

4.3   Composition and chemical properties

The composition and properties of any gas stream depends on the characterization and properties of the hydrocarbons that make up the stream and calculation of the properties of a mixture depends on the properties of its constituents. However, calculation of the property of a mixture based on an average calculation neglects any interactions between the constituents. This makes the issue of modeling properties a difficult one because of the frequent omission of any chemical or physical interactions between the gas stream constituents.

The defining characteristics of the various gas streams in the context of this book are that the gases (1) exist in a gaseous state at room temperature; (2) may contain hydrocarbon constituents with 1−4 carbons, i.e., methane, ethane, propane, and butane isomers; (3) may contain diluents and inert gases; and (4) may contain contaminants in the form of nonhydrocarbon constituents. Each constituent of the gas influences the properties. Thus,

Hydrocarbons	Provide the calorific value of natural gas when it is burned
Diluents/inert gases	Typical gases are carbon dioxide, nitrogen, helium, and argon
Contaminants	Present in low concentrations; may affect processing operations

Many hydrocarbon gases do contain C_5 and C_6 hydrocarbon derivatives and apart from gas streams produced as process by-products in a refinery, the C_{5+} constituents are typically found at lower concentrations (% v/v) in gases than the C_1 to C_4 constituents. There are also a few category members that may contain C_7 and even C_8 hydrocarbons, although such streams would necessarily be at elevated temperature and/or pressure to maintain the heptane (C_7H_{16}) and octane (C_8H_{18}) constituents in the gaseous state. Hydrocarbon compounds such as pentane (C_5H_{12}), hexane (C_6H_{14}), heptane (C_7H_{16}), and octane (C_8H_{18}) derivatives are typically found predominantly in naphtha derived from crude oil.

4.3.1 Composition

Natural gas (predominantly *methane*) denoted by the chemical structure CH_4 is the lowest boiling and least complex of all hydrocarbons. Natural gas from an underground reservoir, when brought to the surface, can contain other higher boiling hydrocarbons and is often referred to as *wet gas*. Wet gas is usually processed to remove the entrained hydrocarbons that are higher boiling than methane and, when isolated, the higher boiling hydrocarbons sometimes liquefy and are called *natural gas condensate*.

Natural gas is found in petroleum reservoirs as free gas (*associated gas*) or in solution with petroleum in the reservoir (*dissolved gas*) or in reservoirs that contain only gaseous constituents and no (or little) petroleum (*unassociated gas*) (Cranmore and Stanton, 2000; Speight, 2014). The hydrocarbon content varies from mixtures of methane and ethane with very few other constituents (*dry* gas) to mixtures containing all of the hydrocarbons from methane to pentane and even hexane (C_6H_{14}) and heptane (C_7H_{16}) (*wet* gas). In both cases some carbon dioxide (CO_2) and inert gases, including helium (He), are present together with hydrogen sulfide (H_2S) and a small quantity of organic sulfur.

The term *petroleum gas(es)* in this context is also used to describe the gaseous phase and liquid phase mixtures comprised mainly of methane to butane (C_1 to C_4 hydrocarbons) that are dissolved in the crude oil and natural gas, as well gases produced during thermal processes in which the crude oil is converted to other products. It is necessary, however, to acknowledge that in addition to the hydrocarbons, gases such as carbon dioxide, hydrogen sulfide, and ammonia are also produced during petroleum refining and will be constituents of refinery gas that must be removed. Olefins are also present in the gas streams of various processes and are not included in liquefied petroleum gas but are removed for use in petrochemical operations (Crawford et al., 1993).

Nonassociated natural gas, which is found in reservoirs in which there is no, or at best only minimal amounts of, petroleum (Chapter 1: History and use). Nonassociated gas is usually richer in methane but is markedly leaner in terms of the higher molecular weight hydrocarbons and condensate. Conversely there is also *associated* natural gas (*dissolved* natural gas) that occurs either as free gas or as gas in solution in the petroleum. Gas that occurs as a solution with the crude petroleum is *dissolved gas*, whereas the gas that exists in contact with the crude petroleum (*gas cap*) is *associated gas* (Chapter 1: History and use) Associated gas is usually leaner in methane than the nonassociated gas but is richer in the higher molecular weight constituents.

The most preferred type of natural gas is the nonassociated gas. Such gas can be produced at high pressure, whereas associated, or dissolved, gas must be separated from petroleum at lower separator pressures, which usually involves increased expenditure for compression. Thus it is not surprising that such gas (under conditions that are not economically favorable) is often flared or vented.

Natural gas is a naturally occurring gas mixture, consisting mainly of methane that is found in porous formations beneath the surface of the Earth, often in association with crude oil but while the gas from the various sources has a similar analysis, it is not entirely the same. In fact, variation in composition varies from field to field and may even vary within a reservoir. In addition, the variation of gas streams from different sources (Chapter 3: Unconventional gas) must also be considered when processing options are being assessed. Thus, because of the lower molecular weight constituents of these gases and their volatility, gas chromatography has been the technique of choice for fixed gas and hydrocarbon speciation and mass spectrometry is also a method of choice for compositional analysis of low molecular weight hydrocarbons (Speight, 2015, 2018; ASTM, 2017). The vapor pressure and volatility specifications will often be met automatically if the hydrocarbon composition is in order with the specification.

Thus natural gas is a combustible mixture of hydrocarbon gases that, in addition to methane, also includes ethane, propane, butane, and pentane. The composition of natural gas can vary widely before it is refined (Table 4.1) (Mokhatab et al., 2006; Speight, 2014). In its purest form, such as the natural gas that is delivered to the consumer is almost pure methane.

The principal constituent of most natural gases is methane with minor amounts of heavier hydrocarbons and certain nonhydrocarbon gases such as nitrogen, carbon dioxide, hydrogen sulfide, and helium (Mokhatab et al., 2006; Speight, 2014, 2018; ASTM, 2017). Methane can be produced in the laboratory by heating sodium acetate with sodium hydroxide and by the reaction of aluminum carbide (Al_4C_3) with water.

$$Al_4C_3 + 12H_2O \rightarrow 4Al(OH)_3 \downarrow + 3CH_4$$

$$CH_3CO_2Na + NaOH \rightarrow CH_4 + Na_2CO_3$$

The members of the hydrocarbon gases are predominantly alkane derivatives (C_nH_{2n+2}, where n is the number of carbon atoms). When inorganic constituents are present in natural gas, they consist of asphyxiant gases such as hydrogen.

Unlike other categories of crude oil products (such as naphtha, kerosene, and the higher boiling products) (Speight, 2014, 2017; ASTM, 2017), the constituents of the various gas streams can be evaluated, and the results of the constituent evaluation can then be used to estimate the behavior of the gas (ASTM, 2017; Speight, 2018). The constituents used to evaluate the behavior of the gas are: (1) the C_1 to C_4 hydrocarbon derivatives; (2) the C_5 to C_6 hydrocarbon derivatives, although in natural gas the C_1 to C_4 constituents predominate; and (3) the asphyxiant gases, i.e., carbon dioxide, nitrogen, and hydrogen. In general, most gas streams used in this text are composed of predominantly the methane (C_1) to butane (C_4) hydrocarbon derivatives, which have extremely low melting points and boiling points.

Each of these gases has a high vapor pressures and low octanol–water partition coefficients—the octanol–water partition coefficient (K_{ow}) is a valuable parameter that represents a measure of the tendency of a chemical to move from the aqueous phase into the organic (octanol) phase. Thus,

$$K_{ow} = C_{op}C_w$$

C_{op} and C_w are the concentrations of the chemical in gram per liter of the chemical in the octanol-rich phase and in the water-rich phase, respectively. In the determination of the partition coefficient at 25°C (77°F), the water-rich phase is essentially pure water (99.99 mol% water) while the octanol-rich phase is a mixture of octanol and water (79.3 mol% octanol).

The aqueous solubility of the various constituents of gas streams varies, but the solubility of most of the hydrocarbon derivatives typically falls within a range of 22 mg/L to several hundred parts per million. There are also a few gas streams that may contain heptane derivatives and octane derivative, although such streams would necessarily be at elevated temperature and/or reduced pressure to maintain the heptane derivatives and the octane derivatives in the gaseous state. Hydrocarbon compounds containing pentane, hexane, heptane, and octane derivatives occur predominantly in low-boiling crude oil naphtha and also occur in gas condensate and natural.

By way of recall, in addition to methane, natural gas contains other constituents that are variously referred to as (1) NGLs, (2) natural gas condensate, and (3) natural gasoline (Chapter 1: History and use). Also, by way of a refresher definition, NGLs are hydrocarbons that occur as gases at ambient conditions (atmospheric pressure and temperature) but as liquids under higher pressure and which can also be liquefied by cooling. The specific pressure and temperature at which the gases liquefy vary by the type of gas liquids and may be described as low-boiling (*light*) or high-boiling (*heavy*) according to the number of carbon atoms and hydrogen atoms in the molecule.

Also, by way of a further reminder, natural-gas condensate (also called *condensate*, or *gas condensate*, or *natural gasoline*) is a low-density mixture of hydrocarbon liquids that are present as gaseous components in the raw natural gas produced from many natural gas fields (Chapter 1: History and use and Chapter 9: Gas condensate). Some gas constituents within the raw (unprocessed) natural gas will condense to a liquid state if the temperature is reduced to below the hydrocarbon dew point temperature at a set pressure. There are many condensate sources, and each has its own unique gas condensate composition.

Within the natural gas family, the composition of associated gas (a by-product of oil production and the oil recovery process) is extremely variable, even within the gas from a petroleum reservoir (Speight, 2014). After the production fluids are brought to the surface, they are separated at a tank battery at or near the production lease into a hydrocarbon liquid stream (crude oil or condensate), a produced water stream (brine or salty water), and a gas stream.

The gas stream traditionally has high proportions of NGLs and is referred to as rich gas. NGLs are the nonmethane constituents such as ethane, propane, butane, and pentanes and higher molecular weight hydrocarbon constituents, which can be separated as liquids during gas processing (Chapter 7: Process classification). The higher molecular weight constituents (i.e., the C_{5+} product) are commonly referred to as gas condensate or natural gasoline. Rich gas will have a high heating value and a high hydrocarbon dew point. When referring to NGLs in the gas stream, the term gallon per thousand cubic feet is used as a measure of high-molecular weight hydrocarbon content. On the other hand, the composition of nonassociated gas (sometimes called well gas) is deficient in NGLs. The gas is produced from geological formations that typically do not contain much, if any, hydrocarbon liquids.

Generally, the hydrocarbon derivatives having a higher molecular weight than methane as well as any acid gases (carbon dioxide and hydrogen sulfide) are removed from natural gas prior to use of the gas as a fuel. However, since the composition of natural gas is never constant, there are standard test methods by which the composition and properties of natural gas can be determined and, thus, prepared for use. It is not the intent to cover the standard test methods in any detail in this text since descriptions of the test methods are available elsewhere (Speight, 2015, 2018; ASTM, 2017).

Organic sulfur compounds and hydrogen sulfide are common contaminants that must be removed prior to most uses. Gas with a significant amount of sulfur impurities, such as hydrogen sulfide, is termed sour gas and often referred to as acid gas. Processed natural gas that is available to end-users is tasteless and odorless. However, before gas is distributed to end-users, it is odorized by adding small amounts of thiols (RSH, also called mercaptans) to assist in leak detection. Processed natural gas is harmless to the human body but natural gas is a simple asphyxiant and can kill if it displaces air to the point where the oxygen content will not support life.

Once the composition of a mixture has been determined it is possible to calculate various properties such as specific gravity, vapor pressure, calorific value, and dew point. In liquefied petroleum gas where the composition is such that the hydrocarbon dew point is known to be low, a dew point method will detect the presence of traces of water.

Typically, natural gas samples are analyzed for molecular composition by gas chromatography and for stable isotopic composition by isotope ratio mass spectrometry. Carbon isotopic composition was determined for methane (CH_4), ethane (C_2H_6), propane (C_3H_8), and butane, particularly *iso*-butane (C_4H_{10}) (ASTM, 2017). Another important property of the gas streams discussed in this text is the hydrocarbon dew point. The hydrocarbon dew point is reduced to such a level that retrograde condensation, i.e., condensation resulting from pressure drop, cannot occur under the worst conditions likely to be experienced in the gas transmission system. Similarly, the water dew point is reduced to a level sufficient to preclude formation of C_1 to C_4 hydrates in the system. Generally, pipeline owners prefer that the specifications for the transmission of natural gas limit the maximum concentration of water vapor allowed. Excess water vapor can cause corrosive conditions, degrading pipelines and equipment. The water can also condense and freeze or form methane hydrates (Chapter 7: Process classification) causing blockages. Water−vapor content also affects the heating value of natural gas, thus influencing the quality of the gas.

The determination of the vapor pressure of liquefied petroleum gas is important for safety reasons to ensure that the maximum operating design pressures of storage, handling, and fuel systems will not be exceeded under normal operating temperature conditions. For liquefied petroleum gases, vapor pressure is an indirect measure of the most extreme low temperature conditions under which initial vaporization can be expected to occur. It can be considered a semiquantitative measure of the amount of the most volatile material present in the product.

In general, gas chromatography will undoubtedly continue to be the method of choice for characterization of low-boiling hydrocarbon derivatives (Speight, 2015, 2018; ASTM, 2017). New and improved detection devices and techniques, such as chemiluminescence, atomic emission, and mass spectroscopy, will enhance selectivity, detection limits, and analytical productivity. Laboratory automation through autosampling, computer control, and data handling will provide improved precision and productivity, as well as simplified method operation.

4.3.2 Chemical properties

In terms of the chemical reactions of natural gas, the most common reaction is combustion process which is represented as chemical reaction between methane and oxygen which results in the production of carbon dioxide (CO_2), water (H_2O) plus the exothermic liberation of energy (heat). Thus,

$$CH_4(g) + 2O_2(g) \rightarrow CO_2(g) + 2H_2O(l)$$

Higher molecular weight hydrocarbon (alkane) constituents will also participate in the combustion reaction. In an unlimited supply of oxygen and assuming that there may be traces of hydrocarbons up to octane in a natural gas stream, the combustion reactions are:

$$C_3H_8 + 5O_2 \rightarrow 3CO_2 + 4H_2O$$

$$2C_4H_{10}(g) + 13O_2(g) \rightarrow 8CO_2(g) + 10H_2O(g)$$

$$C_5H_{12}(g) + 8O_2(g) \rightarrow 5CO_2(g) + 6H_2O(g)$$

$$2C_6H_{14}(l) + 19O_2(g) \rightarrow 12CO_2(g) + 14H_2O(g)$$

$$C_7H_{16}(l) + 11O_2(g) \rightarrow 7CO_2(g) + 8H_2O(g)$$

$$2C_8H_{18}(l) + 25O_2(g) \rightarrow 16CO_2(g) + 18H_2O(g)$$

The balanced chemical equation for the complete combustion of a general hydrocarbon fuel C_xH_y is:

$$C_xH_y + (x + y/4)O_2 \rightarrow xCO_2 + x/2H_2O$$

To the purist, chemical equations do not involve fractions and to balance this final equation, the fractional numbers should be converted to whole numbers.

In an inadequate supply of air, carbon monoxide and water vapor are formed, using methane as the example:

$$2CH_4 + 3O_2 \rightarrow 2CO + 4H_2O$$

In this context of combustion, natural gas is the cleanest of all the fossil fuels. Coal and crude oil are composed of much more complex molecules, with a higher carbon ratio and as well as constituents containing nitrogen and sulfur contents. Thus, when combusted, coal and oil release higher levels of harmful emissions, including a higher ratio of carbon emissions, nitrogen oxides (NO_x), and sulfur dioxide (SO_2), which under the conditions in the atmosphere can be converted to sulfur trioxide (SO_3). Upon further reaction with the water in the atmosphere, the oxides of nitrogen and the oxides of sulfur are converted to acids and, thus, the overall result is the production of acid rain (Chapter 10: Energy security and the environment):

$$SO_2 + H_2O \rightarrow H_2SO_3$$

$$2SO_2 + O_2 \rightarrow 2SO_3$$

$$SO_3 + H_2O \rightarrow H_2SO_4$$

$$2NO + H_2O \rightarrow 2HNO_2$$

$$2NO + O_2 \rightarrow 2NO_2$$

$$NO_2 + H_2O \rightarrow HNO_3$$

Coal and fuel oil also release ash particles into the environment, substances that do not burn but instead are carried into the atmosphere and contribute to pollution. The combustion of natural gas, on the other hand, releases very small amounts of sulfur dioxide and nitrogen oxides, virtually no ash or particulate matter, and lower levels of carbon dioxide, carbon monoxide, and other reactive hydrocarbons.

Substitution reactions will also occur in which the hydrocarbons in natural gas will react with, e.g., chlorine to produce a range of chloro-derivatives:

$$CH_4 + Cl_2 \rightarrow CH_3Cl + HCl$$

$$CH_3Cl + Cl_2 \rightarrow CH_2Cl_2 + HCl$$

$$CH_2Cl_2 + Cl_2 \rightarrow CHCl_3 + HCl$$

$$CHCl_2 + Cl_2 \rightarrow CCl_4 + HCl$$

The reaction of chlorine with ethane may be written similar:

$$C_2H_6 + Cl_2 \rightarrow C_2H_5Cl + HCl$$

$$C_2H_4Cl_2 + Cl_2 \rightarrow C_2H_3Cl_3 + HCl$$

The ultimate product is hexachloroethane. Both of these reactions may be used industrially. As the hydrocarbons increase in molecular size (to propane and butane), the reaction becomes more complex.

In addition to gas streams (particularly natural gas) being used as fuel to produce heat as well as the production of hydrogen (the steam-methane reforming process) and ammonia:

$$CH_4 + H_2O \rightarrow CO + 3H_2 - \text{steam-methane reforming}$$

$$CO + H_2O \rightarrow CO_2 + H_2 - \text{hydrogen production}$$

$$3H_2 + N_2 \rightarrow 2NH_3 - \text{Haber-Bosch process}$$

The steam-methane reforming process is major source of hydrogen for refineries and other industries. In the endothermic process, high-temperature steam (700−1100°C, 1290−2010°F) is used to produce hydrogen from a methane source, such as natural gas at pressures on the order of 45−370 psi. Subsequently, in what is called the *water−gas shift reaction*, the carbon monoxide and steam are reacted using a catalyst to produce carbon dioxide and more hydrogen. In a final process step (*pressure-swing adsorption*), carbon dioxide and other impurities are removed from the gas stream, leaving essentially pure hydrogen.

$$CH_4 + H_2O \rightarrow CO + 3H_2 \quad \text{(steam-methane reforming)}$$

$$CO + H_2O \rightarrow CO_2 + H_2 \quad \text{(water-gas shift reaction)}$$

The other low-molecular weight hydrocarbons in the gas—ethane (C_2H_6), propane (C_3H_8), and the butane isomers (C_4H_{10}), either in the gas phase or liquefied, are also used for heating as well as for motor fuels and as feedstocks for chemical processing. The pentane derivatives (C_5H_{12}) are products of natural gas or crude oil fractionation or refinery operations (i.e., reforming and cracking) that are removed for use as chemical feedstocks (Table 4.4). It is only rarely that olefins occur in natural gas and they are not typically constituents of natural gas stream. However, olefins do occur in biogas produced by thermal methods (Mokhatab et al., 2006; Gary et al., 2007; Speight, 2011, 2014; Hsu and Robinson, 2017).

Because of the wide range of chemical and physical properties, a wide range of tests have been (and continue to be) developed to provide an indication of the means by which a particular gas should be processed although certain of these test methods are in more common use than others (Table 4.2) (Speight, 2015, 2018;

ASTM, 2017). Initial inspection of the nature of the petroleum will provide deductions about the most logical means of refining or correlation of various properties to structural types present and hence attempted classification of the petroleum. Proper interpretation of the data resulting from the inspection of crude oil requires an understanding of their significance.

Having decided what characteristics are necessary for a gas stream, it then remains to describe the product in terms of a specification. This entails selecting suitable test methods to determine the constituents and properties of the gas stream and setting appropriate limits for any variation of the proportion of the constituents and the limits of the variation in the properties.

The hydrocarbon component distribution of liquefied petroleum gases and propene mixtures is often required for end-use sale of this material. Applications such as chemical feedstocks or fuel require precise compositional data to ensure uniform quality. Trace amounts of some hydrocarbon impurities in these materials can have adverse effects on their use and processing. The component distribution data of liquefied petroleum gases and propene mixtures can be used to calculate physical properties such as relative density and vapor pressure. Precision and accuracy of compositional data are extremely important when these data are used to calculate various properties.

An issue that arises during the characterization of liquefied petroleum gas relates to the accurate determination of heavy residues (i.e., higher molecular weight hydrocarbons and even oils) in the gas. Test methods using procedures similar to those employed in gas chromatographic simulated distillation are becoming available. In fact, the presence of any component substantially less volatile than the main constituents of the liquefied petroleum gas will give rise to unsatisfactory performance. It is difficult to set limits to the amount and nature of the *residue* which will make a product unsatisfactory. For example, liquefied petroleum gases that contain certain antiicing additives can give erroneous results by this test method.

Control over the residue content is of considerable importance in end-use applications of liquefied petroleum gases. In liquid feed systems, residues can lead to troublesome deposits and, in vapor withdrawal systems, residues that are carried over can foul regulating equipment. Any residue that remains in the vapor-withdrawal systems will accumulate, can be corrosive, and will contaminate subsequent product. Water, particularly if alkaline, can cause failure of regulating equipment and corrosion of metals. Obviously small amounts of oil-like material can block regulators and valves. In liquid vaporizer feed systems, the gasoline-type material could cause difficulty.

Olefins (ethylene, $CH_2=CH_2$, propylene, $CH_3CH=CH_2$, butylene derivatives, such as $CH_3CH_2CH=CH_2$, and pentylene derivatives, such as $CH_3CH_2CH_2CH=CH_2$) that occur in refinery gas (process gas) have specific characteristics and require specific testing protocols (Speight, 2015, 2018; ASTM, 2017). The amount of ethylene ($CH_2=CH_2$) in a gas stream is limited because it is necessary to restrict the number of unsaturated components to avoid the formation of deposits caused by the polymerization of the olefin constituents. In addition, ethylene (boiling point: $-104°C$, $-155°F$) is more volatile than ethane (boiling point: $-88°C$, $-127°F$) and therefore a product with a substantial proportion of ethylene will have a higher vapor pressure and volatility than one that is predominantly ethane. Butadiene is also undesirable because it may also produce polymeric products that form deposits and cause blockage of lines.

As stated earlier, the amount of ethylene (and other olefins) in the mixture is limited because not only it is necessary to restrict the amount of the unsaturated components (olefins) so as to avoid the formation of deposits caused by the polymerization of the olefin(s) but also to control the volatility of the sample.

Ethylene is one of the highest volume chemicals produced in the world, with global production exceeding 100 million metric tons annually. Ethylene is primarily used in the manufacture of polyethylene, ethylene oxide, and ethylene dichloride, as well as many other lower volume products. Most of these production processes use various catalysts to improve product quality and process yield. Impurities in ethylene can damage the catalysts, resulting in significant replacement costs, reduced product quality, process downtime, and decreased yield.

Ethylene is typically manufactured through the use of steam cracking. In this process, gaseous or light liquid hydrocarbons are combined with steam and heated to 750−950°C (1380−1740°F) in a pyrolysis furnace. Numerous free radical reactions are initiated, and larger hydrocarbons are converted (cracked) into smaller hydrocarbons. In addition, the high temperatures used in steam cracking promote the formation of unsaturated or olefin compounds, such as ethylene. Ethylene feedstocks must be tested to ensure that only high-purity ethylene is delivered for subsequent chemical processing.

Samples of high-purity ethylene typically contain only two minor impurities, methane and ethane, which can be detected in low ppm v/v concentrations. However, steam cracking can also produce higher molecular weight hydrocarbons, especially when propane, butane, or light liquid hydrocarbons are used as starting materials. Although fractionation is used in the final production stages to produce a high-purity ethylene product, it is still important to be able to identify and quantify any other hydrocarbons present in an ethylene sample. Achieving sufficient resolution of all of these compounds can be challenging due to their similarities in boiling point and chemical structure.

4.4 Physical properties

Since the composition of the various gases can vary so widely, no single set of specifications could cover all situations. The requirements are usually based on performances in burners and equipment, on minimum heat content, and on maximum sulfur content. Gas utilities in most states come under the supervision of state commissions or regulatory bodies, and the utilities must provide a gas that is acceptable to all types of consumers and that will give satisfactory performance in all kinds of consuming equipment. However, particularly relevant are the heating values of the various fuel gases and their constituents. For this reason, measurement of the properties of fuel gases is an important aspect of fuel gas technology.

However, the physical properties of unrefined natural gas are variable because the composition of natural gas is never constant. Therefore the properties and behavior of natural gas are best understood by investigating the properties and behavior of the constituents. Thus if the natural gas has been processed (i.e., any constituents such as carbon dioxide and hydrogen sulfide have been removed and the only constituents remaining are hydrocarbons), the properties and behavior of natural gas become a study of the properties and behavior of the relevant.

In addition, the distribution of a set of reservoir fluids depends not only on the characteristics of the rock-fluid system now, but also the history of the fluids, and ultimately their source (Table 4.5). The fundamental forces that drive stabilize, or limit fluid movement are: (1) gravity, which causes separation of gas, oil, and water in the reservoir column; (2) capillary action, which is responsible for the retention of water in microporosity; (3) molecular diffusion, such as small-scale flow acting to homogenize fluid compositions within a given phase; (4) thermal convection, which is convective movement of all mobile fluids, especially gases; and (5) fluid pressure gradients, which is the major force operating during primary production. Although each of these forces and factors vary from reservoir to reservoir, and between lithologies within a reservoir, certain forces are of seminal importance. For

Table 4.5 Examples of factors that affect fluids distribution in the reservoir

Factor	Comment
Depth	The difference in the density of the fluids results in their separation over time due to gravity (i.e., differential buoyancy)
Fluid composition	Important control on its pressure-volume-temperature properties, which define the relative volumes of each fluid in a reservoir. It also affects distribution through the wettability of the reservoir rocks
Reservoir temperature	Exerts a major control on the relative volumes of each fluid in a reservoir
Fluid pressure	Exerts a major control on the relative volumes of each fluid in a reservoir
Fluid migration	Different fluids migrate in different ways depending on their density, viscosity, and the wettability of the rock and the mode of migration helps to define the distribution of the fluids in the reservoir
Trap-type	The effectiveness of the hydrocarbon trap also has a control on fluid distribution (e.g., cap rocks may be permeable to gas but not to oil)
Rock structure	The microstructure of the rock can preferentially accept some fluids and not others through the operation of wettability contrasts and capillary pressure. In addition, the common heterogeneity of rock properties results in preferential fluid distributions throughout the reservoir in all three spatial dimensions

example, it is density (or specific gravity) that ensures that when all three basic fluids types are present in a non-uncompartmentalized reservoir, the stratified order of the fluids (in a perfect density-oriented world) with increasing depth is gas (at the top), oil (in the middle), and water (on the bottom).

The composition of natural gas varies depending on the field, the formation, or the character of the reservoir from which the gas is extracted and that are an artifact of its formation (Chapter 1: History and use and Chapter 2: Origin and production). Also, the properties of other gas streams (Chapter 3: Unconventional gas) vary with the source from which the gas was produced and the process by which the gas was produced. The different hydrocarbons that form the gas streams can be separated using their different physical properties as weight, boiling point, or vapor pressure (Chapter 6: History of gas processing and Chapter 7: Process classification). Depending on its content of higher molecular weight hydrocarbon components, natural gas can be considered as rich (5 or 6 ga or more of recoverable hydrocarbon components per cubic feet) or lean (less than 1 ga of recoverable hydrocarbon components per cubic feet). In terms of chemical behavior, hydrocarbons are simple organic chemicals that contain only carbon and hydrogen. Thus in this section the properties and behavior of hydrocarbon up to and include n-octane (C_8H_{18}) are presented. On the other hand, when natural gas is refined, and any remaining hydrocarbons are removed, the gas is sold to the consumer, the sole component (other than an odorizer) is methane (CH_4) and the properties are constant.

There are two major technical aspects to which gas quality relates: (1) the pipeline specification in which stringent specifications for water content and hydrocarbon dew point are stated along with limits for contaminants such as sulfur—the objective is to ensure pipeline material integrity for reliable gas transportation purpose and (2) the interchangeability specification, which may include analytical data such as calorific value and relative density which are specified to ensure satisfactory performance of end-use equipment.

Gas interchangeability is a subset of the gas quality specification ensuring that gas supplied to domestic users will combust safely and efficiently. The Wobbe number is a common, but not universal, measure of interchangeability and is used to compare the rate of combustion energy output of different composition fuel gases in combustion equipment. For two fuels with identical Wobbe Indices, the energy output will be the same for given pressure and valve settings.

Finally, in terms of properties (and any test methods that are applied to natural gas) (Speight, 2015, 2018; ASTM, 2017), it is necessary to recognize the other constituents of a natural gas stream that is produced from a reservoir as well as any gas streams (Chapter 3: Unconventional gas) that may be blended into the natural gas stream. Briefly, blending is the process of mixing gases for a specific purpose where the composition of the resulting mixture is specified and controlled. Thus NGLs are products other than methane from natural gas: ethane, butane, *iso*-butane, and propane.

Test methods for gaseous fuels have been developed over many years, extending back into the 1930s. Bulk physical property tests, such as density and heating value, as well as some compositional tests, such as the Orsat analysis and the mercuric nitrate method for the determination of unsaturation, were widely used. More recently, mass spectrometry has become a popular method of choice for compositional analysis of low molecular weight and has replaced several older methods. Also, gas chromatography is another method of choice for hydrocarbon identification in gases (Speight, 2015, 2018; ASTM, 2017).

The various gas streams (Chapter 3: Unconventional gas) are generally amenable to analytical techniques and there has been the tendency, and it remains, for the determination of both major constituents and trace constituents than is the case with the heavier hydrocarbons. The complexity of the mixtures that are evident as the boiling point of petroleum fractions and petroleum products increases make identification of many of the individual constituents difficult, if not impossible. In addition, methods have been developed for the determination of physical characteristics such as calorific value, specific gravity, and enthalpy from the analyses of mixed hydrocarbon gases, but the accuracy does suffer when compared to the data produced by methods for the direct determination of these properties.

The different methods for gas analysis include absorption, distillation, combustion, mass spectroscopy, infrared spectroscopy, and gas chromatography. Absorption methods involve absorbing individual constituents one at a time in suitable solvents and recording of contraction in volume measured. Distillation methods depend on the separation of constituents by fractional distillation and measurement of the volumes distilled. In combustion methods, certain combustible elements are caused to burn to carbon dioxide and water, and the volume changes are used to calculate composition. Infrared spectroscopy is useful application and for the most accurate analyses, mass spectroscopy and gas chromatography are the preferred methods.

However, the choice of a particular test to determine any property remains as the decision of the analyst that, then, depends upon the nature of the gas under study. For example, judgment by the analyst is necessary whether or not a test that is applied to a gas stream is suitable for that gas stream insofar as inference from the nonhydrocarbon constituents will be minimal.

The following section presents a brief illustration of the properties of natural gas hydrocarbons from methane up to and including n-octane (C_8H_{18}). This will allow the reader to understand the folly of stating the properties of natural gas as average properties rather than allowing for the composition of the gas mixture and recognition of the properties of the individual constituents.

4.4.1 Behavior

The behavior of natural gas, whether pure methane or a mixture of volatile hydrocarbons and the nonhydrocarbons nitrogen, carbon dioxide, and hydrogen sulfide, must be understood by the engineer designing and operating equipment for its production, processing, and transportation. The constituents of natural gas are most likely to be found in the gaseous state but can occur as liquids and solids (Speight, 2015).

The behavior of a gas is dictated by the gas laws. There are two types of gases: an ideal gas and a nonideal gas. An ideal gas has the following properties: (1) there are no intermolecular forces between the gas particles, (2) the volume occupied by the particles is negligible compared to the volume of the container they occupy, and (3) the only interactions between the particles and the container walls are perfectly elastic collisions. In terms of properties and behavior, natural gas is a nonideal gas and obeys the gas law:

$$PV = nZRT$$

where P is the pressure, V is the volume, T is the absolute temperature (degree Kelvin), Z is the compressibility, n is the number of kilo-moles of the gas, and R is the gas constant. For example, if all other factors remained constant, when the volume of a certain mass of gas is reduced by 50%, the pressure would double and so on. As a gas, it would expand to fill any volume it is in. However, the compressibility, Z, is the factor that differentiates natural gas from an ideal gas. For methane, Z is 1 at 1 atmosphere pressure (14.7 psi).

The kinetic theory of gases treats a gas as crowd of molecules each moving on its own independent path, entirely uncontrolled by forces from the other molecules, although its path may be abruptly altered in both speed and direction whenever it collides with another molecule or strikes the boundary of the containing vessel. In its simplest state, a gas may be considered as composed of particles that has no volume and between which there are no forces.

Compressibility, density, and viscosity of natural gases are necessary in most petroleum engineering calculations. Some of these calculations are gas metering, gas compression, design of processing units, and design of pipeline and surface facilities. Properties of natural gases are also important in calculation of gas flow rate through reservoir rock, material balance calculations, and evaluation of gas reserves.

As stated earlier, an ideal gas is a gas in which all collisions between atoms or molecules are perfectly elastic and in which there are no intermolecular attractive forces. An ideal gas can be characterized by three variables: (1) absolute pressure (P), (2) volume (V), and (3) absolute temperature (T). The relationship between them is the *ideal gas law*:

$$PV = nRT = NkT$$

where n is the number of moles, R is the universal gas constant (8.3145 J/mol K), N is the number of molecules, k is the Boltzmann constant (1.38066×10^{-23} J/K $= 8.617385 \times 10^{-5}$ eV/K), k is equal to R/N_A, N_A is the Avogadro's number ($6.0221 \ 10^{23}$ per mol).

The ideal gas law can arise from the pressure of gas molecules colliding with the walls of a container. And 1 mol of an ideal gas at standard temperature and pressure occupies 22.4 L.

For example, if all other factors remained constant, when the volume of a certain mass of gas is reduced by 50%, the pressure would double and so on. As a gas, it would expand to fill any volume it is in. However, the gas deviation factor, Z, is the factor which differentiates natural gas from an ideal gas. For methane, Z is 1 at 1 bar but decreases to 0.85 at 100 atm, both at 25°C (77°F), that is it compresses to a smaller volume than the proportional relationship.

4.4.2 Compressibility

Natural gas, like any gas, can be pressurized using a compressor in which the volume of the gas is decreased. Typically, natural gas is compressed using pressure on the order of 2900–4300 psi, which gives a 200- to 250-fold reduction in the volume of the gas. The compression factor (also known as the compressibility factor or the real gas factor and given the symbol Z) appears in equations governing volumetric metering. Moreover, the conversion of volume at metering conditions to volume at defined reference conditions can properly proceed with an accurate knowledge of Z at both relevant pressure and relevant temperature conditions.

When gas is compressed, work is done and thus it gets hotter. It is therefore necessary to cool gas during or after compression. Equally when it is expanded, adiabatically (without heat being added) it gets colder. This latter phenomenon is used to cool gas during treatment to remove liquids.

The isothermal gas compressibility (c_g, also called the bulk modulus of elasticity) of natural gas is useful insofar as it is used extensively in determining the compressible properties of the reservoir. Gas usually is the most compressible medium in the reservoir. However, care should be taken so that it is not confused with the gas deviation factor, Z, which is sometimes called the super compressibility factor:

$$c_g = -\frac{1}{V_g}\left(\frac{\partial V_g}{\partial P}\right)$$

where V is the volume, P is the pressure, and T is the absolute temperature. For an ideal gas, the compressibility is defined as:

$$c_g = \frac{1}{P}$$

For a nonideal gas, the compressibility is defined by the equation:

$$c_g = \frac{1}{P} - \frac{1}{Z}\left(\frac{\partial Z}{\partial P}\right)$$

$$c_g\frac{1}{P} - \frac{1}{Z}\left(\frac{\partial Z}{\partial P}\right)_T$$

Gas is difficult to store in the gaseous state outside the reservoir to provide flexibility of supply. Only a small amount of flexibility is provided by the high-pressure gas in pipelines,

4.4.3 Corrosion

Gas streams produced during petroleum recovery, petroleum refining, and natural gas processing, while ostensibly being hydrocarbon in nature, may contain large amounts of corrosive acid gases such as hydrogen sulfide and carbon dioxide resulting in a high potential for corrosion. Although the processing of natural gas is in many respects less complicated than the processing and refining of crude oil, it is equally as necessary to assure that all of the corrosive constituents are removed.

In gas plants handling product streams from refinery units raises the potential for corrosion from moist hydrogen sulfide and cyanide derivatives. When feedstocks are from the visbreaker, the delayed coker, the fluid coker or any other thermal cracking unit, corrosion from hydrogen sulfide and deposits of iron sulfide in the high-pressure sections of gas compressors from ammonium compounds is possible. Furthermore, processing opportunity crudes require refiners to manage greater volumes of corrosive and toxic hydrogen sulfide. In sour-gas streams, the primary corrosion-causing constituents of gas streams are hydrogen sulfide (H_2S) and carbon dioxide (CO_2), with contributions from other corrosive constituents. Streams containing ammonia should be dried before processing. Antifouling additives may be used in absorption oil to protect heat exchangers. Corrosion inhibitors may be used to control corrosion in overhead systems.

As an example, copper corrosion causes difficulties due to the deterioration of copper and copper alloy fittings and connections commonly used in liquefied petroleum gas systems. As little as 1 ppm of hydrogen sulfide can cause a copper strip test failure. Thus the copper strip corrosion test is an extremely sensitive test that will detect virtually all species of corrosive sulfur, including minute traces of hydrogen sulfide. It is critically important that the product being tested does not contain any additives that may diminish the reaction with the copper strip.

Hydrogen sulfide corrosion results in the formation of black iron sulfide scales and is typified by *black water* in the separation units. Underdeposit corrosion frequently occurs beneath the scale layer and can result in forming deep, isolated, or randomly scattered pits. The prime means of removing or reducing the impact of iron sulfide entering an olamine system are to: (1) prevent the corrosion from occurring initially in the piping by using corrosion inhibitors, (2) disperse the iron sulfide particles into the water phase so they can be removed by inlet separation equipment, and (3) remove the iron sulfide from the gas phase upstream of the olamine absorber by use of a suitable filter or by a water wash.

Corrosion by wet carbon dioxide corrosion can result in high corrosion rates, but a carbonate film gives some protection and is more protective at higher temperatures. The carbon dioxide content is often not very high in refinery streams, except in hydrogen reformer plant systems. In addition, one of the major sources of corrosion on carbon steel vessels in sweetening units is heat-stable materials, which are a product of amine degradation. Oxygen plays a major role in amine degradation—the reaction of oxygen and the amine produces organic acids, such as acetic acid, formic acid, and so on.

In addition, the effluent streams from any of the refining processes can contain ammonia (NH_3) and hydrogen sulfide (H_2S) which react to form ammonium bisulfide (NH_4HS), which is highly corrosive to carbon steel and may lead to a catastrophic failure. The severity of ammonium bisulfide-induced corrosion depends upon (1) the concentration of ammonium bisulfide, (2) the fluid velocity and turbulence, (3) wash-water management, as well as (4) piping configuration and temperature of the system.

4.4.4 Density and specific gravity

Density is the mass of a substance contained in a unit volume (simply, density is mass divided by volume). In the SI system of units, the ratio of the density of a substance to the density of water at 15°C is known as the *specific gravity* (*relative density*). Various units of density, such as kg/m^3, lb-mass/ft^3, and g/cm^3, are commonly used. In addition, molar densities or the density divided by the molecular weight is often specified.

Density is a physical property of matter, which is a measure of the relative heaviness of hydrocarbons and other chemicals at a constant volume, and each constituents of natural gas has a unique density associated with it. For most chemical compounds (i.e., those that are solid or liquid), the density is measured relative to water (1.00). For gases, the density is more likely to be compared to the density of air (also given the number 1.00 but this is arbitrary and bears no relationship to the density of water). As a comparison, the density of liquefied natural gas (LNG) is approximately 0.41−0.5 kg/L, depending on temperature, pressure, and composition; in comparison, the density of water is 1.0 kg/L. In terms of composition (Table 4.6), LNG is predominantly methane and it is not surprising that the density of LNG is close to (but not exactly equal to) the density of methane.

The density (or specific gravity) of the various unconventional gases discussed in this chapter may be determined conveniently by any one of several methods and a variety of instruments (Chapter 3: Unconventional gas) (ASTM, 2017; Speight, 2018). Density values (including those of natural gas hydrocarbons) (Fig. 4.1) are

Table 4.6 Composition of liquefied natural gas in various markets

Composition (mole percent)					
Source	Methane	Ethane	Propane	Butane	Nitrogen
Alaska	99.72	0.06	0.0005	0.0005	0.20
Algeria	86.98	9.35	2.33	0.63	0.71
Baltimore	93.32	4.65	0.84	0.18	1.01
New York City	98.00	1.40	0.40	0.10	0.10
San Diego	92.00	6.00	1.00	-	1.00

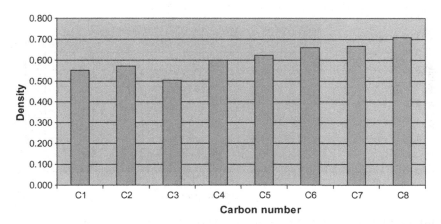

Figure 4.1 Carbon number and density of natural gas hydrocarbons (up to *n*-octane, C_8H_{18}).

given at room temperature unless otherwise indicated by the superscript figure; e.g., 2.487^{15} indicates a density of 2.487 g/cm for the substance at 15°C. A superscript 20 over a subscript 4 indicates a density at 20°C relative to that of water at 4°C. For gases the value of the density is given in grams per liter (g/L).

More specifically, the density of an ideal gas mixture is calculated by simply replacing the molecular weight of the pure component with the average molecular weight of the gas mixture to give:

$$\rho_g = \frac{pM_a}{RT}$$

In this equation, ρ_g is the density of the gas mixture, lb/ft^3, M_a is the average molecular weight, p is the absolute pressure, psia, T is the absolute temperature, and R is the universal gas constant.

The density of low-boiling hydrocarbon derivatives can be determined by several methods including a hydrometer method or by a pressure hydrometer method (ASTM, 2017; Speight, 2018). The specific gravity (relative density) by itself has little significance compared to its use for higher molecular weight liquid petroleum products) and can only give an indication of quality characteristics when combined with values for volatility and vapor pressure. It is important for stock quantity calculations and is used in connection with transport and storage.

Another term, specific gravity is commonly used is relation to the properties of hydrocarbons. The specific gravity of a substance is a comparison of its density to that of water. Both the density of the substance and the density of water should be measured or expressed at the same pressure and temperature. If the behavior of both the gas mixture and the air is described by the ideal gas equation, the specific gravity can then be expressed in the form:

$$\gamma_g = \frac{\rho_g}{\rho_{air}}$$

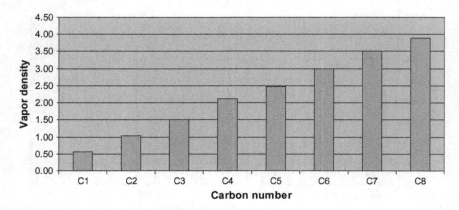

Figure 4.2 Carbon number and vapor density (relative to air = 1.0) of natural gas hydrocarbons (up to n-octane, C_8H_{18}).

Lao:

$$\gamma_g = \frac{\frac{p_{sc}M_a}{RT_{sc}}}{\frac{p_{sc}M_{air}}{RT_{sc}}}$$

$$\gamma_g = \frac{M_a}{M_{air}} = \frac{M_a}{28.96}$$

where γ_g is the specific gravity of the gas, ρ_{air} is the density of the air, M_{air} is the apparent molecular weight of the air which is 28.96, M_a is the apparent molecular weight of the gas, p_{sc} is the standard pressure, psia, and sc is the standard temperature, °R.

The density of any gas compared to the density of air is the vapor density and is a very important characteristic of the constituents of natural gas and natural gas constituents (Fig. 4.2). Put simply, if the constituents of natural gas are less dense (lighter) than air, they will dissipate into the atmosphere whereas if the constituents of natural gas are denser (heavier) than air, they will sink and be less likely to dissipate into the atmosphere. Of the hydrocarbon constituents of natural gas, methane is the only one that is less dense than air.

The density (or specific gravity) can be calculated, but if it is necessary to measure it several pieces of apparatus are available. For determining the density or specific gravity of liquefied petroleum gas in its liquid state there are two methods, using a metal pressure pycnometer.

In fact, relative to air, methane is less dense than air, but the other hydrocarbon constituents of unrefined natural gas (i.e., ethane, propane, and butane) are denser than air (Fig. 4.2). Therefore should a natural gas leak occur in field operations, especially where the natural gas contains constituents other than methane, only methane dissipates readily into the air whereas the other hydrocarbon constituents

Table 4.7 Relative density (specific gravity) of
natural gas hydrocarbons relative to air

Gas	Specific gravity
Air	1.000
Methane, CH_4	0.5537
Ethane, C_2H_6	1.0378
Propane, C_3H_8	1.5219
Butane, C_4H_{10}	2.0061
Pentane, C_5H_{12}	2.487
Hexane, C_6H_{14}	2.973

that are heavier than air do not readily dissipate into the atmosphere. This poses considerable risk if these constituents of natural gas accumulate or pool at ground level when it has been erroneously assumed that natural gas is lighter than air.

The relative density typically relates to the density of natural gas relative to the density of air (Table 4.7) although, in some cases, the density of hydrogen may be used for comparison to the density of natural gas. The relative density, as a measure of gas density relative to air at reference conditions is used for interchangeability specifications to limit the higher hydrocarbon content of the gas. An increased higher hydrocarbon content could lead to combustion problems such as increased carbon monoxide emissions, soot formation, engine knock, or spontaneous ignition on gas turbines even at the same Wobbe index value.

The statement is often made that natural gas is lighter than air. This statement often arises because of the continued insistence by engineers and scientists that the properties of a mixture are determined by the mathematical average of the properties of the individual constituents of the mixture. Such mathematical bravado and inconsistency of thought is detrimental to safety and needs to be qualified. The *relative density* (*specific gravity*) is the ratio of the density (mass of a unit volume) of a substance to the density of a given reference material. Specific gravity usually means relative density (of a liquid) with respect to water.

Relative density = $[\rho(\text{substance})]/[\rho(\text{reference})]$

As it pertains to gases, particularly in relation to safety considerations at commercial and industrial facilities in the United States, the relative density of a gas is usually defined with respect to air, in which air is assigned a *vapor density* of 1 (unity). With this definition, the vapor density indicates whether a gas is denser (greater than 1) or less dense (less than 1) than air. The vapor density has implications for container storage and personnel safety—if a container can release a dense gas, its vapor could sink and, if flammable, collect until it is at a concentration sufficient for ignition. Even if not flammable, it could collect in the lower floor or level of a confined space and displace air, possibly presenting a smothering hazard to individuals entering the lower part of that space.

Gases can be divided into two groups based upon their vapor density: (1) gases which are heavier than air and (2) gases which are as light as air or lighter than air. Gases that have a vapor density greater than 1 will be found in the bottom of storage containers and will tend to migrate downhill and accumulate in low-lying areas. Gases that have a vapor density which is the same or less than the vapor density of air will disperse readily into the surrounding environment. Additionally, chemicals that have the same vapor density as air (1.0) tend to disperse uniformly into the surrounding air when contained and, when released into the open air, chemicals that are lighter than air will travel up and away from the ground.

Methane is the only hydrocarbon constituent of natural gas that is lighter than air (Table 4.7). The higher molecular weight hydrocarbons have a higher vapor density than air and are likely, after a release, to accumulate in low-lying areas and represent a danger to investigator (of the release). However, the other hydrocarbon constituents of unrefined natural gas (i.e., ethane, propane, and butane) are denser than air.

4.4.5 Dew point

The dew point or dew point temperature of a gas is the temperature at which the water vapor or low-boiling hydrocarbon derivatives contained in the gas is transformed into the liquid state. The formed liquid (condensate) exists as a liquid below the dew point temperature but above the dew point the liquid is a gaseous component of the gas.

The hydrocarbon dew point is often considered to be the most important factor when performing any type of gas sampling. In the simplest terms, hydrocarbon dew point is the point at which the gas components begin to change phase from gas to liquid. When phase change occurs, certain components of the gas stream drop out and form liquids thereby making an accurate gas sample impossible to obtain. The hydrocarbon dew point is a function of gas composition and pressure. A hydrocarbon dew point curve is a reference chart that determines the specific pressure and temperature at which condensation occurs. No two hydrocarbon dew point curves are alike due to differing gas compositions. Since dew point can be calculated from composition, direct determination of dew point for a particular liquefied petroleum gas sample is a measure of composition. It is, of course, of more direct practical value and if there are small quantities of higher molecular weight material present, it is preferable to use a direct measurement.

The hydrocarbon dew point is the temperature (at a given pressure) at which the hydrocarbon constituents of any hydrocarbon-rich gas mixture, such as natural gas, will start to condense out of the gaseous phase. The maximum temperature at which such condensation takes place is called the *cricondentherm*. The hydrocarbon dew point is a function of the gas composition as well as the pressure. The hydrocarbon dew point is universally used in the natural gas industry as an important quality parameter, stipulated in contractual specifications (Table 4.8) and enforced throughout the natural gas supply chain, from producers through gas processing (gas cleaning), transmission and distribution companies to the consumer.

Table 4.8 Examples of pipeline specifications for natural gas

Components mol%	Minimum	Maximum
Methane	75	
Ethane		10
Propane		5
Butanes		2
Pentanes plus		0.5
Nitrogen and other inerts		3-4
Carbon dioxide		3-4
Trace components		
Hydrogen sulfide		$0.25-1.0$ g/100 ft^3
Mercaptan sulfur		$0.25-1.0$ g/100 ft^3
Total sulfur		$5-20$ g/100 ft^3
Water vapor		7.0 lbs/mm ft^3
Oxygen		$0.2-1.0$ ppm v/v
Heating value	950	1150 Btu/ft^3

Thus if the hydrocarbon dew point is reduced to such a level that retrograde condensation, i.e., condensation resulting from pressure drop, cannot occur under the worst conditions likely to be experienced in the gas transmission system. Similarly, the water dew point is reduced to a level sufficient to preclude formation of C_1 to C_4 hydrates in the system. The natural gas after appropriate treatment for acid gas reduction, odorization, and hydrocarbon and moisture dew point adjustment would then be sold within prescribed limits of pressure, calorific value, and possibly Wobbe Index (cv/(sp. gr.) (Chapter 8: Gas cleaning processes).

Once the composition of a mixture has been determined it is possible to calculate various properties such as specific gravity, vapor pressure, calorific value, and dew point. Since dew point can be calculated from composition, direct determination of dew point for a liquefied petroleum gas sample is a measure of composition. If there are small quantities of higher molecular weight material present, it is preferable to use a direct measurement.

While the dew point identifies the condition at which vapor first begins to condense to liquid, it provides no information about the quantity of condensation resulting from a small degree of cooling. The condensation rate of liquids in gas transmission lines may vary widely depending on the composition, temperature, and pressure of the system and there needs to be a practical hydrocarbon dew point specification allowing small amounts of liquids that have no significant impact on operations (Bullin et al., 2010).

4.4.6 Flammability

Flammable chemicals are those chemicals that ignite more easily than other chemicals, whereas those that are harder to ignite or burn less vigorously are combustible.

The degree of flammability or combustibility of the constituents of a gas stream in air depends largely upon the chemical composition of the stream which is also related to the volatility of the constituents in the gas stream. Furthermore, volatility is directly related to the boiling point.

The boiling point (boiling temperature) of a substance is the temperature at which the vapor pressure of the substance is equal to atmospheric pressure. At the boiling point, a substance changes its state from liquid to gas. A stricter definition of boiling point is the temperature at which the liquid and vapor (gas) phases of a substance can exist in equilibrium. When heat is applied to a liquid, the temperature of the liquid rises until the *vapor pressure* of the liquid equals the pressure of the surrounding atmosphere (gases). At this point there is no further rise in temperature, and the additional heat energy supplied is absorbed as *latent heat* of vaporization to transform the liquid into gas. This transformation occurs not only at the surface of the liquid (as in the case of *evaporation*) but also throughout the volume of the liquid, where bubbles of gas are formed.

The boiling point of a liquid is lowered if the pressure of the surrounding atmosphere (gases) is decreased. On the other hand, if the pressure of the surrounding atmosphere (gases) is increased, the boiling point is raised. For this reason, it is customary when the boiling point of a substance is given to include the pressure at which it is observed, if that pressure is other than standard, i.e., 760 mm of mercury or 1 atm (STP, Standard Temperature and Pressure).

The boiling points of petroleum fractions are rarely, if ever, distinct temperatures. It is, in fact, more correct to refer to the boiling ranges of the various fractions; the same is true of natural gas. To determine these ranges, the material in question is tested in various methods of distillation, either at atmospheric pressure or at reduced pressure. Thus the boiling points of the hydrocarbon constituents of natural gas increase with molecular weight and the initial boiling point of natural gas corresponds to the boiling point of the most volatile constituents (i.e., methane) (Fig. 4.3).

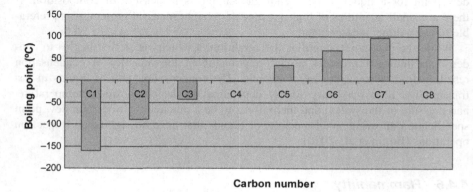

Figure 4.3 Carbon number and boiling point of natural gas hydrocarbons (up to *n*-octane, C_8H_{18}).

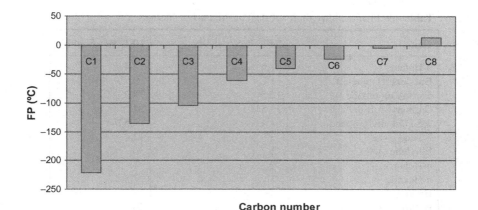

Figure 4.4 Carbon number and flash point of natural gas hydrocarbons (up to *n*-octane, C_8H_{18}).

Purified natural gas is neither corrosive nor toxic, its ignition temperature is high, and it has a narrow flammability range, making it an apparently safe fossil fuel compared to other fuel sources. In addition, purified natural gas (i.e., methane) having a specific gravity (0.60) lower than that of air (1.00) rises if escapes and dissipates from the site of any leak. However, methane is highly flammable, burns easily and almost completely. Therefore natural gas can also be hazardous to life and property through an explosion. When natural gas is confined, such as within a house or in a coal mine, concentration of the gas can reach explosive mixtures that, if ignited, results in blasts that could destroy buildings.

The *flash point* of petroleum or a petroleum product, including natural gas, is the temperature to which the product must be heated under specified conditions to give of sufficient vapor to form a mixture with air that can be ignited momentarily by a specified flame. As with other properties, the flash point is dependent on the composition of the gas and the presence of other hydrocarbon constituents (Fig. 4.4). The *fire point* is the temperature to which the gas must be heated under the prescribed conditions of the method to burn continuously when the mixture of vapor and air is ignited by a specified flame.

From the viewpoint of safety, information about the flash point and fire point is of most significance at or slightly above the maximum temperatures (30−60°C, 86−140°F) that may be encountered in storage, transportation, and use of liquid petroleum products, in either closed or open containers. In this temperature range the relative fire and explosion hazard can be estimated from the flash point. For products with flash point below 40°C (104°F) special precautions are necessary for safe handling. Flash points above 60°C (140°F) gradually lose their safety significance until they become indirect measures of some other quality. The flash point of a petroleum product is also used to detect contamination. A substantially lower flash point than expected for a product is a reliable indicator that a product has

Table 4.9 Flammability limits of the constituents of fuel gases

Gas[a]	LFL % v/v in air	UFL % v/v in air
n-Butane	1.6	8.4
Butylene (1-butene)	1.7	9.7
Carbon monoxide	12.5	74.2
Carbonyl sulfide	12.5	74
2,2-Dimethylpropane	1.4	7.5
Ethane	3	12.4
Ethylene	2.7	36
n-Heptane	1	7
n-Hexane	1	7.5
Hydrogen	4	75
Hydrogen sulfide	4	46
Isobutane	1.82	9.6
Methane	5	15
Methyl mercaptan	3.9	21.8
n-Octane	1	7
iso-Octane	0.6	6
n-Pentane	1.4	7.8
iso-Pentane	1.3	9.2
Propane	2.1	9.5
Propylene	2	11.1

[a]Listed alphabetically.

become contaminated with a more volatile product, such as gasoline. The flash point is also an aid in establishing the identity of a petroleum product.

The flammability range: the range of temperature over which natural gas is flammable. The flammable limits (Table 4.9) are expressed by the lower explosive limit (LEL) and the upper explosive limit (UEL). The LEL is the concentration of natural gas in the air below which the propagation of a flame will not occur on contact with an ignition source. The LEL for natural gas is 5% by volume in air and, in most cases, the smell of gas would be detected well before combustion conditions are met. The UEL is the concentration of natural gas in the air above which the propagation of a flame will not occur on contact with an ignition source. The natural gas UEL is 15% by volume in air.

Explosions caused by natural gas leaks occur a few times each year. Individual homes and small businesses are most frequently affected when an internal leak builds up gas inside the structure. Frequently, the blast will be enough to significantly damage a building but leave it standing. Occasionally, the gas can collect in high enough quantities to cause a deadly explosion, disintegrating one or more buildings in the process.

By way of explanation, before a fire or explosion can occur, three conditions must be met simultaneously. A fuel (i.e., combustible gas) and oxygen (air) must exist in certain proportions, along with an ignition source, such as a spark or flame. The ratio of fuel and oxygen that is required varies with each combustible gas or vapor. The minimum concentration of a particular combustible gas or vapor necessary to support its combustion in air is defined as the LEL for that gas. Below this level, the mixture is too lean to burn. The maximum concentration of a gas or vapor that will burn in air is defined as the UEL. Above this level, the mixture is too rich to burn. The range between the LEL and UEL is known as the flammable range for that gas or vapor. Typically, the values given for the UEL and LEL (Table 4.10) are valid only for the conditions under which they were determined (usually room temperature and atmospheric pressure using a 2-in. tube with spark ignition). The flammability range of most materials expands as temperature, pressure, and container diameter increase.

Table 4.10 Lower explosive limits (LEL) and upper explosive limits (UEL) for various constituents of gases, gas condensate, and natural gasoline

Constituent	LEL	UEL
Benzene	1.3	7.9
1,3-Butadiene	2	12
Butane	1.8	8.4
n-Butanol	1.7	12
iso-Butene	1.6	10
cis-2-Butene	1.7	9.7
trans-2-Butene	1.7	9.7
Carbon monoxide	12.5	74
Carbonyl sulfide	12	29
Cyclohexane	1.3	7.8
Cyclopropane	2.4	10.4
Diethylbenzene	0.8	
2,2-Dimethylpropane	1.4	7.5
Ethane	3	12.4
Ethyl benzene	1	6.7
Ethylene	2.7	36
Gasoline	1.2	7.1
Heptane	1.1	6.7
Hexane	1.2	7.4
Hydrogen	4	75
Hydrogen sulfide	4	44
Isobutane	1.8	8.4
Isobutylene	1.8	9.6
Methane	5	15
3-Methyl-1-butene	1.5	9.1
Methyl mercaptan	3.9	21.8
Pentane	1.4	7.8
Propane	2.1	9.5
Propylene	2.4	11
Toluene	1.2	7.1
Xylene	1.1	6.6

A further aspect of volatility that receives considerable attention is the vapor pressure of petroleum and its constituent fractions. The *vapor pressure* is the force exerted on the walls of a closed container by the vaporized portion of a liquid. Conversely, it is the force that must be exerted on the liquid to prevent it from vaporizing further. The vapor pressure increases with temperature for any given gasoline, liquefied petroleum gas, or other product. The temperature at which the vapor pressure of a liquid, either a pure compound of a mixture of many compounds, equals 1 atm (14.7 psi, absolute) is designated as the boiling point of the liquid.

The vapor density has implications for flammability during storage. If a container can release a dense gas, its vapor could sink and, if flammable, collect until it is at a concentration sufficient for ignition. Even if not flammable, it could collect in the lower floor or level of a confined space and displace air, possibly presenting a smothering hazard to individuals entering the lower part of that space.

Thus the lower and upper limits of flammability indicate the percentage of combustible gas in air below which and above which flame will not propagate (Table 4.6). When flame is initiated in mixtures having compositions within these limits, it will propagate and therefore the mixtures are flammable. A knowledge of flammable limits and their use in establishing safe practices in handling gaseous fuels is important, e.g., when purging equipment used in gas service, in controlling factory or mine atmospheres, or in handling liquefied gases.

The calculation of flammable limits is accomplished by Le Chatelier's modification of the mixture law, which (expressed in the simplest form) is

$$L = 100/\left(p_1/N_1 + p_2/N_2 \ldots + p_n N_n\right)$$

where L is the volume percentage of fuel gas in a limited mixture of air and gas; $p_1, p_2, \ldots, p_n$ are the volume percentages of each combustible gas present in the fuel gas, calculated on an air- and inert-free basis so that $p_1 + p_2 + \ldots p_n = 100$; and $N_1, N_2, \ldots N_n$ are the volume percentages of each combustible gas in a limit mixture of the individual gas and air. The foregoing relation may be applied to gases with inert content of 10% or less without introducing an absolute error of more than 1% or 2% in the calculated limits.

The *rate of flame propagation* (also referred to as the *burning velocity*) in gas—air mixtures is of importance in utilization problems, including those dealing with burner design and rate of energy release. There are several methods that have been used for measuring such burning velocities, in both laminar and turbulent flames. Results by the various methods do not agree, but any one method does give relative values of utility. Maximum burning velocities for turbulent flames are greater than those for laminar flames.

4.4.7 Formation volume factor

Volumetric factors were introduced in petroleum and natural gas calculations to readily relate the *volume* of reservoir fluids that are obtained at the surface (stock tank fluids) to the volume that the fluid(s) occupied when it was (they were) under reservoir pressure and compressed in the reservoir.

The formation volume factor for gas is the ratio of volume of 1 mol of gas at a given pressure and temperature to the volume of 1 mol of gas at standard conditions (Ps and Ts). The gas formation volume factor is used to relate the volume of gas, as measured at reservoir conditions, to the volume of the gas as measured at standard conditions, i.e., 15.5°C (60°F) and 14.7 psia. This gas property is then defined as the actual volume occupied by a certain amount of gas at a specified pressure and temperature, divided by the volume occupied by the same amount of gas at standard conditions. In an equation form, the relationship is expressed as:

$$B_g = \frac{V_{p,T}}{V_{sc}}$$

where B_g is the gas formation volume factor, ft³/scf.; $V_{p,T}$ is the volume of gas at pressure p and temperature, T, in ft³; and V_{sc} is the volume of gas at standard conditions, scf.

The reciprocal of the specific molar volume is the molar density, and thus,

$$B_g = \frac{\tilde{v}_g|_{res}}{\tilde{v}_g|_{sc}} = \frac{\rho_g|_{sc}}{\overline{\rho}_g|_{res}} = \frac{\left(\rho_g/MW_g\right)|_{sc}}{\left(\rho_g/MW_g\right)|_{res}}$$

Introducing the definition for densities in terms of compressibility factor,

$$B_g = \frac{\frac{p_{sc}}{RT_{sc}Z_{sc}}}{\frac{p}{RTZ}}$$

Since $Z_{sc} \approx 1$, the relationship is:

$$B_g = \frac{p_{sc}}{T_{sc}}\frac{ZT}{P} = 0.02827\frac{ZT}{P}\,[\text{RCF/SCF}]$$

where RCF is the reservoir cubic feet.

Gas formation volume factors can be also expressed in terms of [RB/SCF]. In such a case, 1 RB = 5.615 RCF. Thus,

$$B_g = 0.005035\frac{ZT}{P}\,[\text{RCF/SCF}]$$

The formation volume factor of a liquid or condensate (B_o) relates the volume of 1 lb-mol of liquid at reservoir conditions to the volume of that liquid once it has gone through the surface separation facility.

$$B_o = \frac{\text{Volume of 1 lbmol of liquid at reservoir conditions, RB}}{\text{Volume of that lbmol after going through separation, STB}}$$

The total volume occupied by 1 lb-mol of liquid at reservoir conditions $(V_o)_{res}$ can be calculated through the compressibility factor of that liquid, as follows:

$$(V_o)_{res} = \left(\frac{nZ_oRT}{P}\right)_{res}$$

where $n = 1$ lb-mol

Upon separation, some gas is going to be taken out of the liquid stream feeding the surface facility. n_{st} is the moles of liquid leaving the stock tank per mole of feed entering the separation facility. The volume that 1 lb-mol of reservoir fluid (including natural gas) is going to occupy after going through the separation facility is given by:

$$(V_o)_{res} = \left(\frac{n_{st}Z_oRT}{P}\right)_{sc}$$

4.4.8 Heat of combustion

The *heat of combustion* (*energy content*) of natural gas is the amount of energy that is obtained from the burning of a volume of natural gas is measured in British thermal units (Btu). The value of natural gas is calculated by its Btu content. One Btu is the quantity of heat required to raise the temperature of 1 pound of water of 1°F at atmospheric pressure. A cubic foot of natural gas has an energy content of approximately 1031 Btu, but the range of values is between 500 and 1500 Btu depending upon the composition of the gas. The heat value of gases is generally determined at constant pressure in a flow calorimeter in which the heat released by the combustion of a definite quantity of gas is absorbed by a measured quantity of water or air.

For use as heating agents, the relative merits of gases from different sources and having different compositions can be compared readily on the basis of their heating values. Therefore, the heating value is used as a parameter for determining the price of gas in custody transfer as well as an essential factor in calculating the efficiencies of energy conversion devices such as gas-fired turbines. The heating values of a gas depend not only upon the temperature and pressure, but also upon the degree of saturation with water vapor. However, some calorimetric methods for measuring heating values are based upon the gas being saturated with water at the specified conditions.

The heating value of natural gas (or any fuel gas) may be determined experimentally using a calorimeter in which fuel is burned in the presence of air at constant pressure. The products are allowed to cool to the initial temperature and a measurement is made of the energy released during complete combustion. All fuels that contain hydrogen release water vapor as a product of combustion, which is subsequently condensed in the calorimeter. The resulting measurement of the heat released is the higher heating value (HHV), also known as the gross heating value, and includes the heat of vaporization of water. The lower heating value (LHV), also known as the net heating value, is calculated by subtracting the heat of vaporization of water from the measured HHV and assumes that all products of combustion including water remain in the gaseous phase. The US system of measurement uses Btu per pound or Btu per standard cubic foot when expressed on a volume basis. This property is an indicator of the performance and torque potential of the gas for a defined engine configuration.

The principal quality criterion of natural gas is its heating value and the total calorific (heating) value of fuel gas produced or sold in the natural gas range from 900 to 1200 Btu/standard ft^3. In addition, the gas must be readily transportable through high-pressure pipelines and, therefore, the water content, as defined by the water dew point, must be considered to prevent the formation of ice or hydrates in the pipeline. Likewise, the amounts of the entrained hydrocarbons having higher molecular weight than ethane, as defined by the hydrocarbon dew point, should be considered to prevent accumulation of condensable liquids that may block the pipeline.

Thus the energy content of natural gas is variable because natural gas has variations in the amount and types of energy gases (methane, ethane, propane, and butane) it contains; the more noncombustible gases in the natural gas, the lower the energy (Btu). In addition, the volume mass of energy gases which are present in a natural gas accumulation also influences the Btu value of natural gas. The more carbon atoms in a hydrocarbon gas, the higher its Btu value. It is necessary to conduct the Btu analysis of natural gas, which is done at each stage of the supply chain. Gas chromatographic process analyzers are used to conduct fractional analysis of the natural gas streams, separating natural gas into identifiable components. The components and their concentrations are converted into a gross heating value in Btu-cubic foot.

In the United States, at retail, natural gas is often sold in units of Therms (th) (1 Therm = 100,000 Btu. Wholesale transactions are generally done in Decatherms (Dth), or in thousand decatherms (MDth), or in million decatherms (MMDth). A million decatherms is roughly a billion cubic feet of natural gas.

The gross heats of combustion of crude oil and its products are given with fair accuracy by the equation:

$$Q = 12,400 - 2100d^2$$

where d is the 60/60°F specific gravity. Deviation from the formula is generally less than 1%.

Two other terms that need consideration of this point are (1) the gross heating value and (2) the net heating value (Table 4.11). The *gross heating value* is the total energy transferred as heat in an ideal combustion reaction at a standard temperature and pressure in which all water formed appears as liquid. The gross heating is an

Table 4.11 Gross and net heating values of various gases

Gas[a]	Gross heating value		Net heating value	
	(Btu/ft^3)[b]	(Btu/lb)	(Btu/ft^3)	(Btu/lb)
Butane	3225	21,640	2977	19,976
Butylene	3077	20,780	2876	19,420
Carbon monoxide	323	4368	323	4368
Coal gas	149	16,500		
Ethane	1783	22,198	1630	20,295
Ethylene	1631	21,884	1530	20,525
Hexane	4667	20,526	4315	18,976
Hydrogen	325	61,084	275	51,628
Hydrogen sulfide	672	7479		
Methane	1011	23,811	910	21,433
Natural gas	950	19,500	850	17,500
Pentane	3981	20,908	3679	19,322
Propane	2572	21,564	2371	19,834
Propylene	2336	21,042	2185	19,683

[a]Listed alphabetically.
[b]1 Btu/ft^3 = 8.9 kcal/meter^3.

ideal gas property in a hypothetical state since all the water cannot condense to liquid because some of the water would saturate the carbon dioxide in the products. Thus,

$$Hv^{id} = \sum_i y_i Hv_i^{id}$$

where Hv^{id} is the gross heating value per unit volume of ideal gas, MJ/m^3, y_i is the mole fraction in gas phase for component i. Calculation of the ideal energy flow requires multiplication of the gross heating value by the ideal gas volumetric flow rate of gas for the time period. To employ a real gas flow to calculate the ideal energy flow requires converting the real gas flow rate to the ideal gas flow rate by dividing by the Z-factor. Thus Hv^{id}/Z is the ideal gross heating value per unit volume of real gas—the Z-factor must be determined for the natural gas mixture and then divided into the gross heating value of the mixture. Dividing each pure component gross heating value by the pure component Z-factor and then taking the molar average leads to incorrect answers.

The *net heating value* is the total energy transferred as heat in an ideal combustion reaction at a standard temperature and pressure in which all water formed appears as vapor. The net heating is an ideal gas property in a hypothetical state (the water cannot all remain vapor because, after the water saturates the CO_2 in the products, the rest would condense). It is a common misconception that the net heating value applies to industrial operations such as fired heaters and boilers. While the flue gases from these operations do not condense, the net heating value does not apply directly because the gases are not at 15°C (59°F). Where the gases to cool to 15°C (59°F), some of the water would condense while the remainder would saturate the gases. It is possible to use either the gross or net heating value in such situations taking care to utilize the hypothetical state properly.

Typically, a gas with a low calorific value is accompanied by a high inert gas content and a gas with a high calorific value is accompanied by a high content of higher molecular weight hydrocarbons (C_{2+}) content). With respect to the methane number, inert gases are favorable for the methane number level, whereas the higher molecular weight hydrocarbons (C_{2+}) reduce the methane number.

4.4.9 Helium content

Helium is an element that occurs in trace amounts in unprocessed natural gas. Most of the helium that occurs in natural gas is proposed to have been formed by the radioactive decay of uranium and thorium in granite-type rocks that are present in the continental crust of the Earth. As a very light gas, helium is buoyant and seeks to move upward as soon as it is formed. Very few natural gas fields contain enough to justify a helium recovery process. Generally, a natural gas source must contain at least 0.3% v/v helium to be considered as a potential helium source.

The richest helium accumulations are found where three conditions exist: (1) granitoid basement rocks are rich in uranium and thorium, (2) the basement rocks are fractured and faulted to provide escape paths for the helium, (3) porous sedimentary rocks above the basement faults are capped by an impermeable seal of halite or anhydrite. When all three of these conditions are met, helium might accumulate in the porous sedimentary rock layer.

For large-scale use, helium is extracted from the gas stream by fractional distillation (Chapter 7: Process classification). Since helium has a lower boiling point than any other element, low temperature and high pressure are used to liquefy nearly all the other gases, mostly nitrogen and methane (Chapter 7: Process classification and Chapter 8: Gas cleaning processes). The resulting crude helium gas is purified by successive exposures to lowering temperatures, in which almost all of the remaining nitrogen and other gases are precipitated out of the gaseous mixture. Activated charcoal is used as a final purification step, usually resulting in 99.995% pure helium. In a final production step, most of the helium that is produced is liquefied by a cryogenic process that allows more facile transportation of the liquid helium.

4.4.10 Higher molecular weight hydrocarbons

Some utilities may add propane/air mixtures to natural gas during peak demand periods. Propane has a low vapor pressure and if present in significant quantities, it will form a liquid phase at elevated pressures and low temperatures. Fuel variability due to vaporization of this liquid condensate at reduced storage cylinder pressure can lead to difficulty in controlling the air—fuel ratio. In addition, the significant presence of the heavier hydrocarbons in the gas mixture lowers its knock rating and can lead to potential engine damage.

There may also be substantial amounts of oil that can be added to the gas during compression, which can subsequently condense and interfere with the operation of compressed natural gas (CNG) engine components such as gas pressure regulators. On the other hand, a minimum level of carryover oil is required for durable gas injector operation. Various injector manufacturers recommend different minimum oil levels.

Oil in the gas at the compressor outlet is commonly removed by coalescing filters, however, they are insufficient in many cases, as up to 50% of the carryover oil exists in vapor form in the warm (or hot) compressor outlet gas. Additional measures will need to be considered, e.g., by additional cooling of the discharge gas or by using synthetic oil or mineral oil or a combination of mineral oil and a suitable adsorption filter downstream of the coalescing filter (Czachorski et al, 1995).

4.4.11 Measurement

In addition, before natural gas is sold to the consumer, it must be evaluated and measured so that the consumer receives the correct amount of gas (Rhoderick, 2003).

Natural gas can be measured in several different ways (Table 4.12), but it is more often measured by volume at normal temperatures and pressures and the volume is commonly expressed in *cubic feet* at a temperature of 15.5°C (60°F) and an atmospheric pressure of 14.7 pounds per square inch. Thus natural gas is measured (either at the time of production or at the time of delivery to the consumer) in thousands or millions of cubic feet (Mcf or Mscf and MMcf or MMscf, where *scf* is standard cubic feet under prescribed conditions); resources and reserves are calculated in trillions of cubic feet (Tcf). Natural gas is sold in cubic feet (a container 10 ft deep, 10 ft long, and 10 ft wide would hold 1000 ft^3 of natural gas) or in Btu, which is a measure of the heat content or burning properties of natural gas. A *Therm* is equivalent to 100,000 Btu's, or just over 97 ft^3, of natural gas. On the other hand, production and distribution companies commonly measure natural gas in thousands of cubic feet (Mcf), millions of cubic feet (MMcf), or trillions of cubic feet (Tcf).

Table 4.12 Measurement units often applied to natural gas

1 cubic foot (cf or scf standard cubic foot)	=	1027 Btu
100 cubic feet (scf)	=	1 Therm (approximate)
1000 cubic feet (Mcf)	=	1,027,000 Btu (1 MMBtu)
1000 cubic feet (Mcf)	=	1 dekatherm (10 therms)
1 million (1,000,000) cubic feet (MMcf)	=	1,027,000,000 Btu
1 billion (1,000,000,000) cubic feet (Bcf)	=	1.027 trillion Btu
1 trillion (1,000,000,000,000) cubic feet (Tcf)	=	1.027 quadrillion Btu

4.4.12 Mercury content

Mercury is present as a trace component in many natural gas reservoirs and, as an environmentally hazardous element, mercury has to be removed from the gas stream, preferably on the production site (Mussig and Rothman, 1997). The presence of mercury is not caused by human activities. Mercury in natural gas is not a man-made nuisance, it occurs naturally in certain gas formations and is inevitably produced with the gas. The mercury traces are supposed to originate from volcanic rocks, often underlying the gas reservoirs.

As an environmentally hazardous element, mercury has to be removed from the gas stream, preferably on the production site. After the concentration of the metal has been determined by application of standard methods (ASTM, 2017), the most appropriate removal technique has to be applied (Chapter 7: Process classification and Chapter 8: Gas cleaning processes) and all mercury-contaminated areas such as sludge and soil have to be cleaned and the waste has to be disposed of properly and according to the law.

During production and treatment of the gas mercury may be released to the environment and also parts of the plant be contaminated. Mercury may be emitted to air such as by way glycol process overheads, leading to soil contamination together with incidental spillage during maintenance around the treatment facilities. Materials, which have been in contact with mercury, such as sludge from dehydration units and activated carbon filters applied in gas treatment units are contaminated. Regardless of its concentration mercury is adsorbed on any metal surfaces, in scales and corrosion products, therefore, during maintenance, revamp and abandonment activities all these parts might be contaminated and have to be cleaned or treated prior scrapping and disposal or they have to be sent for disposal without affecting the environment.

4.4.13 Methane Number

The main parameter for rating the knock resistance of gaseous fuels is the Methane Number (MN), which is analogous to the Octane Number for gasoline and suitable test methods to determine the Methane Number of a gaseous fuel are being developed. (Malenshek and Olsen, 2009).

Different scales have been used to rate the knock resistance of CNG including the Motor Octane Number (MON) and the Methane Number. The differences in these ratings are the reference fuel blends used for comparison to the natural gas. Methane Number uses a reference fuel blend of methane, with a Methane Number of 100, and hydrogen, with a Methane Number of 0. Correlations have been generated between the reactive hydrogen/carbon ratio (H/C) and the MON and between MON and Methane Number.

$$MON = -406.14 + 508.04 * (H/C) - 173.55 * (H/C)2 + 20.17 * (H/C)3MN$$
$$= 1.624 * MON - 119.1$$

Thus if a gas mixture has a Methane Number of 70, its knock resistance is equivalent to that of a gas mixture of 70% methane and 30% hydrogen.

To ensure safe engine operation the methane number must always be at least equal to the methane number requirement (MNR) of the gas engine. The methane number required by the engine is affected by design and operating parameters, with the adjustment of the MNR being achieved by changing engine operation. Changes in ignition timing, air/fuel ratio, and output are effective measures to reduce the MNR.

4.4.14 Odorization

Odorization is a technique that is applied for safety reasons to the largest consumer product—natural gas. Since natural gas as delivered to pipelines has practically no odor, the addition of an odorant is required by most regulations in order that the presence of the gas can be detected readily in case of accidents and leaks. This odorization is provided by the addition of trace amounts of some organic sulfur compounds to the gas before it reaches the consumer. The standard requirement is that a user will be able to detect the presence of the gas by odor when the concentration reaches 1% of gas in air. Since the lower limit of flammability of natural gas is approximately 5%, this 1% requirement is essentially equivalent to one-fifth the lower limit of flammability. The combustion of these trace amounts of odorant does not create any serious problems of sulfur content or toxicity.

In any form, a minute amount of odorant that has an obvious smell is added to the otherwise colorless and odorless gas, so that leaks can be detected before a fire or explosion. Odorants are considered nontoxic in the extremely low concentrations occurring in natural gas delivered to the end user.

4.4.15 Phase behavior

A multicomponent mixture exhibits an envelope for liquid/vapor phase change in the pressure/temperature diagram, which contains a bubble point line and a dew point line, compared with only a phase change line for a pure component. For a pure substance a decrease in pressure causes a change of phase from liquid to gas at the vapor—pressure line; likewise, in the case of a multicomponent system, a decrease in pressure causes a change of phase from liquid to gas at temperatures below the critical temperature.

The pressure—volume relationships obtained can be plotted on a pressure—volume diagram with the bubble point and dew point locus also included (Fig. 4.5). The bubble point and dew point curves join at a point known as the *critical point*. The region under the bubble point—dew point envelope is the region where the vapor phase and liquid phase can coexist, and hence have an interface (the surface of a liquid drop or of a vapor bubble). The region above this envelope represents the region where the vapor phase and liquid phase do not coexist. The bubble point, dew point, and single-phase regions (Fig. 4.5) are often used to classify reservoirs. At temperatures greater than the cricondentherm, which is the maximum temperature for the formation of two phases, only one phase occurs at any pressure. For instance, if the

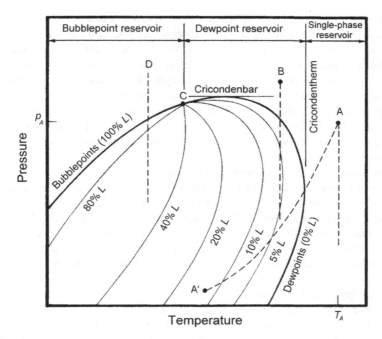

Figure 4.5 A pressure–temperature phase diagram for reservoir systems.

hydrocarbon mixture (Fig. 4.5) was to occur in a reservoir at temperature T_A and pressure p_A (point A), a decline in pressure at approximately constant temperature caused by removal of fluid from the reservoir would not cause the formation of a second phase.

Dry gas is predominantly methane and is in the gas phase under all conditions of pressure and temperature encountered during the production phases from reservoir conditions involving transport and process conditions. In particular, no hydrocarbon-based liquids are formed from the gas although liquid water can condense. Dry gas reservoirs have temperatures above the cricondentherm. During production the fluids are reduced in temperature and pressure. The temperature–pressure path followed during production does not penetrate the phase envelope, resulting in the production of gas at the surface with no associated liquid phase.

A wet gas exists in a pure gas phase in the reservoir but becomes a liquid/gas two-phase mixture in a flow line from the well tube to the separator at the topside platform. During the pressure drop in the flow line, liquid condensate appears in the wet gas. In a wet gas reservoir, the reservoir temperature is just above the cricondentherm. During production the fluids are reduced in temperature and pressure. The temperature–pressure path followed during production just penetrates the phase envelope, resulting in the production of gas at the surface with a small associated liquid phase.

Retrograde gas is the name of a fluid that is gas at reservoir pressure and temperature. However, as pressure and temperature decrease, large quantities of liquids are formed due to retrograde condensation. Retrograde gases are also called retrograde gas condensates, gas condensates, or condensates. In a condensate reservoir, the reservoir temperature is such that it falls between the temperature of the critical point and the cricondentherm. The production path then has a complex history. Initially, the fluids are in an indeterminate vapor phase, and the vapor expands as the pressure and temperature drop. This occurs until the dew point line is reached, whereupon increasing amounts of liquids are condensed from the vapor phase. If the pressures and temperatures reduce further, the condensed liquid may reevaporate, although sufficiently low pressures and temperatures may not be available for this to happen.

In summary, at any given constant low fluid pressure, reduction of fluid volume will involve the vapor condensing to a liquid via the two-phase region, where both liquid and vapor coexist. But at a given constant high fluid pressure (higher than the critical point), a reduction of fluid volume will involve the vapor phase turning into a liquid phase without any fluid interface being generated (i.e., the vapor becomes denser and denser until it can be considered as a light liquid). Thus the critical point can also be viewed as the point at which the properties of the liquid and the gas become indistinguishable (i.e., the gas is so dense that it looks like a low-density liquid and vice versa).

While the fluid in the reservoir remains a single phase, the produced gas splits into two phases as it cools and expands to surface temperature and pressure at point A'. Thus some condensate would be collected at the surface even though only one phase is present in the formation. The amount of condensate collected depends on the operating conditions of the separator. The lower the temperature at a given pressure, the larger the volume of condensate collected.

4.4.16   Residue

The residue in a gas stream is not to be confused with the term residue gas. The residue gas from the NGLs recovery section (Chapter 7: Process classification) is the final, purified sales gas (i.e., methane) which is pipelined to the end-user markets.

On the other hand, the standard test definition for residue in a gas stream is the concentration of contaminants boiling above 37.8°C (100°F) that may be present in the liquefied petroleum gas. The contaminants are usually compressor oils, lubricants from valves, plasticizers from hoses, corrosion inhibitors, and other crude oil products from pumps, pipelines, and storage vessels that are used in multiple service applications. Contaminants, whatever the source, can be particularly troublesome in liquid withdrawal systems such as those used in internal combustion engine fuel systems where the materials accumulate in the vaporizer and will ultimately plug the fuel system.

4.4.17   Sulfur content

Sulfur compounds in natural gas are in the form of mercaptans, hydrogen sulfide, and odorants. The first two are naturally occurring at source (gas fields) and have already been reduced by treatment at the gas processing plant.

The manufacturing processes for liquefied petroleum gas are designed so that the majority, if not all, of the sulfur compounds are removed. The total sulfur level is therefore considerably lower than for other crude oil-based fuels and a maximum limit for sulfur content helps to define the product more completely. The sulfur compounds that are mainly responsible for corrosion are hydrogen sulfide, carbonyl sulfide and, sometimes, elemental sulfur. Hydrogen sulfide and mercaptans have distinctive unpleasant odors. A control of the total sulfur content, hydrogen sulfide and mercaptans ensures that the product is not corrosive or nauseating. Stipulating a satisfactory copper strip test further ensures the control of the corrosion.

Corrosive sulfur compounds can be detected by their effect on copper and the form in which the general copper strip corrosion test (Speight, 2015, 2018; ASTM, 2017) for petroleum products is applied to liquefied petroleum gas. Hydrogen sulfide can be detected by its action on moist lead acetate paper and a procedure is also used as a measure of sulfur compounds.

4.4.18 Viscosity

Knowledge of the viscosity of the reservoir fluids is essential for a study of the dynamic or flow behavior of these fluids through pipes, porous media, or, more generally, wherever transport of momentum occurs in fluid motion. The unit of viscosity is g/cm s, or the poise. The kinematic viscosity is the ratio of the absolute viscosity to the density:

$$\frac{\mu}{\rho} = \frac{\text{centipoise}}{\frac{\text{g}}{\text{cm}^3}} = \text{centistokes}$$

The viscosity of gases near room temperature is in the centipoise (cP) range, so that is a commonly used unit. Gas viscosity is only marginally dependent on pressure near atmospheric pressure and is primarily a function of temperature and can be modeled in terms of temperature with the input of experimental reference measurements.

The viscosity of natural gas is usually several orders of magnitude lower than oil or water, which (fortunately for gas recovery operations) contributes to the higher mobility of the gas the reservoir relative to crude oil or water. The viscosity of gas mixtures at 1 atm and reservoir temperature can be determined from the gas mixture composition. Thus,

$$\mu_{\text{ga}} = \frac{\sum\limits_{i=1}^{N} y_i \mu_i \sqrt{M_{gi}}}{\sum\limits_{i=1}^{N} y_i \sqrt{M_{gi}}}$$

In this equation, μ_{ga} is the viscosity of the gas mixture at the desired temperature and atmospheric pressure, y_i is the mole fraction of the ith component, μ_i is the viscosity of the ith component of the gas mixture at the desired temperature and atmospheric pressure, M_{gi} is the molecular weight of the ith component of the gas mixture, and N is the number of components in the gas mixture.

4.4.19   Volatility and vapor pressure

The vaporization and combustion characteristics of a gas, especially liquefied petroleum gas, are defined for normal applications by volatility, vapor pressure, and to a lesser extent, specific gravity.

Volatility is expressed in terms of the temperature at which 95% of the sample is evaporated and presents a measure of the least volatile component present. Vapor pressure (also called the saturation pressure—the corresponding temperature is called saturation temperature) is, therefore, a measure of the most extreme low temperature conditions under which initial vaporization can take place. By setting limits to vapor pressure and volatility jointly the specification serves to ensure essentially single component products for the butane and propane grades. By combining vapor pressure/volatility limits with specific gravity for propane–butane mixtures, essentially two-component systems are ensured. The residue, i.e., nonvolatile matter, is a measure of the concentration of contaminants boiling above 37.8°C (100°F) that may be present in the gas.

In a closed container, the vapor pressure of a pure compound is the force exerted per unit area of walls by the vaporized portion of the liquid. Vapor pressure is also the pressure at which the vapor phase and the liquid phase of a pure chemical are in equilibrium with each other. In an open air under atmospheric pressure, a liquid at any temperature below its boiling point has its own vapor pressure that is less than 1 atm. When the vapor pressure of a compound reaches 1 atm (14.7 psi) the saturation temperature becomes the normal boiling point. Vapor pressure increases with temperature and the highest value of vapor pressure for a substance is its critical pressure in which the corresponding temperature is the critical temperature.

Vapor pressure is an important thermodynamic property of any chemical and it is a measure of the volatility of a fluid (Table 4.13). Compounds with a higher tendency to vaporize have higher vapor pressures. More volatile compounds are those that have lower boiling points and are called light compounds. For example, propane (C_3) has boiling point less than that of n-butane (nC_4) and as a result it is more volatile. At a fixed temperature, vapor pressure of propane is higher than that of butane. In this case, propane is called the light compound (more volatile) and butane the heavy compound. Generally, more volatile compounds have higher critical pressure and lower critical temperature, and lower density and lower boiling point than those of less volatile (heavier) compounds, although this is not true for the case of some isomeric compounds.

Vapor pressure is a useful parameter in calculations related to hydrocarbon losses and flammability of hydrocarbon vapor in the air. More volatile compounds are more ignitable than heavier compounds. For example, n-butane is added to gasoline to improve its ignition characteristics. Vapor pressure is, therefore, a measure of the most extreme low temperature conditions under which initial vaporization can take place. By setting limits to vapor pressure and volatility jointly the

Table 4.13 Vapor pressure of the hydrocarbon derivatives of various gas streams

Hydrocarbon	Formula	Molecular weight	Vapor pressure[a]
Methane	CH_4	16.043	-5000
Ethane	C_2H_6	30.07	-800
Propane	C_3H_8	44.097	188
n-Butane	C_4H_{10}	58.124	51.54
iso-Butane	C_4H_{10}	58.124	72.39
n-Pentane	C_5H_{12}	72.151	15.575
iso-Pentane	C_5H_{12}	72.151	20.4444
neo-Pentane	C_5H_{12}	72.151	36.66
n-Hexane	C_6H_{14}	86.178	4.96
2-Methylpentane	C_6H_{14}	86.178	6.767
3-Methylpentane	C_6H_{14}	86.178	6.103
neo-Hexane	C_6H_{14}	86.178	9.859
2,3-Dimethylbutane	C_6H_{14}	86.178	7.406
n-Heptane	C_7H_{16}	100.205	1.62
2-Methylhexane	C_7H_{16}	100.205	2.2719
3-Methylhexane	C_7H_{16}	100.205	2.131
3-Ethylpentane	C_7H_{16}	100.205	2.013
2,2-Dimethylpentane	C_7H_{16}	100.205	3.494
2.4-Dimethylpentane	C_7H_{16}	100.205	3.293
3,3-Dimethylpentane	C_7H_{16}	100.205	2.774
n-Octane	C_8H_{18}	114.232	0.537
iso-Octane	C_8H_{18}	114.232	1.709
n-Nonane	C_9H_{20}	128.259	0.1796
n-Decane	$C_{10}H_{22}$	142.286	0.0609
Cyclopentane	C_5H_{10}	70.135	9.914
Methylcyclopentane	C_6H_{12}	84.162	4.503
Cyclohexane	C_6H_{12}	84.162	3.266
Methylcyclohexane	C_7H_{14}	98.189	1.6093
Ethylene	C_2H_4	28.054	
Propylene	C_3H_6	42.081	227.6
1-Butene	C_4H_8	56.108	62.1
cis-2-Butene	C_4H_8	56.108	45.95
trans-2-Butene	C_4H_8	56.108	49.94
iso-Butene	C_4H_8	56.108	63.64
1-Pentene	C_5H_{10}	70.135	19.117
1,3-Butadiene	C_4H_6	54.092	59.4
Benzene	C_6H_6	78.114	3.225
Toluene	$C_6H_5CH_3$	92.141	1.033
Ethyl benzene	C_8H_{10}	106.168	0.376
o-Xylene	C_8H_{10}	106.168	0.263
m-Xylene	C_8H_{10}	106.168	0.325
p-Xylene	C_8H_{10}	106.168	0.3424

[a]psia @100°F.

specification serves to ensure essentially single component products for the butane and propane grades. By combining vapor pressure/volatility limits with specific gravity for propane−butane mixtures, essentially two-component systems are ensured. The residue, i.e., nonvolatile matter, is a measure of the concentration of contaminants boiling above 37.8°C (100°F) that may be present in the gas.

Low-vapor-pressure compounds reduce evaporation losses and chance of vapor lock. Therefore for a fuel there should be a compromise between low and high vapor pressure. However, one of the major applications of vapor pressure is in calculation of equilibrium ratios for phase equilibrium calculations. For pure hydrocarbons, values of vapor pressure are at the reference temperature of 100°F (38°C). For natural gas, the method of Reid is used to measure vapor pressure at 100°F. *Reid vapor pressure* is approximately equivalent to vapor pressure at 100°F (38°C).

Volatility is expressed in terms of the 95% evaporated temperature and is a measure of the amount of least volatile fuel components present in the product. This specification controls the heavy ends in the fuel and is, in effect, a restriction on higher boiling fractions that will not vaporize for use at system temperatures. When coupled with the vapor pressure single component products for propane and butane and two component products for butane−propane mixtures are assured.

Vapor pressure is an important specification property of commercial propane, special duty propane, propane/butane mixtures, and commercial butane that assures adequate vaporization, safety, and compatibility with commercial appliances. Relative density, while not a typical criterion contained in specifications, is necessary for determination of filling densities and custody transfer. The MON is useful in determining the products' suitability as a fuel for internal combustion engines. Precision and accuracy of compositional data are extremely important when these data are used to calculate various properties of these petroleum products.

Simple evaporation tests in conjunction with vapor pressure measurement give a further guide to composition. In these tests a liquefied petroleum gas sample is allowed to evaporate naturally from an open graduated vessel. Results are recorded on the basis of volume/temperature changes, such as the temperature recorded when 95% v/v has evaporated, or volume left at a particular temperature. The evaporation characteristics are a measure of the relative purity of the various types of liquefied petroleum gases and are a benefit in the assurance of volatility performance. The test results can be used to indicate the presence of butane and higher molecular weight constituents in propane-type liquefied petroleum gas, and pentane and higher molecular weight constituents in propane/butane-type fuel gases as well as in butane-type fuel gases. The presence of hydrocarbon compounds less volatile than those of which the liquefied petroleum gas is primarily composed is indicated by an increase in the 95% v/v evaporated temperature. When the type and concentration of higher boiling components is required, chromatographic analysis should be used.

4.4.20   Water content

The use of natural gas with high water content can result in the formation of liquid water, ice particles or hydrates at low operating temperatures and high pressure which will interfere with consistently smooth flow of fuel into the engine and cause problems such as poor drivability or even engine stoppage.

When natural gas contains water, but the component analysis is on a dry basis, the component analysis must be adjusted to reflect the presence of water. The mole fraction of water in the mixture is estimated from the definition of relative humidity (on a one mole basis):

$$y_w = h^g P_W^\sigma / P = n_w/(1 + n_w)$$

where n is the number of moles, n_w is the number of water moles, y_w is the mole fraction of water, P_w is the partial pressure of water in gas phase, kPa, P is the total pressure, kPa, and h_g is the relative humidity.

Thus it is a fundamental requirement that liquefied petroleum gas should not contain free water. Dissolved water may give trouble by forming hydrates and giving moisture vapor in the gas phase. Both of these will lead to blockages. Therefore test methods are available to determine the presence of water using electronic moisture analyzers, dew point temperature, and length-of-stain detector tubes (ASTM, 2017; Speight, 2018).

4.4.21 Wobbe Index

The Wobbe Index gives a measure of the heat input to an appliance through a given aperture at a given gas pressure. Using this as a vertical coordinate and the flame speed factor as the horizontal coordinate a combustion diagram can be constructed for an appliance, or a whole range of appliances, with the aid of appropriate test gases. The concept behind defining the Wobbe Index is to have a measure for the interchangeability of gases, i.e., gases with the same Wobbe Index (under defined pressure conditions) generate the same output during combustion. However, a distinction is made between what is referred to as the higher Wobbe Index and the lower Wobbe Index, depending on whether the superior or the inferior calorific value is used in the formula.

The Wobbe Index (or Wobbe Number = calorific value/(specific gravity)) and the flame speed are usually expressed as a factor or an arbitrary scale on which that of hydrogen is 100. This factor can be calculated from the gas analysis. In fact, calorific value and specific gravity can be calculated from compositional analysis.

The Wobbe Number gives a measure of the heat input to an appliance through a given aperture at a given gas pressure. Using this as a vertical coordinate and the flame speed factor as the horizontal coordinate a combustion diagram can be constructed for an appliance, or a whole range of appliances, with the aid of appropriate test gases. This diagram shows the area within which variations in the Wobbe index of gases may occur for the given range of appliances without resulting in either incomplete combustion, flame lift, or the lighting back of preaerated flames. This method of prediction of combustion characteristics is not sufficiently accurate to eliminate entirely the need for the practical testing of new gases. Another important combustion criterion is the Gas Modulus, $M = P/W$ where P is the gas pressure and W the Wobbe Number of the gas. This must remain constant if a given degree of aeration is to be maintained in a preaerated burner using air at atmospheric pressure.

The natural gas after appropriate treatment for acid gas reduction, odorization, and hydrocarbon and moisture dew point adjustment, would then be sold within prescribed limits of pressure, calorific value, and possibly Wobbe Index (also called the Wobbe index) which equals the calorific value divided by the specific gravity, and the flame speed. Thus,

$$WI = CV/\sqrt{SG}$$

WI is the Wobbe index or Wobbe Index that is typically expressed in Btu/scf, CV is the calorific value (HHV), and SG is the specific gravity.

Gas turbines can operate with a large range of fuels, but the fuel variation that a particular installation can cope with is limited. The Modified Wobbe index (MWI) is used particularly by gas turbine manufacturers because it takes into account the temperature of the fuel. The MWI is the ratio of the LHV to the square root of the product of the specific gravity and the absolute gas temperature:

$$MWI = LHV/\sqrt{SG}_{gas} \times T_{gas}$$

This is equivalent to:

$$MWI = LHV/\sqrt{(MW_{gas}/28.96) \times T_{gas}}$$

where LHV is the lower heating value of the fuel gas (Btu/scf), SG_{gas} is the specific gravity of the fuel gas relative to air, MW_{gas} is the molecular weight of the fuel gas, T_{gas} is the absolute temperature of the fuel gas (in degrees Rankine), and 28.96 is the molecular weight of dry air.

Any change in the heating value of the gas will require a corresponding change in the fuel's flow rate to the machine, incorporation of temperature effects is important in calculating energy flows in turbines where a large input temperature variation is possible. The allowable MWI range is established to ensure that required fuel nozzle pressure ratios are maintained during all combustion/turbine modes of operation. For older, diffusion-type combustors, the gas turbine control system can typically accommodate variations in the MWI as large as $\pm 15\%$. But for newer, dry low NO_x combustors, variations in the MWI of only $\pm 3\%$ could cause problems. Fuel instability can be caused by velocity changes through a precisely sized fuel-nozzle orifice which can cause flame instability, resulting in pressure pulsations and/or combustion dynamics which can, in the worst case, destroy the combustion system.

Also, since the Wobbe Index is an indicator of the interchangeability of fuel gases, it (alone or with other analyses) can be used to control blending of fuel gases. Since the Wobbe Index and the Btu value of fuel gases make similar curves, either could be used to control blending of fuel gases, thereby, controlling the amount of nitrogen in the blended fuel (Segers et al., 2011). Another important combustion criterion is the gas modulus, $M = P/W$ where P is the gas pressure and W the Wobbe Index of the gas. This must remain constant if a given degree of aeration is to be maintained in a preaerated burner using air at atmospheric pressure.

Finally, despite the common acceptance of Wobbe Index as the main interchangeability parameter, a variety of units and reference temperatures are used across the world.

References

API, June 10, 2009. Refinery Gases Category Analysis and Hazard Characterization. Submitted to the EPA by the American Petroleum Institute, Petroleum HPV Testing Group. HPV Consortium Registration # 1100997 United States Environmental Protection Agency, Washington, DC.

ASTM, 2017. Annual Book of Standards. ASTM International, West Conshohocken, PA.

Bullin, J.A., Fitz, C., Dustman, T., 2010. Practical Hydrocarbon Dew Point Specification for Natural Gas Transmission Lines. <https://www.bre.com/PDF/Practical-Hydrocarbon-Dew-Point-Specification-for-Natural-Gas-Transmission-Lines.pdf>.

Cranmore, R.E., Stanton, E., 2000. In: Dawe, R.A. (Ed.), Modern Petroleum Technology. Volume 1: Upstream. John Wiley & Sons Inc, New York (Chapter 9).

Crawford, D.B., Durr, C.A., Finneran, J.A., Turner, W., 1993. Chemicals from natural gas. In: McKetta, J.J. (Ed.), Chemical Processing Handbook. Marcel Dekker Inc, New York, p. 2.

Czachorski, M., Blazek, C., Chao, S., Kriha, K., Koncar, G., 1995. NGV Fueling Station Compressor Oil Carryover Measurement and Control, Report No. GRI-95/0483, 1995. Gas Research Institute, Chicago, IL.

Gary, J.G., Handwerk, G.E., Kaiser, M.J., 2007. Petroleum Refining: Technology and Economics, fifth ed. CRC Press, Taylor & Francis Group, Boca Raton, FL.

Hsu, C.S., Robinson, P.R. (Eds.), 2017. Handbook of Petroleum Technology. Springer International Publishing AG, Cham.

Klimstra, J., 1978. Interchangeability of Gaseous Fuels — The Importance of the Wobbe-Index. Report No. SAE 861578. Society of Automotive Engineers, SAE International, Warrendale, PA.

Liss, W.E., Thrasher, W.R., 1992. Variability of Natural Gas Composition in Select Major Metropolitan Areas of the U.S. Report No. GRI-92/0123. Gas Research Institute, Chicago, IL.

Malenshek, M., Olsen, D.B., 2009. Methane number testing of alternative gaseous Fuels. Fuel 88, 650—656.

Mokhatab, S., Poe, W.A., Speight, J.G., 2006. Handbook of Natural Gas Transmission and Processing. Elsevier, Amsterdam.

Mussig, S., Rothman, B., 1997. Mercury in natural gas — problems and technical solutions for its removal. Paper No. SPE 38088. In: Proceedings. SPE Asia Pacific Oil and Gas Conference and Exhibition, 14-16 April, Kuala Lumpur, Malaysia. April 14-16. Society of Petroleum Engineers, Richardson, TX.

Rhoderick, G.C., 2003. Analysis of natural gas: the necessity of multiple standards for cali-
    bration. J. Chromatogr. A 1017 (1−2), 131−139.
Segers, M., Sanchez, R., Cannon, P., Binkowski, R., Hailey, D., 2011. Blending fuel gas to
    optimize use of off-spec natural gas. In: Proceedings. Presented at ISA Power Industry
    Division 54th Annual I&C Symposium, Concord, North Carolina.
Speight, J.G. (Ed.), 2011. The Biofuels Handbook. The Royal Society of Chemistry, London.
Speight, J.G., 2014. The Chemistry and Technology of Petroleum, fifth ed. CRC Press,
    Taylor & Francis Group, Boca Raton, FL.
Speight, J.G., 2015. Handbook of Petroleum Product Analysis, second ed. John Wiley &
    Sons Inc, New York.
Speight, J.G., 2017. Handbook of Petroleum Refining. Petroleum Refining Processes. CRC
    Press, Taylor & Francis Group, Boca Raton, FL.
Speight, J.G., 2018. Handbook of Natural Gas Analysis. John Wiley & Sons Inc, New York.

Further reading

API, 2017. Refinery Gases Category Analysis and Hazard Characterization. Submitted to the
    EPA by the American Petroleum Institute, Petroleum HPV Testing Group. HPV
    Consortium Registration # 1100997 United States Environmental Protection Agency,
    Washington, DC. June 10.
Drews, A.W., 1998. In: Drews, A.W. (Ed.), Manual on Hydrocarbon Analysis, sixth ed.
    American Society for Testing and Materials, West Conshohocken, PA, Introduction.

Recovery, storage, and transportation

5

5.1 Introduction

Natural gas exists in nature under pressure in rock reservoirs in the crust of the Earth, either in conjunction with and dissolved in crude oil (*associated gas*) or without crude oil (*nonassociated gas*). Reservoirs containing natural gas vary in size, which can be anything from a few hundred yards to miles across in plan, and tens to hundreds of yards thick, with the gas trapped against an impermeable layer similar to crude oil traps (Speight, 2014a). Natural gas is a member of the class of fuels known as gaseous fuels (Chapter 1: History and use) as defined in various standard test methods (ASTM, 2017).

Typically, the gas produced from the reservoir in a manner similar to crude oil (when it occurs in conjunction with crude oil). However, extracting natural gas from deposits deep underground is dependent upon several factors in the underground environment including the (1) pressure of the gas in the reservoir, (2) the composition of the gas, (3) the porosity of the reservoir rock, and (4) the permeability of the reservoir rock. Primary recovery relies on underground pressure to drive the gas though the production well to the surface. When the pressure decreases due to production of the gas, the remaining gas is brought to the surface using artificial lift technologies, such as the horsehead pump at the surface or a downhole pump at the bottom of the production well.

The use of natural has in the past six decades has grown steadily in use—even replacing coal gas in many markets (Speight, 2013)—and, currently, is used extensively in residential as well as in commercial and industrial applications. While natural gas supplies approximately one quarter of the energy used worldwide, and makes up nearly a quarter of electricity generation, as well as playing a crucial role as a feedstock for industry it is, in fact, a predominant source of energy in North America. It is the dominant energy used for domestic heating with slightly more than 50% of the homes in North America heated by natural gas. The use of natural gas is also rapidly increasing in electric power generation with natural gas power-generating facilities replacing coal power-generating facilities and crude oil power-generating facilities. This trend is expected to continue well into the foreseeable future. In addition, whether produced via conventional methods or renewable methods, this clean-burning alternative fuel is also being used as fuel for various vehicles but for such use, the gas must be compressed or liquefied.

Environmentally, natural gas (which is predominantly methane even in the raw unrefined form) (Table 5.1) (Burrus and Ryder, 2003, 2014) is the cleanest burning

Natural Gas. DOI: https://doi.org/10.1016/B978-0-12-809570-6.00005-9

Table 5.1 Constituents of natural gas

Constituent	Formula	% v/v
Methane	CH_4	>85
Ethane	C_2H_6	3−8
Propane	C_3H_8	1−5
n-Butane	C_4H_{10}	1−2
iso-Butane	$C4H_{10}$	<0.3
n-Pentane	C_5H_{12}	1−5
iso-Pentane	C_5H_{12}	<0.4
Hexane, heptane, octane[a]	C_nH_{2n+2}	<2
Carbon dioxide	CO_2	1−2
Hydrogen sulfide	H_2S	1−2
Oxygen	O_2	<0.1
Nitrogen	N_2	1−5
Helium	He	<0.5

[a]Hexane (C_6H_{14}) and higher molecular weight hydrocarbon derivatives up to
octane as well as benzene (C_6H_6) and toluene ($C_6H_5CH_3$).

of the fossil fuels and fossil fuel products and produces (by combustion) primarily carbon dioxide, water vapor, and small amounts of nitrogen oxides.

$$CH_4 + O_2 \rightarrow CO_2 + H_2O$$

Methane is 75% w/w carbon and is not the lowest carbon fuel (Speight, 2013, 2014a) but, in the absence of any contaminants (after gas processing) is the cleanest burning fossil fuel. Thus natural gas is considered to be a versatile fuel and its growth is linked in part to its environmental benefits relative to other fossil fuels, particularly for air quality as well as greenhouse gas emissions. However, the potential to produce as much carbon dioxide as other fossil fuels and the related products is real but whether natural gas has lower life cycle greenhouse gas emissions than coal and oil depends on the assumed leakage rate, the global warming potential of methane over different time frames, the energy conversion efficiency, and other factors.

As a brief introduction, natural gas that is scheduled to be transported and stored must meet specific quality measures so that the pipeline network (or grid) can provide uniform quality gas. Wellhead natural gas will contain other hydrocarbons, inert gases, and contaminants which must be removed before the natural gas can be safely delivered to the high-pressure, long-distance pipelines that transport natural gas to consumers. Natural gas processing can be complex and usually involves several processes, or stages, to remove oil, water, hydrocarbon gas liquids, and other impurities such as sulfur, helium, nitrogen, hydrogen sulfide, and carbon dioxide. The composition of the wellhead natural gas determines the number of stages and the process required to produce pipeline-quality dry natural gas. These processes or stages may be integrated into one unit or operation, be performed in a different order or at alternative locations (lease/plant), or not be required at all.

The stages of natural gas processing/treatment are: (1) gas-oil-water separators, in which pressure relief causes a natural separation of the liquids from the gases in the natural gas; (2) condensate separator, in which gas condensate and/or natural gasoline is removed from the natural gas stream at the wellhead with separators much like gas-oil-water separators; (3) dehydrator, in which water is removed to reduce the potential for corrosion in the pipeline as well as the formation of undesirable hydrates and water condensation in pipelines; (4) contaminant removal units, in which nonhydrocarbon gases such as hydrogen sulfide, carbon dioxide, water vapor, helium, nitrogen, and oxygen are removed from the natural gas stream; (5) nitrogen rejection unit, in which nitrogen is removed and the gas stream is further dehydrated using molecular sieve beds; (6) methane separator, in which methane is removed from the gas stream can occur as a separate operation by cryogenic processing and absorption methods; and (7) fractionator, in which the gas stream is separated into component hydrocarbons (ethane, propane, and butane) using the varying boiling points of the individual hydrocarbons. Each of these unit operations is described in more detail elsewhere in this text (Chapter 4: Composition and properties). Also, there is the need to determine the purity of the gas streams as well as the efficiency of each process unit by means of gas stream analysis, usually by means of online or off-line monitoring of the product streams by means of, e.g., a technique such as gas chromatography (Speight, 2018).

5.2 Recovery

Natural gas is recovered (extracted) from the reservoir through a well (a drilled structure that is not a natural occurrence). However, extraction of natural gas through a well leads to decrease in pressure in the reservoir which, in turn, will lead to decreased rates of gas production from the reservoir.

Once a potential natural gas reservoir has been located, the decision of whether or not to drill a well depends on a variety of factors, not the least of which is the economic characteristics of the gas reservoir. After the decision to drill has been made, the exact placement of the drill site depends on a variety of factors, including the nature of the potential formation to be drilled, the characteristics of the subsurface geology, and the depth and size of the target deposit. During this time, it is also necessary for the drilling company to ensure that they complete all the necessary steps to ensure that they can legally drill in that area. This usually involves securing permits for the drilling operations, establishment of a legal arrangement to allow the natural gas company to extract and sell the resources under a given area of land, and a design for gathering lines that will connect the well to the pipeline.

If the new well does in fact come in contact with a natural gas reservoir, the well is developed to allow for the extraction of the natural gas and is termed a development well or productive well. At this point, with the well drilled and hydrocarbons present, the well may be completed to facilitate its production of natural gas. However, if the exploration team was incorrect in its estimation of the existence of marketable quantity of natural gas at a well site (and this does happen), the well is termed a dry well, and work on the well is terminated.

Gas wells are technically similar to crude oil wells with casing, tubing, and a wellhead and controls at the top (Rojey et al, 1997; Arnold and Stewart, 1999; Speight, 2007). Conventional wells use casings *telescoped* inside each other that is given pressure resistance by cement. However, with *expanded tubulars* each tubular casing is expanded against the previous one by pumping a tool called a *mandrel* down the casing. Thus the well can either be thinner and cheaper for a given size of final tubing, or alternatively, using the conventional 20-in. diameter external casing, there is space to install wider and higher capacity tubing.

When gas is contained in deep low-permeability (*tight*) reservoirs, it is often economic to drill large-diameter wells with horizontal sections through the reservoirs to collect more gas. In addition, gas production can be increased by fracturing the reservoir rock using over-pressure and by pumping in ceramic beads that are maintaining the integrity of the fractures. It is also attractive to install *smart wells* that have measurement gauges permanently installed downhole. This eliminates the need for lowering gauges down the wells during production. With the high gas flow rates, sand can often be pulled into the well with the gas and therefore wearing parts are hardened against sand erosion.

If the natural gas is to be recovered from a crude oil reservoir (as is often the case), the production methods differ somewhat because of the presence of crude oil hydrocarbons in the natural gas. Thus once a crude oil reservoir is discovered and assessed, production engineers begin the task of maximizing the amount of oil or gas that can ultimately be recovered from it. However, before a well can produce crude oil or gas the borehole must also be stabilized with casing that is cemented in place. The casing also serves to protect any fresh water intervals that the well passes through, so that oil cannot contaminate the water. A small-diameter tubing string is centered in the wellbore and held in place with packers. This tubing will carry the crude oil and natural from the reservoir to the surface.

Reservoirs are typically at elevated pressure because of underground forces. To equalize the pressure and avoid the wasteful *gushers* of the early 1900s (and the Hollywood movies), a series of valves and equipment is installed on top of the well. This wellhead (also called a *Christmas tree*) regulates the flow of hydrocarbons out of the well.

Early in its production life, the underground pressure (often referred to as the *reservoir energy*) will push the crude oil and natural gas up the wellbore to the surface and depending on reservoir conditions, this *natural flow* may continue for many years. When the pressure differential is insufficient for the crude oil and natural gas to flow naturally, mechanical pumps must be used to bring the products to the surface (*artificial lift*).

Most wells produce crude oil and natural gas in a predictable pattern (*decline curve*) in production increases for a short period, then peaks, and follow a long, slow decline. The shape of this decline curve, how high the production peaks, and the length of the decline are all driven by reservoir conditions. There are two steps that can be taken to influence the decline curve of a well: (1) perform a periodic *work over*, which cleans out the wellbore to help the crude oil or natural gas move more easily to the surface or (2) fracture or treat the reservoir rock with acid around the bottom of the wellbore to create better pathways for the crude oil and natural gas to move through the subsurface to the producing well.

Throughout their productive life, most crude oil wells produce crude oil, natural gas, and water. This mixture is separated at the surface. However, although natural gas wells usually do not produce crude oil, they do produce varying amounts of liquid hydrocarbons (*natural gas liquids*, NGLs) that are removed in the field or at a gas processing plant (which may remove other impurities as well). NGLs often have significant value as petrochemical feedstocks. Natural gas wells also often produce water, but the volumes are much lower than is typical for oil wells. Once it is produced, onshore natural gas is usually transported through a pipeline, although alternate methods may be more appropriate depending on the location of the well and the destination of the gas.

Just as many crude oil reservoirs require enhancement to continue oil production, gas reservoirs also require application of enhanced production methods to maximize the recovery of the gas. Thus extracting natural gas from deposits deep underground is not merely a matter of drilling and completing a well. Any number of factors in the underground environment—such as the porosity of the reservoir rock—can impede the free flow of product into the well (Sim et al., 2008; Godec et al., 2014). In the past, it was common to recover as little a minority of the gas in the reservoir.

Enhanced gas recovery (EGR) is used to describe the recovery of unconventional, deep, or otherwise difficult-to-recover natural gas (Guo et al., 2014). Unconventional natural gas is the gas contained in geologic formations which have not been exploited traditionally by the oil and gas industry and include shale formation, tight gas sands, formations with active aquifers, and coal bed methane (Chapter 3: Unconventional gas). A number of these geologic formations are known to contain sizable quantities of natural gas. However, EGR is a recent consideration because of the large storage capacity of depleted gas pools. Considering that any additional gas recovered as a result of carbon dioxide injection might offset the cost of the storage of carbon dioxide, a number of experimental and simulation studies have been undertaken to examine displacement efficiency of reservoir gas by gas injection (Mamora and Sea, 2002; Seo and Mamora, 2003; Pooladi-Darvish et al., 2008).

In order to enhance the recovery of natural gas from a gas reservoir, nitrogen is injected into the reservoirs to artificially increase pressure which results in more yield from the well (often referred to as EGR). This process can also be accomplished with other inert gases such as carbon dioxide (CO_2) which sourced from underground wells is an economical alternative.

However, nitrogen can be recovered virtually anywhere from atmospheric air— via an air separation plant. Physically, there are also differences between the two gases: (1) nitrogen requires less compression than carbon dioxide and (2) greater amounts of carbon dioxide are needed to create high pressure in the reservoirs. Thus the EGR process using compressed nitrogen is more energy-efficient than when using carbon dioxide. Another option to maintain the pressure in the reservoir involves injection of part of the recovered natural gas.

The principles of EGR can also be applied to the recovery of coalbed methane which, in the context of this book, is an unconventional source of methane gas (Chapter 3: Unconventional gas). Carbon dioxide has a greater adsorption affinity onto coal than methane and if carbon dioxide is pumped into a coal seam toward the end of a coalbed methane production project, it displaces any remaining methane at the adsorption sites (allowing methane recovery jointly with carbon dioxide storage). However, the low permeability of coal seams means that a large number of wells may be needed to inject sufficient amounts of carbon dioxide to make methane recovery economical. Moreover, the methane in coal represents only a small proportion of the energy value of the coal, and the remaining coal could not be mined or gasified underground without releasing the carbon dioxide to the atmosphere. Also, methane is a far more potent greenhouse gas than carbon dioxide and the necessary precautions would have to be in place to ensure no methane leakage to the atmosphere occurred.

5.3 Storage

As simple as it may seem to many observers, the storage of natural gas is not only a matter of putting the gas into a container without the potential for any adverse consequences. However, there are several analytical measures that must be taken to ensure that the gas is not only stored safely but also recoverable safely and in a usable form.

Typically, in the United States interstate pipeline companies, intrastate pipeline companies, independent storage providers, and local distribution companies are the primary owners and operators of underground natural gas storage facilities. Storage facility owners and/or operators may or may not own the stored natural gas, which can be held under lease with natural gas shipment companies, local distribution centers, or end-users that also own the gas. The entity that owns or operates an underground facility will generally determine how the facility's storage capacity will be used. The individual states primarily regulate natural gas storage facilities involved in intrastate commerce while the Federal Energy Regulatory Commission primarily regulates storage facilities involved with interstate commerce. Thus there are relevant legal issues that must be addressed (Burt, 2016).

Furthermore, because of the geographic and geologic diversity in gas resources and storage operations in the United States, there is no single approach that can be taken without the application of analytical test methods and consideration of the data relating to the (1) potential storage facility, such as a depleted reservoir; (2) the composition and properties of the gas to be stored; and (3) whether the gas is adequate for each and every storage container; in this case, the word container includes natural container such as underground natural storage facilities as described below.

Traditionally, natural gas has been a seasonal fuel and demand is usually higher during the winter, partly because it is used for heat in residential and commercial settings. Therefore the natural gas that reaches its destination is not always needed right away and, fortunately, natural gas can be stored for an indefinite period of time. Storage used to serve only as a buffer between transportation and distribution, to ensure adequate supplies of natural gas were in place for seasonal demand shifts, and unexpected demand surges. Now, in addition to serving those purposes, natural gas storage is also used by industry participants for commercial reasons; storing gas when prices are low, and withdrawing and selling it when prices are high, for instance. The purpose and use of storage has been closely linked to the regulatory environment of the time.

Gas storage facilities largely contribute to the reliability of gas supplies to consumers and enable the gas providers to level off daily gas consumption fluctuations and meet the peak demand in winter. Natural gas in storage also serves as insurance against any unforeseen accidents, natural disasters, or other occurrences that may affect the production or delivery of natural gas.

Storage for base load requirements (base load storage) is gas that is used to meet seasonal demand increases and the facilities are capable of holding enough natural gas to satisfy long-term seasonal demand requirements. Typically, the turn-over rate for natural gas in these facilities is a year; natural gas is generally injected during the summer (nonheating season) and withdrawn during the winter (heating season). These reservoirs are larger, but their delivery rates are relatively low, meaning the natural gas that can be extracted each day is limited. Instead, these facilities provide a prolonged, steady supply of natural gas. Depleted gas reservoirs are the most common type of base load storage facility.

On the other hand, storage for peak load requirements (peak load storage) is designed to have high deliverability for short periods of time during which the natural gas can be withdrawn from storage quickly as the need arises. Peak load facilities that are intended to meet sudden, short-term demand increases cannot hold as much natural gas as base load facilities. However, peak load facilities can deliver smaller amounts of gas more quickly and can also be replenished in a shorter amount of time than base load facilities. While base load facilities have long-term injection and withdrawal seasons, turning over the natural gas in the facility about once per year, peak load facilities can have turnover rates as short as a few days or weeks.

Stored natural gas plays a vital role in ensuring that any excess supply delivered during the summer months is available to meet the increased demand for gas during the winter months. In addition, the recent trend toward electricity generation using natural gas as the fuel had caused demand to increase during the summer months due to the need for electricity to power air conditioners and the like. Natural gas in storage also serves as insurance against any unforeseen accidents, natural disasters, or other occurrences that may affect the production or delivery of natural gas. In addition, natural gas storage plays a vital role in maintaining the reliability of supply needed to meet the demands of consumers. The increased use of natural gas and the need to provide a plentiful supply of this fuel is causing an expansion of the need for natural gas storage facilities so that this domestically produced gaseous fuel will be readily available through the utility infrastructure. Storage is the main link between the three segments of the natural gas industry that are involved in delivering natural gas from the wellhead to the consumer: (1) the production company explores, drills and extracts the natural gas from the ground; (2) the transportation company operates the pipelines that link the gas fields to major consuming areas; and (3) the distribution company—in the form of local utility companies that deliver natural gas to the customer.

Underground natural gas storage fields grew in popularity shortly after World War II. At that time, the natural gas industry noted that seasonal demand increases could not feasibly be met by pipeline delivery alone. In order to meet seasonal demand increases, the deliverability of pipelines (and thus their size) had to increase dramatically but the technology required for constructing such large pipelines to consuming regions was, at the time, unattainable and unfeasible. At that time, underground storage fields were the only option.

Any underground storage facility is reconditioned before injection, to create a sort of storage vessel underground. Natural gas is injected into the formation, building up pressure as more natural gas is added. In this sense, the underground formation becomes a sort of pressurized natural gas container. As with newly drilled wells, the higher the pressure in the storage facility, the more readily gas may be extracted. Once the pressure drops to below that of the pressure at the wellhead, there is no pressure differential left to push the natural gas out of the storage facility. This means that, in any underground storage facility, there is a certain amount of gas that may never be extracted. This is known as physically unrecoverable gas; it is permanently embedded in the formation.

In addition to this physically unrecoverable gas, underground storage facilities contain what is known as base gas or cushion gas. This is the volume of gas that must remain in the storage facility to provide the required pressurization to extract the remaining gas. In the normal operation of the storage facility, this cushion gas remains underground. However, a portion of it may be extracted using specialized compression equipment at the wellhead.

For example, in the case of underground storage facilities, the operators check for weak points and leakage, and investigate suspicious indications through a variety of downhole logging techniques, including formation evaluation tools (such as neutron logging), fluid flow and movement indicators (noise and temperature surveys), casing inspections (magnetic flux leakage and ultrasonic methods), mechanical calipers, downhole cameras, and cathodic protection profile surveys. Multiple methods to make decisions on the mitigation of any adverse effects require continuous maintenance work that must ensure the integrity of the storage facility. Of necessity, storage facility operators are using risk-based assessments by collection of analytical data, data analysis, and recognition of the indicator of the integrity of the storage area, to drive the underground storage management programs. In addition to tool-based assessment methods, the owner of a storage facility industry uses pressure tests and pressure monitoring as methods to assess the integrity of the facility. Common approaches used by operators include (1) shut in pressure monitoring at each facility, (2) annulus pressure or flow monitoring, and (3) mechanical integrity tests.

Facility owners may also employ a risk-based approach that considers risks and threats specific to each facility when selecting well-integrity practices. Commonly assessed risks or threats include, but are not limited to the following: (1) the physical properties of the casing, such as diameter, weight, and grade; (2) presence of atmospheric or external corrosion at or near the surface; (3) known metal loss indications from casing inspection surveys; (4) the presence of annulus pressure or flow; (5) water production in the well; (6) the presence of corrosive hydrogen sulfide and bacteria; (7) or naturally corrosive zones; (8) the flow potential of the well; and (9) the work history of the well and its reliability in providing natural gas service. Each category requires the production of data by application of the relevant test methods. In addition, there are several volumetric measures that are used to quantify the fundamental characteristics of an underground storage facility and the gas contained within it. Indeed, it is important to distinguish between the characteristic of a facility, such as the *capacity* of the facility and the characteristic of the natural gas within the facility, such as the actual *inventory level*.

By way of definition, the injection capacity (or injection rate) is the complement of the deliverability or withdrawal rate and is the amount of natural gas that can be injected into a storage facility on a daily basis (or unit-time basis). As with deliverability, injection capacity is usually expressed in millions of standard cubic feet per day (MMcf/day), although dekatherms/day is also used. A further definition that is necessary at this point, a dekatherm (dth) is a unit of energy used primarily to measure natural gas and was developed in about 1972 by the Texas Eastern Transmission Corporation—a natural gas pipeline company. The dekatherm is equal to 10 Therms or 1,000,000 British thermal units (MMBtu) or, using the SI system, 1.055 gigajoules (GJ). The dekatherm is also approximately equal to one thousand cubic feet (Mcf, Mft^3) of natural gas or exactly one thousand cubic feet of natural gas with a heating value of 1000 Btu/ft^3.

The total natural gas storage capacity of an underground storage facility is the maximum volume of natural gas that can be stored in the facility in accordance with the design of the facility, which comprises (1) the physical characteristics of the facility, such as if it is a depleted reservoir, (2) the installed equipment, and (3) the operating procedures that are particular to the site. Following from this, the total gas in storage is the volume of natural gas in the underground facility at a particular time. Thus the injection capacity of a storage facility is also variable and is dependent on factors comparable to those that determine the deliverability of the natural gas. In contrast, or as can be anticipated, the injection rate varies inversely with the total amount of gas in storage: the injection rate is at its lowest when the facility is full, but the injection rate increases as gas is withdrawn.

The base gas (or cushion gas) is the volume of natural gas intended as permanent inventory in a storage facility to maintain adequate pressure and deliverability rates throughout the withdrawal season. On the other hand, the working gas is the volume of gas in the storage facility above the level of base gas and is, simply, the natural gas that is available for withdrawal and sales. In relation to the definition of working gas, the working gas capacity is the total gas storage capacity minus base gas. The deliverability of the gas from the storage facility (also referred to as the deliverability rate, withdrawal rate, or withdrawal capacity) is often expressed as a measure of the amount of gas that can be delivered (withdrawn) from a storage facility on a daily basis.

Gas deliverability is usually expressed in terms of million cubic feet per day (MMcf/day). Occasionally, deliverability is expressed in terms of equivalent heat content of the gas withdrawn from the facility, most often in dekatherms per day (1 Therm is equal 100,000 Btu, which is roughly equivalent to 100 ft^3 of natural gas; a dekatherm is the equivalent of about one thousand cubic feet, Mcf). The deliverability of a given storage facility is variable, and it depends on factors such as, but not limited to: (1) the amount of natural gas in the storage facility, (2) the pressure within the storage facility, (3) the compression capability available to the storage facility, and (4) the configuration and capabilities of surface facilities associated with the underground storage facility. In general, the deliverability rate of a storage facility will vary directly with the total amount of natural gas in the facility. Thus gas deliverability of the gas is at its highest when the facility is full and declines as working gas is withdrawn.

In addition, the mineralogy of the underground storage facility is important since the presence of certain minerals may enhance the adsorption (and nonrecovery) of certain constituents of the gas which can result in changes in the composition of the gas. Also, other mineralogical constituents of the storage facility may cause adverse chemical changes to the gas. Thus in addition to use of geological, geophysical, and core analysis methods, presence of discontinuities in the caprock (e.g., because of presence of faults or fractures) is often tested using long water pump tests. In these tests, water is either withdrawn or is injected into the target formation, and pressure (or water level) is monitored in the formation (aquifer) above the caprock (Katz and Coats, 1968; Crow et al., 2008).

Because of the potential variability in gas composition (Mokhatab et al., 2006; Burrus and Ryder, 2003, 2014; Speight, 2014a, 2017) and mineralogical characteristics of the storage facility, none of the aforementioned measures for any given storage facility are fixed or absolute. The injection rate and the withdrawal rate can (will) change as the level of natural gas varies within the facility. In practice, a storage facility may be able to exceed certificated total capacity in some circumstances by exceeding certain operational parameters. The total capacity of the storage facility can also vary, temporarily or permanently, as its defining parameters vary. Measures of base gas, working gas, and working gas capacity can also change from time to time. These changes occur, e.g., when a storage operator reclassifies one category of natural gas to the other, often as a result of new wells, equipment, or operating practices (such a change generally requires approval by the appropriate regulatory authority and is subject to data derived by chemical and physical analysis of the gas). Finally, storage facilities can withdraw base gas for supply to market during times of particularly heavy demand, although by definition, this gas may not be intended for that use.

5.4   Storage facilities

Natural gas may be stored in several different ways. In the modern world, natural gas is most commonly held in inventory underground under pressure in three main types of facilities. These underground facilities are (1) depleted reservoirs in oil and/or natural gas fields, (2) aquifers, and (3) salt cavern formations. Other potential types of underground storage facilities such as an abandoned mine or a hard rock cavern are not included here being the less likely potentially usable storage facilities. Natural gas is also stored in liquid or gaseous form in above-ground storage tanks, which were the conventional method of storing coal gas in the early-to-mid 20th century (Speight, 2013), and such storage facilities may still be seen at some sites.

More pertinent to underground storage facilities, each type of facility has its own physical characteristics (porosity, permeability, and retention capability) and economics (site preparation and maintenance costs, deliverability rates, and cycling capability), which govern its suitability for storage applications. In relation to the type of facility, two important characteristics of an underground storage facility are (1) the capacity to hold natural gas for future use and (2) the rate at which gas inventory can be withdrawn (the deliverability rate).

It is important to recognize, that no matter which type of underground storage method is used, the caverns need to be prepared to receive the natural gas. This often includes adding pipes and valves, as well as sealing any cracks that might have occurred with the original drilling. These costs, gas accessibility, ease of gas extraction, or proximity to a gas distribution center must be given due consideration in order to evaluate the ability of the reservoir or the cavern as a viable option for natural gas storage.

5.4.1   Depleted reservoirs

Most existing natural gas storage in the United States is in depleted natural gas reservoirs or depleted crude oil reservoirs that are close to consumption centers. A benefit of the use of such reservoirs is that conversion of a gas field or crude oil field from production to storage duty takes advantage of existing wells, gathering systems, and pipeline connections. Because of their wide availability, depleted crude oil reservoirs and depleted natural gas reservoirs are the most commonly used underground of all storage facilities.

By definition, depleted reservoirs are those formations that have already been tapped of all their recoverable natural gas and/or crude oil. This leaves an underground formation with porosity and permeability that is capable of holding natural gas. In addition, using an already developed reservoir for storage purposes allows the use of the extraction and distribution equipment left over from when the field was productive. Having this extraction network in place reduces the cost of converting a depleted reservoir into a storage facility. The factors that determine whether or not a depleted reservoir will make a suitable storage facility are both geographic and geologic. Geographically, depleted reservoirs must be relatively close to consuming regions and they must also be close to transportation infrastructure, including pipelines, and distribution systems.

To be suitable for storage of natural gas, a depleted reservoir formation must have high porosity and high permeability. The porosity of the formation determines the amount of natural gas that it may hold, while the permeability of the formation determines the rate at which natural gas flows through the formation, which in turn determines the rate of injection and withdrawal of working gas. However, a depleted reservoir where withdrawal of the crude oil or natural gas has caused a decrease in the porosity and/or a decrease in the permeability may not be acceptable for natural gas storage. In certain instances, the formation may be stimulated to increase permeability (Speight, 2016).

In some cases, and in order to maintain pressure in depleted reservoirs, approximately 50% v/v of the natural gas in the formation must be kept as cushion gas. However, a depleted reservoir that has already been used as a source of filled with natural gas may not require the injection of what will become physically unrecoverable gas (cushion gas) since that gas already exists in the formation.

In regions without depleted reservoirs, like the upper Midwest of the United States, one of the other two storage options is required—either an aquifer or a salt cavern. Of the approximately 400 active underground storage facilities in the United States, the majority (ca. 79%) are depleted natural gas or depleted crude oil reservoirs. Conversion of a crude oil or a natural gas reservoir from production to storage takes advantage of existing infrastructure such as wells, gathering systems, and pipeline connections. Depleted crude oil and natural gas reservoirs are the most commonly used underground storage sites because of their relatively wide availability.

5.4.2   Aquifers

An aquifer is an underground highly porous and permeable rock formation that can act as natural water reservoirs. Natural aquifers may be suitable for gas storage if the water-bearing sedimentary rock formation is overlaid with an impermeable cap rock. Also, the aquifer should not be a part of drinking water system of aquifers and make up only approximately 10% of the natural gas storage facilities. In some cases, an aquifer may be reconditioned and used as a facility for natural gas storage.

Typically, an aquifer is costlier to develop than a depleted reservoir and, consequent, these aquifer-derived storage facilities are used only in areas where there are no nearby (and cheaper to develop) depleted reservoirs. As might be anticipated since aquifers have not always been explored to the same extent as crude oil or natural gas reservoirs, seismic testing must be performed, much the same as is performed for the exploration of potential natural gas formations and application of text method to determine the mineralogy of the aquifer would also be a benefit, perhaps even a necessity. The area of the formation, the composition and porosity of the formation itself, and the existing formation pressure as well as the capacity for gas storage must all be determined prior to development of the aquifer for gas storage. Typically, an aquifer is suitable for gas storage if the water-bearing sedimentary rock formation is overlaid with an impermeable cap rock. Although the geology of aquifers is similar to depleted production formations, their use for natural gas storage usually requires more base gas (cushion gas) and allows less flexibility in gas injection and gas withdrawal.

Furthermore, in order to develop a natural aquifer into an effective natural gas storage facility, all of the associated infrastructure must also be developed. This includes installation of wells, extraction equipment, pipelines, dehydration facilities, and possibly compression equipment. Since aquifers are naturally full of water, in some instances powerful injection equipment must be used, to allow sufficient injection pressure to push down the resident water and replace it with natural gas.

While natural gas being stored in aquifers has already undergone all of its processing, upon extraction from a water-bearing aquifer formation the gas typically requires further dehydration prior to transportation, which would, more than likely, require the installation of specialized equipment at or near to the wellhead. In addition, the aquifer may not have the same natural gas retention capabilities as a depleted crude oil reservoir or a natural gas reservoir and a part of the injected natural gas that may escape from the formation and must be gathered and extracted by *collector wells* that are specifically designed to collect any gas that has escaped from the primary aquifer formation.

In addition, an aquifer formation may typically require cushion gas that is required in a depleted crude oil reservoir or a depleted natural gas reservoir. Since there is no naturally occurring gas in the aquifer formation to begin with, a portion of the natural gas that is injected will ultimately prove to be physically unrecoverable. While it is possible to extract cushion gas from depleted reservoirs, extracting cushion gas from an aquifer formation could have negative effects, among which formation damage might be extensive. As such, most of the cushion gas that is injected into an aquifer formation may remain unrecoverable, even after the storage facility is shut down and abandoned.

Nevertheless, in spite of these issues, in some areas of the United States, most notably the Midwestern United States, natural aquifers have been converted to storage reservoirs for natural gas. Deliverability rates may be enhanced by the presence of an active water drive, which supports the reservoir pressure through the injection and production cycles.

5.4.3 Salt caverns

Essentially, salt caverns are formed out of existing salt deposits. These underground salt deposits may exist in two possible forms: salt domes and salt beds. Salt domes are thick formations created from natural salt deposits that, over time, leach up through overlying sedimentary layers to form large dome-type structures. They can be as large as a mile in diameter, and 30,000 ft in height. Typically, salt domes used for natural gas storage are between 6000 and 1500 ft beneath the surface, although in certain circumstances they can come much closer to the surface. Salt beds are shallower, thinner formations. These formations are usually no more than 1000 ft in height. Because salt beds are wide, thin formations, once a salt cavern is introduced, they are more prone to deterioration, and may also be more expensive to develop than salt domes.

Once a suitable salt dome or salt bed deposit is discovered, and deemed suitable for natural gas storage, it is necessary to develop a "salt cavern" within the formation. Essentially, this consists of using water to dissolve and extract a certain amount of salt from the deposit, leaving a large empty space in the formation. This is done by drilling a well down into the formation and cycling large amounts of water through the completed well. This water will dissolve some of the salt in the deposit, and be cycled back up the well, leaving a large empty space that the salt used to occupy—this process is known as "salt cavern leaching." Some of the salt is dissolved leaving a void and the water, now saline, is pumped back to the surface. The process continues until the cavern is the desired size. Once created, a salt cavern offers an underground natural gas storage vessel with very high deliverability.

Salt cavern leaching is used to create caverns in both types of salt deposits and can be quite expensive. However, once created, a salt cavern offers an underground natural gas storage vessel with very high deliverability. In addition, cushion gas requirements are the lowest of all three storage types, with salt caverns only requiring about 33% of total gas capacity to be used as cushion gas.

These underground salt caverns offer another option for natural gas storage. These formations are well suited to natural gas storage in that salt caverns, once formed, allow little injected natural gas to escape from the formation unless specifically extracted. The walls of a salt cavern also have the structural strength of steel, which makes it very resilient against degradation over the life of the storage facility. Salt caverns allow very little of the injected natural gas to escape from storage unless specifically extracted. The walls of a salt cavern are strong and impervious to gas over the lifespan of the storage facility.

Salt cavern storage facilities are primarily located along the Gulf Coast, as well as in the northern states, and are best suited for peak load storage. Salt caverns are typically much smaller than depleted gas reservoirs and aquifers, in fact underground salt caverns usually take up only one-hundredth of the acreage taken up by a depleted gas reservoir. As such, salt caverns cannot hold the volume of gas necessary to meet base load storage requirements. However, deliverability from salt caverns is typically much higher than for either aquifers or depleted reservoirs. Therefore natural gas stored in a salt cavern may be more readily (and quickly) withdrawn, and caverns may be replenished with natural gas more quickly than in either of the other types of storage facilities. Moreover, salt caverns can readily begin flowing gas on as little as one hour's notice, which is useful in emergency situations or during unexpected short-term demand surges. Salt caverns may also be replenished more quickly than other types of underground storage facilities.

Typically, a salt cavern will provide a high withdrawal rate and a high injection rate relative to the working gas capacity of the cavern. Most salt cavern storage facilities have been developed in salt dome formations located in the Gulf Coast states of the United States. Salt caverns have also been formed (by a leaching process) in bedded salt formations in Northeastern, Midwestern, and Southwestern states.

Salt formation storage facilities make up about 10% of the natural gas storage facilities. These subsurface salt formations provide very high withdrawal and injection rates.

5.4.4 Gas holders

In addition to underground storage of natural gas, there are also facilities for above-ground storage of the gas. This may be the situation if there are no regional underground storage facilities or if it is more convenient, because of the lower volume of gas to be stored, and the need for the gas is immediate. In the case of above-ground storage, the gas is stored in specially fabricated tanks above ground which do allow for easy access to the gas and complete control of gas extraction from storage. However, while the costs for above-ground storage options are typically less than underground, tanks can store only a fraction of the natural gas that underground caverns can. Another style of above-ground storage is the transportable tank which are typically used to maximize the amount of liquefied natural gas (LNG) being moved. Transportable tanks can be loaded onto train cars, used with 18-wheeler trucks for road transport, or moved onto barges for an overseas journey.

A gas holder (also called a gasometer) is a large container in which natural gas (or town gas) can be stored at or near atmospheric pressure and at ambient temperature. Gas holders were initially developed in the early part of the 20th century for the storage of coal gas (Speight, 2013). The volume of a gas holder follows the quantity of stored gas, with pressure coming from the weight of a movable cap. Typical volumes for large gas holders are approximately 1,800,000 ft^3, with 200 ft diameter structures. Gas holders tend to be used for balancing purposes (making sure gas pipes can be operated within a safe range of pressures) rather than for actually storing gas for later use.

A gas holder can provide on-site storage for purified gas and can act as a buffer by removing the need for continuous gas processing. The weight of the gas holder lift (cap) controlled the pressure of the gas in the mains and provided back pressure for the gas-making plant. A water-sealed gas holder consisted of two parts: a deep tank of water that was used to provide a seal, and a vessel that rose above the water as the gas volume increased. In the well-known telescoping type of gas holder, the tank floats in a circular or annular water reservoir, held up by the roughly constant pressure of a varying volume of gas, the pressure determined by the weight of the structure, and the water providing the seal for the gas within the moving walls.

The rigid waterless gas holder is a design that neither expands or contracts. The modern version of the waterless gas holder is the dry-seal type (membrane type) of gas holder that consists of a static cylindrical shell, within which a piston rises and falls. As it moves, a grease seal, tar/oil seal, or a sealing membrane which is rolled out and in from the piston keeps the gas from escaping.

The benefit of using a gas holder (although possibly limited in storage capacity) is that the gas holder can store gas at district pressure and can provide extra on-site gas very quickly and at peak times. Furthermore, the gas holder is the only storage method that can maintain the gas at the required pressure, which is the pressure required in local gas lines, and thus the gas holder may hold a large advantage over other methods of storage.

5.5 Transportation

The efficient and effective movement of natural gas from the production site to the consumers requires an extensive and elaborate transportation system. In many instances, natural gas produced from at the wellhead will have to travel a considerable distance to reach its point of use. The typical transportation system for natural gas consists of a complex network of pipelines, designed to quickly and efficiently transport natural gas from its origin, to areas of high natural gas demand. Transportation of natural gas is closely linked to storage insofar as if the natural gas being transported is not required when it reaches the end of the pipeline (or transportation system) it can be put into storage facilities to be released at a time when it is needed.

Natural gas, as it is used by consumers, is much different from the natural gas that is brought from underground up to the wellhead. In some cases, constituents of the gas may remain in the reservoir because of various property constraints or because of interactions with the reservoir rock. Furthermore, although the processing (treating and refining) of natural gas is, in many respects, less complicated than the processing and refining of crude oil (Chapter 4: Composition and properties), it is equally as necessary before its use by end-users (Parkash, 2003; Mokhatab et al., 2006; Gary et al., 2007; Speight, 2007, 2014a, 2017; Riazi et al., 2013; Faramawy et al., 2016; Hsu and Robinson, 2017). The natural gas used by consumers is composed almost entirely of methane.

However, natural gas produced at the wellhead, although still composed primarily of methane, is by no means as pure and includes several varieties of impurities such as higher molecular weight hydrocarbons, carbon dioxide, and hydrogen sulfide (including mercaptans). Raw natural gas comes from three types of wells: oil wells, gas wells, and condensate wells. Natural gas that comes from oil wells is typically termed *associated gas* which can separate from oil in the formation (free gas) that occur with crude oil in the reservoir or gas that dissolved in the crude oil (dissolved gas). Natural gas from gas and condensate wells, in which there is little or no crude oil, is termed *nonassociated gas* (Chapter 1: History and use).

The natural gas used by consumers is composed almost entirely of methane but the natural gas that is produced at the wellhead, although still composed primarily of methane, is by no means as pure. Whatever the source of the natural gas (Chapter 1: History and use), it commonly exists in mixtures with other hydrocarbons; principally ethane, propane, butane, and pentanes. In addition, raw natural gas contains water vapor, hydrogen sulfide (H_2S), carbon dioxide, helium, nitrogen, and other compounds.

Natural gas processing (Chapter 4: Composition and properties) consists of separating all of the various hydrocarbons and fluids from the pure natural gas, to produce pipeline-quality dry natural gas. Major transportation pipelines usually impose restrictions on the make-up of the natural gas that is allowed into the pipeline. That means that before the natural gas can be transported it must be purified. The associated hydrocarbons, which include ethane, propane, butane, *iso*-butane, and natural gasoline, are often referred to as NGLs can be very valuable by-products of natural gas processing. These NGLs are sold separately and have a variety of uses including enhancing oil recovery in oil wells, providing raw materials for oil refineries or petrochemical plants, and as sources of energy. If the natural gas has a significant helium content, the helium may be recovered by fractional distillation. Natural gas may contain as much as 7% helium and is the commercial source of this noble gas (Ward and Pierce, 1973).

While some of the needed processing can be accomplished at or near the wellhead (field processing), the complete processing of natural gas takes place at a processing plant (Chapter 4: Composition and properties). The extracted natural gas is transported to these processing plants through a network of gathering pipelines, which are small-diameter, low-pressure pipes. In addition to processing performed at the wellhead and at centralized processing plants, some final processing is also sometimes accomplished at *straddle extraction plants* which are located on major pipeline systems. Although the natural gas that arrives at these straddle extraction plants is already of pipeline quality, in certain instances there still exist small quantities of NGLs, which are extracted at the straddle plants.

Major transportation pipelines usually impose restrictions on the make-up of the natural gas that is allowed into the pipeline. Natural gas processing consists of separating all of the various hydrocarbons and fluids from the pure natural gas to produce what is known as "pipeline quality" dry natural gas. Associated hydrocarbons (NGLs) can be very valuable by-products of natural gas processing. These liquids include ethane, propane, butane, isobutane, and natural gasoline and are sold separately. NGLs have a variety of uses including enhancing oil recovery in oil wells and provide raw materials for oil refineries or petrochemical plants, and as sources of energy.

Thus the raw natural gas must be purified to meet the quality standards specified by the major pipeline transmission and distribution companies. These quality standards vary from pipeline to pipeline and are usually a function of a design of the pipeline system and the markets that are served by the natural gas. In general, the standards specify that the natural gas should:

1. Be within a specific range of heating value (caloric value). For example, in the United States, the heating value should be on the order of $1035 \pm 5\%$ Btu per cubic foot of gas at 1 atm and 15.6°C (60°F).
2. Be delivered at or above a specified hydrocarbon dew point temperature (below which some of the hydrocarbons in the gas might condense at pipeline pressure forming liquid

slugs that could damage the pipeline). The dew point adjustment serves the reduction of the concentration of water and heavy hydrocarbons in natural gas to such an extent that no condensation occurs during the ensuing transport in the pipelines.
3. Be free of particulate solids and liquid water to prevent erosion, corrosion, or other damage to the pipeline.
4. Be dehydrated (water removed) sufficiently to prevent the formation of methane hydrates within the gas processing plant or subsequently within the sales gas transmission pipeline. A typical water content specification in the United States is that gas must contain no more than 7 pounds of water per million standard cubic feet (MMSCF) of gas.
5. Contain no more than trace amounts of components such as hydrogen sulfide, carbon dioxide, mercaptans, and nitrogen. The most common specification for hydrogen sulfide content is 4 ppm of hydrogen sulfide per 100 ft^3 of gas. The specifications for the allowable amount of carbon dioxide typically limit the content to less than 3% by volume.
6. Maintain mercury at less than detectable limits (approximately 0.001 parts per billion, ppb, (0.001×10^{12}) by volume) primarily to avoid damaging equipment in the gas processing plant or the pipeline transmission system from mercury amalgamation and embrittlement of aluminum and other metals.

Furthermore, there is need for caution when condensate mixtures are to be transported. The transport of gas-condensate mixtures of various compositions has been found to be accompanied by a slight increase in viscosity in the coldest period when ground temperatures at depth of a condensate pipeline reach −4 to 0°C (25 to 32°F). The decrease in the temperature of reservoir oil fluids under study to minus −30 to 10°C (−22 to 50°F) is accompanied by a sharp increase in all structural and rheological parameters of the mixture. As a result, cloud and pour point of a mixture falls, its amount decreases, the structure of paraffin deposits changes (Loskutova et al., 2014).

5.5.1 Pipelines

After extraction from the reservoir, natural gas must be transported to different places to be processed, stored, and then finally delivered to the end consumer and this can occur by means of a pipeline or ship. Thus once natural gas is extracted from onshore and offshore sites, it is transported to consumers, typically by way of pipelines. Before it reaches the pipelines, however, it needs to be purified into the state it will be in when it enters homes and businesses. This requires the separation of various hydrocarbons and fluids from the pure natural gas to produce "pipeline quality" dry gas. Restrictions are placed on the quality of natural gas that is allowed to enter pipelines. Natural gas can also be stored in large underground areas because demand is higher in different seasons of the year. From the large pipelines, the gas goes into smaller pipelines called mains, and then further into even smaller pipes called services that lead directly into homes and buildings to be heated. Natural gas can also be cooled to a very cold temperature and stored as a liquid. Changing the phase of the natural gas from gas to liquid allows for easier storage because it takes up less space. Then, when it needs to be distributed, it is returned to its original state and sent through pipelines.

There are essentially three major types of pipelines along the natural gas transportation route: (1) the gathering pipeline system, (2) the transmission pipeline system, sometimes referred to as the interstate pipeline system, and (3) the distribution system.

Gathering systems, primarily made up of small-diameter, low-pressure pipelines, move raw natural gas from the wellhead to a natural gas processing plant or to an interconnection with a larger mainline pipeline. Transmission pipelines are typically wide-diameter, high-pressure transmission pipelines that transport natural gas from the producing and processing areas to storage facilities and distribution centers. Compressor stations (or pumping stations) on the pipeline network keep the natural gas flowing forward through the pipeline system. Local distribution companies deliver natural gas to consumers through small-diameter, lower pressure service lines. Whatever the pipeline system, there is the need to determine if the natural gas received at the wellhead has a high sulfur content and a high carbon dioxide content (sour gas), a specialized sour gas gathering pipe must be installed. Sour gas is extremely corrosive and dangerous, thus its transportation from the wellhead to the sweetening plant must be done carefully (Speight, 2014b).

Thus an issue that always arises at the wellhead when natural gas is transported by pipeline is the degree of processing at the wellhead to remove potential corrosive contaminants that would seriously affect the integrity (corrosivity) of the pipeline. While carbon dioxide and hydrogen sulfide are often considered to be noncorrosive in the dry state, the presence of water in the natural gas can render these two gases extremely corrosive (Speight, 2014b). This emphasizes the need for compositional analysis of the natural gas as it exits the production well. As a result of the compositional analysis, the pipeline operators can make the decision about the extent of the wellhead treating such as separation of (1) hydrocarbon gas liquids, (2) nonhydrocarbon gases, and (3) water from the natural gas before the (treated) gas is delivered into a mainline transmission system.

Major transportation pipelines usually impose restrictions on the make-up of the natural gas that is allowed into the pipeline and natural gas must be processed to produce "pipeline quality" dry natural gas. Some field processing can be accomplished at or near the wellhead; however, the complete processing of natural gas takes place at a processing plant, usually located in a natural gas producing region. Thus, from the wellhead, natural gas is transported to processing plants through a network of small-diameter, low-pressure gathering pipelines which may consist of a complex gathering system can consist of thousands of miles of pipes, interconnecting the processing plant to upwards of 100 wells in the area.

Thus it is essential that the composition of the gas be known so that natural-gas processing can begin at the wellhead. The composition of the raw natural gas extracted from producing wells depends on the type, depth, and location of the underground deposit and the geology of the area. A natural gas processing plant is a facility designed to clean raw natural gas by separating impurities and various nonmethane hydrocarbons and fluids to produce what is known as *pipeline quality* dry natural gas. A gas processing plant is also used to recover NGLs (condensate, natural gasoline, and liquefied petroleum gas (LPG)) and sometimes other substances such as sulfur-containing constituents and should be (at least) checked for sulfur content and for residues to ensure that the LPG meets the specification (Speight, 2015, 2018).

Four basic types of LPGs are provided to cover the common use applications, particularly LPGs consisting of propane, propene (propylene), butane, and mixtures of these materials that are intended for use as domestic, commercial and industrial heating, and engine fuels. However, care must be taken to in sampling of the liquefied gases to ensure that the sample is representative otherwise the test results may not be significant (Speight, 2018). All four types of LPGs should conform to the specified requirements for vapor pressure, volatile residue, residue matter, relative density, and corrosion. There is also a further series of standard test methods that can be used to provide information about the composition and properties of domestic and industrial fuel gases (Speight, 2018).

While some of the needed processing can be accomplished at or near the wellhead (field processing), the complete processing of natural gas takes place at a processing plant, usually located in a natural gas producing region. The extracted natural gas is transported to these processing plants through a network of gathering pipelines, which are small-diameter, low-pressure pipes. A complex gathering system can consist of thousands of miles of pipes, interconnecting the processing plant to upwards of 100 wells in the area. In addition to processing done at the wellhead and at centralized processing plants, some final processing is also sometimes accomplished at straddle extraction plants. These plants are located on major pipeline systems. Although the natural gas that arrives at these straddle extraction plants is already of pipeline quality, in certain instances there still exist small quantities of NGLs, which are extracted at the straddle plants.

The actual practice of processing natural gas to pipeline dry gas quality levels can be quite complex, but usually involves four main processes to remove the various impurities: (1) oil and condensate removal, (2) water removal, (3) separation of NGLs, and (4) hydrogen sulfide removal, and (5) carbon dioxide removal (Chapter 4: Composition and properties) (Mokhatab et al., 2006; Speight, 2007, 2014a). If mercury is present in the gas, typically as trace amounts, opinions differ whether or not the mercury should be removed at the wellhead. The presence of mercury can cause corrosion of aluminum heat exchangers as well as cause environmental pollution. If needed, there are two forms of removal processes: (1) regenerative processes and (2) nonregenerative processes. The regenerative process uses sulfur-activated carbon or alumina, while nonregenerative processes use silver on a molecular sieve (Mokhatab et al., 2006; Speight, 2007, 2014a).

In addition, hydrogen sulfide and carbon dioxide can be removed by olamine scrubbing (Mokhatab et al., 2006; Speight, 2007, 2014a; Kidnay et al., 2011) and heaters and scrubbers are installed, usually at or near the wellhead. The scrubbers serve primarily to remove sand and other large-particle impurities. The heaters ensure that the temperature of the gas does not drop too low. With natural gas that contains even low quantities of water, natural gas hydrates (NGHs) tend to form when temperatures drop. These hydrates are solid or semisolid compounds, resembling ice-like crystals and when gas hydrates accumulate, they can impede the passage of natural gas through valves and gathering systems. After processing, the pipeline quality natural gas is injected into gas transmission pipelines and transported to the end-users. This often involves transportation of the gas over hundreds of miles, as the location of gas production is generally not the location where the gas is used.

Nevertheless, whatever of the extent of the gas processing operations (at the wellhead or in a processing facility) it is essential that the composition of the gas be known through analysis in order to ensure a (relatively) pure product for transportation and that the gas meets sales specification goals.

5.5.2   Liquefied natural gas

LNG is the liquid form of natural gas and is used principally for transporting natural gas to markets, where it is regasified and distributed as pipeline natural gas. The temperature required to condense natural gas depends on its precise composition, but it is typically between $-120°C$ and $-170°C$ ($-184°F$ and $-274°F$). The advantage of LNG is that it offers an energy density comparable to petrol and diesel fuels, extending range and reducing refueling frequency. The disadvantage is the high cost of cryogenic storage on vehicles and the major infrastructure requirement of LNG dispensing stations, production plants and transportation facilities.

LNG is composed predominantly of methane since the liquefaction process requires the removal of the nonmethane components like carbon dioxide, water, butane, pentane, and heavier components from the produced natural gas. LNG is odorless, colorless, noncorrosive, and nontoxic. When vaporized it burns only in concentrations of 5%−15% when mixed with air. In terms of composition (Table 5.2), LNG is predominantly methane and it is not surprising that the density of LNG is close to (but not exactly equal to) the density of methane.

If gas is produced at lower pressures than typical sales pipeline pressure (approximately 700−1000 psi), it is compressed to sales gas pressure (Mokhatab et al., 2006, Chapter 8: Gas cleaning processes). Transport of sales gas is done at high pressure to reduce pipeline diameter. Pipelines may operate at very high pressures (above 1000 psig) to keep the gas in the dense phase thus preventing condensation and two-phase flow. Compression typically requires two to three stages to attain

Table 5.2 Composition of liquefied natural gas in various markets

Composition (mole percent)					
Source	Methane	Ethane	Propane	Butane	Nitrogen
Alaska	99.72	0.06	0.0005	0.0005	0.20
Algeria	86.98	9.35	2.33	0.63	0.71
Baltimore	93.32	4.65	0.84	0.18	1.01
New York City	98.00	1.40	0.40	0.10	0.10
San Diego	92.00	6.00	1.00	−	1.00

sales gas pressure. As stated previously, processing may be done after the first or second stage, prior to sales compression.

Compression is used in all aspects of the natural gas industry, including gas lift, reinjection of gas for pressure maintenance, gas gathering, gas processing operations (circulation of gas through the process or system), transmission and distribution systems, and reducing the gas volume for shipment by tankers or for storage. In recent years, there has been a trend toward increasing pipeline-operating pressures. The benefits of operating at higher pressures include the ability to transmit larger volumes of gas through a given size of pipeline, lower transmission losses due to friction, and the capability to transmit gas over long distances without additional boosting stations. In gas transmission, two basic types of compressors are used: reciprocating and centrifugal compressors. Reciprocating compressors are usually driven by either electric motors or gas engines, whereas centrifugal compressors use gas turbines or electric motors as drivers.

Thus when natural gas is cooled to a temperature of approximately $-160°C$ (approximately $-260°F$) at atmospheric pressure, it condenses to a liquid (LNG). One volume of this liquid takes up about 1/600th the volume of natural gas. LNG weighs less than one-half that of water, approximately 45% as much. LNG is odorless, colorless, noncorrosive, and nontoxic. When vaporized it burns only in concentrations of 5%−15% when mixed with air (Section 2.4). Neither LNG, nor its vapor, can explode in an unconfined environment. Since LNG takes less volume and weight, is presents more convenient options for storage and transportation.

The task of gas compression is to bring gas from a certain suction pressure to a higher discharge pressure by means of mechanical work. The actual compression process is often compared to one of three ideal processes: (1) isothermal, (2) isentropic, and (3) polytropic compression.

Isothermal compression occurs when the temperature is kept constant during the compression process. It is not adiabatic because the heat generated in the compression process must be removed from the system. The compression process is isentropic or adiabatic reversible if no heat is added to or removed from the gas during compression and the process is frictionless. The polytropic compression process is, like the isentropic cycle, reversible but it is not adiabatic. It can be described as an infinite number of isentropic steps, each interrupted by isobaric heat transfer. This heat addition guarantees that the process will yield the same discharge temperature as the real process.

Contrary to some loose and inaccurate definitions, LNG is not the same as LPG (often referred to as *propane*) which is determined by a standard test method (Speight, 2015, 2018; ASTM 2017).

LNG can be used in natural gas vehicles, although it is more common to design vehicles to use compressed natural gas (CNG). The relatively high cost of production of LNG and the need to store the liquid in expensive cryogenic tanks have prevented its widespread use in commercial applications. Prior to and during transportation, CNG is stored on the vehicle in high-pressure tanks, usually on the order of 3000−3600 psi.

The production of LNG process involves removal of certain components, such as dust, acid gases (such as hydrogen sulfide and carbon dioxide), helium, water vapor, and higher molecular weight hydrocarbon derivatives, which could cause difficulty downstream of the wellhead. The natural gas is then condensed into a liquid at close to atmospheric pressure by cooling it to approximately $-162°C$ ($-260°F$), maximum transport pressure is set at around 4 psi. One volume of this liquid takes up approximately 1/600th the volume of natural gas. LNG weighs less than one-half that of water and it is colorless, odorless, and noncorrosive. When vaporized, LNG burns only in concentrations of 5−15% v/v when mixed with air. Since LNG takes less volume and weight that natural gas, liquefaction enhances ease of storing and transporting. Hazards include flammability after vaporization into a gaseous state, freezing and asphyxia.

Facilities for liquefying natural gas require complex machinery with moving parts and special refrigerated ships for transporting the LNG to market. The costs of building LNG plant have lowered over the past 25 years because of greatly improved thermodynamic efficiencies so that LNG is becoming a major gas export method worldwide and many plants being extended, or new ones built in the world.

LNG is natural gas stored as a super-cooled (cryogenic) liquid—the temperature required to condense natural gas depends on its precise composition, but it is typically between $-120°C$ and $-170°C$ ($-184°F$ and $-274°F$). Large cryogenic tanks are needed to store the LNG which, typically, may be 230 ft in diameter, 145 ft high and hold over 26,400,000 ga of LNG. At the consumer end of the transportation process, an infrastructure for handling the reprocessing of vast quantities of natural gas from LNG is required, which is also expensive and vulnerable to sabotage.

The advantage of LNG is that it offers an energy density comparable to petrol and diesel fuels, extending range and reducing the frequency of refueling. The disadvantage, however, is the high cost of cryogenic storage on vehicles and the major infrastructure requirement of dispensing station for the LNG, production, plants and transportation facilities. For transportation, the LNG is loaded onto double-hulled ships which are used for both safety and insulating purposes. Once the ship arrives at the receiving port, the LNG is typically off-loaded into well-insulated storage tanks. Regasification is used to convert the LNG back into its gas form, which enters the domestic pipeline distribution system and is ultimately delivered to the end-user. The current largest specially built refrigerated tankers can carry $135,000 \ m^3$ LNG, which is approximately $4,767,480 \ ft^3$ of gas.

Within the United States and many other countries, LNG must meet heating value specifications and the gas can contain only moderate quantities of NGLs. If LNG is shipped with NGLs, the NGLs must be removed upon receipt or blended with lean gas or nitrogen before the natural gas can enter the transportation system (especially the pipeline system).

5.5.3 Liquefied petroleum gas

LPG is the term applied to certain specific hydrocarbons and their mixtures, which exist in the gaseous state under atmospheric ambient conditions but can be converted to the liquid state under conditions of moderate pressure at ambient temperature. Typically, fuel gas with four or less carbon atoms in the hydrogen−carbon combination have boiling points that are lower than room temperature and these products are gases at ambient temperature and pressure. LPG also called liquid petroleum gas (LP gas), also referred to as simply propane or butane, are flammable mixtures of hydrocarbon gases used as fuel in heating appliance and vehicles. Propylene and butylene derivatives as well as various other hydrocarbons are also usually present in small concentrations.

LPG is prepared at a crude oil refinery or at a gas processing plant and is almost entirely derived from fossil fuel sources being (1) manufactured during the refining of crude oil or (2) extracted from crude oil or natural gas streams as they emerge from the ground. As its boiling point is below room temperature, LPG will evaporate quickly at normal temperature and pressure (STP) and is usually supplied in pressurized steel vessels. The pressure at which LPG becomes liquid is the vapor pressure which varies depending on composition and temperature. LPG is heavier than air and, unlike methane, will flow along floors and tend to settle in low spots, such as basement or depressions in the Earth. This can lead to explosion if the mixture of the LPG and air is within the explosive limits and there is an ignition source. Also, LPG can cause suffocation when it displaces air, causing a decrease in oxygen concentration.

LPG is a hydrocarbon mixture containing propane ($CH_3 \cdot CH_2 \cdot CH_3$) and butane ($CH_3 \cdot CH_2 \cdot CH_2 \cdot CH_3$). To a lesser extent, *iso*-butane [$CH_3 \cdot CH(CH_3) \cdot CH_3$] may also be present. The most common commercial products are propane, butane, or a specific of the two gases (Table 5.3) and are generally extracted from natural gas or crude oil. The propane and butane can be derived from natural gas or from refinery operations but, in this latter case, substantial proportions of the corresponding olefins will be present and need to be separated. The hydrocarbons are normally liquefied under pressure for transportation and storage.

Propylene and butylene isomers result from cracking other hydrocarbons in a crude oil refinery and are two important chemical feedstocks. The presence of propylene and butylenes in LPG used as fuel gas is not critical. The vapor pressures of these olefins are slightly higher than those of propane and butane and the flame speed is substantially higher, but this may be an advantage since the flame speeds of propane and butane are slow. However, one issue that often limits the amount of the olefins in LPG is the propensity of the olefins to form soot.

As already noted, the compositions of natural, manufactured, and mixed gases can vary so widely, no single set of specifications could cover all situations (Chapter 3: Unconventional gas). The requirements are usually based on performances in burners and equipment, on minimum heat content, and on maximum sulfur content. Gas utilities in most states come under the supervision of state commissions or regulatory bodies and the utilities must provide a gas that is acceptable to all types of consumers and that will give satisfactory performance in all kinds of consuming equipment.

The specific gravity of product gases, including LPG, may be determined conveniently by a number of methods and a variety of instruments (Speight, 2015, 2018; ASTM 2017). This test method covers the determination of relative density of gaseous fuels, including LPGs, in the gaseous state at normal temperatures and

Table 5.3 Properties of liquefied petroleum gas

Constituent	Propane	Butane
Formula	C_3H_8	C_4H_{10}
Boiling point (°F)	−44°	32°
Specific gravity of the gas (air = 1.00)	1.53	2.00
Specific gravity of the liquid (water = 1.00)	0.51	0.58
Lbs./gallon: liquid @ 60°F	4.24	4.81
BTU/gallon: gas @ 60°F	91,690	102,032
BTU/lb: gas	21,591	21,221
BTU/ft^3: gas @ 60°F	2516	3280
Ft^3 of vapor @ 60°F./gal. of liquid, 60°F	36.39	31.26
Ft^3 of vapor @ 60°F./lb. of liquid, 60°F	8.547	6.506
Latent heat of vaporization @ boiling point BTU/gal	785.0	808.0
Combustion data		
Flash point (°F)	−155	−76
Autoignition temperature (°F)	878	761
Maximum flame temperature in air (°F)	3595	3615
Flammability limits, % v/v of gas in air mixture:		
Lower limit (%)	2.4	1.9
Upper limit (%)	9.6	8.6
Octane number (*iso*-octane = 100)	100 +	92

pressures. The methods specified as subsections of the test method are sufficiently varied in nature so that one or more may be used for laboratory, control, reference (quality control), gas measurement, or in fact, for any purpose in which it is desired to know the relative density of gas or gases as compared to the density of dry air at the same temperature and pressure.

The *heat value* of a fuel gas is generally determined at constant pressure in a flow calorimeter in which the heat released by the combustion of a definite quantity of gas is absorbed by a measured quantity of water or air.

The lower and upper limits of *flammability* indicate the percentage of combustible gas in air below which and above which flame will not propagate. When flame is initiated in mixtures having compositions within these limits, it will propagate and therefore the mixtures are flammable. A knowledge of flammable limits and their use in establishing safe practices in handling gaseous fuels is important, such as (1) when purging equipment is used in gas service, (2) in controlling the atmospheric composition in a factory or in a mine, or (3) in handling liquefied gases.

Many factors enter into the experimental determination of flammable limits of gas mixtures, including the diameter and length of the tube or vessel used for the test, the temperature and pressure of the gases, and the direction of flame propagation-upward or downward. For these and other reasons, great care must be taken in the application of the data. In monitoring closed spaces where small amounts of gases enter the atmosphere, often the maximum concentration of the combustible gas is limited to one-fifth of the concentration of the gas at the lower limit of flammability of the gas–air mixture.

LPG can be transported in a number of ways, including by ship, rail, tanker trucks, intermodal tanks, cylinder trucks, pipelines, and local gas reticulation systems. However, a challenge with LPG is that it can vary widely in composition, leading to variable engine performance and cold starting performance. At normal temperatures and pressures, LPG will evaporate, and, because of this tendency, LPG is stored in pressurized steel bottles. Unlike natural gas, LPG is heavier than air, and thus will flow along floors and tend to settle in low spots, such as basements or, if outside, in depressions in the Earth. Such accumulations can cause explosion hazards and are the reason that LPG fueled vehicles are prohibited from indoor parking area in some jurisdictions.

In addition, LPG is usually available in different grades (usually specified as: (1) commercial propane, (2) commercial butane, (3) commercial propane–butane (P-B) mixtures, and (4) special duty propane). During the use of LPG, the gas must vaporize completely and burn satisfactorily in the appliance without causing any corrosion or producing any deposits in the system.

Commercial propane consists predominantly of propane and/or propylene, while *commercial butane* is mainly composed of butanes and/or butylenes. Both must be free from harmful amounts of toxic constituents and free from mechanically entrained water (that may be further limited by specifications). *Commercial propane–butane* mixtures are produced to meet sales specifications such as volatility, vapor pressure, specific gravity, hydrocarbon composition, sulfur and its compounds, corrosion of copper, residues, and water content. These mixtures are used as fuels in areas and at times where low ambient temperatures are less frequently encountered. Analysis by gas chromatography is possible (Speight, 2015, 2018; ASTM, 2017). *Special duty propane* is intended for use in spark-ignition engines and the specification includes a minimum *motor octane number* to ensure satisfactory antiknock performance. Propylene ($CH_3CH = CH_2$) has a significantly lower octane number than propane, so there is a limit to the amount of this component that can be tolerated in the mixture. Analysis by gas chromatography is possible (Speight, 2015, 2018; ASTM, 2017).

LPG and LNG can share the facility of being stored and transported as a liquid and then vaporized and used as a gas. In order to achieve this, LPG must be maintained at a moderate pressure but at ambient temperature. The LNG can be at ambient pressure but must be maintained at a temperature of roughly −1 to 60°C (30 to 140°F). In fact, in some applications it is actually economical and convenient to use LPG in the liquid phase. In such cases, certain aspects of gas composition (or quality such as the ratio of propane to butane and the presence of traces of heavier hydrocarbons, water, and other extraneous materials) may be of lesser importance compared to the use of the gas in the vapor phase.

For normal (gaseous) use, the contaminants of LPG are controlled at a level at which they do not corrode fittings and appliances or impede the flow of the gas. For example, hydrogen sulfide (H_2S) and carbonyl sulfide (COS) should be absent.

Organic sulfur to the level required for adequate odorization, or *stenching*, is a normal requirement in LPG, dimethyl sulfide (CH_3SCH_3) and ethyl mercaptan (C_2H_5SH) are commonly used at a concentration of up to 50 ppm. Natural gas is similarly treated possibly with a wider range of volatile sulfur compounds.

The presence of water in LPG (or in natural gas) is undesirable since it can produce hydrates that will cause, e.g., line blockage due to the formation of hydrates under conditions where the water *dew point* is attained. If the amount of water is above acceptable levels, the addition of a small methanol will counteract any such effect.

In addition to other gases, LPG may also be contaminated by higher boiling constituents (residua) such as the constituents of middle distillates to lubricating oil. These contaminants become included in the gas during handling and must be prevented from reaching unacceptable levels. Olefins and especially diolefins are prone to polymerization and should be removed.

Control over the amount of residue in LPG is essential in end-use applications of LPG. In fact, an oily residue in LPG is a contamination that can lead to problems during production, transportation, storage, or when in use. For example, when the liquefied gas is used for automotive fuel application, a residue can lead to troublesome deposits that will accumulate and corrode or plug the LPG fuel filter, the low-pressure regulators, the fuel mixer or the control solenoids. LPG can be contaminated with oily residues during its production or transport. Transport contamination can be a result of shared pipelines, valves, and trucks used for the distribution of other products. Production sources such as the desulfurization process may contribute sulfur absorbent oil to the LPG stream. Commercial LPG, especially for automotive applications, should comply with current fuel specifications.

Fuel specifications for oily residue in LPG use a method known as the oil stain method (Speight, 2015, 2018; ASTM, 2017). In the method 100 mL of the LPG are evaporated and the remaining volume of residue is read from the glass evaporation tube. In addition, the residue is dissolved in a solvent and the resulting solution is slowly dripped on the adsorption paper. The size and persistence of the stain which remains on the paper after the solvent evaporates is the other, empirical, quantification of the oily residue in the LPG sample. The accuracy of both quantifications is subject to question.

On the other hand, the liquefied gas injector method, in which a dedicated sampler (the liquefied gas injector) is used to inject the liquefied gas at room temperature directly on to the gas-chromatographic column. This test method is based on evaporation of large sample volumes followed by visual or gravimetric estimation of residue content. In addition, this method provides enhanced sensitivity in measurements of heavier (oily) residues, with a quantification limit of 10 mg/kg total residue. This test method gives both quantitative results and information about contaminant composition such as boiling point range and fingerprint, which can be very useful in tracing the source of a particular contaminant. The method covers the determination, by gas chromatography, of soluble hydrocarbon materials, sometimes called "oily residue," which can be present in LPG and which are substantially less volatile than the LPG itself. Also, the method offers quantitative data over the range of 10−600 mg/kg (ppm w/w), the residue with a boiling point between 174°C (345°F) and 522°C (970°F) (decane to tetracontane, i.e., C_{10} to C_{40}) in LPG. Higher boiling materials, or materials that adhere permanently to the chromatographic column, will not flow though the column and, therefore, will not be detected.

Finally, most methods for the compositional analysis of LPG recommend the use of a liquid sampling valve for sample introduction into the split inlet of the gas chromatographic instrument.

5.5.4 Compressed natural gas

CNG (methane stored at high pressure) is a fuel which can be used in place of gasoline, diesel fuel, and LPG. The composition and properties of compressed natural are dependent on the properties and composition of the original natural gas feedstock and test methods for application to the compressed natural should be designed based on the test methods applied to the gaseous (noncompressed) feedstock.

CNG is often confused with LNG—the main difference between the two (CNG and LNG) is that CNG is stored at ambient temperature and high pressure, while LNG is stored at low temperature and approximately ambient pressure. In their respective storage conditions, LNG is a liquid, while CNG is a supercritical fluid (a fluid at a temperature and pressure above the critical point where distinct liquid and gas phases do not exist). Also, CNG does not require an expensive cooling process and cryogenic tanks but CNG does require a much larger volume to store the energy equivalent of gasoline and the use of very high pressures (3000−4000 psi). As a consequence of this, LNG is often used for transporting natural gas over large distances, in ships, trains, or pipelines, where the gas is converted into CNG before distribution to the consumers.

The combustion of CNG produces fewer undesirable gases than many other fuels. Also, CNG is safer than other fuels in the event of a spill because natural gas is lighter than air and disperses quickly when released. CNG may be found above oil deposits or may be collected from landfills or wastewater treatment plants where it is known as biogas. CNG is produced by compressing natural gas to less than 1% of the volume it occupies at standard atmospheric pressure. It is stored and distributed in hard containers at a pressure on the order of 2900−3600 psi, usually in cylindrical-shaped or spherical-shaped containers. Also, gas can be transported in containers at high pressures, typically 1800 psig for a rich gas (significant amounts of ethane, propane, etc.) to roughly 3600 psig for a lean gas (mainly methane). Gas at these pressures is termed CNG.

CNG has been used as a transportation fuel, mostly in public transit as an alternative to conventional fuels (gasoline or diesel). CNG, which is compressed at over 3000 psi to 1% of the volume the gas would occupy at normal atmospheric pressure, can be burned in an internal combustion engine that has been appropriately modified. Compared to gasoline, CNG vehicles emit far less carbon monoxide, nitrogen oxides (NO_x), and particulates. The main disadvantage of CNG is its low energy density compared with liquid fuels. A gallon of CNG has only a quarter of the energy in a gallon of gasoline. CNG vehicles therefore require big, bulky fuel tanks, making CNG practical mainly for large vehicles such as buses and trucks. The filling stations can be supplied by pipeline gas, but the compressors needed to get the gas to 3000 psig can be expensive to purchase, maintain, and operate.

An alternative approach has dedicated transport ships carrying straight long, large diameter pipes in an insulated cold storage cargo package. The gas has to be dried, compressed, and chilled for storage onboard. By careful control of temperature, more gas should be transported in any ship of a given payload capacity, subject to volume limitation and amount and weight of material of the pipe (pressure and safety considerations). Suitable compressors and chillers are needed but would be much less expensive than a natural gas liquefier, and would be standard, so that costs could be further minimized. According to the proposers, the terminal facilities would also be simple and hence would be of low cost.

5.5.5 Gas-to-solid

Gas can be transported as a solid, with the solid being the gas hydrate (Børrehaug and Gudmundsson, 1996; Gudmundsson, 1996; Gudmundsson, and Børrehaug, 1996; Gudmundsson et al., 1997, 1998; Taylor et al., 2003). NGH is the product of mixing natural gas with liquid water to form a stable water crystalline ice-like substance (Chapter 1: History and use and Chapter 9: Gas condensate). The transport of NGHs is (with due consideration of the safety issues) a viable alternative to LNG or pipelines for the transportation of natural gas from source to demand. Consideration of the various aspects leads to the conclusion that transporting natural gas as the solid gas hydrates can be done at a higher temperature and at a lower pressure than the transportation of LNG and the risk of ignition in transport is much lower remembering that gas hydrates may decompose explosively with simultaneous ignition.

The gas-to-solids process involves three stages: (1) production, (2) transportation, and (3) regasification. NGHs are created when certain small molecules, particularly methane, ethane, and propane, stabilize the hydrogen bonds within water to form a three-dimensional cage-like structure with the gas molecule trapped within the cages. A cage is made up of several water molecules held together by hydrogen bonds and the solid gas hydrate has a snow-like appearance. Hydrates are formed from natural gas in the presence of liquid water provided the pressure is above and the temperature is below the equilibrium line of the phase diagram of the gas and liquid water.

For the most part, in the crude oil industry and in the natural gas industry, gas hydrates are a pipeline nuisance and safety hazard, and require considerable care by the operators to ensure that they do not form as they can block pipelines if precautions, such as methanol injection, are not taken. On the other hand, vast quantities of gas hydrate have been found in permafrost and at the seabed in depths below 1500 ft (500 m), and if properly exploited could become the major energy source in the next 30 years.

For gas transport, NGHs can be deliberately formed by mixing natural gas and water at approximately 1175–1500 psi and 2–10°C (35–50°F). If the slurry is refrigerated to approximately −15°C (5°F), it decomposes very slowly at atmospheric pressure, so that the hydrate can be transported by ship to market in simple insulated under near adiabatic conditions—conditions that occur without transfer of heat or matter between a thermodynamic system and its surroundings.

At the market, the slurry is melted back to gas and water by controlled warming for use after appropriate drying in electricity power generation stations or other requirements. The hydrate mixture yields up to 5600 ft³ (approximately 160 m³) of natural gas per ton of hydrate, depending on the manufacturing process. The manufacture of the hydrate could be carried out using mobile equipment for onshore and ship for offshore using a floating production, storage and offloading vessel with minimal gas processing (cleaning, etc.) prior to hydrate formation, which is attractive commercially.

The water can be used at the destination if there is water shortage, or returned as ballast to the hydrate generator and, since it is saturated with gas, will not take more gas into solution. Process operability of continuous production of hydrate in a large-scale reactor, long-term hydrate storage, and controlled regeneration of gas from storage has all been demonstrated.

The hydrate mixture can be stored at normal temperatures (0 to −10°C; 32 to 14°F) and pressures (10–1 atm) where 1 m³ of hydrate should contain about 160 m³ gas per m³ of water. This concentration of gas is attractive as it is easier to produce, safer and cheaper to store compared to the 200 m³ per 1 m³ of compressed gas (high pressure ca. 3000 psig) or the 637 m³ gas per 1 m³ of LNG (low temperatures of −162°C, −260°F).

Gas storage in hydrate form becomes especially efficient at relatively low pressures where substantially more gas per unit volume is contained in the hydrate than in the free state or in the compressed state when the pressure has dropped. When compared to the transportation of natural gas by pipeline or as LNG, the hydrate concept has lower capital and operating costs for the movement of quantities of natural gas over adverse conditions.

Thus gas hydrate is very effective for gas storage and transport as it eliminates low temperatures and the necessity of compressing the gas to high pressures. To recall, dry hydrate pellets yield approximately 160 volumes of gas at standard conditions from 1 volume of hydrate compared to the approximately 637 volumes of gas per volume of LNG. This is a considerable volume penalty (and hence transport cost) if considered in isolation, with the cheaper ships for hydrate transport the process could be economic.

5.5.6 Gas-to-power

Electric power can be an intermediate product, such as in the case of mineral refining in which electricity is used to refine bauxite into aluminum; or it can be an end product that is distributed into a large utility power grid. Thus the concept of gas-to-power (GTP, sometimes referred to as *gas-to-grid*) is not a new thought but is certainly worthy of consideration in this section.

Currently, much of the transported gas destination is fuel for electricity generation. Electricity generation at or near the storage facility and transportation by cable to the destination(s) (GTP) is possible. Thus for instance offshore or isolated gas could be used to fuel an offshore power plant (may be sited in less hostile waters), which would generate electricity for sale onshore or to other offshore customers. Unfortunately, installing high-power lines to reach the shoreline appear to be almost as expensive as pipelines, so that GTP could be viewed as defeating the purpose of an alternative cheaper solution for transporting gas. There is significant energy loss from the cables along the long-distance transmission lines, more so if the power is AC rather than DC; additionally, losses also occur when the power is converted to DC from AC and when it converted from the high voltages used in the transmission to the lower values needed by the consumers.

Some observers consider having the energy as gas at the consumers' end gives greater flexibility and better thermal efficiencies, because the waste heat can be used for local heating and desalination. This view is strengthened by the economics as power generation uses approximately 1 million scf/day of gas for every 10 MW of power generated, so that even large generation capacity would not consume much of the gas from larger fields, and thus not generate large revenues for the gas producers. Nevertheless, GTP has been an option much considered in the United States for getting energy from the Alaskan gas and oil fields to the populated areas.

There are other practical considerations to note such as if the gas is associated gas, then if there is generator shutdown and no other gas outlet, the whole oil production facility might also have to be shut down, or the gas released to flare. Also, if there are operational problems within the generation plant the generators must be able to shut down quickly (on the order of 60 seconds, or less) to keep a small incident from escalating to a major incident. Additionally, the shutdown system itself must be safe so that any plant that has complicated processes that requires a purge cycle or a cool-down cycle before it can shut down is clearly unsuitable (Ballard, 1965). Finally, if the plant cannot shut down easily and/or be able to start up again quickly (perhaps in an hour), operators will be hesitant to ever shut down the process, for fear of financial retribution from the power distributors.

5.5.7 Gas-to-liquids

In gas-to-liquids (GTL) transport, the natural gas is converted to a liquid (Fig. 5.1), such as methanol (Fig. 5.2), and transported as such (Knott, 1997; Skrebowski, 1998; Thomas, 1998; Gaffney Cline and Associates, 2001). In the process, methane is first mixed with steam and converted to syngas (mixtures of carbon monoxide and hydrogen, $CO + H_2$) by one of a number of routes using suitable new catalyst technology. Thus:

In the steam-methane reforming process, methane reacts with steam under pressure (45−360 psi) in the presence of a catalyst to produce hydrogen, carbon monoxide, and a relatively small amount of carbon dioxide. The process is endothermic

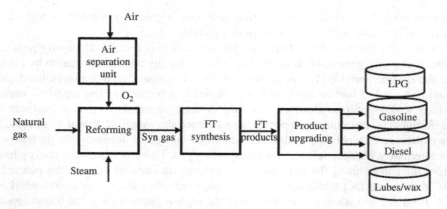

Figure 5.1 Production of liquids from natural gas.

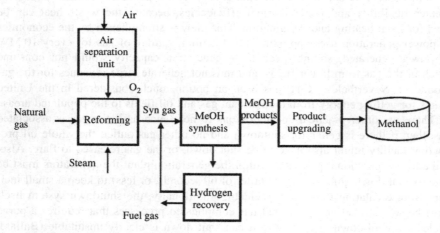

Figure 5.2 Production of methanol from natural gas.

insofar as heat must be supplied to the process for the reaction to proceed. Subsequently, in what is referred to as the *water-gas shift reaction*, the carbon monoxide and steam are reacted using a catalyst to produce carbon dioxide and more hydrogen. In a final process step (pressure-swing adsorption step) carbon dioxide and other impurities are removed from the gas stream, leaving essentially pure hydrogen. Steam reforming can also be used to produce hydrogen from other fuels, such as ethanol, propane, or even gasoline.

$$CH_4 + H_2O \rightarrow CO + 3H_2 \text{ (steam-methane reforming reaction)}$$

$$CO + H_2O \rightarrow CO_2 + H_2 \text{ (+ small amount of heat) (water } - \text{ gas shift reaction)}$$

$$2H_2 + CO \rightarrow CH_3OH \text{ (methanol synthesis)}$$

Also:

$$2CH_3OH \rightarrow CH_3OCH_3 + H_2O$$

$$CH_3OCH_3 \rightarrow C_2H_4 + H_2O$$

The ethylene is polymerized and hydrogenated to give gasoline with hydrocarbon constituents having five or more carbon atoms that constitute approximately making up 80% w/w of the fuel.

In the partial oxidation process, methane and other hydrocarbons in natural gas react with a limited amount of oxygen (typically from air) that is not enough to completely oxidize the hydrocarbons to carbon dioxide and water. With less than the stoichiometric amount of oxygen available, the reaction products contain primarily hydrogen and carbon monoxide (and nitrogen, if the reaction is carried out with air rather than pure oxygen), and a relatively small amount of carbon dioxide and other compounds. Subsequently, in a water−gas shift reaction, the carbon monoxide reacts with water to form carbon dioxide and more hydrogen.

Partial oxidation is an exothermic process insofar as heat is evolved during the process. It is, typically, a much more rapid reaction than steam reforming and requires a smaller reactor vessel. As can be seen in chemical reactions of partial oxidation, this process initially produces less hydrogen per unit of the input fuel than is obtained by steam reforming of the same fuel.

$$CH_4 + \tfrac{1}{2}O_2 \rightarrow CO + 2H_2 \text{ (Partial oxidation of methane)}$$

$$CO + H_2O \rightarrow CO_2 + H_2 \text{ (water − gas shift reaction)}$$

The synthesis gas (syngas) is then converted into a liquid using a Fischer−Tropsch process (in the presence of a catalyst) or an oxygenation method (mixing synthesis gas with oxygen in the presence of a suitable catalyst).

$$CO + nH_2 \rightarrow H(-CH_2-)_xH + H_2O$$

The produced liquid can be a fuel, usually a clean burning motor fuel (syncrude) or lubricant, or ammonia or methanol or some precursor for plastics manufacture (e.g., urea, dimethyl ether (DME), which is also used as (1) a transportation fuel, (2) a LPG substitute or power generation fuel as well as (3) a chemical feedstock).

DME (CH_3OCH_3) is a colorless gas at ambient temperature, chemically stable, with a boiling point of $-25°C$ ($-13°F$). As its vapor pressure is approximately 0.6 mPa at 25°C (77°F), dimethyl ether is easily liquefied. Liquid dimethyl ether is colorless. The viscosity of the liquid is on the order of 0.12−0.15 kg/ms, which is almost equivalent to the viscosity of liquid propane or liquid butane. Dimethyl ether has a Wobbe index (ratio of calorific value and flow resistance of gaseous fuel) 52−54 that of natural gas, cooking stove for natural gas can be used for dimethyl ether without any modification. The thermal efficiency and emissions with dimethyl ether are almost same as with natural gas. Its physical properties are so similar to those of LPG that dimethyl ether can be distributed and stored.

The reaction formulas and reaction heat concerning the exothermic synthesis of dimethyl ether is:

$$3CO + 3H_2 \rightarrow CH_3OCH_3 + CO_2$$

$$2CO + 4H_2 \rightarrow CH_3OCH_3 + H_2O$$

$$2CO + 4H_2 \rightarrow 2CH_3OH$$

$$2CH_3OH \rightarrow CH_3OCH_3 + H_2O$$

$$CO + H_2O \rightarrow CO_2 + H_2$$

It is more important to control the reaction temperature than in the case of methanol synthesis, because the higher equilibrium conversion of dimethyl ether synthesis could give higher reaction heat, and hot spot in the reactor could damage the catalyst. Synthesis gas produced by high temperature gasification of coal, oil residue and woody biomass has a composition of hydrogen-to-carbon monoxide ratio of approximately $0.5-1$, which should be adjusted to $H_2/CO = 2$ before the dimethyl ether synthesis by shift converter.

Methanol is a GTL option that has been in commission since the mid-1940s. While methanol produced from gas was originally a relatively inefficient conversion process, optimized technology has improved the efficiency. Methanol can be used in internal combustion engines as a fuel, but the current market for methanol as a fuel is limited, although the development of fuel cells for motor vehicles may change this. Methanol is best used as a basic chemical feedstock for the manufacture of plastics.

Other GTL processes are being developed to produce clean fuels, e.g., syncrude, diesel, or many other products including lubricants and waxes, from gas but require complex (expensive) chemical plant with novel catalyst technology.

5.5.8 Gas-to-commodity

The gas-to-commodity (GTC) concept involves use of the components in natural gas, methane, ethane, propane, n- and iso-butanes and pentanes that useful in their own right. The higher paraffin derivatives are particularly valuable for a wealth of chemicals and polymer precursors such as acetic acid, formaldehyde, olefin derivatives, polyethylene, polypropylene, acrylonitrile, ethylene glycol, etc., as well as portable premium fuels, such as propane. In addition, methane can be converted via syngas to methanol, ammonia, syncrude, lubricant, or some precursor for chemicals manufacture, e.g., DME and urea, and then used to make chemicals for export.

Thus, when using the GTC concept, the gas is converted to thermal or electrical power, which is then used in the production of the commodity, which is then sold, on the open market. It is the energy from the gas, heat via electricity or direct combustion, and not the components of the GTL concept that is used. The gas energy is, in essence, transported via the commodity.

In some cases, the gas feedstock may be suitable for hydrogen production which can be obtained through various thermochemical methods utilizing methane (natural gas), liquefied petroleum gas, coal gasification, or biomass (biomass gasification), from electrolysis of water, or by a thermolysis process. Hydrogen or H_2 gas is highly flammable and will burn at concentrations as low as 4% H_2 in air. For automotive applications, hydrogen is generally used in two forms: internal combustion or fuel cell conversion. In combustion, it is essentially burned as conventional gaseous fuels are, whereas a fuel cell uses the hydrogen to generate electricity that in turn is used to power electric motors on the vehicle. Hydrogen gas must be produced and is therefore is an energy storage medium, not an energy source. The energy used to produce it usually comes from a more conventional source. Hydrogen holds the promise of very low vehicle emissions and flexible energy storage but the technical challenges (such as the flammability of the gas) required to realize these benefits may delay the widespread implementation for several decades (Rigden, 2003). However, there are disadvantages that must be given serious consideration.

For example, hydrogen poses a number of hazards to human safety, from potential detonation and fires when mixed with air to causing an asphyxia. In addition, liquid hydrogen is a cryogenic material and does present dangers (such as frostbite). Also, hydrogen can dissolve in many metals, and, in addition to leaking out, may have adverse effects such as hydrogen embrittlement which can lead to cracks and explosions. Hydrogen gas leaking into external air may spontaneously ignite.

In fact, on occasions when the use of hydrogen as a fuel is proposed, the memory of the Hindenburg disaster always arises. At 7.25 p.m. (local time) on May 6, 1937, the German passenger airship LZ 129 Hindenburg caught fire and was destroyed during its attempt to dock with its mooring mast at the Lakehurst Naval Air Station in Manchester Township, New Jersey, United States. The airship had 97 people on board (36 passengers and 61 crewmen) of which there were 36 fatalities (13 passengers, 22 crewmen, and 1 worker on the ground).

References

ASTM, 2017. Annual Book of Standards. ASTM International, West Conshohocken, PA.
Arnold, K., Stewart, M., 1999. Surface Production Operations, Volume 2: Design of Gas-Handling Systems and Facilities, second ed. Gulf Professional Publishing, Houston, TX.
Ballard, D., 1965. How to operate quick-cycle plants. Hydrocarbon Process. Crude oil Refiner 44 (4), 131.
Burruss, R.C., Ryder, R.T., 2003. Composition of Crude Oil and Natural Gas Produced from 14 Wells in the Lower Silurian "Clinton" Sandstone and Medina Group, Northeastern Ohio and Northwestern Pennsylvania. Open-File Report 03−409, United States Geological Survey, Reston, VA.

Burruss, R.C., Ryder, R.T., 2014. Composition of natural gas and crude oil produced from 10 wells in the lower Silurian "Clinton" Sandstone, Trumbull County, Ohio. In: Ruppert, L. F., Ryder, R.T. (Eds.), Coal and Crude Oil Resources in the Appalachian Basin; Distribution, Geologic Framework, and Geochemical Character. United States Geological Survey, Reston, VA, Professional Paper 1708.

Burt, S.L., 2016. Who Owns the Right to Store Gas: A Survey of Pore Space Ownership in U.S. Jurisdictions. <http://www.duqlawblogs.org/joule/wp-content/uploads/2016/07/Who-Owns-the-Right-to-Store-Gas-A-Survey-of-Pore-Space-Ownership-in-U.S.-Jurisdictions-.pdf>.

Børrehaug, A., Gudmundsson, J.S., 1996. Gas Transportation in Hydrate Form, EUROGAS 96, 3—5 June, Trondheim, pp. 35—41.

Crow, W., Williams, B., Carey, J.W., Celia, M.A., Gasda, S. 2008. Wellbore integrity analysis of a natural CO_2 producer. In: Proceedings. 9th International Conference on Greenhouse Gas Control Technologies, Washington, D.C., November 16—20, 2008, Elsevier, New York.

Faramawy, S., Zaki, T., Sakr, A.A.-E., 2016. Natural gas origin, composition, and processing: a review. J. Nat. Gas Sci. Eng. 34, 34—54.

Gaffney Cline and Associates, 2001. GTL Discussion Paper. Prepared for the Gas to Liquids Taskforce, Australian Dept. of Industry, Science & Resources, Commonwealth of Australia, June.

Gary, J.G., Handwerk, G.E., Kaiser, M.J., 2007. Crude oil Refining: Technology and Economics, fifth ed. CRC Press, Taylor & Francis Group, Boca Raton, FL.

Godec, M., Koperna, G., Petrusak, R., Oudino, A., 2014. Enhanced gas recovery and CO_2 storage in gas shales: a summary review of its status and potential. Energy Procedia 63, 5849—5857.

Gudmundsson, J.S., 1996. Method for Production of Gas Hydrate for Transportation and Storage, U.S. Patent No. 5,536,893.

Gudmundsson, J.S., Børrehaug, A., 1996. Frozen hydrate for transport of natural gas. In: Proc. 2nd International Conf. Natural Gas Hydrates, June 2—6, Toulouse, pp. 415—422.

Gudmundsson, J.S., Andersson, V., Levik, O.I., 1997. Gas Storage and Transport Using Hydrates, Offshore Mediterranean Conference, Ravenna, March 19—21.

Gudmundsson, J.S., Andersson, V., Levik, O.I., Parlaktuna, M., 1998. Hydrate concept for capturing associated gas. In: Proceedings. SPE European Crude oil Conference, The Hague, The Netherlands, 20—22 October 1998.

Guo, P., Jing, S., Peng, C., 2014. Technologies and countermeasures for gas recovery enhancement. Nat. Gas Industry B1, 96—102.

Hsu, C.S., Robinson, P.R. (Eds.), 2017. Handbook of Crude oil Technology, C.S. Springer International Publishing AG, Cham.

Katz, D.L., Coats, K.H., 1968. Underground Gas Storage of Fluids. Ulrich Books Inc, Ann Arbor, MI.

Kidnay, A., McCartney, D., Parrish, W., 2011. Fundamentals of Natural Gas Processing. CRC Press, Taylor & Francis Group, Boca Raton, FL.

Knott, D., 1997. Gas-to-liquids projects gaining momentum as process list grows. Oil Gas J. June 23, 16—21.

Loskutova, Yu.V., Yadrevskaya, N.N., Yudina, N.V., Usheva, N.V., 2014. Study of viscosity-temperature properties of oil and gas-condensate mixtures in critical temperature ranges of phase transitions. Procedia Chem. 10, 343—348.

Mamora, D.D., Seo, J.G., 2002. Enhanced gas recovery by carbon dioxide sequestration in depleted gas reservoirs. SPE Paper No. 77347. In: Proceedings. SPE Annual Technical

Conference and Exhibition, Houston, Texas. September 29—October 2. Society of Petroleum Engineers, Richardson, TX.

Mokhatab, S., Poe, W.A., Speight, J.G., 2006. Handbook of Natural Gas Transmission and Processing. Elsevier, Amsterdam.

Parkash, S., 2003. Refining Processes Handbook. Gulf Professional Publishing, Elsevier, Amsterdam.

Pooladi-Darvish, M., Hong, H., Theys, S., Stocker, R., Bachu, S., Dashtgard, S., 2008. CO_2 injection for enhanced gas recovery and geological storage of CO_2 in the Long Coulee Glauconite F Pool, Alberta. SPE Paper No. 115789, SPE Annual Technical Conference and Exhibition, Denver, Colorado. September 21—24. Society of Petroleum Engineers, Richardson, TX.

Riazi, M., Eser, S., Agrawal, S., Peña Díez, J., 2013. Crude Oil Refining and Natural Gas Processing. Manual 58 MNL58. ASTM International, West Conshohocken, PA.

Rigden, J.S., 2003. Hydrogen: The Essential Elements. Harvard University Press, Cambridge, MA.

Rojey, A., Jaffret, C., Cornot-Gandolph, S., Durand, B., Jullin, S., Valais, M., 1997. Natural Gas Production, Processing, Transport. Editions Technip, Paris.

Seo, J.G., Mamora, D.D., March 10—12, 2003. Enhanced gas recovery by carbon dioxide sequestration in depleted gas reservoirs. SPE Paper No. 81200. In: Proceedings. SPE/EPA/DOE Exploration and Production Environmental Conference, San Antonio, Texas.

Sim, S.S.K., Turta, A.T., Signal, A.K., Hawkins, B.F., June 17—19, 2008. Enhanced gas recovery: factors affecting gas-gas displacement efficiency. CIMPC Paper No. 2008—145. In: Proceedings. Canadian International Petroleum Conference/SPE Gas Technology Symposium 2008 Joint Conference, Calgary, Alberta, Canada.

Skrebowski, C., 1998. Gas-to-Liquids or LNG? Crude oil Review. January, pp. 38—39.

Speight, J.G., 2007. Natural Gas: A Basic Handbook. GPC Books, Gulf Publishing Company, Houston, TX.

Speight, J.G., 2013. The Chemistry and Technology of Coal, fifth ed. CRC Press, Taylor & Francis Group, Boca Raton, FL.

Speight, J.G., 2014a. The Chemistry and Technology of Crude Oil, fifth ed. CRC Press, Taylor & Francis Group, Boca Raton, FL.

Speight, J.G., 2014b. Oil and Gas Corrosion Prevention. Gulf Professional Publishing, Elsevier, Oxford.

Speight, J.G., 2015. Handbook of Crude oil Product Analysis, second ed. John Wiley & Sons Inc, Hoboken, NJ.

Speight, J.G., 2016. Handbook of Hydraulic Fracturing. John Wiley & Sons Inc, Hoboken, NJ.

Speight, J.G., 2017. Handbook of Crude Oil Refining. CRC Press, Taylor & Francis Group, Boca Raton, FL.

Speight, J.G., 2018. Handbook of Natural Gas Analysis. John Wiley & Sons Inc, Hoboken, NJ.

Taylor, M., Dawe, R.A., Thomas, S., 2003. Fire and ice: gas hydrate transportation — a possibility for the Caribbean Region. Paper no. SPE 81022. In: Proceedings. SPE Latin American and Caribbean Crude oil Engineering Conference. Port-of-Spain, Trinidad, West Indies. April 27—30. Society of Crude oil Engineers, Richardson, TX.

Thomas, M., 1998. Water into Wine: Gas-to-Liquids Technology the Key to Unlocking Future Reserves, Euroil, May 17—21.

Ward, D.E., Pierce, A.P., 1973 Helium. Professional Paper No. 820. In: United States Mineral Resources, US Geological Survey, Reston, VA, pp. 285—290.

Further reading

CFR, October 2017. Part 192: Transportation of Natural and Other Gas by Pipeline: Minimum Federal Safety Standards. Code of Federal Regulations. <https://www.ecfr.gov/cgi-bin/text-idx?SID = bc6ba2aedb111021352940bb1e5e9811&mc = true&node = pt49.3.192&rgn = div5>.

Gudmundsson, J.S., Hveding, F., Børrehaug, A., 1995. Transport of natural gas as frozen hydrate. In: Proc. 5th International Offshore and Polar Engineering Conf., The Hague, June 11−16, vol. I, pp. 282−288.

Manning, F.S., Thompson, R.E., 1991. Oil Field Processing of Crude oil, Volume 1: Natural Gas. Pennwell Publishing Company, Tulsa, OK.

US EPA. 2014. Interim Chemical Accident Prevention Advisory − Design of LPG Installations at Natural Gas Processing Plants. EPA 540-F-14-001 United States Environmental Protection Agency, Washington, DC.

Part II

Gas Processing

History of gas processing

6

6.1 Introduction

Although naturally occurring gas has been known since ancient times (Chapter 1: History and use), gas processing (gas cleaning) is a relatively modern innovation that commenced with the Industrial Revolution in the late 18th century. In the modern world, natural gas is recognized as an important component of the world energy supply. In fact, natural gas currently supplies more than one-half of the energy consumed by residential and commercial customers and contributes approximately 41% of the energy used by industries in the United States.

Along with the increased use of natural gas, distribution companies have always been subject to regulation by state and local governments. In 1938, however, with the growing importance of natural gas, concern over the heavy concentration of the natural gas industry, and the monopolistic tendencies of interstate pipelines to charge higher than competitive prices due to their market power, the government of the United States began to regulate the interstate natural gas industry with passage of the Natural Gas Act. The Act was intended to protect consumers from possible abuses such as unreasonably high prices. The Act gave the Federal Power Commission (FPC) jurisdiction to regulate the transportation and sale of natural gas in interstate commerce. The FPC was charged with regulating the rates that were charged for interstate natural gas delivery and with certifying new interstate pipeline construction if it was consistent with the public convenience and necessity. Along with this regulation came the recognition of the need for cleaning the gas of the impurities that accompanied the gas from the well. Furthermore, the need to protect the environment has brought about the need for further regulations (Chapter 10: Energy security and the environment).

Natural gas, as transported to domestic and industrial consumers, is much different from the natural gas that appears at the wellhead (Chapter 3: Unconventional gas). Although the processing of natural gas is in many respects less complicated than the processing and refining of crude oil, processing (refining) natural gas is equally as necessary before sales. Natural gas used by consumers is composed almost entirely of methane (usually >95% by volume). However, natural gas found at the wellhead, although still composed primarily of methane (usually >65% by volume), is by no means as pure as required by sales specifications.

Raw (impure) natural gas comes from three types of wells: (1) crude oil wells, (2) gas wells, and (3) condensate wells. Natural gas that is produced from crude oil wells is typically termed *associated gas* and can exist separate from the crude oil in the reservoir (*free gas*) or dissolved in the crude oil (*dissolved gas*). Natural gas from gas and condensate wells, in which there is little or no crude oil, is

Natural Gas. DOI: https://doi.org/10.1016/B978-0-12-809570-6.00006-0

nonassociated gas. Gas wells typically produce only natural gas, while condensate wells produce natural gas along with a liquid hydrocarbon *condensate*. Whatever the source of the natural gas, once separated from crude oil (if present) it commonly exists in mixtures with other hydrocarbons, principally ethane, propane, butane, and pentanes. In addition, raw natural gas contains water vapor (H_2O), hydrogen sulfide (H_2S), carbon dioxide (CO_2), helium (He), nitrogen (N_2), and other miscellaneous compounds.

Natural gas processing (refining) consists of separating all of the various hydrocarbons and fluids from the pure natural gas, to produce *pipeline quality* dry natural gas (i.e., gas that meets a specified analytical composition). Pipeline companies usually impose restrictions on the make-up of the natural gas that is allowed into the pipeline. Thus before the natural gas can be transported it must be purified and while the ethane, propane, butane, and pentanes must be removed from the methane, they are by no means waste products. These higher molecular weight hydrocarbons, once extracted, are termed *natural gas liquids* (NGLs) and are used in other products. But whatever the composition of the raw gas, it must be processed before transportation and use.

As already noted (Chapter 1: History and use), the history of natural gas extends into antiquity, but the history of gas processing is somewhat more recent. In fact, the history of gas processing is, of course, carefully intertwined with the development of gas use and gas technology (Murphy et al., 2005; Speight, 2013). Therefore any treatise relating to the history of gas processing must refer to the evolution of the use and development of gas production technology. This must, of necessity, include reference to the original commercial gas industry that involved the production of gas from coal. It is from such an industry that the modern gas processing industry evolved.

6.2 Coal gas

By way of introduction to this important topic, coal was known to be used in Britain during the 4000−5000 years ago where it has been detected as forming part of the composition of funeral pyres. It was also commonly used as a fuel in the later period of the Roam occupation. Evidence of trade in coal is plentiful and shows that coal was in common use as a source of fuel in dwellings. Carbon forms more than 60% w/w of coal and is dependent on coal *rank*, with higher rank coals containing less hydrogen, oxygen, and nitrogen, until 95% purity of carbon is achieved at anthracite (Speight, 2013). Methane gas (coal bed methane, Chapter 1: History and use) is another component of coal and is dangerous, as it can cause coal seam explosions, especially in underground mines, and may cause the coal to spontaneously combust.

Throughout history, especially since the time of the Industrial Revolution, the combustion of coal has been the recognized fuel for the generation of heat and power. Combustion is the conversion of primary chemical energy contained in fuels such as coal into heat (secondary energy) through the process of oxidation. Combustion therefore is the technical term for the chemical reaction of oxygen with the combustible components of fuels including the release of energy.

Prior to the development of natural gas supplies and transmission in the United States during 1940s and 1950s, virtually all fuel and lighting gas was manufactured, and the by-product coal tars were, at times, an important chemical feedstock for the chemical industries. The development of manufactured gas paralleled that of the industrial revolution and urbanization. Thus, before natural gas (and by inference the natural gas industry), there was coal gas (also called town gas) (Sugg, 1884; Stewart, 1958; Higman and Van Der Burgt, 2003). Town gas is a gas manufactured from coal produced for sale to consumers and municipalities. The terms manufactured gas, syngas (synthetic natural gas), and Hygas are also common. Depending on the processes used for gas production, the gas is a mixture of different gases of different heat value, i.e., hydrogen, carbon monoxide, methane, and higher molecular weight volatile hydrocarbons with small amounts of gases such carbon dioxide and nitrogen that detract from the heat value of the mixture.

Coal gas is the gaseous mixture (mainly hydrogen, methane, and carbon monoxide) produced by the destructive distillation (i.e., heating in the absence of air) of bituminous coal. Sometimes steam is added to react with the hot coke, thus increasing the yield of gas; coal tar and coke are obtained as by-products. Thus, coal gas is a flammable gaseous fuel that was obtained when coal is heated strongly in the absence of air and was supplied to the consumer by way of a piped distribution system. The term *town gas* is a more general term referring to manufactured gaseous fuels produced for sale to consumers and municipalities. The facilities where the gas was produced were often known as a manufactured gas plant (MGP) or a gasworks—most towns and cities of any size had at least one gasworks.

Originally created as a by-product of the coke-producing process, the use for coal gas developed during the 19th century and the early 20th century. By-products from the production process included coal tar and ammonia, which were important chemical feedstock for the dyestuffs industry and for the evolving chemical industry with a wide range of artificial dyes being made from coal gas and coal tar.

Combustion occurs in a combustion chamber; other control units are required for fuel supply and fuel distribution, combustion air supply, heat transfer, exhaust gas cleaning and for discharge of exhaust gases and combustion residues (ash, slag). Solid fuels are fired on a fixed or fluidized bed or in a flue dust/air mixture. Liquid fuels are fed to the burning chamber together with the combustion air as mist. Gaseous fuels are mixed with combustion air already in the burner. Combustion processes proceed with high temperatures (up to 1000°C, 1800°F) and above). The oxygen required for the combustion is supplied as part of the combustion air fed to the process. From that a considerable volume of exhaust gas (flue gas and off gas) is produced together with, in the case of coal combustion, considerable quantities of residues, such as slag and ash.

The exhaust gases of combustion plants contain the reaction products of fuel and combustion air and residual substances such as particulate matter (PM, dust), sulfur oxides, nitrogen oxides, and carbon monoxide. When burning coal, HCl and HF may be present in the flue gas as well as hydrocarbons and heavy metals in case of incineration of waste materials.

In many countries, as part of a national environmental protection program, exhaust gases must comply with strict governmental regulations regarding the limit values of pollutants such as dust, sulfur and nitrogen oxides, and carbon monoxide. To meet these limit values combustion plants are equipped with flue gas cleaning systems such as gas scrubbers and dust filters. In all cases, knowledge of the fuel composition is important for an optimum and economical combustion process. Increasing percentage of incombustible (inert) fuel components reduces the gross and net calorific value of the fuel and increases contamination of the furnace walls. Increasing water content raises the water dew point and consumes energy to evaporate water in the flue gas. The sulfur contained in the fuel is burnt (oxidized) to sulfur dioxide and sulfur trioxide, which, at temperatures below the dew point, may lead to the formation of aggressive sulfurous and sulfuric acids.

6.2.1 History

Coal gas is produced when coal is heated in the absence of air in an enclosed chamber. When bituminous coal is heated to a temperature of about 400°C it softens and coalesces, giving off water vapor, rich gas, and tar. As the temperature is raised to 1000°C (1800°F) the remaining volatile matter, ultimately hydrogen, is almost entirely driven off leaving coke residue. The gas produced by this process consists largely of hydrogen, carbon monoxide, and methane. But in its raw state it also contains condensable products such as tar and ammonia, which are removed in the purification process.

The early history of gas production by carbonization starts with the Flemish scientist Jan Baptista van Helmont (1577–1644) who discovered that a wild spirit escaped from heated wood and coal, and, thinking that it differed little from the chaos of the ancients, he named it gas in his Origins of Medicine (published c.1609). Among several others who carried out similar experiments, were Johann Becker of Munich (in, or about, 1681) and about 3 years later John Clayton of Wigan England, the latter amusing his friends by lighting, what he called, Spirit of the Coal.

In other examples of the use of gas in demonstrations, Prof. Jan Pieter Minckelers lit his lecture room at the University of Louvain in 1783 and Lord Dundonald lit his house at Culross, Scotland, in 1787, the gas being carried in sealed vessels from the local tar works. In France, Phillipe Lebon patented a gas-burning stove in 1799 and demonstrated street lighting in 1801. Other demonstrations followed in France and in the United States, but, it is generally recognized that the first commercial gas works was built by the London and Westminster Gas Light and Coke Company in Great Peter Street in 1812 laying wooden pipes to illuminate Westminster Bridge with gas lights on New Year's Eve in 1813. In 1816, Rembrandt Peale and four others established the Gas Light Company of Baltimore, the first manufactured gas company in the United States. In 1821, natural gas was being used commercially in Fredonia, New York, through the auspices of the

Fredonia Gas Light Company—the first American natural gas company. On the other side of the Atlantic Ocean, the first German gas works was built in Hannover in 1825 and by 1870 there were 340 gas works in Germany making town gas from coal, wood, peat, and other materials.

Working conditions in the Gas Light and Coke Company's Horseferry Road Works, London, in the 1830s were described by a French visitor, Flora Tristan, in her Promenades Dans Londres (Walks in London) (Tristan, 1840). Thus:

> *Two rows of furnaces on each side were fired up; the effect was not unlike the description of Vulcan's forge, except that the Cyclops were animated with a divine spark, whereas the dusky servants of the English furnaces were joyless, silent and benumbed. . . . The foreman told me that stokers were selected from among the strongest, but that nevertheless they all became consumptive after seven or eight years of toil and died of pulmonary consumption. That explained the sadness and apathy in the faces and every movement of the hapless men.*

The first public piped gas supply was to 13 gas lamps, each with three glass globes along the length of Pall Mall, London in 1807. The credit for this goes to the inventor and entrepreneur Fredrick Winsor and the plumber Thomas Sugg who made and laid the pipes. Digging up streets to lay pipes required legislative action and this delayed the development of street lighting and gas for domestic use. Meanwhile William Murdoch and his pupil Samuel Clegg were installing gas lighting in factories and work places, encountering no such impediments.

Gas clean up varied from processes to process but, in many processes, volatile matter driven from the coal mass passes upward through cast iron goosenecks into a common horizontal steel pipe (called the collecting main), which connects all the ovens in series. This unpurified foul gas contained water vapor, tar, light oils, solid particulates (coal dust), heavy hydrocarbons, and complex carbon compounds. The condensable materials were removed from the exhaust gas to obtain purified coke oven gas. As it left the heating chamber, the gas was initially cleaned with a weak ammonia spray, which condenses some tar and ammonia from the gas stream. This liquid condensate flowed down the collecting main until it reached a settling tank. Collected ammonia was used in the weak ammonia spray, while the rest was pumped to ammonia still. Collected coal tar was pumped to a storage tank to await sales or use as a fuel. The remaining gas was cooled as it passed through a condenser and then compressed by an exhauster unit. Any remaining coal tar was removed by a tar extractor, either by impingement against a metal surface or collection by an electrostatic precipitator (ESP). Further amounts of ammonia were removed by passing the gas through a saturator containing a 5%−10% solution of sulfuric acid where the ammonia reacted with sulfuric acid to form ammonium sulfate, which was crystallized and removed. The gas was further cooled, resulting in the condensation of naphthalene. The light oils were removed in an absorption tower containing water mixed with straw oil (a high boiling petroleum fraction). The straw oil acted as an absorbent for the light oils and was later heated to release the light oils for recovery and refinement. The last cleaning step was the removal of hydrogen sulfide from the gas that was usually achieved in a scrubbing tower (Fig. 6.1) after which the gas was suitable for use as fuel for the coke ovens, other plant combustion processes, or sold.

The more traditional term *coal gasification* refers to a process through which solid coal is converted into a low-to-medium energy fuel gas. Coal gasification has its origin in 1780 where it was used to produce a product referred to variously as town gas or blue gas. Coal gasification is the process by which coal is converted into a fuel gas. The process was first developed in about 1780 and was widely commercialized by the early 1900s. Before natural gas became widely available in the 1940s, many North American and European cities used coal gas as a heating and lighting fuel. It was referred to variously as *blue gas*, *producer gas*, *water gas*, *town gas*, or *fuel gas* depending on the method of production, gas properties, and gas use. Often using the same low-pressure mains for distribution, natural gas replaced fuel gas in most uses by the 1950s because of its greater heating value and lack of contaminants. In the modern world, gasification units have been installed at many refineries and the feedstock is not always coal but can be any carbonaceous material that is suitable for the task of producing a usable gaseous product (Speight, 2014b).

Nevertheless, the first process used to produce gaseous products from coal was the carbonization and partial pyrolysis of coal (Speight, 2013). The off gases liberated in the high-temperature carbonization (coking) of coal in coke ovens were collected, scrubbed, and used as fuel. Depending on the goal of the plant, the desired

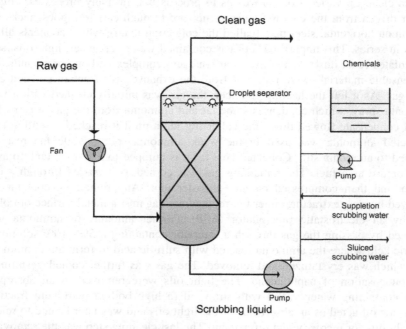

Figure 6.1 Representation of a scrubbing tower.

product was either a high-quality coke for metallurgical use, with the gas being a side product or the production of a high-quality gas with coke being the side product. Coke plants are typically associated with metallurgical facilities such as smelters and blast furnaces, while gas works typically served urban areas.

In the early years of MGP operations, the goal of a utility gas works was to produce the greatest amount of highly illuminating gas. The illuminating power of a gas was related to amount of soot-forming hydrocarbon derivative (illuminants) dissolved in it. These hydrocarbons gave the gas flame its characteristic bright yellow color. Gas works would typically use oily bituminous coals as feedstock and these coals would give off large amounts of volatile hydrocarbons into the coal gas, but would leave behind a crumbly, low-quality coke not suitable for metallurgical processes. Coal or coke oven gas typically had a heat value (CV) between 250 and 550 Btu/ft^3.

Coal gas was used for lighting, heating, and cooking. In 1802 William Murdoch (later spelled as Murdock; 1754−1839) installed gas burners at James Watt's (steam engine) factory, near Manchester. In 1812 gas lighting was used in the streets of London. In England objectors to gaslight argued that it was destroying the whaling industry, because demand for whale oil for lighting was declining. Murdoch is also reputed to have heated coal in his mother's teapot to produce gas. From this beginning, he discovered new ways of making, purifying, and storing gas; illuminating his house at Redruth (or his cottage at Soho) in 1792, the entrance to the premises of the Manchester Police Commissioner in 1797, the exterior of the factory of Boulton and Watt in Birmingham, England, and a large cotton mill in Salford, Lancashire in 1805.

Coal gas was produced using a retorting operation in which the coal, inside airtight retorts, was heated by a coke furnace. Air was forced through the coke to produce carbon monoxide which burned around the retorts. The temperature was high enough to drive off all volatiles over a period of hours. Tar and water (with dissolved ammonia) were formed in the air-cooled condensers. These liquids were pumped to the liquid separator. The gaseous impurities were hydrogen cyanide (HCN, 0.1% v/v), hydrogen sulfide (H_2S, 1.3% v/v), and carbon disulfide (CS_2, 0.04% v/v). The gas purification section of the plant produced gas suitable for heating and lighting. The gas was scrubbed to remove the last traces of ammonia.

Carbon disulfide was removed by reaction with hydrogen using a nickel catalyst in an atmosphere of hydrogen:

$$CS_2 + 2H_2 \rightarrow 2H_2S + C$$

$$CS_2 + 4H_2 \rightarrow 2H_2S + CH_4$$

In the following step, the gas was passed through iron oxide (ferric oxide, Fe_2O_3) to remove hydrogen sulfide:

$$Fe_2O_3 + 3H_2S \rightarrow Fe_2S_3 + 3H_2O$$

The hydrogen cyanide was removed (as Prussian blue, $Fe_7(CN)_{18}$) by washing with water. Prussian blue is a dark blue pigment produced by oxidation of ferrous ferrocyanide salts. The chemical formula is complex and is conventionally represented as $Fe_7(CN)_{18}$. Another name for the color is Berlin blue or, in painting, as Parisian blue or Paris blue. Prussian blue was the first modern synthetic pigment and is prepared as a fine colloidal dispersion because the compound is not soluble in water. It contains variable amounts of other ions and its appearance depends sensitively on the size of the colloidal particles. The remaining material was sold to sulfuric acid plants.

The ferric oxide was regenerated by reacting the ferric sulfide with an excess of oxygen:

$$2Fe_2S_3 + 9O_2 \rightarrow 2Fe_2O_3 + 6SO_2$$

The liquid separator produced ammonia solution and coal tar. Nearly all the ammonia was converted into fertilizer (ammonium sulfate).

$$2NH_3(aq) + H_2SO_4(aq) \rightarrow (NH_4)_2SO_4(aq)$$

Benzene, naphthalene, and other aromatic organics were distilled from the tar. The remaining viscous liquid was used for road making. In the gas holders, the gas was stored till required. This gas consisted of hydrogen 48%, methane 32%, carbon monoxide 8%, and ethylene (ethene) 2%. Overall, the goal of gas cleaning was tar removal and there are indications that gas manufacturer did not check the quality of the gas after the tar had been removed.

Fuel gas for industrial use was made using what was known as producer gas technology in which the gas was made by blowing air through an incandescent fuel bed (commonly coke or coal) in a gas producer. The reaction of fuel with insufficient air for total combustion produced carbon monoxide (CO) and the reaction, being exothermic, was self-sustaining. The addition of steam to the input air of a gas producer would result in an increase in the calorific value of the fuel gas by enriching it with carbon monoxide and hydrogen. However, the producer gas has a relatively low calorific value ($99-150$ Btu/ft^3), because the high-calorific gases (carbon monoxide and hydrogen) were diluted with inert nitrogen (from the air) and carbon dioxide (from the combustion):

$$2C_{coal} + O_2 \rightarrow 2CO \quad \text{exothermic producer gas reaction}$$

$$C_{coal} + H_2O \rightarrow CO + H_2 \quad \text{endothermic water gas reaction}$$

$$C_{coal} + 2H_2O \rightarrow CO_2 + 2H_2 \text{ endothermic}$$

$$CO + H_2O \rightarrow CO_2 + H_2 \quad \text{exothermic water gas shift reaction}$$

The problem of nitrogen dilution was overcome by the Siemens blue water gas (BWG) process that was developed in the 1850s. The incandescent fuel bed would be alternately blasted with air followed by steam. The air reactions during the blow cycle are exothermic, heating up the bed, while the steam reactions during the make cycle are endothermic and cool down the bed. The products from the air cycle contain noncalorific nitrogen and are exhausted out the stack, while the products of the steam cycle are kept as BWG. The product gas stream was composed almost entirely of carbon monoxide and hydrogen (with a calorific value of the order of 300 Btu/ft^3).

However, the BWG lacked illuminants; it would not burn with a luminous flame in a simple fishtail gas jet as existed prior to the invention of the gas mantle in the 1890s. Various attempts were made to enrich BWG with illuminants from gas oil in the 1860s. In 1875, the carbureted water gas process came into being and revolutionized the manufactured gas industry and was the standard technology until the end of manufactured gas era. A carbureted water gas generating set consisted of three elements: a producer (generator), a carburetor, and a superheater connected in series with gas pipes and valves. In the process, steam would be passed through the generator to make BWG. From the generator the hot water gas would pass into the top of the carburetor, where light petroleum oils would be injected into the gas stream. The light oils would be thermally cracked as they came in contact with the white-hot checker-work fire bricks inside the carburetor. The hot enriched gas would then flow into the superheater, where the gas would be further cracked by more hot fire bricks.

The production of commercial amounts of gas coal gasification can be traced to the 1850s. Gas producers were invented, and the water gas process was discovered. Mond Gas: In the 1850s, the Europeans also discover that using coal instead of coke in a producer results in producer gas that contains ammonia and coal tar. The year 1875 saw the invention of the carbureted water gas process by Prof. T.S.C. Lowe and the carbureted water gas technology became the dominant technology from 1880s until 1950s, replacing coal gasification. In addition, the golden age of gas light developed with the invention and commercial use of the Welsbach mantle.

In the process, coke was produced in the retorts as a by-product of coal gas manufacture. It was used in part to heat the retorts, some was sold as a solid fuel, and some was used to produce water gas. Thus the gas was produced at high temperatures in a reactor (deep brick generator) filled with coke. Blowing air through the hot coke for about 2 minutes raised the temperature and made the coke incandescent whereupon the air was shut off and steam was blown through. This reacted with the coke (carbon) to produce water gas, a mixture of hydrogen and carbon monoxide. After approximately 1½ minutes the steam was shut off and air blown through again to reheat the coke. This air-steam cycle was maintained automatically. The water gas was then passed through a second heated brick chamber (the carburetor) in which a low-boiling petroleum liquid (a naphtha fraction) was sprayed onto the brickwork where it was cracked to produce gas which boosted the energy content of the water gas which could then be mixed with coal gas from the retorts.

In the earliest days of the town gas industry, coal tar was the major waste from which the gas had to be cleaned. The tar considered a waste and often disposed into the environment in and around the plant locations. While uses for coal tar developed by the late-1800s, the market for tar varied, and plants that could not sell tar at a given time could either store tar for future use, attempt to burn it as fuel for the boilers, or dump the tar as waste. Waste tars were often disposed of in old gas holders, adits, or even mine shafts (if present). Over time, the waste tars degrade with phenol derivatives, benzene derivatives (including other mono-aromatics—toluene and xylene), polycyclic aromatic hydrocarbon derivatives released as pollutant plumes that can escape into the surrounding environment.

The shift to the carbureted water gas process initially resulted in a reduced output of water gas tar as compared to the volume of coal tars. The advent of automobiles reduced the availability of naphtha for carburetion oil, as that fraction was desirable as motor fuel. Gas manufacturing plants that shifted to heavier grades of oil often experienced problems with the production of tar—water emulsions, which were difficult, time consuming, and costly to break. The production of large volumes of tar—water emulsions quickly filled up available storage capacity at MGPs and the waste was often dumped in pits, from which they may or may not have been later reclaimed. Even if the tar was reclaimed, the environmental damage from placing tars in unlined pits remained. The increased use of natural gas removed the necessity of dealing with the tar by-products that are a product of coal gasification. However, as was soon discovered, natural gas usually contains substances that can have a detrimental effect on distribution systems, such as water and sulfurous substances that can cause corrosion.

High moisture can result in clogged pipelines because water and methane form solid hydrates under certain pressures and temperatures. In addition, extracted natural gas contains dust that can cause defects in compressor or regulation stations. After extraction, natural gas undergoes a process during which it is dried and solid particles (dust) are removed. If necessary, the process includes the removal of higher hydrocarbons and any sulfurous substances that may be present. The refining technology depends on composition.

The advent of incandescent lighting in factories, homes, and in the streets, replacing oil lamps and candles with steady clear light, almost matching daylight in its color, turned night into day for many and made night-shift work possible in industries—such as the spinning and weaving industries—where light was very important for garment making. The social significance of this change is difficult for generations brought up with lighting after dark available at the touch of a switch to appreciate. Not only was industrial production accelerated, but streets were made safe, social intercourse facilitated, and reading and writing made more widespread. Gas works were built in almost every town, main streets were brightly illuminated, and gas was piped in the streets to the majority of urban households. The invention of the gas meter and the prepayment meter in the late 1880s played an important role in selling town gas to domestic and commercial customers.

During the World War I, gas industry by-products such as phenol (C_6H_5OH), toluene ($C_6H_5CH_3$), and ammonia (NH_3) as well as sulfur-containing chemicals were valuable ingredients for the manufacture of explosives. Much of the coal for the gas works was shipped by sea and was vulnerable to enemy attack. The gas industry was a large employer of clerks, mainly male before the war. But the advent of the typewriter brought about another important social change that was to have long-lasting effects.

The interwar years were marked by the development of the continuous vertical retort, which replaced many of the batch fed horizontal retorts. There were improvements in storage, especially the waterless gas holder, and distribution with the advent of 2- to 4-in. steel pipes to convey gas at up to 50 psi as feeder mains to the traditional cast iron pipes working at an average of 2- to 3-in. water gauge. Benzole (a product almost equivalent to aromatic naphtha) as a vehicle fuel and coal tar as the main feedstock for the emerging organic chemicals industry provided the gas industry with substantial revenues. Following from this, crude oil supplanted coal tar as the primary feedstock of the organic chemical industry after World War II and the loss of this market contributed to the economic problems of the gas industry after the war.

The advent of electric lighting forced utilities to search for other markets for manufactured gas. MGPs that once produced gas almost exclusively for lighting shifted their efforts toward supplying gas primarily for heating and cooking, and even refrigeration and cooling. As a result, during the 1920s the chemical industry began to explore gasification to synthesize chemicals, but it was not until World War II that the real potential for gasification was realized. In the face of scarce oil supplies and for several years thereafter, gasification was used extensively by Germany (as well as Britain and France) to produce liquid fuels from coal. In the face of this increasing use of coal for gasification purposes, tar was another product of the coal gasification process and had to be separated from the gas before the gas was sent to the consumer.

By the 1960s, manufactured gas, compared with its main rival in the energy market, electricity, was considered nasty, smelly, dirty, and dangerous (to quote market research of the time) and seemed doomed to lose market share still further, except for cooking where its controllability gave it marked advantages over both electricity and solid fuel. The development of more efficient gas-fueled fires (for domestic heating and industrial use) assisted gas to resist competition in the market for room heating. Concurrently a new market for whole house central heating by hot water was being developed by the crude oil industry and the gas industry followed suit.

Gas warm-air heating found a market niche in new housing areas built by local authorities where low installation costs gave it an advantage. These developments, the realignment of managerial thinking away from commercial management (selling what the industry produced) to marketing management (meeting the needs, wants and desires of customers) and the lifting of an early moratorium preventing nationalized industries from using television advertising, saved the gas industry for long enough to provide a viable market for what was to come.

The slow death of the town gas industry was signaled, even hastened, by the repeated discoveries of natural gas and the advantage was taken of this new indigenous energy source. As a result, there was a "rush to gas" for use in peak load electricity generation and in low-grade uses in industry. The effects on the coal industry were very significant; not only did coal lose its market for town gas production, it came to be displaced from much of the bulk energy market also.

6.2.2 Modern aspects

The gasification of coal or a derivative (i.e., char produced from coal) is, essentially, the conversion of coal (by any one of a variety of processes) to produce combustible gases (Radović et al., 1983; Radović and Walker, 1984; Garcia and Radović, 1986; Calemma and Radović, 1991; Kristiansen, 1996). With the rapid increase in the use of coal from the 15th century onwards (Nef, 1957; Taylor and Singer, 1957), it is not surprising concept of using coal to produce a flammable gas, especially the use of the water and hot coal, became common-place (Elton, 1958). In fact, the production of gas from coal has been a vastly expanding area of coal technology, leading to numerous research and development programs. As a result, the characteristics of rank, mineral matter, particle size, and reaction conditions are all recognized as having a bearing on the outcome of the process, not only in terms of gas yields but also on gas properties (Massey, 1974; Hanson et al., 2002).

Gasification has been considered for many years as an alternative to combustion of solid or liquid fuels. It is easier to clean gaseous mixtures than it is to clean solid or high-viscosity liquid fuels. Clean gas can be used in internal combustion-based power plant that would suffer from severe fouling or corrosion if solid or low-quality liquid fuels were burnt inside them. In some cases, process complexities arise because of the need for recovery of the materials used to remove the contaminants or even recovery of the contaminants in the original, or altered, form.

The purpose of preliminary cleaning of gases which arise from coal utilization is the removal of materials such as mechanically carried solid particles (either process products and/or dust) as well as liquid vapors (i.e., water, tars, and aromatics such as benzene derivatives and/or naphthalene derivatives); in some instances, preliminary cleaning might also include the removal of ammonia gas. For example, cleaning of town gas is the means by which the crude "tarry" gases from retorts or coke-ovens are (first), in a preliminary step, freed from tarry matter, condensable aromatics (such as naphthalene) and (second) purified by removal of materials such as hydrogen sulfide, other sulfur compounds, and any other unwanted components that will adversely affect the use of the gas (Speight, 2007, 2013).

In more general terms, gas cleaning is divided into removal of particulate impurities and gaseous impurities. For the purposes of this chapter, the latter operation includes the removal of hydrogen sulfide, carbon dioxide, sulfur dioxide, and the like. There was also need for subdivision of these two categories as dictated by needs and process capabilities: (1) coarse cleaning whereby substantial amounts of unwanted impurities are removed in the simplest, most convenient manner, (2) fine cleaning for the removal of residual impurities to a degree sufficient for the majority of normal chemical plant operations, such as catalysis or preparation of normal commercial products; or cleaning to a degree sufficient to discharge an effluent gas to atmosphere through a chimney, (3) ultra-fine cleaning where the extra step (as well as the extra expense) is justified by the nature of the subsequent operations or the need to produce a particularly pure product. To make matters even more complicated, a further subdivision of the processes, which applies particularly to the removal of gaseous impurities, is by process character insofar as there are processes which rely upon chemical and physical properties/characteristics of the gas stream to enhance separation of the constituents (Chapter 7: Process classification) (Speight, 2013, 2014a).

Since coal is a complex, heterogeneous material, there is a wide variety of constituents that are not required in a final product and must be removed during processing. Coal composition and characteristics vary significantly; there are varying amounts of sulfur, nitrogen, and trace-metal species which must be disposed of in the correct manner. Thus, whether the process be gasification to produce a fuel gas or gas cleaning before ejection into the atmosphere, the stages required during this processing are numerous and can account for a major portion of a gas cleaning facility.

Generally, the majority of the sulfur that occurs naturally in the coal is driven into the product gas. Thermodynamically, the majority of the sulfur should exist as hydrogen sulfide, with smaller amounts of carbonyl sulfide (COS) and carbon disulfide (CS_2). However, data from some operations (coke ovens) shows higher than expected (from thermodynamic considerations) concentrations of carbonyl sulfide and carbon disulfide.

The existence of mercaptans, thiophenes, and other organic sulfur compounds in (gasifier) product gas will probably be a function of the degree of severity of the process, contacting schemes, and heat-up rate. Those processes that tend to produce tars and oils may also tend to drive off high molecular weight organic sulfur compounds into the raw product gas.

In general terms, the gaseous emissions from coal combustion and gasification facilities may be broadly classed as those originating from four processing steps: (1) pretreatment, (2) conversion, and (3) upgrading, as well as those from (4) ancillary processes. In conventional coal combustion power plants, pulverized coal is burned in a boiler, where the heat vaporizes water in steam tubes. The resulting steam turns the blades of a turbine, and the mechanical energy of the turbine is converted to electricity by a generator. Waste gases produced in the boiler during combustion, among them, sulfur dioxide, nitrogen oxide(s), and carbon dioxide, flow from the boiler to a particulate removal device and then to the stack and the air.

Recent developments in gas turbine technology have resulted in combined cycle units with efficiencies close to 60% when generating electricity from natural gas. Gas turbine improvements lead to a number of power plants where "dirty" fuels (usually coal, residual oil, or petroleum coke) are gasified, the gas is cleaned and used in a combined cycle gas turbine power plant. Such power plants generally have higher capital cost, higher operating cost, and lower availability than conventional combustion and steam cycle power plant on the same fuel. Efficiencies of the most sophisticated plants have been broadly similar to the best conventional steam plants with losses in gasification and gas cleaning being balanced by the high efficiency of combined cycle power plants. Environmental aspects resulting from the gas cleaning before the main combustion stage have often been excellent, even in plants with exceptionally high levels of contaminants in the feedstock fuels.

Power plants under about 350 MWe cannot use the latest high-efficiency combined cycle technologies. Those below about 250 MWe cannot use particularly high-efficiency steam turbines because of friction losses and leaks in small dimension gas paths. Those below about 100 MWe cannot economically use reheat steam cycles, giving a further efficiency drop. Moving further down in size gives a steady reduction in efficiency of the gas turbine, whichever manufacturer is selected. The scale effect of gas turbine efficiencies is due to flow paths and pressure drops and can only be partly compensated with additional components such as intercoolers or reheaters.

At smaller sizes, reciprocating engines become relatively more attractive compared with rotating machinery. Their electricity generation efficiency is higher for power generation unit sizes of a few tens of MWe and less. Their major disadvantage is often the frequent and expensive maintenance required.

These technical considerations indicate some of the incentives for large unit size of power plant. Labor requirements per unit of installed capacity provide yet another driver toward large unit size. With the exception of power plants using easily handled fuels (generally natural gas) where there is a significant heat demand as well as a power demand, the trend has been toward larger power plants.

Gasification and pyrolysis processes can be classified as entrained gasifiers, fluidized-bed gasifiers (bubbling bed or circulating, atmospheric, or pressurized), small industrial-scale gasifiers (fixed-bed or grate, which can be up-draught or down-draught), and hybrid systems.

Gasification agents are normally air, oxygen-enriched air, or oxygen. Steam is sometimes added for temperature control, heating value enhancement, or to permit the use of external heat (allothermal gasification). The major chemical reactions break and oxidize hydrocarbons to give a product gas of carbon monoxide, carbon dioxide, hydrogen, and water. Other important components include hydrogen sulfide, various compounds of sulfur and carbon, ammonia, light hydrocarbons, and heavy hydrocarbons (tars).

The products from the gasification of coal may be of low, medium, or high heat-content (high-Btu) as dictated by the process as well as by the ultimate use for the gas (Fryer and Speight, 1976; Mahajan and Walker, 1978; Anderson and Tillman, 1979; Cavagnaro, 1980; Bodle and Huebler, 1981; Probstein and Hicks, 1990; Lahaye and Ehrburger, 1991).

6.2.3 Coal gas cleaning

Contrary to the general belief of some scientists and engineers, all gas cleaning systems are *not* alike and having a good understanding of the type of gaseous effluents from coal-based processes is necessary to implementing the appropriate solution. The design of a gas cleaning system must always consider the operation of the upstream installations, since every process will have a specific set of requirements. In some cases, the application of a dry dusting removal unit may not be possible and thus requires a special process design of the wet gas cleaning plant. Thus the gas cleaning process must always be of optimal design one for both the upstream and downstream processes.

Flue and waste gases from power plants and other industrial operations where coal is used as a feedstock and invariably contain constituents that are damaging to the climate or environment—these will be constituents such as carbon dioxide (CO_2), nitrogen oxides (NO_x), sulfur oxides (SO_x), dust and particles, and toxins such as dioxin and mercury. The processes that have been developed for gas cleaning vary from a simple once-through wash operation to complex multistep systems with options for recycle of the gases and are direct descendants of the early processes used to clean coal gas.

In the early years of coal gas production cleaning was a means to remove tarry products and any other liquid products from the gas stream to enable shipping and use without too much difficulty. However, as distributed, coal gas contained varying proportions of methane (CH_4), hydrogen (H), carbon monoxide (CO), and simple hydrocarbon illuminants, including ethylene (C_2H_4, at the time called olefiant gas) and acetylene (C_2H_2). In addition, prior to treatment, the coal gas contained coal tars (complex aliphatic and aromatic hydrocarbons), ammoniacal liquor (gaseous ammonia, NH_3), aqueous ammonia (NH_4OH), hydrogen sulfide (H_2S, also called the sulfuret of hydrogen), and carbon disulfide (CS_2, also called the sulfuret of carbon).

Thus, from the retort, the gas would first pass through a tar/water "trap" (similar to a trap in plumbing) called a hydraulic main, where a considerable fraction of coal tar was given up and the gas was significantly cooled. Then, it would pass through the main out of the retort house into an atmospheric or water-cooled condenser, where it would be cooled to the temperature of the atmosphere or the water used. At this point, the gas entered the exhauster house and passes through an "exhauster," an air pump which maintains the hydraulic mains and, consequently, the retorts at a negative pressure (with a zero pressure being atmospheric). The gas would then be washed in a "washer" by bubbling it through water, to extract any remaining tars. After this, it would enter a purifier. The gas would then be ready for distribution and pass into a gas holder for storage.

Scrubbers which utilized water were designed in the 25 years after the foundation of the industry. It was discovered that the removal of ammonia from the gas depended upon the way in which the gas to be purified was contacted by water. This was found to be best performed by the tower scrubber, which consisted of a tall cylindrical vessel, which contained trays or bricks which were supported on grids. The water, or weak gas liquor, trickled over these trays, thereby keeping the exposed surfaces thoroughly wetted. The gas to be purified was run through the tower to be contacted with the liquid.

In addition, the chemical industries demanded coal tar, and the gas-works could provide it for them; and the coal tar was stored on site in large underground tanks. As a rule, these were single wall metal tanks—i.e., if they were not porous masonry. At the time, underground tar leaks were seen as merely a waste of tar; out of sight was truly out of mind; and such leaks were generally addressed only when the loss of revenue from leaking tar wells, as these were sometimes called, exceeded the cost of repairing the leak.

In a modern coal gas cleaning system, the first step in gas cleaning is usually a device to remove large particles of coal and other solid materials. This is followed by cooling, quenching, or washing, to condense tars and oils and remove dust and water-soluble materials—phenols, chlorides, ammonia, hydrogen cyanide, thiocyanate, and perhaps some sulfur compounds from the gas stream. Water washing is desirable for simplicity in gas cleaning; however, the purification of this water is not simple.

Cleanup steps and their sequence can be affected by the type of gas produced and its end use (Speight, 2007, 2008). The minimum requirement in this respect would be the application of low heat-value (low-Btu) gas produced from low-sulfur anthracite coal as a fuel gas. The gas may pass directly from the gasifier to the burners and, in this case, the burners are the cleanup system. Many variations on this theme are possible and, in addition, the order of the cleanup stages may be varied.

Finally, the detrimental effects of PM on the atmosphere have been of some concern for several decades. In fact, the total output of PM into the atmosphere has increased in Europe since medieval times and, although the sources are various, there is special concern because of the issue of PM from fossil fuel use. Species such as mercury, selenium, and vanadium which can be ejected into the atmosphere from fossil fuel combustion (Kothny, 1973; Lakin, 1973; Zoller et al., 1973) are particularly harmful to the flora and fauna mercury. Thus there is the need to remove such materials from gas streams that are generated during fossil fuel processing.

There are many types of particulate collection devices in use and they involve a number of different principles for removal of particles from gas streams. However, the selection of an appropriate particle removal device must be based upon equipment performance as anticipated/predicted under the process conditions. To enter into a detailed description of the various devices available for particulate removal is well beyond the scope of this text. However, it is essential for the reader to be aware of the equipment available for particulate removal and the means by which this might be accomplished. Thus by use of: (1) cyclones, (2) ESPs, (3) granular-bed filters, and (4) wet scrubbers.

The selection of a particular process-type for a gas cleaning operation is not a simple choice. Many factors have to be considered, not the least of which is the constitution of the gas stream that requires treatment. Indeed, process selectivity indicates the preference with which the process will remove one acid gas component relative to (or in preference to) another. For example, some processes remove both hydrogen sulfide and carbon dioxide, while other processes are designed to remove hydrogen sulfide only.

Most gas systems employ two stages: one for fly ash removal and the other for removal of sulfur dioxide. Attempts have been made to remove both the fly ash and sulfur dioxide in one scrubbing vessel but these systems tended to experience severe maintenance problems and low removal efficiency. In wet scrubbing systems, the gas normally passes first through a fly ash removal device, either an ESP or a baghouse, and then into the sulfur dioxide absorber. However, in dry injection or spray drying operations, the sulfur dioxide is first reacted with lime, and then the flue gas passes through a particulate control device.

Another important design consideration associated with wet gas systems is that the flue gas exiting the absorber is saturated with water and still contains some sulfur dioxide. These gases are highly corrosive to any downstream equipment such as fans, ducts, and stacks. Two methods that may minimize corrosion are: (1) reheating the gases to above the dew point or (2) using materials of construction and designs that allow equipment to withstand the corrosive conditions.

Gas cleaning by sorption by a liquid or solid sorbent is one of the most widely applied operations for cleaning flue gases generated by the chemical and process industries (Biondo and Marten, 1977). Some processes have the potential for sorbent regeneration but, in a few cases, the process is applied in a nonregenerative manner. The interaction between sorbate and sorbent may either be physical in nature or consist of physical sorption followed by chemical reaction. Other gas stream treatments use the principle of chemical conversion of the contaminants with the production of "harmless" (noncontaminant) products or substances which can be removed much more readily than the impurities from which they are derived (Speight, 2007, 2008).

Since sulfur dioxide is an acid gas, the typical sorbent slurries or other materials used to remove the sulfur dioxide from the flue gases are alkaline. The reaction taking place in wet scrubbing using a limestone ($CaCO_3$) slurry produces calcium sulfite ($CaSO_3$). Thus

$$CaCO_{3(s)} + SO_{2(g)} \rightarrow CaSO_{3(s)} + CO_{2(g)}$$

When wet scrubbing with a hydrated lime [$Ca(OH)_2$] slurry, the reaction also produces calcium sulfite:

$$Ca(OH)_{2(s)} + SO_{2(g)} \rightarrow CaSO_{3(s)} + H_2O_{(l)}$$

When wet scrubbing with a magnesium hydroxide [$Mg(OH)_2$] slurry, the reaction produces magnesium sulfite ($MgSO_3$):

$$Mg(OH)_{2(s)} + SO_{2(g)} \rightarrow MgSO_{3(s)} + H_2O_{(l)}$$

Some process options, particularly dry sorbent injection systems, further oxidize the calcium sulfite to produce marketable gypsum ($CaSO_4 \cdot 2H_2O$) that can be of the high quality required for use in wallboard and other products. Thus

$$CaSO_{3(aq)} + 2H_2O_{(l)} + \tfrac{1}{2}O_{2(g)} \rightarrow CaSO_4 \cdot 2H_2O_{(s)}$$

A natural alkaline usable to absorb sulfur dioxide is seawater—the sulfur dioxide is absorbed in the water and, when oxygen is added, reacts to form sulfate ions

(SO_4^-) and free H^+. The surplus of H^+ is offset by the carbonate derivatives in the seawater which move the carbonate equilibrium to release carbon dioxide:

$$SO_{2(g)} + H_2O_{(l)} + \tfrac{1}{2}O_{2(g)} \rightarrow SO_4^{2-}{}_{(aq)} + 2H^+$$

$$HCO_3^- + H^+ \rightarrow H_2O_{(l)} + CO_{2(g)}$$

In industry sodium hydroxide (caustic or caustic soda, NaOH) is often used to remove sulfur dioxide from gas streams thereby producing sodium sulfite:

$$2NaOH_{(aq)} + SO_{2(g)} \rightarrow Na_2SO_{3(aq)} + H_2O_{(l)}$$

6.2.3.1   Wet scrubbers

Scrubber systems are a diverse group of devices that are used to control air pollution, and can be used to remove some PM and/or gases from industrial exhaust-gas streams. Traditionally, the term *scrubber* has referred to pollution control devices that use liquid to wash unwanted pollutants from a gas stream but, more recently with the evolution of gas cleaning operations, the term has also been used to describe systems that inject a dry reagent or slurry into a dirty exhaust stream to remove acid gases, such as carbon dioxide and hydrogen sulfide (Chapter 7: Process classification and Chapter 8: Gas cleaning processes) which elevate scrubbers to be one of the primary devices that control gaseous emissions, especially acid gases.

To promote maximum gas—liquid surface area and residence time, a number of wet scrubber designs have been used, including spray towers, venturis, plate towers, and mobile packed beds. Because of scale buildup, plugging, or erosion, which affect sulfur dioxide removal dependability and absorber efficiency, the trend is to use simple scrubbers such as spray towers instead of more complicated ones. The configuration of the tower may be vertical or horizontal, and flue gas can flow concurrently, countercurrently, or cross-currently with respect to the liquid. The chief drawback of spray towers is that they require a higher liquid-to-gas ratio requirement for equivalent removal of sulfur dioxide than other absorber designs.

6.2.3.2   Venturi scrubbers

A venturi scrubber is a device that is designed to effectively use the energy from the inlet gas stream to atomize the liquid being used to scrub the gas stream. This type of technology is a part of the group of air pollution controls collectively referred to as wet scrubbers. Although venturi devices have been used successfully for over 100 years to measure fluid flow, it was not until the late 1940s, that it was discovered the venturi configuration could be used to remove particles from gas streams from which followed the use of the venturi scrubbers as a regular part of gas processing operations.

A venturi scrubber consists of three sections: (1) a converging section, (2) a throat section, and (3) a diverging section. The inlet gas stream enters the converging section and, as the area decreases, gas velocity increases. Liquid is introduced either at the throat or at the entrance to the converging section. When the liquid stream is injected at the throat, which is the point of maximum velocity, the turbulence caused by the high gas velocity atomizes the liquid into small droplets, which creates the surface area necessary for mass transfer to take place. The inlet gas, forced to move at extremely high velocities in the small throat section, shears the liquid from its walls, producing an enormous number of very tiny droplets. Particle removal and gas removal occur in the diverging section as the inlet gas stream mixes with the fog of tiny liquid droplets. The inlet stream then exits through the diverging section, where it is forced to slow down.

For simultaneous removal of sulfur dioxide and fly ash, venturi scrubbers can be used. In fact, many of the industrial sodium-based throwaway systems are venturi scrubbers originally designed to remove PM. These units were slightly modified to inject a sodium-based scrubbing liquor. Although removal of both particles and sulfur dioxide in one vessel can be economic, the problems of high pressure drops and finding a scrubbing medium to remove heavy loadings of fly ash must be considered. However, in cases where the particle concentration is low, such as from oil-fired units, it can be more effective to remove particulate and sulfur dioxide simultaneously.

6.2.3.3 Packed-bed scrubbers

A packed-bed scrubber consists of a tower with packing material inside. This packing material can be in the shape of saddles, rings, or some highly specialized shapes designed to maximize contact area between the dirty gas and liquid. Like the venturi scrubber, the packed-bed technology is also a part of the group of air pollution controls collectively referred to as wet scrubbers and are suitable for the removal of air pollutants by inertial impact or diffusional impact, reaction with a sorbent or reagent slurry, or absorption into liquid solvent. When used to control inorganic gases, they may also be referred to as *acid gas scrubbers*.

The technology is suitable for the removal of pollutants such as: inorganic fumes, vapors, and gases (such as chromic acid, hydrogen sulfide, ammonia, chlorides, fluorides, and sulfur dioxide), volatile organic compounds (VOCs); and PM, including PM that is less than or equal to $10 \mu m$ in aerodynamic diameter (PM_{10}), PM less than or equal to $2.5 \mu m$ in aerodynamic diameter (PM_{25}), and hazardous air pollutants (HAPs) in particulate form (PM_{HAP}).

The packed-bed scrubber operates using the principle of absorption, which is widely used as a raw material and/or product recovery technique in separation and purification of gaseous streams containing high concentrations of VOCs, especially water-soluble compounds such as methanol, ethanol, isopropanol, butanol, and acetone. Hydrophobic VOCs can be absorbed using an amphiphilic block copolymer dissolved in water. However, as an emission control technique, it is much more commonly employed for controlling inorganic gases than for VOCs.

When using absorption (Chapter 7: Process classification) as the primary control technique for organic vapors, the spent solvent must be easily regenerated or disposed of in an environmentally acceptable manner. When used for removal of PM, caution is advised since a high concentration of PM in the gas stream can clog the bed, which is often cited as a reason for application of the packed-bed technology to gas streams with a low loading of PM.

Packed towers typically operate at much lower pressure drops than venturi scrubbers and are therefore cheaper to operate. They also typically offer higher sulfur dioxide removal efficiency. The drawback is that they have a greater tendency to plug up if particles are present in excess in the exhaust air stream.

6.2.3.4 Spray towers

A spray (spray column, spray chamber) tower is the simplest type of scrubber and is essentially a gas—liquid that is used to achieve mass and heat transfer between a gas phase (that can contain dispersed solid particles) and a dispersed liquid phase. The tower consists of an empty cylindrical vessel (typically, a steel vessel) with nozzles that spray liquid into the vessel. The nozzles are placed across the tower at different heights to spray all of the gas as it moves up through the tower. This type of technology also falls into the category of wet scrubbers.

The gas stream usually enters at the bottom of the tower and moves upward, while the liquid is sprayed downward from one or more levels (countercurrent floe, countercurrent contact). Countercurrent flow exposes the outlet gas with the lowest concentration of pollutants to the freshest scrubbing liquid. The positioning of the nozzles allows a maximum number of the fine droplets to interact with the pollutant particles and to provide a large surface area for absorption.

Spray towers are typically used when circulating a slurry. The high speed of a venturi would cause erosion problems, while a packed tower would plug up if it tried to circulate a slurry. Countercurrent packed towers are infrequently used because they tend to become plugged by collected particles or to scale when lime or limestone scrubbing slurries are used.

Although spray towers are used for gas absorption, they are not as effective as packed towers (or plate towers). Spray towers can be very effective in removing pollutants from gas streams if the pollutants are highly soluble or if a chemical reagent is added to the liquid.

6.2.3.5 Scrubbing reagents

Traditionally, the term *scrubber* has referred to pollution control devices that use liquid to wash unwanted pollutants from a gas stream. Recently, the term has also been used to describe systems that inject a dry reagent or a slurry into a gas stream to *wash out* any acid gases. Thus the reagent used in any scrubber is of paramount importance for the efficiency of the operation. Generally, the scrubbing processes can be classed as those involving (1) dry sorbent injection or (2) slurry injection.

Dry sorbent injection involves the addition of an alkaline chemical, such as hydrated lime [$Ca(OH)_2$], soda ash (Na_2CO_3), or sodium bicarbonate ($NaHCO_3$) into the gas stream to react with the acid gases. The acid gases react with the alkaline sorbent to form solid salts, which are removed in a particulate control device. On the other hand, in a slurry injection process, the acid gases are contacted with a finely atomized alkaline slurry and are absorbed by the slurry mixture and react to form solid salts, which are also removed in a particulate control device.

Depending on the application, the two most important reagents are lime (CaO; hydrated lime is CaO plus H_2O to give $Ca(OH)_2$) and sodium hydroxide (caustic soda, $NaOH$). Lime is typically used on large coal- or oil-fired boilers as found in power plants, as it is very much less expensive than caustic soda but often results in a slurry being circulated through the scrubber instead of a solution. The use of lime results in a slurry of calcium sulfite ($CaSO_3$) that must be disposed of. Fortunately, calcium sulfite can be oxidized to produce by-product gypsum ($CaSO_4 \cdot 2H_2O$), which is marketable for use in the building products industry. The use of caustic soda is limited to smaller combustion units because it is more expensive than lime, but it has the advantage that it forms a solution rather than a slurry. The product is a spent caustic solution of sodium sulfite/bisulfite (depending on the pH), or sodium sulfate that must be sent for environment-friendly disposal.

It is possible to scrub sulfur dioxide by using a cold solution of sodium sulfite (Na_2SO_3), which forms a solution of sodium hydrogen sulfite ($NaHSO_3$). By heating this solution, it is possible to reverse the reaction to form sulfur dioxide and the sodium sulfite solution.

$$2NaHSO_3 \rightarrow Na_2SO_3 + SO_2 + H_2O$$

Since the sodium sulfite solution is not consumed, it is a regenerative process— the application of this reaction is used in the Wellman–Lord process (Chapter 8: Gas cleaning processes).

The chemistry of the various reagents can be represented by a set of simple equations that have a high probability of success but if the chemical and the engineering planning steps are not given detailed consideration that is necessary, then Murphy's Law is operative, viz. *whatever can go wrong will go wrong* to which the addendum *Murphy was an optimist* is often cited The chemical reactions are as follows:

1. *Using calcium carbonate:*

$$CaCO_{3(s)} + SO_{2(g)} \rightarrow CaSO_{3(s)} + CO_{2(g)}$$

2. *Using hydrated lime:*

$$Ca(OH)_{2(s)} + SO_{2(g)} \rightarrow CaSO_{3(s)} + H_2O_{(l)}$$

3. *Using magnesium hydroxide:*

$$Mg(OH)_{2(s)} + SO_{2(g)} \rightarrow MgSO_{3(s)} + H_2O_{(l)}$$

4. *Using caustic soda:*

$$2NaOH_{(aq)} + SO_{2(g)} \rightarrow Na_2SO_{3(aq)} + H_2O_{(l)}$$

5. *Using seawater:*

$$SO_{2(g)} + H_2O_{(l)} + \tfrac{1}{2}O_{2(g)} \rightarrow SO_4^{2-}{}_{(aq)} + 2H^+$$

$$HCO_3^- + H^+ \rightarrow H_2O_{(l)} + CO_{2(g)}$$

An alternative to removing sulfur-containing compounds from the flue gas after burning is to remove the sulfur from the fuel before or during combustion. For example, the hydrodesulfurization process (Speight, 2013, 2014a, 2017) has been used for treating fuel oil before use and the fluidized-bed combustion process often adds lime (CaO) to the fuel during combustion. The lime reacts with the sulfur dioxide to form calcium sulfate, which become part of the combustion ash (Speight, 2013). This elemental sulfur is then separated and finally recovered at the end of the process for further usage in, e.g., agricultural products.

In fact, there are many variables in gas cleaning and the precise area of application of a given process is difficult to define although there are several factors that need to be considered: (1) the types and concentrations of contaminants in the gas, (2) the degree of contaminant removal desired, (3) the selectivity of acid gas removal required, (4) the temperature, pressure, volume, and composition of the gas to be processed, (5) the carbon dioxide to hydrogen sulfide ratio in the gas, and (6) the desirability of sulfur recovery due to process economics and/or environmental issues.

6.3 Natural gas

The process for gas drilling begins with locating the type of rock that will likely contain gas (Chapter 2: Origin and production and Chapter 3: Unconventional gas). Drilling can be performed on land (onshore) or in the ocean (offshore) (Speight, 2015). Whatever the site, it is preferable that natural gas processing begins at the wellhead, but this is not as straightforward as it may seem.

The composition of the raw natural gas extracted from producing wells depends on the depth, and location of the reservoir, the geology of the area, and whether the gas is associated or nonassociated (Chapter 3: Unconventional gas). Gas processing at the wellhead is used to ensure that the gas to be shipped (say, by pipeline) meets the necessary specification put in place by the transportation company. In addition to wellhead processing done and processing at centralized processing plants, some final processing is also sometimes accomplished at straddle extraction plants, which are located on major pipeline systems. Although the natural gas that arrives at these straddle extraction plants is already of pipeline quality, in certain instances, there still exist small quantities of NGLs, which are extracted at the straddle plants.

However, more specifically, gas processing (Speight, 2014a, 2017) consists of separating all of the various hydrocarbons and other nonmethane compounds for the raw natural gas stream (Table 6.1) (Fig. 6.2). Major transportation pipelines usually impose restrictions on the make-up of the natural gas that is allowed into the pipeline. That means that before the natural gas can be transported it must be purified. While the ethane, propane, butane, and pentanes must be removed from natural gas, this does not mean that they are all waste products. Gas processing is necessary to ensure that the natural gas intended for use is as clean and pure as possible, making it the clean burning and environmentally sound energy choice. Thus, natural gas, as it is used by consumers, is much different from the natural gas that is brought from underground up to the wellhead. Although the processing of natural gas is in many respects less complicated than the processing and refining of crude oil, it is equally as necessary before its use by end-users. The natural gas used by consumers is composed almost entirely of methane. However, natural gas found at the wellhead, although still composed primarily of methane, is by no means as pure.

Raw natural gas comes from three types of wells: (1) crude oil wells, (2) gas wells, and (3) gas condensate wells. Natural gas that comes from oil wells is typically termed "associated gas." This gas can exist separate from oil in the formation (free gas) or dissolved in the crude oil (dissolved gas). Natural gas from gas and condensate wells, in which there is little or no crude oil, is termed *nonassociated gas*. Gas wells typically produce raw natural gas by itself, while condensate wells produce free natural gas along with a semiliquid hydrocarbon condensate. Whatever the source of the natural gas, once separated from crude oil (if present), it commonly exists in mixtures with other hydrocarbons, principally ethane, propane, butane, and pentanes. In addition, raw natural gas contains water vapor, hydrogen sulfide (H_2S), carbon dioxide, helium, nitrogen, and other compounds. In fact, associated hydrocarbons, known as NGLs can be very valuable by-products of natural gas processing. NGLs include ethane, propane, butane, *iso*-butane, and natural gasoline that are sold separately and have a variety of different uses, including enhancing oil recovery in oil wells, providing raw materials for oil refineries or petrochemical plants, and as sources of energy.

Most natural gas production contains, to varying degrees, small (two to eight carbons) hydrocarbon molecules in addition to methane. Although they exist in a gaseous state at underground pressures, these molecules will become liquid (condense) at normal atmospheric pressure. Collectively, they are called condensates or NGLs. The natural gas extracted from coal reservoirs and mines (coal bed methane) is the primary exception, being essentially a mix of mostly methane and carbon dioxide (about 10%).

6.3.1 History

Commercial production of natural gas and transportation in the United States dates back in 1859 when Edwin Drake struck oil near Titusville, PA. Natural gas that

Table 6.1 Steps[a] in the production of sales-quality gas

1. *Gas—oil separators*

   Gas—oil separators are commonly closed cylindrical shells, horizontally mounted with inlets at one end, an outlet at the top for removal of gas, and an outlet at the bottom for removal of oil. Separation is accomplished by alternately heating and cooling (by compression) the flow stream through multiple steps. Some water and condensate, if present, will also be extracted as the process proceeds.

2. *Condensate separator*

   Condensates are most often removed from the gas stream at the wellhead through the use of mechanical separators. In most instances, the gas flow into the separator comes directly from the wellhead, since the gas—oil separation process is not needed. Extracted condensate is routed to on-site storage tanks.

3. *Dehydration*

   Dehydration is the removal of this water from the produced natural gas and is accomplished by several methods. Among these is the use of ethylene glycol (glycol injection) systems as an absorption mechanism to remove water and other solids from the gas stream. Alternatively, adsorption dehydration may be used, utilizing dry-bed dehydrators towers, which contain desiccants such as silica gel and activated alumina, to perform the extraction.

4. *Contaminant removal*

   Removal of contaminants includes the elimination of hydrogen sulfide, carbon dioxide, water vapor, helium, and oxygen. The most commonly used technique is to first direct the flow through a tower containing an olamine solution which absorbs sulfur compounds. After desulfurization, the gas flow is directed to the next section, which contains a series of filter tubes. As the velocity of the stream reduces in the unit, primary separation of remaining contaminants occurs due to gravity.

5. *Nitrogen extraction*

   Once the hydrogen sulfide and carbon dioxide are processed to acceptable levels, the stream is routed to a nitrogen rejection unit; it is routed through a series of passes through a column and a brazed aluminum plate fin heat exchanger. Helium, if any, can be extracted from the gas stream through membrane diffusion in a pressure swing adsorption unit.

6. *Methane separation*

   Cryogenic processing and absorption methods are some of the ways to separate methane from natural gas liquids. The cryogenic method is better at extraction of the lighter liquids, such as ethane, than is the alternative absorption method and consists of lowering the temperature of the gas stream to approximately $-120°F$. The absorption method, on the other hand, uses a "lean" absorbing oil to separate the methane from the natural gas liquids. While the gas stream is passed through an absorption tower, the absorption oil soaks up a large amount of the natural gas liquids. The "enriched" absorption oil, now containing NGLs, exits the tower at the bottom. The enriched oil is fed into distillers where the blend is heated to above the boiling point of the natural gas liquids, while the oil remains fluid. The oil is recycled, while the natural gas liquids are cooled and directed to a fractionator tower. Another absorption method that is often used is the refrigerated oil absorption method where the lean oil is chilled rather than heated, a feature that enhances recovery rates somewhat.

7. *Fractionation*

   Fractionation, the process of separating the various natural gas liquids present in the remaining gas stream, uses the varying boiling points of the individual hydrocarbons in the stream. The process occurs in stages as the gas stream rises through several towers where heating units raise the temperature of the stream, causing the various liquids to separate and exit into specific holding tanks.

[a]The number of steps and the type of techniques used in the process of creating pipeline-quality natural gas depends upon the source and makeup of the wellhead production stream. While natural gas liquids, such as propane and butane, are the by-products most often related to the natural gas recovery process, several other products (such as hydrogen sulfide, carbon dioxide, and helium) are also extracted from natural gas at field or gas treatment facilities.

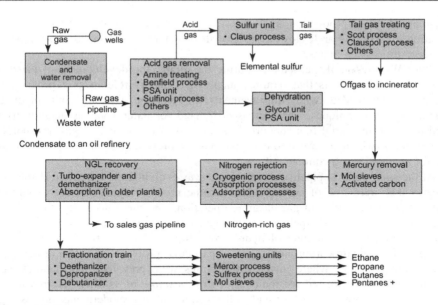

Figure 6.2 General layout of a gas processing plant.

accompanied the production of crude oil was an undesirable by-product of petroleum production efforts that was often simply vented or flared. Soon however, small volumes of natural gas began to be piped to local municipalities to displace manufactured gas volumes which were used in early distribution systems for lighting and fuel.

Various types of processing plants have been utilized since the mid-1850s to extract liquids, such as natural gasoline, from crude oil produced from a reservoir. However, for many years, natural gas was not a sought-after fuel. Prior to the early 20th century, most of it was flared or simply vented into the atmosphere, primarily because the available pipeline technology permitted only very short-distance transmission.

An early recorded example of the transmission of natural gas dates to 1872 when a 2-in. iron pipeline (approximately 5 miles in length) was constructed between Titusville (Pennsylvania) and Newton (Pennsylvania), which operated at a pressure of 80 pounds per square inch. The first natural gas pipeline that spanned a distance of more than 100 miles was constructed in 1891 and connected gas fields in central Indiana to Chicago (Illinois). After the success of these pipelines operations, larger diameter pipelines were installed and operated in Pennsylvania, New York, Kentucky, West Virginia, and Ohio.

It was not until the early 1920s, when reliable pipe welding techniques were developed, that a need for natural gas processing arose. It is debatable which came first: the need for the gas or the more-developed pipeline technology. Whichever came first, a preliminary network of relatively long-distance natural gas pipelines was in place by the 1930s, and some natural gas processing plants were installed upstream in major production areas. However, the depression of the 1930s and the duration of World War II slowed the growth of natural gas demand and the need for more processing plants.

After World War II, particularly during the 1950s, the development of plastics and other new products that required natural gas and petroleum as a production component coincided with improvements in pipeline welding and pipeline manufacturing techniques. The increased demand for natural gas as an industrial feedstock and industrial fuel supported the growth of major natural gas transportation systems, which in turn improved the marketability and availability of natural gas for residential and commercial use.

At the same time as pipeline development was evolving, the requirement for gas streams that were free of contaminants became more stringent and the evolution and installation of efficient gas cleaning operations became a necessity. For example, many of these gas pipelines soon began to experience operational problems with plugging from a mixture of hydrocarbon liquids that was referred to as *drip gasoline*. These hydrocarbon liquids posed particular problems in colder conditions and stream crossings where lower temperatures were prevalent. The natural gas stream began to be treated for removal of *drip gasoline* or as often referred to as casing head gasoline, which provided the origins of the modern gas processing industry (Speight, 2013, 2014a, 2017). While initially there was very little demand for gasoline, the advent of the automobile in the early 1900s forever changed this landscape. Automobiles were soon being built faster than the supply of natural gasoline could keep up with. Motor vehicles numbered around 2 million in 1915 and roughly doubled in the next 2 years. Demand and prices for gasoline soared.

As the natural gas market developed and pipeline technology advanced, producers began to explore for nonassociated gas fields and together with associated gas began to condition the produced gas and crude oil for transportation. Natural gas contains mostly methane and ethane, but as produced will often contain higher molecular weight hydrocarbon components such as propane, butane derivatives, and pentane derivatives along with water, carbon dioxide, hydrogen sulfide, helium nitrogen, and other trace elements (Chapter 3: Unconventional gas). Conditioning generally consisted of the removal of water and any free liquids, partial or full dehydration of the gas stream and *sweetening of the gas* with chemical agents to offset carbon dioxide and hydrogen sulfide concentrations were required to make the gas merchantable and suitable for pipeline transportation.

The development of the natural gas processing industry was now maturing from the era of simple collection of casing head gasoline (natural gasoline). Beginning in the 1920s to about 1940 natural gasoline production was accomplished through early absorption plant technology. As this technology advanced in the 1940s, new process for the extraction of lower molecular weight constituents emerged. Liquefied petroleum gas (LPG or LP Gas) made up of a combination of butane derivatives and propane, which were removed through more advanced lean oil absorption technology.

Subsequently, propane emerged as the dominant NGL product, which by the mid-1950s exceeded production of natural gasoline. Again, technological progress coupled with economics allowed further advances in the industry, which began to focus on the production of ethane in the late 1950s and early 1960s. Ethane production continued to grow and exceeded propane production by the early 1990s. It is interesting to note that the early economic drivers for ethane production included controlled wellhead natural gas prices and volumetric based pricing, i.e., per Mcf, which largely ignored the heating content value of the residual gas stream. Most of the market demand for these products evolved as the refining and petrochemical industries developed.

Prior to regulation in the United States, commercial natural gas quality characteristics evolved over time and reflected generally accepted operating practices, the state of processing technology, and market conditions. With the advancement and standardization of gas processing technology in large capacity "straddle" plants along major trunk lines, the conditioning and processing of the natural gas resulted in an ever increasingly stable, consistent, and functional end-use product. Thus one of the first steps in improving the quality of the natural gas was the removal of the NGLs.

6.3.2   Modern aspects

Initially, raw natural gas is commonly collected from a group of adjacent wells and is first processed at that collection point for removal of free liquid water and natural gas condensate. The condensate is usually then transported to an oil refinery and the water is disposed of as wastewater. The raw gas is then pipelined to a gas processing plant where the initial purification is usually the removal of acid gases (hydrogen sulfide and carbon dioxide).

There are many processes that are available for that purpose as shown in the flow diagram, but amine treating is the process that was historically used (Chapter 7: Process classification and Chapter 8: Gas cleaning processes) (Speight, 2014a, 2017). However, due to a range of performance and environmental constraints of the amine process, processes based on the use of polymeric membranes to separate the carbon dioxide and hydrogen sulfide from the natural gas stream has gained increasing acceptance. These processes are attractive since no reagents are consumed but the efficiency of a membrane can deteriorate in the presence of acid gases.

Acid gases, if present, are removed by membrane or amine treating can then be routed into a sulfur recovery unit, which converts the hydrogen sulfide in the acid gas into either elemental sulfur or sulfuric acid. Of the processes available for these conversions, the Claus process is by far the most well-known for recovering elemental sulfur, whereas the conventional contact process and the wet sulfuric acid process are the most used technologies for recovering sulfuric acid (this chapter 6 and Chapter 7: Process classification). The residual gas from the Claus process (commonly called *tail gas*) is then processed in a tail gas treating unit to recover and recycle residual sulfur-containing compounds back into the Claus unit (Chapter 7: Process classification). There are a number of processes available for treating the Claus unit tail gas and for that purpose a wet sulfuric acid process is also very suitable since it can work autothermally on tail gases.

The next step in the gas processing plant is to remove water vapor from the gas using either the regenerable absorption using (1) in liquid triethylene glycol, commonly referred to as glycol dehydration, (2) deliquescent chloride desiccants, and or (3) a pressure swing adsorption unit which is regenerable adsorption using a solid adsorbent. Other newer processes such as those based on the use of membranes may also be considered. Mercury is then removed by using adsorption processes (as shown in the flow diagram) such as activated or regenerable molecular sieves (Chapter 7: Process classification).

Although not common, nitrogen is sometimes removed and rejected using one of the three processes (1) a cryogenic process, often referred to as a nitrogen rejection unit, which employs low-temperature distillation—this process can be modified to also recover helium; (2) an absorption process, which uses lean oil or a special solvent as the absorbent; and (3) an adsorption process, using activated carbon or molecular sieves as the adsorbent—this process may have limited applicability because it can involve the loss of butane derivatives and higher molecular weight hydrocarbon derivatives.

The next step is to recover the NGLs for which most large, modern gas processing plants use another cryogenic low-temperature distillation process involving expansion of the gas through a turbo-expander unit followed by distillation in a fractionating column to remove any methane (known as a demethanizing column or demethanizing unit). Some gas processing plants use a lean oil absorption process rather than the cryogenic turbo-expander process.

The recovered stream of NGLs can be processed through a fractionation train consisting of three distillation towers in series: (1) a deethanizer, (2) a depropanizer, and (3) a debutanizer. The overhead product from the deethanizer is ethane and the bottoms are fed to the depropanizer. The overhead product from the depropanizer is propane and the bottoms are fed to the debutanizer. The overhead product from the debutanizer is a mixture of normal and iso-butane, and the bottoms product is a C_{5+} mixture.

The recovered streams of propane, butanes, and C_{5+} may be sweetened (any sour gas removed) in a Merox process unit to convert undesirable mercaptans into disulfide derivatives and, along with the recovered ethane, are the final NGLs byproducts from the gas processing plant.

By way of explanation, Merox is a shortened trade name derived from the term *mercaptan oxidation*. It is a proprietary catalyst chemical process developed that is used in crude oil refineries and natural gas processing plants to remove mercaptan derivatives from LPG, propane, butane derivatives, low-boiling naphtha streams, kerosene, and jet fuel by converting them to liquid hydrocarbon disulfide derivatives. The process requires an alkaline environment which, in some process versions, is provided by an aqueous solution of sodium hydroxide (NaOH), a strong base, commonly referred to as caustic. In other versions of the process, the alkalinity is provided by ammonia, which is a weak base. The catalyst in some versions of the process is a water-soluble liquid. In other versions, the catalyst is impregnated onto charcoal granules.

Currently, most cryogenic plants do not include fractionation for economic reasons, and the NGL stream is instead transported as a mixed product to standalone fractionation complexes located near refineries or chemical plants that use the components for feedstock. In case laying pipeline is not possible for geographical reason, or the distance between source and consumer exceed 3000 km, natural gas is then transported by ship as liquefied natural gas (LNG) and again converted into its gaseous state in the vicinity of the consumer.

The residue gas from the LNG recovery section is the final, purified sales gas which is pipelined to the end-user markets. Rules and agreements are made between buyer and seller regarding the quality of the gas. These usually specify the maximum allowable concentration of carbon dioxide, hydrogen sulfide, and water as well as requiring the gas to be commercially free from objectionable odor and materials, and dust or other solid or liquid matter, waxes, gums, and gum-forming constituents, which might damage or adversely affect operation of the buyers' equipment.

References

Anderson, L.L., Tillman, D.A., 1979. Synthetic Fuels from Coal: Overview and Assessment. John Wiley & Sons Inc, New York.

Biondo, S.J., Marten, J.C., 1977. A history of flue gas desulphurization systems since 1850. J. Air Pollut. Control Assoc. 27 (10), 948−949.

Bodle, W.W., Huebler, J., 1981. In: Meyers, R.A. (Ed.), Coal Handbook. Marcel Dekker Inc, New York (Chapter 10).

Calemma, V., Radović, L.R., 1991. On the gasification reactivity of Italian Sulcis Coal. Fuel 70, 1027.

Cavagnaro, D.M., 1980. Coal Gasification Technology. National Technical Information Service, Springfield, VA.

Elton, A., 1958. In: Singer, C., Holmyard, E.J., Hall, A.R., Williams, T.I. (Eds.), A History of Technology. Clarendon Press, Oxford, vol. IV (Chapter 9).

Fryer, J.F., Speight, J.G., 1976. Coal Gasification: Selected Abstract and Titles. Information Series No. 74. Alberta Research Council, Edmonton.

Garcia, X., Radović, L.R., 1986. Gasification reactivity of Chilean coals. Fuel 65, 292.

Hanson, S., Patrick, J.W., Walker, A., 2002. The effect of coal particle size on pyrolysis and steam gasification. Fuel 81, 531−537.

Higman, C., Van Der Burgt, M., 2003. Gasification. Butterworth Heinemann, Oxford.

Kothny, E.L., 1973. In: Kothny, E.L. (Ed.), Trace Metals in the Environment. Advances in Chemistry Series No. 123. American Chemical Society, Washington, DC (Chapter 4).

Kristiansen, A., 1996. IEA Coal Research Report IEACR/86. Understanding Coal Gasification. International Energy Agency, London.

Lahaye, J., Ehrburger, P. (Eds.), 1991. Fundamental Issues in Control of Carbon Gasification Reactivity. Kluwer Academic Publishers, Dordrecht.

Lakin, H.W., 1973. In: Kothny, E.L. (Ed.), Trace Metals in the Environment. Advances in Chemistry Series No. 123. American Chemical Society, Washington, DC.

Mahajan, O.P., Walker Jr., P.L., 1978. In: Karr Jr, C. (Ed.), Analytical Methods for Coal and Coal Products. Academic Press Inc, New York, vol. II (Chapter 32).

Massey, L.G. (Ed.), 1974. Coal Gasification. Advances in Chemistry Series No. 131. American Chemical Society, Washington, DC.

Murphy, B.L., Sparacio, T., Walter, J., Shields, W.J., 2005. Environ. Forens. 6 (2), 161.

Nef, J.U., 1957. In: Singer, C., Holmyard, E.J., Hall, A.R., Williams, T.I. (Eds.), A History of Technology. Clarendon Press, Oxford, vol. III (Chapter 3).

Probstein, R.F., Hicks, R.E., 1990. Synthetic Fuels. pH Press, Cambridge, MA (Chapter 4).

Radović, L.R., Walker Jr., P.L., 1984. Reactivities of chars obtained as residues in selected coal conversion processes. Fuel Process. Technol. 8, 149.

Radović, L.R., Walker Jr., P.L., Jenkins, R.G., 1983. Importance of carbon active sites in the gasification of coal chars. Fuel 62, 849.

Speight, J.G., 2013. The Chemistry and Technology of Coal, third ed. CRC Press, Taylor & Francis Group, Boca Raton, FL.

Speight, J.G., 2014a. The Chemistry and Technology of Petroleum, fifth ed. CRC Press, Taylor & Francis Group, Boca Raton, FL.

Speight, J.G., 2014b. Gasification of Unconventional Feedstocks. Gulf Professional Publishing, Elsevier, Oxford.

Speight, J.G., 2015. Handbook of Offshore Oil and Gas Operations. Gulf Professional Publishing, Elsevier, Oxford.

Speight, J.G., 2017. Handbook of Petroleum Refining. CRC Press, Taylor & Francis Group, Boca Raton, FL.

Stewart, E.G., 1958. Town Gas: Its Manufacture and Distribution. H.M. Stationery Office, London.

Sugg, W.T., 1884. The Domestic Uses of Coal Gas. Walter King, London.

Taylor, F.S., Singer, C., 1957. In: Singer, C., Holmyard, E.J., Hall, A.R., Williams, T.I. (Eds.), A History of Technology. Clarendon Press, Oxford, vol. II (Chapter 10).

Tristan, F., 1840. Promenades Dans Londres (D. Palmer, G. Pincetl, Trans.) (1980) Flora Tristan's London Journal, A Survey of London Life in the 1830s George Prior, Publishers, London. Extract Worse than the slave trade in Appendix 1, Barty-King, H., 1985.

Zoller, W.H., Gordon, G.E., Gladney, E.S., Jones, A.G., 1973. In: Kothny, E.L. (Ed.), Trace Metals in the Environment. Advances in Chemistry Series No. 123. American Chemical Society, Washington, DC (Chapter 3).

Further reading

Massey, L.G., 1979. In: Wen, C.Y., Lee, E.S. (Eds.), Coal Conversion Technology. Addison-Wesley Publishers Inc, Reading, MA, p. 313.

Process classification

7

7.1 Introduction

There is a wide variety of fuels currently in widespread use (Chapter 3: Unconventional gas) of which the simplest in composition is natural gas that consists primarily of methane but includes several other constituents from ethane to butane (collectively known as natural gas liquids, NGLs) as well as higher boiling hydrocarbons from pentane to octane (C_8H_{18}) or, in some cases, to dodecane ($C_{12}H_{26}$) decane (collectively known as gas condensate and generally referred to as the C_{5+} fraction). In addition, process gas, biogas, and landfill gas are also gases of interest in this present content.

Natural gas, as it is used by consumers, is composed almost entirely of methane and is much different from the natural gas that exists in the reservoir and that is brought up to the wellhead. Gas processing cleaning (also called *gas cleaning* and *gas refining*) is a necessary procedure to produce the product that meets the various specifications. Gas processing is, in fact, an integrated system of unit processes that are used to remove objectionable products such as acid gases (e.g., carbon dioxide and hydrogen sulfide) and to separate natural gas into other useful gas streams. Thus gas processing is instrumental in ensuring that the natural gas intended for use is as clean and pure as possible, making it the clean burning and an environmentally sound energy choice.

Major transportation pipelines usually impose restrictions on the make-up of the natural gas that is allowed into the pipeline. That means that before the natural gas can be transported it must be purified. This can commence at the wellhead where efforts are made to remove (at least) water, carbon dioxide, and hydrogen sulfur to prevent damage (corrosion) to the pipeline and associated equipment (Speight, 2014b). The goal of this first stage of natural gas processing consists of separating (some or all) the various hydrocarbon derivatives (C_{2+} as well as natural gasoline) which, depending upon the dew point of the hydrocarbon and the conditions in the pipeline, could separate in the pipeline and cause liquid blockage.

Although the processing of natural gas is in many respects less complicated than the processing and refining of crude oil, it is equally as necessary before its use by end-users. The natural gas used by consumers is composed almost entirely of methane, but the natural gas produced at the wellhead, although still composed primarily of methane, is by no means as pure. Raw (untreated) natural gas is produced from three types of wells: (1) wells that produce crude oil and natural gas, (2) wells that produce natural gas only, and (3) wells that produce gas condensate and natural gas. In addition, raw natural gas contains water vapor, hydrogen sulfide (H_2S), carbon dioxide (CO_2), helium (He), nitrogen (N_2), and a variety of other compounds, often in trace amounts (typically $<1\%$ v/v in total) (Table 7.1).

Natural Gas. DOI: https://doi.org/10.1016/B978-0-12-809570-6.00007-2

Table 7.1 Constituents of natural gas

Constituent	Formula	% v/v
Methane	CH_4	>85
Ethane	C_2H_6	3−8
Propane	C_3H_8	1−5
n-Butane	C_4H_{10}	1−2
iso-Butane	$C4H_{10}$	<0.3
n-Pentane	C_5H_{12}	1−5
iso-Pentane	C_5H_{12}	<0.4
Hexane, heptane, octane[a]	C_nH_{2n+2}	<2
Carbon dioxide	CO_2	1−2
Hydrogen sulfide	H_2S	1−2
Oxygen	O_2	<0.1
Nitrogen	N_2	1−5
Helium	He	<0.5

[a]Hexane (C_6H_{14}) and higher molecular weight hydrocarbon derivatives up to octane as well as benzene (C_6H_6) and toluene ($C_6H_5CH_3$).

Briefly, gas processing (Mokhatab et al., 2006; Speight, 2014a) consists of separating all of the various hydrocarbons and any nonhydrocarbon constituents from the methane using a variety of integrated unit processes (Fig. 7.1). Gas processing is necessary to ensure that the natural gas intended for use is as clean and pure as possible that meets the specifications which guarantee that the gas will be the clean-burning and environmentally sound energy choice. While the necessary preprocessing of natural gas can be accomplished at or near the wellhead (field processing), the complete processing of natural gas takes place at a processing plant, usually located in a natural gas producing region.

In addition to processing done at the wellhead and at centralized processing plants, some final processing is also sometimes accomplished at *straddle extraction plants*. These plants are located on major pipeline systems and although the natural gas that arrives at these straddle extraction plants is already of pipeline quality, in certain instances there still exist small quantities of NGLs, which are extracted at the straddle plants.

Sulfur and a variety of sulfur-containing compounds (such as hydrogen sulfide, H_2S, mercaptans, RSH—also called thiols—and carbonyl sulfide, COS) occur in gas streams. These compounds are detrimental (especially in the presence of water) by causing corrosion to the pipeline and can cause deactivation of catalysts used in gas-treating systems. The sulfur levels in the natural gas (or, for that matter, in any gas stream) are an important parameter for gas treatment. Therefore it is essential that sulfur-containing compounds such as hydrogen sulfide (as well acid gases such as carbon dioxide) be removed from a gas stream and converted to more useful products such as sulfur or sulfuric acid (Bartoo, 1985; Mokhatab et al., 2006; Speight, 2014a).

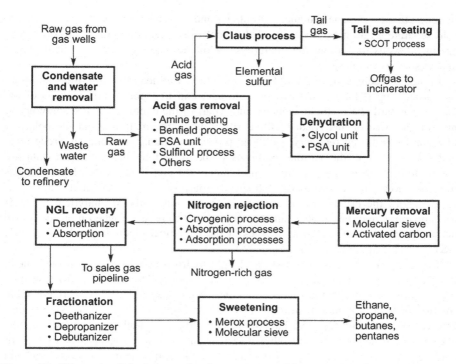

Figure 7.1 Schematic of the integrated unit processes used for gas processing.

It is also worthy of note that, because of the variation of gas volume with changes in temperature and changes in pressure, it is necessary to use one of the following alternatives for describing a concentration value: (1) additional specification of gas temperature and pressure values existing during measurement or (2) conversion of the measured concentration value into the corresponding value at standard zero conditions.

7.2 Gas streams

Gas streams are highly variable in composition (Chapter 1: History and use)—even natural gas from the same reservoir can vary in composition over time and/or is dependent upon the placement of the well. Thus the object of gas processing is to even out these variables and produce a gas stream that meets the designated specifications for transportation, storage, and use.

Thus gas processing consists of separating all of the various hydrocarbons and fluids from the pure natural gas, to produce what is known as specification-grade dry gas (generally, sometimes referred to as pipeline-quality dry natural gas). In fact, the NGLs can be very valuable by-products of natural gas processing. NGLs include ethane, propane, butane, *iso*-butane, and natural gasoline. These NGLs are sold separately and have a variety of different uses; including enhancing oil recovery in oil wells, providing raw materials for oil refineries or petrochemical plants, and as sources of energy.

In fact, at each stage of natural gas production, wellhead treating, transportation, and processing (Chapter 5: Recovery, storage, and transportation), analysis of the gas by standard test methods is an essential part of the chemistry and technology of natural gas. Use of analytical methods offers vital information about the behavior of natural gas during recovery, wellhead processing, transportation, gas processing, and use. However, the various purposes to be served by the analytical data are listed, and the necessary accuracy with which each constituent of each type of gas must be known in order to serve each specific purpose is then estimated. These estimates afford the criteria by means of which the suitability of analytical methods and apparatus may be judged, but they are subject to revision when more is known about the limiting attainable accuracies of the analytical methods (Nadkarni, 2005; ASTM, 2017; Speight, 2018).

Typically, natural gas is a mixture of light hydrocarbon gases rich in methane—the methane content of natural gases is usually above 75% with C_{5+} fraction (i.e., the fraction commonly known as *gas condensate*) less than 1% v/v. If the mole fraction of hydrogen in a natural gas is less than 4 ppm v/v it is called "sweet" gas. Dry gases contain no C_{5+} and have more than 90 mol% methane. The main difference between natural gas and other reservoir fluids is that the amount of C_{5+} or even C_{5+} in the mixture is quite low and the main components are light paraffinic hydrocarbons.

Another type of reservoir fluid that is in gaseous phase under reservoir conditions is gas condensate systems (the C_{5+} fraction). The C_{5+} fraction of a mixture should be treated as an undefined fraction and its properties may be determined by a series of standard test methods that are different to the test methods used for various gases as well as the liquid constituents of gas streams (Part II). The C_{5+} content is more than that of natural gases and it is about few percent, while its methane content is less than that of natural gas. However, these reservoir fluids generally contain components such as carbon dioxide (CO_2), hydrogen sulfide (H_2S), or nitrogen (N_2). Presence of such compounds affects the properties of the gas mixture.

In the description of the composition of natural gas, the C_{5+} (condensate) fraction is a relatively narrow boiling range fraction and characterization methods are available that can be used to determine the various properties of this fraction. However, contrary to the C_{5+} fraction of natural gas, the standard test methods applied to this condensate are not the typical standard test methods applied to crude oil liquids, but should be selected with discretion from the test methods to be applied to low-boiling naphtha which, even when selectively chosen, may require some modification for application to the condensate.

Information on the composition of natural gas (including the condensate fraction) is generally much more available than the information for liquid fuels through application of a series of standard test methods (ASTM, 2017). Rarely is the molecular composition of liquid fuels known with any degree of certainty (e.g., there are many suppositions that go into describing the molecular composition of, say, fuel oil) because of the complexity of the liquid fuels contain a large (sometime undefinable) number of hydrocarbon species. The most commonly reported composition data is derived from the *ultimate analysis*, which consists of measurements of the elemental composition of the fuel, generally presented as mass fractions (% w/w) of carbon, hydrogen, sulfur, oxygen, nitrogen, and ash, where appropriate as well as the potential for low-boiling (even gaseous) hydrocarbons which, with suitable choice of the column and instrumental parameters can also include determination of olefin constituents (Nadkarni, 2005; ASTM, 2017; Speight, 2018).

Generally, volume and mole fractions are used interchangeably for all types of gas mixtures. Composition of gas mixtures is rarely expressed in terms of weight fraction and this type of composition has very limited application for gas systems. Whenever composition in a gas mixture is expressed only in percentage it should be considered as mole % or volume % (i.e., % v/v). Thus by applying the appropriate standard test methods to the gas stream, it is possible to determine the distribution of the constituents from methane to the constituents of the condensate from which the behavior of the reservoir, including the gas composition from different wells within a reservoir, can be assessed.

While some of the needed processing can be accomplished at or near the wellhead (field processing), the complete processing of natural gas takes place at a processing plant, usually located in a natural gas producing region. The extracted natural gas is transported to these processing plants through a network of gathering pipelines, which are small-diameter, low-pressure pipes. A complex gathering system can consist of thousands of miles of pipes, interconnecting the processing plant to upwards of 100 wells in the area.

The actual practice of processing the various gas streams to either pipeline dry gas at the wellhead or specification-quality gas in a gas processing plant levels can be quite complex, but usually involves four main processes to remove the impurities. Natural gas streams while ostensibly being hydrocarbon in nature may contain large amounts of acid gases such as hydrogen sulfide and carbon dioxide. Most commercial plants employ hydrogenation to convert organic sulfur compounds into hydrogen sulfide. Hydrogenation is achieved by means of recycled hydrogen-containing gases or external hydrogen over a nickel molybdate or cobalt molybdate catalyst (Parkash, 2003; Gary et al., 2007; Speight, 2014a, 2017; Hsu and Robinson, 2017).

In summary, refinery process gas, in addition to hydrocarbons, may contain other contaminants, such as carbon oxides (CO_x, where $x = 1$ and/or 2), sulfur oxides (SO_x, where $x = 2$ and/or 3), as well as ammonia (NH_3), mercaptans (RSH), and carbonyl sulfide (COS) as well as olefins (Chapter 2: Origin and production).

In addition to hydrogen sulfide (H_2S) and carbon dioxide (CO_2), natural gas (and other gases: processes gases, biogases, and landfill gases) may contain other contaminants, such as mercaptans (RSH) and carbonyl sulfide (COS). Higher amounts of these contaminants may occur in biogas and landfill, depending upon the source of these gases and the process parameters that were operative when they were produced. The presence of these impurities may eliminate some of the sweetening (acid gas removal) processes, since some processes remove large amounts of acid gas but not to a sufficiently low concentration. On the other hand, there are those processes that are not designed to remove (or are incapable of removing) large amounts of acid gases. However, these processes are also capable of removing the acid gas impurities to very low levels when the acid gases are there in low to medium concentrations in the gas.

Gas processing involves the use of several different types of processes, but there is always overlap between the individual processing concepts. In addition, the terminology used for gas processing can often be confusing and/or misleading because of the overlap (Nonhebel, 1964; Curry, 1981; Maddox, 1982; Mokhatab et al., 2006; Speight, 2014a). For example, at power plants, flue gas (Chapter 8: Gas cleaning processes) is often treated with a series of chemical processes and scrubbers, which remove pollutants. Most flue gas desulfurization systems employ two stages: (1) one for fly ash removal and (2) the other for sulfur dioxide removal. For example, as environmental regulations regarding sulfur dioxide emissions have been enacted and enforced in many countries, sulfur dioxide is removed from flue gases by a variety of methods. Common methods used: (1) wet scrubbing, which uses a slurry of alkaline sorbent, usually limestone or lime, or seawater to scrub the gases; (2) spray-dry scrubbing, which uses similar sorbent slurries as described in the first category; (3) a wet sulfuric acid process, which allows the recovery of sulfur in the form of commercial quality sulfuric acid; (4) a process commonly referred to as a SNOX flue gas desulfurization process, which removes sulfur dioxide, nitrogen oxides, and particulate matter from flue gases; and (5) a dry sorbent injection process, which introduces powdered hydrated lime or other sorbent material into the exhaust ducts to eliminate sulfur dioxide and sulfur trioxide from the process emissions.

In wet scrubbing systems, the flue gas normally passes first through a fly ash removal device, either an electrostatic precipitator or a baghouse, and then into absorber that removes sulfur dioxide. However, in dry injection or spray drying operations, the sulfur dioxide is first reacted with the lime, and then the flue gas passes through a particulate control device. Another important design consideration associated with wet flue gas desulfurization systems is that the flue gas exiting the absorber is saturated with water and still contains some sulfur dioxide. These gases are highly corrosive to any downstream equipment such as fans, ducts, and stacks and two methods that may minimize corrosion are: (1) reheating the gases to above the dew point or (2) using materials of construction and designs that allow equipment to withstand the corrosive conditions. Nitrogen oxides are treated either by modifications to the combustion process to prevent their formation, or by high temperature or catalytic reaction with ammonia (NH_3) or urea (H_2NCONH_2). In either case, the aim is to produce nitrogen gas, rather than nitrogen oxides.

Technologies based on regenerative capture by amines (usually called *olamine processes*) for the removal of carbon dioxide from flue gas have been deployed to provide high-purity carbon dioxide for further industrial use as well as for enhanced oil recovery (EOR, Speight, 2014a).

Because of the variable number of constituents and the variable composition of the gas stream, there are many variables in treating natural gas. The precise area of application of a given process is difficult to define. Several factors (not necessarily in order of importance) must be considered: (1) the types of contaminants in the gas, (2) the concentrations of contaminants in the gas, (3) the degree of contaminant removal desired, (4) the selectivity of acid gas removal required, (5) the temperature of the gas to be processed, (6) the pressure of the gas to be processed, (7) the volume of the gas to be processed, (8) the composition of the gas to be processed, (9) the carbon dioxide–hydrogen sulfide ratio in the gas, and (10) the desirability of sulfur recovery due to process economics or environmental issues. Nevertheless, there are four general processes used for emission control (often referred to in another, more specific context as flue gas desulfurization): (1) adsorption, (2) absorption, (3) catalytic oxidation, and (4) thermal oxidation (Soud and Takeshita, 1994; Mokhatab et al., 2006; Speight, 2014a).

Process selectivity indicates the preference with which the process removes one acid gas component relative to (or in preference to) another. For example, some processes remove both hydrogen sulfide and carbon dioxide, other processes are designed to remove hydrogen sulfide only. It is important to consider the process selectivity for, say, hydrogen sulfide removal compared to carbon dioxide removal that ensures minimal concentrations of these components in the product, thus the need for consideration of the carbon dioxide to hydrogen sulfide in the gas stream.

7.2.1 Natural gas streams

Natural gas is also capable of producing emissions that are detrimental to the environment. While the major constituent of natural gas is methane, there are components such as carbon dioxide (CO), hydrogen sulfide (H_2S), and mercaptans (thiols; RSH), as well as trace amounts of sundry other emissions such as carbonyl sulfide (COS). The fact that methane has a foreseen and valuable end-use makes it a desirable product, but in several other situations it is considered a pollutant, having been identified a greenhouse gas.

A sulfur removal process must be very precise, since natural gas contains only a small quantity of sulfur-containing compounds that must be reduced several orders of magnitude. Most consumers of natural gas require less than 4 ppm in the gas. Also, a characteristic feature of natural gas that contains hydrogen sulfide is the presence of carbon dioxide (generally in the range of 1%–4% by volume). In cases where the natural gas does not contain hydrogen sulfide, there may also be a relative lack of carbon dioxide.

In practice, heaters and scrubbers are installed, usually at or near the wellhead. The scrubbers serve primarily to remove sand and other large-particle impurities and the heaters ensure that the temperature of the gas does not drop too low. With natural gas that contains even low quantities of water, natural gas hydrates tend to form when temperatures drop. These hydrates are solid or semisolid compounds, resembling ice-like crystals. If the hydrates accumulate, they can impede the passage of natural gas through valves and gathering systems (Zhang et al., 2007).

To reduce the occurrence of hydrates, small natural gas-fired heating units are typically installed along the gathering pipe wherever it is likely that hydrates may form.

While natural gas hydrates are usually considered as possible nuisances in the
development of oil fields and gas fields, mainly in deep-water drilling operations
and if multiphase production and transportation technologies are to be examined.
On the other hand, hydrates can be used for the safe and economic storage of natural gas, mainly in cold countries. In remote offshore areas, the use of hydrates for
natural gas transportation is also presently considered as an economic alternative to
the processes based either on liquefaction or on compression (Lachet and Béhar,
2000).

7.2.2 Crude oil gas streams

In order to process and transport associated dissolved natural gas, it must be separated from the oil in which it is dissolved. This separation of natural gas from oil is
most often done using equipment installed at or near the wellhead. The actual process used to separate oil from natural gas, as well as the equipment that is used, can
vary widely. Although dry pipeline quality natural gas is virtually identical across
different geographic areas, raw natural gas from different regions will vary in composition (Chapter 1: History and use) and therefore separation requirements may
emphasize or deemphasize the optional separation processes.

In many instances, natural gas is dissolved in oil underground primarily due to
the formation pressure. When this natural gas and oil is produced, it is possible that
it will separate on its own, simply due to decreased pressure; much like opening a
can of soda pop allows the release of dissolved carbon dioxide. In these cases, separation of oil and gas is relatively easy, and the two hydrocarbons are sent separate
ways for further processing. The most basic type of separator is known as a conventional separator. It consists of a simple closed tank, where the force of gravity
serves to separate the heavier liquids like oil, and the lighter gases, like natural gas.

In certain instances, however, specialized equipment is necessary to separate oil
and natural gas. An example of this type of equipment is the low-temperature separator. This is most often used for wells producing high-pressure gas along with light
crude oil or condensate. These separators use pressure differentials to cool the wet
natural gas and separate the oil and condensate. Wet gas enters the separator, being
cooled slightly by a heat exchanger. The gas then travels through a high-pressure
liquid vessel (often referred to as a *knockout pot*), which serves to remove any
liquids into a low-temperature separator. The gas then flows into this low-
temperature separator through a choke mechanism, which expands the gas as it
enters the separator. This rapid expansion of the gas allows for the lowering of the
temperature in the separator. After liquid removal, the dry gas then travels back
through the heat exchanger and is warmed by the incoming wet gas. By varying the
pressure of the gas in various sections of the separator, it is possible to vary the
temperature, which causes the oil and some water to be condensed out of the wet
gas stream. This basic pressure—temperature relationship can work in reverse as
well, to extract gas from a liquid oil stream.

From an environmental viewpoint, it is not means by which these gases can be utilized but it is the effects of these gases on the environment when they are introduced into the atmosphere. Thus in addition to the corrosion of equipment of acid gases, the escape into the atmosphere of sulfur-containing gases can eventually lead to the formation of the constituents of acid rain, i.e., the oxides of sulfur (SO_2 and SO_3). Similarly, the nitrogen-containing gases can also lead to nitrous and nitric acids (through the formation of the oxides NO_x, where $x = 1$ or 2), which are the other major contributors to acid rain. The release of carbon dioxide and hydrocarbons as constituents of refinery effluents can also influence the behavior and integrity of the ozone layer.

Finally, another acid gas, hydrogen chloride (HCl), although not usually considered to be a major emission, is produced from mineral matter and the brine that often accompany crude oil during production and is gaining increasing recognition as a contributor to acid rain. However, hydrogen chloride may exert severe local effects, because it does not need to participate in any further chemical reaction to become an acid. Under atmospheric conditions that favor a buildup of stack emissions in the areas where hydrogen chloride is produced, the amount of hydrochloric acid in rain water could be quite high.

7.2.3 Other gas streams

Gas processing removes one or more components from the produced gas to prepare it for use. Common components removed to meet pipeline, safety, environmental, and quality specifications include hydrogen sulfide, carbon dioxide, nitrogen, higher molecular weight hydrocarbons, and water. The technique employed to process the gas varies with the components to be removed as well as with the properties of the gas stream (e.g., temperature, pressure, composition, and flow rate).

When biogas, or any other unconventional gas stream, is sent to processing for impurity removal, the goal, e.g., is to produce a gas that can be blended with a natural; gas stream without the potential for incompatibility so that the producer of the unconventional gas stream can utilize the local gas distribution networks. Water (H_2O), hydrogen sulfide (H_2S), and particulates are removed if present at high levels or if the gas is to be completely cleaned. Carbon dioxide is less frequently removed, but it must also be separated to achieve pipeline quality gas. If the gas is to be used without extensively cleaning, it is sometimes cofired with natural gas to improve combustion. Unconventional gas that has been processed (cleaned) to meet the specifications of pipeline quality gas may be referred to as processed natural gas. In this form the gas can be used in any application in which natural gas is used.

On the other hand, raw (untreated) biogas produced from digestion is approximately roughly 60% v/v methane and 29% v/v carbon dioxide with trace elements of hydrogen sulfide. Landfill gas is produced by wet organic waste decomposing under anaerobic conditions in a digester and may contain the same constituents as biogas but, like biogas, in different proportions depending upon the composition of the source material and the process parameters under which the gas was produced

(Chapter 1: History and use) and there are processes for converting biogas to bio-methane (Ryckebosch et al., 2011).

Just as for natural gas from the reservoir, unconventional gas streams must also be treated to remove impurities, condensate, and particulates. However, the treatment system depends on the end-use: (1) minimal treatment is needed for the direct use of gas in boiler, furnaces, or kilns and (2) using the gas in electricity generation typically requires more in-depth treatment. Thus treatment systems are divided into primary and secondary treatment processing. Primary processing systems remove moisture and particulates. Gas cooling and compression are common in primary processing. Secondary treatment systems employ multiple cleanup processes, physical and chemical, depending on the specifications of the end-use. Two constituents that may need to be removed are siloxane derivatives, and sulfur compounds, which are damaging to equipment and significantly increase maintenance cost. Adsorption and absorption are the most common technologies used in secondary treatment processing, such as for the removal of siloxane derivatives (Yao et al., 2016).

The presence of these impurities may eliminate some of the sweetening processes, since some processes remove large amounts of acid gas but not to a sufficiently low concentration. On the other hand, there are those processes not designed to remove (or incapable of removing) large amounts of acid gases whereas they are capable of removing the acid gas impurities to very low levels when the acid gases are present only in low-to-medium concentration in the gas (Katz, 1959; Mokhatab et al., 2006; Speight, 2014a). The processes that have been developed to accomplish gas purification vary from a simple once-through wash operation to complex multistep recycling systems (Mokhatab et al., 2006; Speight, 2014a). In many cases, the process complexities arise because of the need for recovery of the materials used to remove the contaminants or even recovery of the contaminants in the original, or altered, form (Katz, 1959; Kohl and Riesenfeld, 1985; Newman, 1985; Mokhatab et al., 2006; Speight, 2014a).

Acid-gas removal is commonly by absorption of the hydrogen sulfide and carbon dioxide into aqueous olamine solutions. This method is adequate for high-pressure gas streams and those with moderate to high concentrations of the acid-gas component. Physical solvents such as methanol or the Selexol process may also be used in some cases. However, if the concentration of carbon dioxide in the gas stream is high, such as in the gas stream from carbon dioxide-flooded reservoirs, membrane technology affords bulk carbon dioxide as a preprocessing step before other methods are applied. For minimal amounts of hydrogen sulfide in a gas stream, scavengers can be a cost-effective approach to hydrogen sulfide removal.

The use of absorption as the primary control technique for organic vapors is subject to several limiting factors. One factor is the availability of a suitable solvent. The volatile organic compound must be soluble in the absorbing liquid and even then, for any given absorbent liquid, only volatile organic compounds that are soluble can be removed. Some common solvents that may be useful for volatile organic compounds include water, mineral oils, or other nonvolatile petroleum oils. Another factor that affects the suitability of absorption for organic emissions control is the

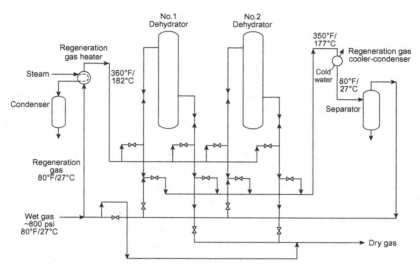

Figure 7.2 The glycol dehydration process.

availability of vapor−liquid equilibrium data for the specific organic/solvent system in question.

Gas streams that are saturated with water require dehydration to increase the heating value of the gas and to prevent pipeline corrosion and the formation of solid hydrates. In most cases, dehydration with a glycol is employed and the water-rich glycol can be regenerated by reducing the pressure and applying heat (Fig. 7.2). Another possible dehydration method is use of molecular sieves that contact the gas with a solid adsorbent to remove the water. Molecular sieves can remove the water down to the extremely low levels required for cryogenic separation processes.

Distillation uses the different boiling points of heavier hydrocarbons and nitrogen for separation. Cryogenic temperatures, required for separation of nitrogen and methane, are achieved by refrigeration and expansion of the gas through an expander. Removal of any higher molecular weight hydrocarbons is dictated by pipeline quality requirements, while deep removal is based on the economic and process parameters required for production of NGLs.

7.3 Process variations

Gas processing is a complex sequence of unit processes designed to clean contaminated gas streams by separating impurities and various nonmethane gases (in the context of this book) nonmethane hydrocarbon derivatives and contaminants fluids to produce a specification-grade product. When processing takes place at the wellhead, the goal is to produce a pipeline-quality dry gas stream so that the gas can be transported (by pipeline) to the nearest processing facility.

Wellhead processing is a necessary aspect of gas technology because the composition of the raw natural gas extracted from one or more producing wells depends on the type, depth, and location of the underground deposit and the geology of the area. If the wells also produced crude oil, the natural gas produced the wells (associated or dissolved gas), the natural gas will contain a higher proportion of low-molecular weight hydrocarbon derivatives than nonassociated gas from a natural gas reservoir or from a gas condensate reservoir. In addition, to common contaminants such as water, carbon dioxide (CO_2), and hydrogen sulfide (H_2S), some of the nonmethane constituents have economic value and are processed to high purity and or sold. A fully operational gas-processing plant delivers pipeline-quality dry natural gas that can be used as fuel by residential, commercial, and industrial consumers.

By way of introduction and by way of definition, throughout this chapter there are two terms used frequently: (1) absorption and (2) adsorption.

Absorption is an approach in which the absorbed gas is ultimately distributed throughout the absorbent (liquid). The process depends only on physical solubility and may include chemical reactions in the liquid phase (*chemisorption*). Common absorbing media used are water, aqueous amine solutions, caustic, sodium carbonate, and nonvolatile hydrocarbon oils, depending on the type of gas to be absorbed. Usually, the gas—liquid contactor designs which are employed are plate columns or packed beds.

Packed-bed scrubbers consist of a chamber containing layers of variously shaped packing material, such as Raschig rings, spiral rings, or saddles, that provides a large surface area for liquid—particle contact. The packing is held in place by wire mesh retainers and supported by a plate near the bottom of the scrubber and the scrubbing liquid is evenly introduced above the packing and flows down through the bed. The liquid coats the packing and establishes a thin film. The pollutant to be absorbed must be soluble in the fluid. In vertical designs (packed towers), the gas stream flows up the chamber (countercurrent to the liquid). In some cases, the packed beds are designed horizontally for gas flow across the packing (cross-current flow).

In packed-bed scrubbers, the gas stream is forced to follow a circuitous path through the packing material, on which much of the particulate matter contacts. The liquid on the packing material collects the particulate matter and flows down the chamber toward the drain at the bottom of the tower. A mist eliminator (demister) is typically positioned above/after the packing and scrubbing liquid supply. Any scrubbing liquid and wetted particulate matter entrained in the exiting gas stream will be removed by the mist eliminator and returned to drain through the packed bed.

However, some caution is advised since a high concentrations of particulate matter can clog the bed, hence the limitation of these devices to streams with relatively low dust loadings. Plugging is a serious problem for packed-bed scrubbers, because the packing is more difficult to access and clean than other scrubber designs. Mobile-bed scrubbers are available that are packed with low-density plastic spheres that are free to move within the packed bed and are less susceptible to plugging because of the increased movement of the packing material.

Absorption is achieved by dissolution in a liquid (a physical phenomenon) or by reaction with a reagent (a chemical phenomenon, sometime referred to as chemisorption) (Barbouteau and Dalaud, 1972; Ward, 1972). Chemical adsorption processes adsorb sulfur dioxide onto a carbon surface where it is oxidized (by oxygen in the flue gas) and absorbs moisture to give sulfuric acid impregnated into and on the adsorbent.

Physical absorption depends on properties of the gas stream and liquid solvent, such as density and viscosity, as well as specific characteristics of the pollutant(s) in the gas and the liquid stream (such as diffusivity and equilibrium solubility). These properties are temperature dependent, and lower temperatures generally favor absorption of gases by the solvent. Absorption is also enhanced by (1) high contact surface area, (2) high liquid—gas ratios, and (3) high concentrations in the gas stream. Chemical absorption may be limited by the rate of reaction, although the rate-limiting step is typically the physical absorption rate, not the chemical reaction rate.

Adsorption differs from *absorption*, in that it is a physical—chemical phenomenon in which the gas is concentrated on the surface of a solid or liquid to remove impurities. Usually, carbon is the adsorbing medium, which can be regenerated upon *desorption* (Fulker, 1972; Speight, 2014a). The quantity of material adsorbed is proportional to the surface area of the solid and, consequently, adsorbents are usually granular solids with a large surface area per unit mass. Subsequently, the captured gas can be desorbed with hot air or steam either for recovery or for thermal destruction.

Because of the variation in the composition and quality of the gas streams entering a gas cleaning facility (Chapter 2: Origin and production and Chapter 3: Unconventional gas), various types of processing plants have been utilized since the mid-1850s to extract liquids, such as natural gasoline, from produced crude oil. However, for many years, natural gas was not a sought-after fuel. Prior to the early 20th century, most of it was flared or simply vented into the atmosphere, primarily because the available pipeline technology permitted only very short-distance transmission.

As the gas processing industry evolved (Chapter 5: Recovery, storage, and transportation), the natural gas received and transported by the major intrastate and interstate mainline transmission systems in the United States had to meet the quality standards specified by pipeline companies. These quality standards vary from pipeline to pipeline and are usually a function of a pipeline system's design, its downstream interconnecting pipelines, and its customer base. However, generally speaking, these standards specify that the natural gas must: (1) be within a specific Btu content range (1035 Btu per cubic feet, $\pm$ 50 Btu); (2) be delivered at a specified hydrocarbon dew point temperature level (Chapter 9: Gas condensate), below which any vaporized gas liquid in the mix will tend to condense at pipeline pressure); (3) contain no more than trace amounts of elements such as hydrogen sulfide, carbon dioxide, nitrogen, water vapor, and oxygen, and (4) be free of particulate solids and liquid water that could be detrimental to the pipeline or its ancillary operating equipment.

Furthermore, several different water dew point specifications exist for natural gas. The water dew point specification is specified in sales gas contracts or given by requirements for transport, processing or storage. For production of liquefied natural gas the water specification needs to be very stringent and a specification of <0.1 ppm (mole) water in the gas is normally used. Due to the low solubility of water in hydrocarbon liquids, the water specification for liquefied petroleum gas (LPG) products need to be low.

Gas processing equipment, whether in the field or at processing/treatment plants, assures that these requirements can be met. While in most cases processing facilities extract contaminants and higher molecular weight hydrocarbons (NGLs) from the gas stream. However, in some cases, the higher molecular weight hydrocarbons may be blended into the gas stream to bring it within acceptable heat content (Btu levels). Whatever the situation, there is the need to prepare the gas for transportation and use in domestic and commercial furnaces. Thus natural gas processing begins at the wellhead and since the composition of the raw natural gas extracted from producing wells depends on the type, depth, and location of the underground deposit and the geology of the area, processing must offer several options (even though each option may be applied to a different degree) to accommodate the difference in composition of the extracted gas (Fig. 7.1).

In those few cases where pipeline-quality natural gas is produced at the wellhead or field facility, the natural gas is moved directly to the pipeline system. In other instances, especially in the production of nonassociated natural gas, field or lease facilities referred to as *skid-mount plants* are installed nearby to dehydrate (remove water) and decontaminate (remove dirt and other extraneous materials) raw natural gas into acceptable pipeline-quality gas for direct delivery to the pipeline system. The *skids* are often specifically customized to process the type of natural gas produced in the area and are a relatively inexpensive alternative to transporting the natural gas to distant large-scale plants for processing.

Gas processing (Mokhatab et al., 2006) consists of separating all of the various hydrocarbons, nonhydrocarbons (such as carbon dioxide and hydrogen sulfide), and fluids from the methane (Table 7.1). Major transportation pipelines usually impose restrictions on the make-up of the natural gas that is allowed into the pipeline. That means that before the natural gas can be transported it must be purified. While the ethane, propane, butanes, and pentanes must be removed from natural gas, this does not mean that they are all waste products.

Gas processing (gas refining) is necessary to ensure that the natural gas intended for use is clean-burning and environmentally acceptable. Natural gas used by consumers is composed almost entirely of methane, but natural gas that emerges from the reservoir at the wellhead is by no means as *pure* (Chapter 3: Unconventional gas). Although the processing of natural gas is in many respects less complicated than the processing and refining of crude oil, it is equally as necessary before its use by end-users.

Raw natural gas comes from three types of wells: oil wells, gas wells, and condensate wells (Chapter 2: Origin and production and Chapter 4: Composition and properties). *Associated gas* (Chapter 1: History and use), i.e., gas from petroleum wells, can exist separate from oil in the formation (free gas), or dissolved in the crude oil (dissolved gas). *Nonassociated gas*, i.e., gas from gas wells or condensate wells is *free* natural gas along with a semiliquid hydrocarbon condensate. Whatever the source of the natural gas, once separated from crude oil (if present) it commonly exists in mixtures with other hydrocarbons; principally ethane, propane, butane, and pentanes. In addition, raw natural gas contains water vapor, hydrogen sulfide (H_2S), carbon dioxide, helium, nitrogen, and other compounds. In fact, the associated hydrocarbons (NGLs) can be very valuable by-products of natural gas processing (Table 7.2). NGLs include ethane, propane, butane, *iso*-butane, and natural gasoline that are sold separately and have a variety of different uses; including enhancing oil recovery in oil wells, providing raw materials for oil refineries or petrochemical plants, and as sources of energy.

The actual practice of processing natural gas to high-quality pipeline gas for the consumer usually involves four main processes to remove the various impurities (1) water removal, (2) liquids removal, (3) enrichment, (4) fractionation, and (5) the process by which hydrogen sulfide is converted to sulfur (the Claus process) (Chapter 8: Gas cleaning processes).

In many instances pressure relief at the wellhead will cause a natural separation of gas from oil (using a conventional closed tank, where gravity separates the gas hydrocarbons from the higher boiling crude oil). In some cases, however, a

Table 7.2 Natural gas liquids: formula and uses

NGL	Chemical formula	Uses	Other uses
Ethane	C_2H_6	Ethylene production Power generation	Plastics Antifreeze Detergents
Propane	C_3H_8	Heating fuel Transportation petrochemical feedstock	Plastics
Butane derivatives (*n*-butane, *iso*-butane)	C_4H_{10}	Petrochemical feedstock Refinery feedstock Blend stock for gasoline	Plastics Synthetic rubber
Condensate	C_5H_{12} and higher boiling hydrocarbons	Petrochemical feedstock Additive to gasoline Diluent for heavy crude oil	Solvents

multistage gas−oil separation process is needed to separate the gas stream from the crude oil. These *gas−oil separators* are commonly closed cylindrical shells, horizontally mounted with inlets at one end, an outlet at the top for removal of gas, and an outlet at the bottom for removal of oil. Separation is accomplished by alternately heating and cooling (by compression) the flow stream through multiple steps; some water and condensate, if present, will also be extracted as the process proceeds.

Dependent on temperature, pressure, and water vapor concentration, the maximum water precipitation temperature will correspond either to the water dew point, the frost point, or the hydrate point. In a conservative design the specification of water vapor concentration should be based on this maximum precipitation temperature, and not the traditional water dew point temperature. Accurate conversion between water vapor concentration and precipitation temperature is therefore crucial.

At some stage of the processing, the gas flow is directed to a unit that contains a series of filter tubes. As the velocity of the stream reduces in the unit, primary separation of remaining contaminants occurs due to gravity. Separation of smaller particles occurs as gas flows through the tubes, where they combine into larger particles which flow to the lower section of the unit. Further, as the gas stream continues through the series of tubes, a centrifugal force is generated which further removes any remaining water and small solid particulate matter.

7.4   Solids removal

Historically, *particulate matter control* (*dust control*) (Mody and Jakhete, 1988) has been one of the primary concerns of industries, since the emission of particulate matter is readily observed through the deposition of fly ash and soot as well as in impairment of visibility. Particulate matter consists of microscopic solid particles or liquid droplets which are small enough to enter the lungs and cause health problems. Both nitrogen oxides (NO_x) and sulfur oxides (SO_x) are associated with the formation of particulate matter, but other processes can contribute to their formation. Aside from health concerns, particulates cause reduced visibility and haze when released in the atmosphere.

Ash is formed in coal combustion and gasification from inorganic impurities in the coal (Chapter 3: Unconventional gas). Some of these impurities react to form microscopic solids which can be suspended in the exhaust gases in the case of combustion, or the syngas produced by gasification. In the latter case, raw syngas leaving the gasifier contains fine ash and/or slag that needs to be removed prior to sending the gas downstream for further processing. The bulk of the particulate matter is removed using dry particulate removal systems such as filters and/or cyclones.

Differing ranges of control can be achieved by use of various types of equipment. Upon proper characterization of the particulate matter emitted by a specific process, the appropriate piece of equipment can be selected, sized, installed, and performance tested. The general classes of control devices for particulate matter are as follows: (1) cyclone collectors, (2) electrostatic precipitators, (3) fabric filters, (4) sand-bed filters, (5) wet scrubbing, and (6) dry scrubbing

Cyclone collectors are the most common of the inertial collector class and are effective in removing coarser fractions of particulate matter and are dependent on mass for the removal. The gas stream and particles are introduced tangentially into a cylinder so that a rotational movement is obtained. Centrifugal forces carry the particles toward the wall of the cylinder, to the vortex chamber and then to the dust collection chamber. Cyclones and other mass force separators can be characterized by the particle cut diameter. A mass force separator has a removal efficiency of 50% for the stated cut particle diameter. Particles with diameters larger than the particle cut diameter are removed with higher rates and smaller particles, with less mass, are removed with lower efficiency.

In the operation of the unit, the particle-laden gas stream enters an upper cylindrical section tangentially and proceeds downward through a conical section. Particles migrate by centrifugal force generated by providing a path for the carrier gas to be subjected to a vortex-like spin. The particles are forced to the wall and are removed through a seal at the apex of the inverted cone. A reverse-direction vortex moves upward through the cyclone and discharges through a top center opening.

Cyclones are highly efficient for the coarse fraction but have minor effect on the submicron particles. The main application for cyclones in many gas-treating plants is as precollectors. A special application is the cyclones at about $900°C$ ($1650°F$) in circulating fluidized bed boilers. An alternative to cyclones as a precollector would be a Louvre-type separator, which operates with a sharp controlled deflection. Some small-diameter high-efficiency cyclones are utilized. The equipment can be arranged either in parallel or in series to both increase efficiency and decrease pressure drop. These units for particulate matter operate by contacting the particles in the gas stream with a liquid. In principle the particles are incorporated in a liquid bath or in liquid particles which are much larger and therefore more easily collected.

Electrostatic precipitators or *fabric filters* remove particulate matter and flue gas desulfurization units capture the sulfur dioxide produced by burning fossil fuels, particularly coal. Electrostatic precipitators operate on the principle of imparting an electric charge to particles in the incoming air stream, which are then collected on an oppositely charged plate across a high-voltage field. Charging of particles is good for particles bigger than approximately 1 μm-field charging—for particles smaller than approximately 0.2 μm, diffusion charging is the preferred method. This phenomenon is one of the reasons why electrostatic precipitators often have a lower removal efficiency for about 0.5-μm particles. Dry electrostatic precipitator units are sensitive to dust with high resistivity as flashes can occur in the dust layer on the collector, but high-voltage pulsing decreases this risk.

Particles of high resistivity create the most difficulty in collection. Conditioning agents such as sulfur trioxide (SO_3) have been used to lower resistivity. Important parameters include design of electrodes, spacing of collection plates, minimization of air channeling, and collection-electrode rapping techniques (used to dislodge particles). Techniques under study include the use of high-voltage pulse energy to enhance particle charging, electron-beam ionization, and wide plate spacing. Electrical precipitators are capable of efficiencies >99% under optimum conditions, but performance is still difficult to predict in new situations.

Fabric filters are typically designed with nondisposable filter bags. The most common adhesion separator, fabric filter, normally has a large area of woven or needled fabric which the flue gas has to flow through. During passage, the particles are removed by deflection (mass forces), interception, diffusion (adhesion forces), and electrical forces. It is the combination of these forces which gives this type of filter its unique performance of almost 100% removal efficiency for most kinds of dust, independent of particle size (correctly designed, operated, and maintained).

As the dusty emissions flow through the filter media (typically cotton, polypropylene, Teflon, or fiberglass), particulate matter is collected on the bag surface as a dust cake. Fabric filters are generally classified on the basis of the filter bag processing mechanism employed. Fabric filters operate with collection efficiencies up to 99.9% although other advantages are evident. Fabric filters can operate up to 200−250°C (390−480°F) with common fabrics, higher temperatures require special material. The dust cake which is formed on the fabric is regularly removed by shaking, pulsing, or by reversing the gas flow.

Sand-bed filters, which are robust and can tolerate extreme conditions, also use mass forces for removal of particles. When the particles are deflected during passage through the sand bed, they stick on the surface and are collected. The particle cut diameter for a well-designed sand-bed filter is around 1 μm and it can then remove the main part of the coarse particles. The fractional efficiency curve is normally steep so the removal of particles smaller than 1 μm is limited. By introducing an electrostatic field in the sand bed, the removal of small harmful particles can be improved.

Wet scrubbing (Chapter 6: History of gas processing) involves the use of devices in which a countercurrent spray liquid is used to remove particles from an air stream. Device configurations include plate scrubbers, packed beds, orifice scrubbers, venturi scrubbers, and spray towers, individually or in different combinations. It is very important to *saturate* the flue gas with water vapor so that particles can absorb water and then increase in size and weight. *Condensation* after saturation gives another improvement when the water vapor condenses on the small particles. By optimizing the design and high flue gas humidity the removal of small particles can be much improved at the same time as low-temperature heat is recovered.

Dry scrubbing (more correctly called dry absorption) involves sorption and a chemical reaction between impurities in gas stream and an absorbent which is injected into the gas. The reaction products are dry and are removed in a dust collector. Compared to wet scrubbing, dry absorption is a relatively recent development. Adsorption on materials such as alumina (Al_2O_3) and charcoal has been known since the early days of the gas processing industry, but absorption in dry and/or semidry phase where the reaction products penetrate into the inner part of the absorbent was not practiced until the early part (or the mid-point at the latest) of the 20th century. The dry absorption principle also includes variants such as semidry, spray-drying, wet−dry, and dry−dry processes. The principle of dry absorption is relatively simple insofar as a filter is precoated with an absorbent with a large specific surface, the gas stream is passed through the precoated filter, unwanted impurities react with the absorbent, particles are removed simultaneously, remove the filter cake, and precoated repeat use.

If the gas stream contains hydrogen chloride, hydrogen fluoride, or sulfur dioxide (as occurs in many biogas streams), there are two predominant absorbents can be used: (1) calcium hydroxide, $Ca(OH)_2$, also called slaked lime and (2) sodium bicarbonate, $NaHCO_3$, also called baking soda. Calcium hydroxide obtains a rather large specific surface already during the slaking process. The sodium bicarbonate has a low specific surface but when it is heated to 150−160°C (300−320°F) and water is driven off leaving sodium carbonate (Na_2CO_3) that has three with many pores and a large specific surface. Thus:

$$2NaHCO_3 \rightarrow Na_2CO_3 + CO_2 + H_2O$$

For good absorption calcium hydroxide is dependent on a certain temperature in relation to the humidity in the flue gas, because the calcium hydroxide must be able to absorb water either added as a slurry, wetted powder or absorbed from the humidity in the flue gas. Sodium carbonate, which is generated from the added sodium bicarbonate, is very reactive and not dependent on water nor humidity. It can therefore be used to absorb sulfur dioxide, hydrogen chloride, or hydrogen fluoride as well as a mixture of these gases in a wide temperature range of 150−300°C (300−570°F) with and without water in the gas stream.

Other methods include use of high-energy input *venturi scrubbers* or electrostatic scrubbers where particles or water droplets are charged, and flux force/condensation scrubbers where a hot humid gas is contacted with cooled liquid or where steam is injected into saturated gas. In the latter scrubber the movement of water vapor toward the cold-water surface carries the particles with it (*diffusiophoresis*), while the condensation of water vapor on the particles causes the particle size to increase, thus facilitating collection of fine particles. The *foam scrubber* is a modification of a wet scrubber in which the particle-laden gas is passed through a foam generator, where the gas and particles are enclosed by small bubbles of foam.

Some relatively new processes for dehydration involve applying isentropic cooling and separation using high centrifugal forces in supersonic gas flow. To estimate the limits of such techniques it is important to have models that can calculate water vapor concentration in equilibrium with hydrate, ice, and water in natural gas at operational temperature and pressure. Briefly, by way of clarification an isentropic process is an idealized thermodynamic process that is both adiabatic and reversible.

7.5 Water removal

Typically, reservoirs containing natural gas reservoirs always have water associated with them and thus, the gas in the reservoir is water-saturated. When the gas is produced water is produced as well, although some of the water is produced directly from the reservoir. Other water produced with the gas is water of condensation formed because of the changes in pressure and temperature during production. In fact, water is a common impurity in gas streams, and removal of water is necessary to prevent condensation of the water and the formation of ice or gas hydrates $(C_nH_{2n+2} \cdot xH_2O)$.

Most of the water associated with natural gas is removed by simple separation methods at or near the wellhead. However, the removal of the water vapor that exists in solution in natural gas requires a more complex treatment. This treatment consists of *dehydrating* the natural gas, which usually involves one of two processes: either absorption or adsorption. It is worthy of note that water can also occur in synthesis gas (syngas) and must also be removed before the gas is used for synthesis of hydrocarbon derivatives or methanol (Sharma et al., 2013)

Water is a common impurity in gas streams, and removal of water is necessary to prevent condensation of the water and the formation of ice or gas hydrates $(C_nH_{2n+2} \cdot xH_2O)$. Water in the liquid phase causes corrosion or erosion problems in pipelines and equipment, particularly when carbon dioxide and hydrogen sulfide are present in the gas. Water in natural gas can create problems during transportation and processing, of which the most severe is the formation of gas hydrates which are representatives of a class of compounds known as clathrates or inclusion compounds. Natural gas and crude oil normally reside in reservoir in contact with connate water. Water can combine with low-molecular weight natural gases to form a solid, hydrate, even if the temperature is above water freezing point. Hydrates can block transmission lines, plug blowout preventers, jeopardize the foundation of deep water platforms and pipelines, cause tubing and casing collapses, and foul process heat exchangers, valves, and expanders. Also, corrosion of materials in contact with natural gas and condensed water, particularly when carbon dioxide and hydrogen sulfide are present in the gas, is also a common problem in the oil and gas industry. The simplest method of water removal (refrigeration or cryogenic separation) is to cool the gas to a temperature at least equal to or (preferentially) below the dew point, as in the glycol dehydration process (Fig. 7.2) or the glycol refrigeration process (Fig. 7.3) (Geist, 1985).

In addition to separating crude oil and some condensate from the wet gas stream, it is necessary to remove most of the associated water. Absorption occurs when the water vapor is taken out by a dehydrating agent whereas adsorption occurs when the water vapor is condensed from the gas stream and collected on the surface.

Thus the first step in evaluating a gas stream for a dehydration system is to determine the water content of the gas by a recognized standard test method (Nadkarni, 2005; ASTM, 2017; Speight, 2018). This data is most important when one designs sour-gas dehydration facilities and estimates water production with sour gas in the plant inlet separator. When natural gas is sent to sweetening units (i.e., units that remove carbon dinopid and hydrogen sulfide), aqueous solvents are typically employed. The sweetened gas, with the carbon dioxide and hydrogen sulfide removed, is saturated with water. In addition, the acid gas byproduct of the sweetening is also saturated with water. In the transmission of natural gas further

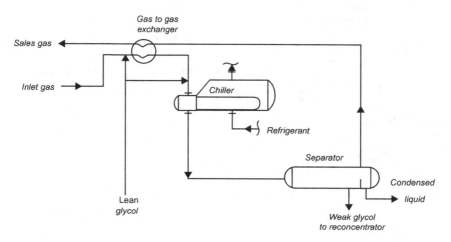

Figure 7.3 The glycol refrigeration process.

condensation of water is problematic, since water can increase pressure drop in the line and often leads to corrosion problems. Thus water should be removed from the natural gas before it is sold to the pipeline company or these reasons, the water content of natural gas and acid gas is an important engineering consideration.

Hydrates form when a gas or liquid containing free water experiences specific temperature/pressure conditions. Dehydration is the removal of this water from the produced natural gas and is accomplished by several methods. Among these is the use of ethylene glycol (glycol injection) systems as an absorption process. The process removes water and other solids from the gas stream. Alternatively, adsorption dehydration may be used, utilizing dry-bed dehydrators towers, which contain desiccants such as silica gel and activated alumina, to perform the extraction (Fig. 7.4).

Dehydration is the process by which water is to remove natural gas and NGLs and is required to (1) prevent formation of gas hydrates and condensation of free water in processing and transportation facilities, (2) meet a water content specification, and (3) prevent corrosion. In a majority of cases of natural gas dehydration, cooling alone is insufficient and, for the most part, impractical for use in field operations. Other water removal processes use (1) *hygroscopic* liquids, such as diethylene glycol (DEG) or triethylene glycol (TEG)—in fact, ethylene glycol can be directly injected into the gas stream in refrigeration plants and (2) solid adsorbents or desiccants, of which alumina, Al_2O_3, silica gel, SiO_2, and molecular sieves.

The formation of hydrates can be prevented by dehydrating the gas or liquid to eliminate the formation of a condensed water (liquid or solid) phase. In some cases, however, dehydration may not be practical or economically feasible. In these cases, chemical inhibition can be an effective method of preventing hydrate formation and uses injection of thermodynamic inhibitors or low dosage hydrate inhibitors (LDHIs).

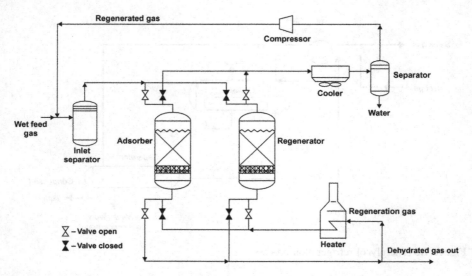

Figure 7.4 Solid-desiccant dehydration process (GPSA, 1998).

Thermodynamic inhibitors are the traditional inhibitors (i.e., one of the glycols or methanol), which lower the temperature of hydrate formation. LDHIs are either kinetic hydrate inhibitors or antiagglomerants, which do not lower the temperature of hydrate formation but do diminish the effect of the hydrates. For example, kinetic hydrate inhibitors lower the rate of hydrate formation, which inhibits its development for a defined duration while antiagglomerants do not inhibit the formation of hydrates but restrict the crystals to submillimeter size.

However, to be effective, the inhibitor must be present at the very point where the wet gas is cooled to its hydrate temperature. For example, in refrigeration plants glycol inhibitors are typically sprayed on the tube-sheet faces of the gas exchangers so that it can flow with the gas through the tubes. As water condenses, the inhibitor is present to mix with the water and prevent hydrates. Injection must be in a manner to allow efficient distribution of the inhibitor to every tube or plate surface in the chillers and heat exchangers that are set to operate below the gas hydrate temperature.

Kinetic hydrate inhibitors inhibit hydrate formation at a concentration range of 0.1%−1.0% w/w polymer in the free water phase. At the maximum recommended dosage, the current inhibition capabilities are −2°C (29°F) of subcooling in a gas system. For relative comparison, methanol or glycol typically may be required at concentrations ranging 20%−50% w/w, respectively, in the water phase.

Three of the most common methods for dehydration of natural gas are physical absorption using glycols, adsorption on solids (e.g., molecular sieve/silica gel) and condensation by a combination of cooling and chemical injection (ethylene glycol/ methanol). TEG dehydration is the most frequent method used to meet pipeline sales specifications. Adsorption processes are used to obtain very low water vapor concentration (0.1 ppm or less) required in low-temperature processing.

Several methods have been developed to dehydrate gases on an industrial scale. The three major methods of dehydration are (1) direct cooling, (2) adsorption, and (3) absorption. Molecular sieves (zeolites), silica gel (SiO_2), and bauxite are the desiccants used in adsorption processes. Bauxite consists mostly of the aluminum minerals gibbsite, [$Al(OH)_3$], boehmite [γ-$AlO(OH)$], and diaspore [α-$AlO(OH)$], mixed with the two iron oxides goethite [α-$FeOH(OH)$] and hematite (Fe_2O_3). In absorption processes, the most frequently used desiccants are diethylene and triethylene glycols. Usually, the absorption/stripping cycle is used for removing large amounts of water, and adsorption is used for cryogenic systems to reach low moisture contents.

In summary techniques for dehydrating natural gas, associated gas condensate and NGLs include: (1) absorption using liquid desiccants, (2) adsorption using solid desiccants, (3) dehydration with calcium chloride, $CaCl_2$, (4) dehydration by refrigeration, (5) dehydration by membrane permeation, (6) dehydration by gas stripping, and (7) dehydration by distillation.

Among these various gas dehydration processes, absorption is the most common technique, where the water vapor in the gas stream becomes absorbed in a liquid solvent stream. Glycols are the most widely used absorption liquids as they approximate the properties that meet commercial application criteria. Several glycols have been found suitable for commercial application. TEG is by far the most common liquid desiccant used in natural gas dehydration as it exhibits most of the desirable criteria of commercial suitability.

In a majority of cases, cooling alone is insufficient and, for the most part, impractical for use in field operations. Other more convenient water removal options use (1) *hygroscopic* liquids (e.g., diethylene glycol or triethylene glycol) and (2) solid adsorbents or desiccants (e.g., alumina, silica gel, and molecular sieves). Ethylene glycol can be directly injected into the gas stream in refrigeration plants.

7.5.1 Absorption

Absorption differs from *adsorption*, in that it is not a physical—chemical surface phenomenon, but an approach in which the absorbed gas is ultimately distributed throughout the absorbent (liquid). The process depends only on physical solubility and may include chemical reactions in the liquid phase (*chemisorption*). Common absorbing media used are water, aqueous amine solutions, caustic, sodium carbonate, and nonvolatile hydrocarbon oils, depending on the type of gas to be absorbed. Usually, the gas—liquid contactor designs which are employed are plate columns or packed beds (Mokhatab et al., 2006; Speight, 2014a). Thus absorption is achieved by dissolution (a physical phenomenon) or by reaction (a chemical phenomenon) (Barbouteau and Dalaud, 1972; Ward, 1972; Mokhatab et al., 2006; Speight, 2014a). Chemical adsorption processes adsorb sulfur dioxide onto a carbon surface where it is oxidized (by oxygen in the flue gas) and absorbs moisture to give sulfuric acid impregnated into and on the adsorbent.

Liquid absorption processes (which usually employ temperatures below 50°C (120°F)) are classified either as *physical solvent processes* or *chemical solvent processes*. The former processes employ an organic solvent, and absorption is enhanced by low temperatures, or high pressure, or both. Regeneration of the solvent is often accomplished readily (Staton et al., 1985; Mokhatab et al., 2006; Speight, 2014a). In chemical solvent processes, absorption of the acid gases is achieved mainly by use of alkaline solutions such as amines or carbonates (Kohl and Riesenfeld, 1985). Regeneration (desorption) can be brought about by use of reduced pressures and/or high temperatures, whereby the acid gases are stripped from the solvent.

Solvents used for emission control processes should have: (1) a high capacity for acid gas, (2) a low tendency to dissolve hydrogen, (3) a low tendency to dissolve low-molecular weight hydrocarbons, (4) low vapor pressure at operating temperatures to minimize solvent losses, (5) low viscosity, (6) low thermal stability, (7) absence of reactivity toward gas components, (8) a low tendency for fouling, (9) a low tendency for corrosion, and (10) be economically acceptable (Mokhatab et al., 2006; Speight, 2014a).

Amine washing of gas emissions involves chemical reaction of the amine with any acid gases with the liberation of an appreciable amount of heat and it is necessary to compensate for the absorption of heat. Amine derivatives such as ethanolamine (monoethanolamine, MEA), diethanolamine (DEA), triethanolamine (TEA), methyldiethanolamine (MDEA), diisopropanolamine (DIPA), and diglycolamine (DGA) have been used in commercial applications (Katz, 1959; Kohl and Riesenfeld, 1985; Maddox et al., 1985; Polasek and Bullin, 1985; Jou et al., 1985; Pitsinigos and Lygeros, 1989; Mokhatab et al., 2006; Speight, 2014a).

The chemistry can be represented by simple equations for low partial pressures of the acid gases:

$$2RNH_2 + H_2S \rightarrow (RNH_3)_2S$$

$$2RHN_2 + CO_2 + H_2O \rightarrow (RNH_3)_2CO_3$$

At high acid gas partial pressure, the reactions will lead to the formation of other products:

$$(RNH_3)_2S + H_2S \rightarrow 2RNH_3HS$$

$$(RNH_3)_2CO_3 + H_2O \rightarrow 2RNH_3HCO_3$$

The reaction is extremely fast, the absorption of hydrogen sulfide being limited only by mass transfer; this is not so for carbon dioxide. Regeneration of the solution leads to near complete desorption of carbon dioxide and hydrogen sulfide. A comparison between monoethanolamine, diethanolamine, and diisopropanolamine shows that monoethanolamine is the cheapest of the three but shows the highest heat of reaction and corrosion; the reverse is true for diisopropanolamine.

An example of absorption dehydration is known as *glycol dehydration* and DEG ($HOCH_2CH_2CH_2CH_2OH$), the principal agent in this process, has a chemical affinity for water and removes water from the gas stream. In this process, a liquid desiccant dehydrator serves to absorb water vapor from the gas stream. In the process, glycol dehydration involves using a glycol solution, usually either DEG or TEG ($HOCH2CH2CH2CH_2CH_2CH_2OH$), which is brought into contact with the wet gas stream in a *contactor*. The glycol solution absorbs the water from the wet gas and, once absorbed, the glycol-wager mix increases in density (becomes heavier) and sinks to the bottom of the contactor where it (the mixture) is removed. The natural gas, having been stripped of most of its water content, is then transported out of the dehydrator. The glycol solution, bearing all of the water stripped from the natural gas, is put through a specialized boiler designed to vaporize only the water out of the solution. The boiling point differential between water (100°C, 212°F) and glycol (204°C, 400°F) makes it relatively easy to remove water from the glycol solution, allowing it to be reused in the dehydration process.

As well as absorbing water from the wet gas stream, the glycol solution occasionally carries with it small amounts of methane and other compounds found in the wet gas. In the past, this methane was simply vented out of the boiler. In addition to losing a portion of the natural gas that was extracted, this venting contributes to air pollution and the greenhouse effect. In order to decrease the amount of methane and other compounds that are lost, flash tank separator-condensers work to remove these compounds before the glycol solution reaches the boiler. Essentially, a flash tank separator consists of a device that reduces the pressure of the glycol solution stream, allowing the methane and other hydrocarbons to vaporize (*flash*).

After dehydration, the glycol solution then travels to the boiler, which may also be fitted with air or water-cooled condensers, which serve to capture any remaining organic compounds that may remain in the glycol solution. The regeneration (stripping) of the glycol is limited by temperature: diethylene glycol and triethylene glycol decompose at or before their respective boiling points. Also, triethylene glycol is more easily regenerated in an atmospheric stripper because of its high boiling point and decomposition temperature. Such techniques as stripping of hot triethylene glycol with dry gas or vacuum distillation are recommended. In practice, absorption systems recover 90−99% v/v of methane that would otherwise be flared into the atmosphere.

Water may be removed from gas streams at the same time as hydrogen sulfide is removed. Moisture removal is necessary to prevent harm to anhydrous catalysts and to prevent the formation of hydrocarbon hydrates (e.g., $C_3H_8 \cdot 18H_2O$) at low temperatures. A widely used dehydration and desulfurization process is the glycolamine process, in which the treatment solution is a mixture of ethanolamine and a large amount of glycol. The mixture is circulated through an absorber and a reactivator in the same way as ethanolamine is circulated in the Girbotol process. The glycol absorbs moisture from the hydrocarbon gas passing up the absorber; the ethanolamine absorbs hydrogen sulfide and carbon dioxide. The treated gas leaves the top of the absorber; the spent ethanolamine-glycol mixture enters the reactivator tower, where heat drives off the absorbed acid gases and water.

7.5.2 *Adsorption*

Removal of water by *adsorption* of the water on to a solid adsorbent (often referred to as solid-desiccant dehydration) is another process option for the dehydration of gas streams. Adsorption is a physical—chemical phenomenon in which the gas is concentrated on the surface of a solid or liquid to remove impurities and is the underlying principle of solid desiccant dehydration systems. Adsorption involves a form of adhesion between the surface of the solid desiccant and the water vapor in the gas. The water forms an extremely thin film that is held to the desiccant surface by forces of attraction, but there is no chemical reaction.

The process typically consists of two or more adsorption towers, which are filled with a solid desiccant. Typical desiccants include activated alumina or a granular silica gel material. Wet natural gas is passed through these towers, from top to bottom. As the wet gas passes around the particles of desiccant material, water is retained on the surface of these desiccant particles. Passing through the entire desiccant bed, almost all of the water is adsorbed onto the desiccant material, leaving the dry gas to exit the bottom of the tower.

Solid-desiccant dehydrators are typically more effective than glycol dehydrators. However, in order to reduce the size of the solid-desiccant dehydrator, a glycol dehydration unit is often used for bulk water removal. The glycol unit can reduce the water content of the gas stream from approximately 100 ppm v/v to approximately 60 ppm v/v, which is a means of reducing the amount of solid desiccant necessary for final drying. These types of dehydration systems are best suited for large volumes of gas under very high pressure and are thus usually located on a pipeline downstream of a compressor station. Two or more towers are required due to the fact that after a certain period of use, the desiccant in a particular tower becomes saturated with water. To "regenerate" the desiccant, a high-temperature heater is used to heat gas to a very high temperature. Passing this heated gas through a saturated desiccant bed vaporizes the water in the desiccant tower, leaving it dry and allowing for further natural gas dehydration.

The solid-desiccant process is conducted alternately and periodically, with each bed going through successive steps of adsorption and desorption. During the adsorption step, the gas to be processed is sent on the adsorbent bed, which selectively retains the water. When the bed is saturated, hot gas is sent to regenerate the adsorbent. After regeneration and before the adsorption step, the bed must be cooled. This is achieved by passing through cold gas. After heating, the same gas can be used for regeneration. In these conditions, four beds are needed in practice in cyclic operation to dry the gas on a continuous. Two beds operate simultaneously in adsorption or gas drying cycle, one bed in cooling cycle and one bed in regeneration cycle. In the simplest case, one bed operates in the adsorption mode, while the second bed operates in the desorption mode, and both beds are periodically switched (Rojey et al., 1997).

Adsorber units are widely used to increase a low gas concentration prior to incineration unless the gas concentration is very high in the inlet air stream. Adsorption also is employed to reduce problem odors from gases. There are several limitations to the use of adsorption systems, but it is generally felt that the major one is the requirement for minimization of particulate matter and/or condensation of liquids (e.g., water vapor) that could mask the adsorption surface and drastically reduce its efficiency (Table 7.1). Thus, in any gas processing plant, it is necessary to know not only the composition of the gas entering the plan but also the composition of the gas entering (and exiting) each of the unit processes that constitute the gas processing plant and methods involving gas chromatographic analysis are particularly useful (Nadkarni, 2005; ASTM, 2017; Speight, 2018). This type of analytical data can prevent overloading the unit process and (through the removal of corrosive constituents) mitigate the potential for equipment corrosion.

Solid adsorbent or solid-desiccant dehydration is the primary form of dehydrating natural gas using adsorption, and usually consists of two or more adsorption towers, which are filled with a solid desiccant. Typical desiccants include activated alumina or a granular silica gel material. Wet natural gas is passed through these towers, from top to bottom. As the wet gas passes around the particles of desiccant material, water is retained on the surface of these desiccant particles. Passing through the entire desiccant bed, almost all of the water is adsorbed onto the desiccant material, leaving the dry gas to exit the bottom of the tower.

Typical desiccants include activated alumina or a granular silica gel material. Wet natural gas is passed through these towers, from top to bottom. As the wet gas passes around the particles of desiccant material, water is retained on the surface of these desiccant particles. Passing through the entire desiccant bed, almost all of the water is adsorbed onto the desiccant material, leaving the dry gas to exit the bottom of the tower.

A variety of solid desiccants are available in the market for specific applications. Some are good only for dehydrating the gas, while others can perform both dehydration and removal of heavy hydrocarbon components. For solid desiccants used in gas dehydration, the following properties are desirable (Daiminger and Lind, 2004):

1. The desiccant must have a high adsorption capacity at equilibrium, which lowers the required adsorbent volume, allowing for the use of smaller vessels.
2. The desiccant must have high selectivity, which minimizes the undesirable removal of valuable components and reduces overall operating expenses.
3. The desiccant must be capable of easy regeneration, which involves a relatively low regeneration temperature to minimize overall energy requirements.
4. The desiccant must have good mechanical properties (such as high crush strength, low attrition, low dust formation, and high stability against aging) which serve to reduce the frequency of adsorbent replacement.
5. The desiccant must be noncorrosive, nontoxic, chemically inert, high bulk density, and no significant volume changes upon adsorption and desorption of water.

Of the common desiccants, silica gel (SiO_2) and alumina (Al_2O_3) have good capacities for water adsorption (up to 8% by weight). Bauxite (crude alumina, Al_2O_3) adsorbs up to 6% by weight water, and molecular sieves adsorb up to 15% by weight water. Silica is usually selected for dehydration of sour gas because of its high tolerance to hydrogen sulfide and to protect molecular sieve beds from plugging by sulfur. As a desiccant, silica gel is a widely used and is characterized by: (1) more easily regenerated than molecular sieves and (2) has a high capacity for water and can adsorb up to 45% of its own weight in water.

Alumina *guard beds* (which serve as protectors by the act of attrition and may be referred to as an *attrition catalyst*) (Speight, 2000) may be placed ahead of the molecular sieves to remove the sulfur compounds. Downflow reactors are commonly used for adsorption processes, with an upward flow regeneration of the adsorbent and cooling in the same direction as adsorption.

Solid desiccant units generally cost more to buy and operate than glycol units. Therefore their use is typically limited to applications such as gases having a high hydrogen sulfide content, very low water dew point requirements, simultaneous control of water, and hydrocarbon dew points. In processes where cryogenic temperatures are encountered, solid-desiccant dehydration is usually preferred over conventional methanol injection to prevent hydrate and ice formation (Kindlay and Parrish, 2006).

7.5.3 Molecular sieve processes

Molecular sieves belong to a class of aluminosilicates which produce the lowest water dew point, and which can be used to simultaneously sweeten dry gases and liquids (Maple, and Williams 2008)—are commonly used in dehydrators ahead of plants designed to recover ethane and other NGLs. These plants operate at very cold temperatures and require very dry feed gas to prevent formation of hydrates. Dehydration to $-100°C$ ($-148°F$) dew point is possible with molecular sieves. Water dew points less than $-100°C$ ($-148°F$) can be accomplished with special design and definitive operating parameters (Mokhatab et al., 2006; Speight, 2014a).

Molecular sieves are highly selective for the removal of hydrogen sulfide (as well as other sulfur compounds) from gas streams and over continuously high absorption efficiency. They are also an effective means of water removal and thus offer a process for the simultaneous dehydration and desulfurization of gas. However, gas that has excessively high content of water that may require upstream dehydration.

The *molecular sieve process* is similar to the iron oxide process (Chapter 8: Gas cleaning processes). Regeneration of the bed is achieved by passing heated clean gas over the bed. As the temperature of the bed increases, it releases the adsorbed hydrogen sulfide into the regeneration gas stream. The sour effluent regeneration gas is sent to a flare stack, and up to 2% of the gas seated can be lost in the regeneration process. A portion of the natural gas may also be lost by the adsorption of hydrocarbon components by the sieve.

In this process, unsaturated hydrocarbon components, such as olefins and aromatics, tend to be strongly adsorbed by the molecular sieve. Molecular sieves are susceptible to poisoning by such chemicals as glycols and require thorough gas cleaning methods before the adsorption step. Alternatively, the sieve can be offered some degree of protection by the use of *guard beds* in which a less expensive catalyst is placed in the gas stream before contact of the gas with the sieve, thereby protecting the catalyst from poisoning. This concept is analogous to the use of guard beds or attrition catalysts in the petroleum industry (Speight, 2000).

Alumina *guard beds* (which serve as protectors by the act of attrition and may be referred to as an *attrition catalyst*) (Speight, 2000) may be placed ahead of the molecular sieves to remove the sulfur compounds. Downflow reactors are commonly used for adsorption processes, with an upward flow regeneration of the adsorbent and cooling in the same direction as adsorption.

Although two-bed adsorbent treaters have become more common (while one bed is removing water from the gas, the other undergoes alternate heating and cooling), on occasion, a three-bed system is used: one bed adsorbs, one is being heated, and one is being cooled. An additional advantage of the three-bed system is the facile conversion of a two-bed system so that the third bed can be maintained or replaced, thereby ensuring continuity of the operations and reducing the risk of a costly plant shutdown.

Molecular sieves are versatile adsorbents because they can be manufactured for a specific pore size, depending on the application. The sieves are (1) capable of dehydration to less than 1 ppm water content, (2) excellent for hydrogen sulfide removal, (3) carbon dioxide removal, (4) dehydration, and (5) removal of higher molecular weight hydrocarbon derivatives.

7.5.4 Membrane processes

Membrane separation process are very versatile and are designed to process a wide range of feedstocks and offer a convenient solution for the removal and recovery of higher boiling hydrocarbons (NGLs) from natural gas (Foglietta, 2004) as well as cleaning other gases such as biogas (Schweigkofler and Niessner, 2001; Popat and Deshusses, 2008; Deng and Hagg, 2010; Matsui and Imamura, 2010.). Synthetic membranes are made from a variety of polymers including polyethylene, cellulose acetate, polysulfone, and polydimethylsiloxane (Table 7.3) (Isalski, 1989; Robeson, 1991). The material from which the membrane is manufactured plays an important role in the ability of the membrane to provide the desired performance characteristics (Table 7.4). For process optimization, the membrane should have a high permeability as well as sufficient selectivity. It is also important to match the membrane properties to that of the system operating conditions (e.g., pressures and gas composition).

The membrane material can be polymeric, inorganic, or a mixture of those previously mentioned. The different materials have their advantages and disadvantages. Polymeric membranes are widely used as they are easier to fabricate at a lower cost compared to inorganic membranes (Scholes et al., 2012). However, for most polymeric membranes there is a trade-off between the two major performance parameters, permeability and selectivity, known as the Robeson upper bound (Robeson, 2008, 1991).

A huge number of different membrane materials have been investigated for CO_2/CH_4 separation. The performances of most of the materials developed before 1991 come under the 1991 upper bound. Great efforts have been made to improve the gas permeability and selectivity during the last two decades, and the new

Table 7.3 Examples of polymer structures used for the production of membranes

Polyethylene

Cellulose acetate

Polysulfone

Polydimethylsiloxane

Table 7.4 General classification of membranes

Material (synthetic)	Bulk structure
Polymeric	Symmetrical
Inorganic	Asymmetrical
Ceramic	Composite
Metallic	
Zeolite	
Carbon	
Hybrid	
Nanocomposite	
Mixed matrix	

materials developed after 1991 made the upper bound move forward and reach a new plot known as the 2008 Robeson upper bound. There are some exceptions of membranes with an even higher performance that surpass the Robeson upper bound, such as mixed-matrix membranes, facilitated transport fixed-site-carrier membranes, thermally rearranged polymers and high free volume polymers.

Particular membrane materials of interest include nanocomposites, mixed organic/inorganic composites, and chemically inert materials. Particular processes/ systems of interest include membranes for the separation of biobased products, membranes for hydrogen separation and purification, and membranes for industrial applications. Gas separation membranes rely on the difference in chemical or physical interaction between the components present in the gas mixture and the membrane material. This difference causes one of the components to permeate faster through the membrane than the other (Fig. 7.3). Gas absorption membranes are used as contacting devices between a gas flow and a liquid flow. The absorption liquid on one side of the membrane causes the separation by selectively removing certain components from the gas stream on the other side of the membrane (Sanchez et al., 2001; Mokhatab and Towler, 2007).

The separation process is based on high-flux membranes that selectively permeates higher boiling hydrocarbons (compared to methane) and are recovered as a liquid after recompression and condensation. Polymeric membranes are a common option for the separation of carbon dioxide from flue gas because of the maturity of the technology in a variety of industries, namely the petrochemical industry. The ideal polymer membrane has both a high selectivity and permeability. Polymer membranes are examples of systems that are dominated by the solution-diffusion mechanism. The membrane is considered to have holes in which the gas can dissolve (solubility) and the molecules can move from one cavity to the other (diffusion).

Silica membranes can be produced with high uniformity (the same structure throughout the membrane) and the high porosity of these membranes is accompanied by high permeability. Synthesized membranes have smooth surfaces and can be modified on the surface to drastically improve selectivity. For example, silica membrane surfaces that have been functionalized by inclusion of an amine (on the surface) allows the membranes to separate carbon dioxide from flue gas streams more effectively (Jang et al., 2011).

Zeolites (crystalline aluminosilicate minerals) with a regular repeating structure of molecular-sized pores also can be used to produce workable membranes. These membranes selectively separate molecules based on pore size and polarity and are thus highly tunable to specific gas separation processes. In general, smaller molecules and those with stronger zeolite-adsorption properties are adsorbed onto zeolite membranes with larger selectivity. The capacity to discriminate based on both molecular size and adsorption affinity makes zeolite membranes an attractive candidate for carbon dioxide separation from natural gas.

Silica membranes can be produced with high uniformity (the same structure throughout the membrane) and the high porosity of these membranes is accompanied by high permeability. Synthesized membranes have smooth surfaces and can be modified on the surface to drastically improve selectivity. For example, silica membrane surfaces that have been functionalized by inclusion of an amine (on the surface) allow the membranes to separate carbon dioxide from flue gas streams more effectively (Jang et al., 2011).

Zeolites (crystalline aluminosilicate minerals) with a regular repeating structure of molecular-sized pores also can be used to produce workable membranes. These membranes selectively separate molecules based on pore size and polarity and are thus highly tunable to specific gas separation processes. In general, smaller molecules and those with stronger zeolite-adsorption properties are adsorbed onto zeolite membranes with larger selectivity. The capacity to discriminate based on both molecular size and adsorption affinity makes zeolite membranes an attractive candidate for carbon dioxide separation from natural gas.

7.6 Liquids removal

Natural gas coming directly from a well contains many NGLs that are commonly removed. In most instances, NGLs have higher value as separate added-value products, and it is thus economical to remove them from the gas stream. The removal of NGLs usually takes place in a relatively centralized processing plant and uses techniques similar to those used to dehydrate natural gas. Recovery of the liquid hydrocarbons can be justified either because it is necessary to make the gas salable or because economics dictate this course of action. The justification for building a liquid recovery (or a liquid removal) plant depends on the price differential between the enriched gas (containing the higher molecular weight hydrocarbons) and lean gas with the added value of the extracted liquid.

Most natural gas is processed to remove the heavier hydrocarbon liquids from the natural gas stream. These heavier hydrocarbon liquids, commonly referred to as NGLs, include ethane, propane, butanes, and natural gasoline (condensate). Recovery of NGLs in gas not only may be required for hydrocarbon dew point control in a natural gas stream (to avoid the unsafe formation of a liquid phase during transport), but also yields a source of revenue as NGLs normally have significantly greater value as separate marketable products than as part of the natural gas stream.

Lower boiling constituents of NGLs, such as ethane, propane, and butanes, can be sold as fuel or feedstock to refineries and petrochemical plants, while the heavier portion can be used as gasoline-blending stock. The price difference between selling NGLs as a liquid and as fuel, commonly referred to as the "shrinkage value" often dictates the recovery level desired by the gas processors. Regardless of the economic incentive, however, gas usually must be processed to meet the specification for safe delivery and combustion. The removal of NGLs usually takes place in a relatively centralized processing plant, where the recovered NGLs are then treated to meet commercial specifications before moving into the NGLs transportation infrastructure.

There are two basic steps to the treatment of NGLs in the natural gas stream. First, the liquids must be *extracted* from the natural gas. Second, these NGLs must be separated themselves, down to their base components. These two processes account for approximately 90% of the total production of NGLs. There are three principle techniques for extracting NGLs from the natural gas stream: (1) the absorption method, (2) the cryogenic expander process, and (3) a membrane process. The extraction of NGLs from the natural gas stream produces both cleaner, purer natural gas, as well as the valuable hydrocarbons that are the NGLs themselves.

7.6.1 Gas–crude oil separation

In order to process and transport associated dissolved natural gas, the first step must be the separation of the gas from the crude oil in which the gas is dissolved. This separation of natural gas from oil is most often done using equipment installed at or near the wellhead. The actual process used to separate oil from natural gas, as well as the equipment that is used, can vary widely. Although dry pipeline quality natural gas may be identical (or almost identical) across different geographic areas, raw natural gas from different regions may have different compositions and separation requirements (Chapter 1: History and use). The most basic type of separator is known as a conventional separator which consists of a closed tank, where the force of gravity serves to separate the heavier liquids like oil, and the lighter gases, like natural gas.

An example of the type of separator is the *low-temperature separator* which is often used for wells producing high-pressure gas along with light crude oil or condensate. These separators use pressure differentials to cool the wet natural gas and separate the oil and condensate.

In the process, wet gas enters the separator, being cooled slightly by a heat exchanger. The gas then travels through a high-pressure liquid *knockout pot*, which serves to remove any liquids into a low-temperature separator. The gas then flows into this low-temperature separator through a choke mechanism, which expands the gas as it enters the separator. This rapid expansion of the gas allows for the lowering of the temperature in the separator. After liquid removal, the dry gas then travels back through the heat exchanger and is warmed by the incoming wet gas. By varying the pressure of the gas in various sections of the separator, it is possible to vary the temperature, which causes the oil and some water to be condensed out of the wet gas stream.

7.6.2 Extraction

There are two basic steps to the treatment of NGLs in the natural gas stream: (1) the liquids must be extracted from the natural gas and (2) the NGLs must be separated themselves, down to their base components. These two processes account for approximately 90% v/v of the total production of NGLs.

There are two principle techniques for removing NGLs from the natural gas stream: (1) the absorption method and (2) the cryogenic expander process.

7.6.2.1 Absorption process

The absorption method of extraction is very similar to using absorption for dehydration. The main difference is that, in the absorption of NGLs, absorbing oil is used as opposed to glycol. This absorbing oil has an affinity for NGLs in much the same manner as glycol has an affinity for water. Before the oil has picked up any NGLs, it is termed *lean* absorption oil. As the natural gas is passed through an absorption tower, it is brought into contact with the absorption oil which soaks up a high proportion of the NGLs. The rich absorption oil, now containing NGLs, exits the absorption tower through the bottom. It is now a mixture of absorption oil, propane, butanes, pentanes, and other heavier hydrocarbons. The rich oil is fed into lean oil stills, where the mixture is heated to a temperature above the boiling point of the NGLs, but below that of the oil. This process allows for the recovery of around 75% v/v butane derivatives, and 85–90% v/v pentane derivatives as well as higher molecular weight hydrocarbons from the natural gas stream.

The *oil absorption process* involves the countercurrent contact of the lean (or stripped) oil with the incoming wet gas with the temperature and pressure conditions programmed to maximize the dissolution of the liquefiable components in the oil. The *rich* absorption oil (sometimes referred to as *fat* oil), containing NGLs, exits the absorption tower through the bottom. It is now a mixture of absorption oil, propane, butanes, pentanes, and other higher boiling hydrocarbons. The rich oil is fed into lean oil stills, where the mixture is heated to a temperature above the boiling point of the NGLs but below that of the oil. This process allows for the recovery of around 75% by volume of the butanes, and 85%–90% by volume of the pentanes and higher boiling constituents from the natural gas stream.

The basic absorption process above can be modified to improve its effectiveness, or to target the extraction of specific NGLs. In the refrigerated oil absorption method, where the lean oil is cooled through refrigeration, propane recovery can be upwards of 90% by volume and approximately 40% by volume of the ethane can be extracted from the natural gas stream. Extraction of the other, higher boiling NGLs can be close to 100% by volume using this process.

The absorption method, on the other hand, uses an absorbing oil to separate the methane from the NGLs. While the gas stream is passed through an absorption tower, the absorption oil (*lean oil*) soaks up a large amount of the NGLs. The absorption oil (*enriched oil*), now containing NGLs, exits the base of the tower after which the enriched oil is fed into distillers where the blend is heated to above the boiling point of the NGLs, while the oil remains fluid. The absorption oil is recycled, while the NGLs are cooled and directed to a fractionator tower (see Section 7.5).

Another absorption method that is often used is the refrigerated oil absorption method, where the lean oil is chilled rather than heated, a feature that has the potential to enhance recovery rates.

7.6.2.2 Cryogenic expander process

Cryogenic processes are also used to extract NGLs from natural gas. While absorption methods can extract almost all of the higher molecular weight NGLs, the lighter hydrocarbons, such as ethane, are often more difficult to recover from the natural gas stream. In certain instances, it is economic to simply leave the lower molecular weight NGLs in the natural gas stream. However, if it is economic to extract ethane and other lighter hydrocarbons, cryogenic processes are required for high recovery rates. Essentially, cryogenic processes consist of dropping the temperature of the gas stream to approximately −85°C (−120°F). The rapid temperature drop condenses ethane and other hydrocarbon derivatives in the gas stream, but methane remains in gaseous form.

In the *cryogenic expander process*, a turboexpander is used to produce the necessary refrigeration and very low temperatures and high recovery of light components, such as ethane and propane, can be attained. Essentially, cryogenic processing consists of lowering the temperature of the gas stream to approximately −85°C (−120°F). There are several ways to perform this function but the turboexpander process (in which external refrigerants are used to chill the gas stream) is the most effective. The quick drop in temperature is that the expander can condense the hydrocarbons in the gas stream but maintains methane in its gaseous form.

In the process, the natural gas is first dehydrated using a molecular sieve followed by cooling (Fig. 7.5). The separated liquid containing most of the heavy fractions is then demethanized, and the cold gases are expanded through a turbine that produces the desired cooling for the process. The expander outlet is a two-phase stream that is fed to the top of the demethanizer column. This serves as a separator in which: (1) the liquid is used as the column reflux and the separator vapors combined with vapors stripped in the demethanizer are exchanged with the feed gas and (2) the heated gas, which is partially recompressed by the expander compressor, is further recompressed to the desired distribution pressure in a separate compressor. This process allows for the recovery of about 90%−95% by volume of the ethane originally in the gas stream. In addition, the expansion turbine is able to convert some of the energy released when the natural gas stream is expanded into recompressing the gaseous methane effluent, thus saving energy costs associated with extracting ethane. The cryogenic method is better at extraction of the lower boiling constituents, such as ethane, than is the alternative absorption method.

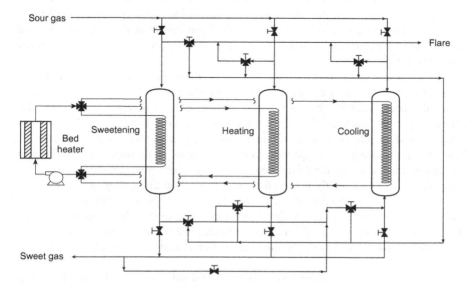

Figure 7.5 The molecular sieve dehydration process.

The process allows for the recovery of about 90−95% v/v of the ethane originally in the gas stream. The extraction of NGLs from the natural gas stream produces both cleaner, purer natural gas, as well as the valuable hydrocarbons that are
the NGLs themselves.

7.6.3 Fractionation

Fractionation is the process of separating the various NGLs present in the remaining
gas stream by using the varying boiling points of the individual hydrocarbons in the
gas stream. Fractionation is a unit operation in which the difficulty of a separation
is directly related to the relative volatility of the components and the required purity
of the product streams. The process occurs in stages as the gas stream rises through
several towers where heating units raise the temperature of the stream, causing the
various liquids to separate and exit into specific holding tanks.

After separation of the NGLs from the natural gas stream, the mixture must be
broken down into their base components to be useful. That is, the mixed stream of
different NGLs must be separated. The process used to accomplish this task is
called fractionation, which is based on the different boiling points of the different
hydrocarbons in the NGLs stream. Essentially, fractionation occurs in stages consisting of the boiling off of hydrocarbons one by one.

The name of a particular fractionator gives an idea as to its purpose, as it is conventionally named for the hydrocarbon that is boiled off. The entire fractionation
process is broken down into steps, starting with the removal of the lighter NGLs
from the stream. The fractionators are used in the following order: (1) the deethanizer, which separates the ethane from the NGLs stream, (2) the depropanizer, which
separates propane from the deethanized stream, (3) the debutanizer, which removes
the butane isomers from the deethanized−depropanized stream and leaves the pentane derivatives and higher molecular weight hydrocarbon derivatives in what reasons of the NGLs stream. There is also a butane splitter (deisobutanizer) which
separates the iso and normal butanes.

Fractionation processes are very similar to those processes classed as *liquids
removal* processes but often appear to be more specific in terms of the objectives,
hence the need to place the fractionation processes into a separate category. The
fractionation processes are those processes that are used (1) to remove the more significant product stream first or (2) to remove any unwanted light ends from the
heavier liquid products. The process occurs in stages consisting of the boiling off
the hydrocarbons one by one. The name of a fractionator gives an idea as to its purpose, as it is conventionally named for the hydrocarbon that is boiled off. The fractionators are used in the following order: (1) the *deethanizer* separates the ethane
from the stream of NGLs, (2) the *depropanizer* separates the propane from the
deethanized stream, (3) the *debutanizer* separates the butanes from the higher boiling hydrocarbons, and (4) the *butane splitter* or *deisobutanizer* separates the *iso*-
butane and *n*-butane.

In the process (Fig. 7.6), heat is introduced to distillation vessel to produce stripping vapors which rise through the column contacting the descending liquid. The

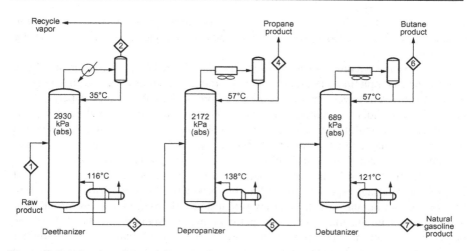

Figure 7.6 A fractionation train.

vapor leaving the top of the column enters the condenser where heat is removed by some type of cooling medium. Liquid is returned to the column as reflux to limit the loss of heavy components overhead. Internals such as trays or packing promote the contact between the liquid and vapor streams in the column. Intimate contact of the vapor and liquid phases is required for efficient separation. Vapor entering a separation stage will be cooled, which results in some condensation of heavier components. The liquid phase will be heated, which results in some vaporization of the lighter components. Thus the higher molecular weight constituents are concentrated in the liquid phase and eventually become the bottom product. The vapor phase is continually enriched in the light components which will make up the overhead product.

The vapor leaving the top of the column may be totally or partially condensed. In a total condenser system, all vapor entering the condenser is condensed to liquid and the reflux returned to the column has the same composition as the distillate or overhead product. In a partial condenser, only a portion of the vapor entering the condenser is condensed to liquid. In most partial condensers, only sufficient liquid will be condensed to serve as reflux for the tower. In some cases, however, more liquid will be condensed than is required for reflux and there will be two overhead products, one a liquid having the same composition as the reflux and the other a vapor product which is in equilibrium with the liquid reflux.

The three-tower system most commonly produces commercial propane, commercial butane, and natural gasoline as products. In this system also, the deethanizer must work properly to remove all constituents that cannot be sold in one of the three products. However, regardless of how the fluids are removed from natural gas and/or gasoline, fractionation is necessary if products that meet any kind of rigid specification are to be made. The number of fractionating columns required depends on the number of products to be made and the character of the liquid, which serves as feed. The single tower system ordinarily produces one specification product from the bottom stream, with all other components in the feed passing overhead.

For a system containing several towers, the split desired in each column should be made before completing analysis of any one product which assures that the splits set up for the different towers produce satisfactory products in all streams. A perfect separation between adjacent components cannot be specified, since this will lead to a situation that is difficult to achieve in an actual column. For example, in the production of propane, it may be that a small amount of ethane and butane is in the propane but, in such a case, the propane must still meet the necessary purity specifications.

The purification of hydrocarbon gases by any of these processes is an important part of gas refining, especially regarding the production of LPG. This is a mixture of propane and butane, which is an important domestic fuel, as well as an intermediate material in the manufacture of petrochemicals (Speight, 2014a, 2017). The presence of ethane in LPG must be avoided because of the inability of this lighter hydrocarbon to liquefy under pressure at ambient temperatures and its tendency to register abnormally high pressures in the LPG containers. In addition, the presence of pentane in LPG must also be avoided, since this hydrocarbon (a liquid at ambient temperatures and pressures) may separate into a liquid state in the gas lines.

Various types of trays are used in fractionation columns: bubble cap trays, sieve trays, and valve trays. Due to the rise in the bubble cap, it is the only tray which can be designed to prevent liquid from weeping (escaping) through the vapor passage. The sieve or valve trays control weeping by vapor velocity. The bubble cap tray has the highest turndown ratio, with designs of 8:1 to 10:1 ratio being common and bubble cap trays are almost always used in glycol dehydration columns.

Typically, many fractionation columns in gas processing plants were equipped with trays. An option to trayed columns is to use packing and, with packed columns, contact between the vapor and liquid phases is achieved throughout the column rather than at specific levels. There are generally three types of packed columns: (1) random packing wherein discrete pieces of packing are dumped in a random manner into a column shell—the packings are of a variety of designs and each design has particular surface area, pressure drop, and efficiency characteristics; (2) structured packing, in which a specific geometric configuration is achieved— these types of packing can either be the knitted-type mesh packing or sectionalized beds; and (3) grids which are systematically arranged packing which use an open lattice structure—these types of packings have found application in vacuum operation and low pressure drop applications, although little use of these types of packings are seen in high pressure services.

7.7 Nitrogen removal

Nitrogen may often occur in sufficient quantities in natural gas and, consequently, lower the heating value of the gas. Thus several plants for *nitrogen removal* from natural gas have been built, but it must be recognized that nitrogen removal requires liquefaction and fractionation of the entire gas stream, which may affect process economics. In many cases the nitrogen-containing natural gas is blended with a gas having a higher heating value and sold at a reduced price depending upon the thermal value (Btu/ft^3).

A significant fraction of many natural gas reserves is subquality (low heating value) due to the high-nitrogen content of the gas. Gas containing more than about 6% nitrogen must be treated to remove the nitrogen. In many cases where nitrogen occurs in natural gas, the reserves cannot currently be exploited because of the lack of suitable nitrogen removal technology.

The separation of methane and nitrogen is challenging for any technology because these gases are similar in size, boiling point, and chemical nature. Conventional processes such as cryogenic distillation and pressure swing adsorption (PSA) are in use (Table 7.5), but application of these technologies is not widespread because the costs of nitrogen removal from subquality gas increase the capital and operating costs of natural gas processing to a point at which the economics are too poor. Most plants that practice nitrogen rejection were built for dual use, such as production of helium and production of carbon dioxide for EOR applications.

Once the hydrogen sulfide and carbon dioxide are processed to acceptable levels, the stream is routed to a nitrogen rejection unit (NRU), where it is further dehydrated using molecular sieve beds. In the NRU, the gas stream is routed through a series of passes through a column and a brazed aluminum plate fin heat exchanger, where the nitrogen is cryogenically separated and vented.

Another type of NRU involves separation of separates methane and heavier hydrocarbons from nitrogen using an absorbent solvent. The absorbed methane and heavier hydrocarbons are flashed off from the solvent by reducing the pressure on the processing stream in multiple gas decompression steps. The liquid from the flash regeneration step is returned to the top of the methane absorber as lean solvent. Helium, if any, can be extracted from the gas stream by various techniques, which include: (1) membrane diffusion, (2) PSA, and (3) cryogenic technique. Among these techniques, cryogenic processes are the most economical method and

Table 7.5 Processes currently suggested for the removal of nitrogen from natural gas (Membrane Technology and Research, Inc., 1999)

Process	Method of separation	Application
Cryogenic distillation	Condensation and distillation at cryogenic temperatures	Typically, high-flow-rate applications
Pressure swing adsorption	Adsorption of methane	Generally small to medium flow rates
Lean oil absorption	Absorption of methane in chilled hydrocarbon liquid	Suitable for high-nitrogen streams
Nitrogen absorption	Selective absorption of nitrogen in chelating solvent	No methane recompression needed

have been commonly used to produce helium at high recovery and/or purity from natural gas or other streams containing low purity helium (Froehlich and Clausen, 2008). Generally, nitrogen removal from natural gas requires liquefaction and fractionation of the entire gas stream, which may affect process economics. In many cases the nitrogen-containing natural gas is blended with a gas having a higher heating value and sold at a reduced price depending upon the thermal value (Btu/ft^3).

A membrane system can produce pipeline-quality gas and nitrogen-rich fuel from raw natural gas. The process relies on proprietary membranes that are significantly more permeable to methane, ethane, and other hydrocarbons than to nitrogen. Gas containing 8%−18% nitrogen is compressed and passed across a first set of membrane modules. The permeate, which contains 4% nitrogen, is sent to the pipeline; the nitrogen-rich residue gas is passed to a second set of membrane modules. These modules produce a residue gas containing 50% nitrogen and a nitrogen-depleted permeate containing about 9% nitrogen. The residue gas is used as fuel; the permeate is mixed with the incoming feed gas for further recovery. The membrane process divides the gas into two streams. The first stream is product gas containing less than 4% nitrogen, which is sent to the pipeline; the second stream, which contains 30%−50% nitrogen, is used to fuel the compressor engines. In some cases, a third stream is produced, this stream, which contains 60%−85% nitrogen, is flared or reinjected.

In a typical two-step process, the feed gas containing 10%−15% nitrogen is compressed to 800−1200 psi using a gas-powered compressor. The gas then passes through the first set of methane-permeable membrane modules. The product gas, which contains 4% nitrogen, is sent to the pipeline. The residue is sent to a second set of modules but the permeate from these modules is too rich in nitrogen to be delivered to the pipeline, and the gas is recirculated to the front of the compressor for further treatment. The residue gas from the second set of modules, which contains about 50% nitrogen, is used to fuel the compressor engines. The process achieves 80%−90% recovery of the fuel gas Btu value in the pipeline product. Recovery values as high as 95% or higher can be achieved depending on the composition of the inlet gas (http://www.mtrinc.com/Pages/NaturalGas/ng.html).

7.8 Acid gas removal

In addition to water and NGLs removal, one of the most important parts of gas processing involves the removal of hydrogen sulfide and carbon dioxide. Natural gas from some wells contains significant amounts of hydrogen sulfide and carbon dioxide and is usually referred to as *sour gas*. Sour gas is undesirable because the sulfur compounds it contains can be extremely harmful, even lethal, to breath and the gas can also be extremely corrosive. The process for removing hydrogen sulfide from sour gas is commonly referred to as *sweetening* the gas. Acid gas removal (i.e., removal of carbon dioxide and hydrogen sulfide from natural gas streams) is achieved by application of one or both of the following process types: (1) absorption and (2) adsorption.

For example, carbon dioxide capture from natural gas can be performed by several techniques like solvent scrubbing, membranes, or cryogenic distillation. Carbon dioxide capture by adsorption process has a potential in reducing energy requirement and operational costs due to smaller energy consumption and low maintenance requirements. Physical adsorbents and PSA process are considered to be suitable for carbon dioxide capture at higher partial pressure of carbon dioxide. The high pressure and high flows used in this application are not similar to other applications where PSA is widely used (Grande et al., 2017).

As currently practiced, acid gas removal processes involve the selective absorption of the contaminants into a liquid, which is passed countercurrent to the gas. Then the absorbent is stripped of the gas components (regeneration) and recycled to the absorber. The process design will vary and, in practice, may employ multiple absorption columns and multiple regeneration columns.

Liquid absorption processes (which usually employ temperatures below 50°C (120°F) are classified either as *physical solvent processes* or *chemical solvent processes*. The former processes employ an organic solvent, and absorption is enhanced by low temperatures, or high pressure, or both. Regeneration of the solvent is often accomplished readily (Staton et al., 1985). In chemical solvent processes, absorption of the acid gases is achieved mainly by use of alkaline solutions such as amines or carbonates (Kohl and Riesenfeld, 1985). Regeneration (desorption) can be brought about by use of reduced pressures and/or high temperatures, whereby the acid gases are stripped from the solvent.

Adsorber units are widely used to increase a low gas concentration prior to incineration unless the gas concentration is very high in the inlet air stream. Adsorption also is employed to reduce problem odors from gases. There are several limitations to the use of adsorption systems, but it is generally felt that the major one is the requirement for minimization of particulate matter and/or condensation of liquids (e.g., water vapor) that could mask the adsorption surface and drastically reduce its efficiency.

7.8.1 Olamine processes

The primary process (Fig. 7.7) for sweetening sour natural gas is quite similar to the processes of glycol dehydration and removal of NGLs by absorption. In this case, however, amine (*olamine*) solutions are used to remove the hydrogen sulfide (the *amine process*). The sour gas is run through a tower, which contains the olamine solution. There are two principle amine solutions used, MEA and DEA. Either of these compounds, in liquid form, will absorb sulfur compounds from natural gas as it passes through. The effluent gas is virtually free of sulfur compounds, and thus loses its sour gas status. Like the process for the extraction of NGLs and glycol dehydration, the amine solution used can be regenerated for reuse.

The primary process for sweetening sour natural gas is quite similar to the processes of glycol dehydration and removal of NGLs by absorption. In this case, however, amine (*olamine*) solutions are used to remove the hydrogen sulfide (the *amine process*). The sour gas is run through a tower, which contains the olamine solution.

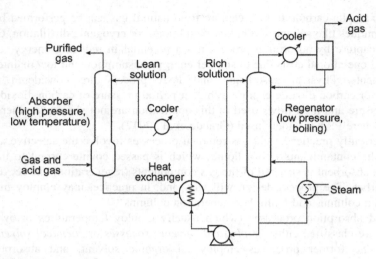

Figure 7.7 The amine (olamine) process.

There are two principle amine solutions used, MEA) and DEA. Either of these compounds, in liquid form, will absorb sulfur compounds from natural gas as it passes through. The effluent gas is virtually free of sulfur compounds, and thus loses its sour gas status. Like the process for the extraction of NGLs and glycol dehydration, the amine solution used can be regenerated for reuse. Although most sour gas sweetening involves the amine absorption process, it is also possible to use solid desiccants like iron sponge (*q.v.*) to remove hydrogen sulfide and carbon dioxide.

Amine derivatives such as ethanolamine (MEA), DEA, TEA, MDEA, DIPA, and DGA are the most widely used *olamines* in commercial applications (Table 7.6) (Katz, 1959; Kohl and Riesenfeld, 1985; Maddox et al., 1985; Polasek and Bullin, 1985; Jou et al., 1985; Pitsinigos and Lygeros, 1989; Mokhatab et al., 2006; Speight, 2014a). They are selected according to their relative ability to interact with and remove carbon dioxide and/or hydrogen sulfide.

The chemistry can be represented by simple equations for low partial pressures of the acid gases:

$$2RNH_2 + H_2S \rightarrow (RNH_3)_2S$$

$$2RHN_2 + CO_2 + H_2O \rightarrow (RNH_3)_2CO_3$$

At high acid gas partial pressure, the reactions will lead to the formation of other products:

$$(RNH_3)_2S + H_2S \rightarrow 2RNH_3HS$$

$$(RNH_3)_2CO_3 + H_2O \rightarrow 2RNH_3HCO_3$$

Table 7.6 Olamines used for gas processing

Olamine	Formula	Derived name	Molecular weight	Specific gravity	Melting point (°C)	Boiling point (°C)	Flash point (°C)	Relative capacity (%)
Ethanolamine (monoethanolamine)	$HOC_2H_4NH_2$	MEA	61.08	1.01	10	170	85	100
Diethanolamine	$(HOC_2H_4)_2NH$	DEA	105.14	1.097	27	217	169	58
Triethanolamine	$(HOC_2H_4)_3N$	TEA	148.19	1.124	18	335[a]	185	41
Diglycolamine (hydroxyethanolamine)	$H(OC_2H_4)_2NH_2$	DGA	105.14	1.057	-11	223	127	58
Diisopropanolamine	$(HOC_3H_6)_2NH$	DIPA	133.19	0.99	42	248	127	46
Methyldiethanolamine	$(HOC_2H_4)_2NCH_3$	MDEA	119.17	1.03	-21	247	127	51

[a]With decomposition.

The reaction is extremely fast, the absorption of hydrogen sulfide being limited only by mass transfer; this is not so for carbon dioxide.

Regeneration of the solution leads to near complete desorption of carbon dioxide and hydrogen sulfide. A comparison between monoethanolamine, diethanolamine, and diisopropanolamine shows that monoethanolamine is the cheapest of the three but shows the highest heat of reaction and corrosion; the reverse is true for diisopropanolamine.

The processes using ethanolamine and potassium phosphate are now widely used. The ethanolamine process, known as the *Girbotol* process, removes acid gases (hydrogen sulfide and carbon dioxide) from liquid hydrocarbons as well as from natural and from refinery gases. The Girbotol process uses an aqueous solution of ethanolamine ($H_2NCH_2CH_2OH$) that reacts with hydrogen sulfide at low temperatures and releases hydrogen sulfide at high temperatures. The ethanolamine solution fills a tower called an absorber through which the sour gas is bubbled. Purified gas leaves the top of the tower, and the ethanolamine solution leaves the bottom of the tower with the absorbed acid gases. The ethanolamine solution enters a reactivator tower where heat drives the acid gases from the solution. Ethanolamine solution, restored to its original condition, leaves the bottom of the reactivator tower to go to the top of the absorber tower, and acid gases are released from the top of the reactivator.

7.8.2 Carbonate washing and water washing processes

In *chemical conversion processes*, contaminants in gas emissions are converted to compounds that are not objectionable or that can be removed from the stream with greater ease than the original constituents. For example, a number of processes have been developed that remove hydrogen sulfide and sulfur dioxide from gas streams by absorption in an alkaline solution.

Carbonate washing is a *chemical conversion processes*, contaminants in natural gas are converted to compounds that are not objectionable or that can be removed from the stream with greater ease than the original constituents. In the carbonate processes, hydrogen sulfide and sulfur dioxide are removed from gas streams by reaction with an alkaline solution (Mokhatab et al., 2006; Speight, 2014a) and uses the principle that the rate of absorption of carbon dioxide by potassium carbonate increases with temperature. It has been demonstrated that the process works best near the temperature of reversibility of the reactions:

$$K_2CO_3 + CO_2 + H_2O \rightarrow 2KHCO_3$$

$$K_2CO_3 + H_2S \rightarrow KHS + KHCO_3$$

Water washing, in terms of the outcome, is analogous to washing with potassium carbonate (Kohl and Riesenfeld, 1985). Acid gas removal is purely physical and there is also a relatively high absorption of hydrocarbons, which are liberated at the same time as the acid gases during the regeneration step. Typical water-soluble gases contained in some gas streams (especially in biogas streams) are: (1) sulfur dioxide, (2) hydrogen chloride, (3) hydrogen fluoride, and (4) ammonia. Other water-soluble gases may also be present but in very low concentrations. These soluble gases can be removed with wet or dry methods.

Absorption of water-soluble gases in wet phase is an old technology. Suitable equipment is wet scrubbers with a large gas/liquid contact area such as spray, tray, column, or plate scrubbers, that is more or less the same as for the removal of supermicron particles. Absorption can be physical or chemical but even if the absorption can be pure physical, such as for hydrogen chloride, chemical reactions are always involved during neutralization.

The process using potassium phosphate is known as phosphate desulfurization, and it is used in the same way as the Girbotol process to remove acid gases from liquid hydrocarbons as well as from gas streams. The treatment solution is a water solution of potassium phosphate (K_3PO_4), which is circulated through an absorber tower and a reactivator tower in much the same way as the ethanolamine is circulated in the Girbotol process; the solution is regenerated thermally.

Other processes include the *Alkazid process* for removal of hydrogen sulfide and carbon dioxide using concentrated aqueous solutions of amino acids. The hot potassium carbonate process decreases the acid content of natural and refinery gas from as much as 50% to as low as 0.5% and operates in a unit similar to that used for amine treating. The *Giammarco-Vetrocoke* process is used for hydrogen sulfide and/or carbon dioxide removal. In the hydrogen sulfide removal section, the reagent consists of sodium or potassium carbonates containing a mixture of arsenite derivatives and arsenate derivatives; the carbon dioxide removal section utilizes hot aqueous alkali carbonate solution activated by arsenic trioxide or selenous acid or tellurous acid.

7.8.3   Metal oxide processes

Treatment of gas to remove the acid gas constituents (hydrogen sulfide and carbon dioxide) is most often accomplished by contact of the natural gas with an alkaline solution. The most commonly used treating solutions are aqueous solutions of the ethanolamine or alkali carbonates, although a considerable number of other treating agents have been developed in recent years (Mokhatab et al., 2006; Speight, 2014a). Most of these newer treating agents rely upon physical absorption and chemical reaction. When only carbon dioxide is to be removed in large quantities or when only partial removal is necessary, a hot carbonate solution or one of the physical solvents is the most economical selection.

The most well-known hydrogen sulfide removal process is based on the reaction of hydrogen sulfide with iron oxide (often also called the iron sponge process or the dry box method) in which the gas is passed through a bed of wood chips impregnated with iron oxide.

The most well-known hydrogen sulfide removal process is based on the reaction of hydrogen sulfide with iron oxide (*iron sponge process*, *dry box method*) in which the gas is passed through a bed of wood chips impregnated with iron oxide. The process is one of the several metal oxide-based processes that scavenge hydrogen sulfide and organic sulfur compounds (mercaptans) from gas streams through reactions with the solid-based chemical adsorbent (Kohl and Riesenfeld, 1985; Mokhatab et al., 2006; Speight, 2014a). The process is governed by the reaction of a metal oxide with hydrogen sulfide to form the metal sulfide. For regeneration, the metal oxide is reacted with oxygen to produce elemental sulfur and the regenerated metal oxide.

The iron oxide process (also known as the iron sponge process) is the oldest and still the most widely used batch process for sweetening natural gas and NGLs (Duckworth and Geddes, 1965; Anerousis and Whitman, 1984; and Zapffe, 1963). The process was implemented during the 19th century. In the process (Fig. 7.8) the sour gas is passed down through the bed. In the case where continuous regeneration is to be utilized a small concentration of air is added to the sour gas before it is processed. This air serves to continuously regenerate the iron oxide, which has reacted with hydrogen sulfide, which serves to extend the on-stream life of a given tower but probably serves to decrease the total amount of sulfur that a given weight of bed will remove.

The process is usually best applied to gases containing low to medium concentrations (300 ppm) of hydrogen sulfide or mercaptans. This process tends to be highly selective and does not normally remove significant quantities of carbon dioxide. As a result, the hydrogen sulfide stream from the process is usually high purity. The use of iron sponge process for sweetening sour gas is based on adsorption of the acid gases on the surface of the solid sweetening agent followed by chemical reaction of ferric oxide (Fe_2O_3) with hydrogen sulfide:

$$2Fe_2O_3 + 6H_2S \rightarrow 2Fe_2S_3 + 6H_2O$$

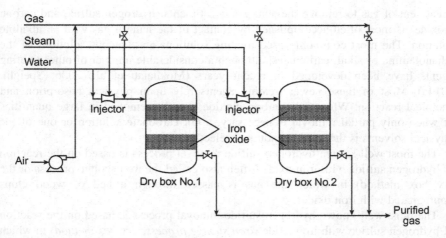

Figure 7.8 The iron oxide (iron sponge) process.

The reaction requires the presence of slightly alkaline water and a temperature below 43°C (110 °F) and bed alkalinity (pH +8−10) should be checked regularly, usually on a daily basis. The pH level is maintained through the injection of caustic soda with the water. If the gas does not contain sufficient water vapor, water may need to be injected into the inlet gas stream.

The ferric sulfide produced by the reaction of hydrogen sulfide with ferric oxide can be oxidized with air to produce sulfur and regenerate the ferric oxide:

$$2Fe_2S_3 + 3O_2 \rightarrow 2Fe_2O_3 + 6S$$

$$S_2 + 2O_2 \rightarrow 2SO_2$$

The regeneration step is exothermic, and air must be introduced slowly so the heat of reaction can be dissipated. If air is introduced quickly the heat of reaction may ignite the bed. Some of the elemental sulfur produced in the regeneration step remains in the bed. After several cycles this sulfur will form a cake over the ferric oxide, decreasing the reactivity of the bed. Typically, after 10 cycles the bed must be removed, and a new bed introduced into the vessel.

Removal of larger amounts of hydrogen sulfide from gas streams requires a continuous process, such as the *Ferrox* process or the Stretford process. The *Ferrox process* is based on the same chemistry as the iron oxide process except that it is fluid and continuous. The *Stretford* process employs a solution containing vanadium salts and anthraquinone disulfonic acid (Maddox, 1974). Most hydrogen sulfide removal processes (Fig. 7.9) return the hydrogen sulfide unchanged, but if the

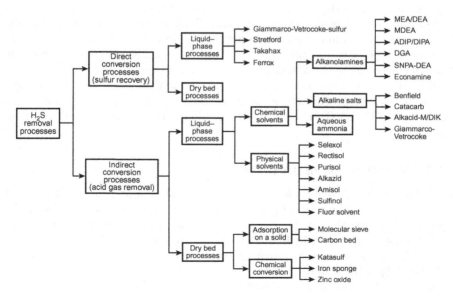

Figure 7.9 Example of hydrogen sulfide removal processes.

quantity involved does not justify installation of a sulfur recovery plant (usually a Claus plant) it is necessary to select a process that directly produces elemental sulfur.

The iron oxide process is one of several metal oxide-based processes that scavenge hydrogen sulfide and organic sulfur compounds (mercaptans) from gas streams through reactions with the solid-based chemical adsorbent (Kohl and Riesenfeld, 1985; Mokhatab et al., 2006; Speight, 2014a). They are typically nonregenerable, although some are partially regenerable, losing activity upon each regeneration cycle. Most of the processes are governed by the reaction of a metal oxide with hydrogen sulfide to form the metal sulfide. For regeneration, the metal oxide is reacted with oxygen to produce elemental sulfur and the regenerated metal oxide. In addition, to iron oxide, the primary metal oxide used for dry sorption processes is zinc oxide.

7.8.4 Catalytic oxidation

Catalytic oxidation is a chemical conversion process that is used predominantly for destruction of volatile organic compounds and carbon monoxide. These systems operate in a temperature regime of 205–595°C (400–1100°F) in the presence of a catalyst. Without the catalyst, the system would require higher temperatures. Typically, the catalysts used are a combination of noble metals deposited on a ceramic base in a variety of configurations (e.g., honeycomb-shaped) to enhance good surface contact.

Catalytic systems are usually classified on the basis of bed types such as *fixed bed* (or *packed bed*) and *fluid bed* (*fluidized bed*). These systems generally have very high destruction efficiencies for most volatile organic compounds, resulting in the formation of carbon dioxide, water, and varying amounts of hydrogen chloride (from halogenated hydrocarbons). The presence in emissions of chemicals such as heavy metals, phosphorus, sulfur, chlorine, and most halogens in the incoming air stream act as poison to the system and can foul up the catalyst.

Thermal oxidation systems, without the use of catalysts, also involve chemical conversion (more correctly, chemical destruction) and operate at temperatures in excess of 815°C (1500°F), or 220–610°C (395–1100°F) higher than catalytic systems.

7.8.5 Molecular sieve processes

Molecular sieves are highly selective for the removal of hydrogen sulfide (as well as other sulfur compounds) from gas streams and over continuously high absorption efficiency. They are also an effective means of water removal and thus offer a process for the simultaneous dehydration and desulfurization of gas. However, gas that has a high content of water may require further (upstream) dehydration (Mokhatab et al., 2006; Speight, 2014a).

The *molecular sieve process* is similar to the iron oxide process. Regeneration of the bed is achieved by passing heated clean gas over the bed. As the temperature of the bed increases, it releases the adsorbed hydrogen sulfide into the regeneration gas stream. The sour effluent regeneration gas is sent to a flare stack, and up to 2% v/v of the gas seated can be lost in the regeneration process. A portion of the natural gas may also be lost by the adsorption of hydrocarbon components by the sieve (Mokhatab et al., 2006; Speight, 2014a).

In this process, unsaturated hydrocarbon components, such as olefins and aromatics, tend to be strongly adsorbed by the molecular sieve. Molecular sieves are susceptible to poisoning by such chemicals as glycols and require thorough gas processing methods before the adsorption step. Alternatively, the sieve can be offered some degree of protection by the use of *guard beds* in which a less expensive catalyst is placed in the gas stream before contact of the gas with the sieve, thereby protecting the catalyst from poisoning. This concept is analogous to the use of guard beds or attrition catalysts in the crude oil industry (Speight, 2000).

7.9 Enrichment

The purpose of *enrichment* is to produce natural gas for sale and enriched tank oil. The tank oil contains more light hydrocarbon liquids than natural crude oil, and the residue gas is drier (leaner, i.e., has lesser amounts of the higher molecular weight hydrocarbons). Therefore, the process concept is essentially the separation of hydrocarbon liquids from the methane to produce a lean, dry gas.

Crude oil enrichment is used when there is no separate market for light hydrocarbon liquids or when the increase in API gravity of the crude provides a substantial increase in the price per unit volume as well as volume of the stock tank oil. A very convenient method of enrichment involves manipulation of the number and operating pressures of the gas—oil separators (traps). However, it must be recognized that alteration or manipulation of the separator pressure affects the gas compression operation as well as influences other processing steps.

One method of removing light ends involves the use of a pressure reduction (vacuum) system. Generally, stripping of light ends is achieved at low pressure, after which the pressure of the stripped crude oil is elevated so that the oil acts as an absorbent. The crude oil, which becomes enriched by this procedure, is then reduced to atmospheric pressure in stages or using fractionation (rectification).

7.10 Other constituents

Natural gas is a mixture of gases containing primarily hydrocarbon gases. It is colorless and odorless in its pure form. It is the cleanest fossil fuel with the lowest carbon dioxide emissions and, therefore, is an important fuel source as well as a major feedstock for fertilizers and petrochemicals. Processing natural gas to isolate the hydrocarbon constituents in specification-quality forms is a challenge that has been met with good success. In addition, many gas streams also contain helium, mercury, vestiges of natural-occurring radioactive materials (NORMs). These constituents offer further challenges because, if not removed, and are left to occur in the products of gas processing there is potential harm to the environment and human health.

7.10.1 Helium

Helium occurs frequently in gas streams (Rogers, 1921) and the recovery of helium from natural gas can generally be considered to occur in two distinct processes, which may take place at the same physical location. Crude helium (i.e., 50%–80% by volume) from the natural gas stream is extracted in a first stage. In the second stage, the crude helium is purified to commercial grades in gaseous and liquid states which are the Grade A (99.996%) helium gas or liquid helium product. Helium is often separated from natural gas as part of a process of removing nitrogen. Nitrogen is removed in order to improve the heating value of the natural gas. Helium can be extracted from a natural gas stream by membrane diffusion in a PSA unit.

Due to the boiling temperature difference between helium ($-268.9°C$, $-452.1°F$), nitrogen ($-195.8°C$, $-320.4°F$), hydrogen ($-252.9°C$, $-423.2°F$), methane ($-164.0°C$, $-263.2°F$), and other components, either condensation-based, distillation-based, or the integration of condensation and distillation-based cryogenic process to extract helium from natural gas streams. Through the distillation process, most methane and other hydrocarbons are recovered as bottom product, and the left overhead gas, which is the mixture of nitrogen, helium and hydrogen, can be further separated by cryogenic condensation process (Xiong et al., 2017).

Purification methods required for helium are also important in producing the other major helium commodity: cryogenic liquid helium. Before liquefaction, helium is purified by using either (1) activated charcoal absorbers at liquid-nitrogen temperatures and high pressure or (2) PSA processes (Chapter 8: Gas cleaning processes). Low-temperature adsorption can produce helium purities of 99.9999%, while a PSA processes can recover helium at better than, but close to, 99.99% purity. A PSA unit can be relatively cheap for gaseous helium, but it is usually the more expensive option if liquefied helium production is required.

7.10.2 Mercury

Mercury is a highly toxic element commonly found in various gas streams and wet scrubbers are only effective for removal of soluble mercury species, such as oxidized mercury (Hg^{2+}). Mercury vapor in its elemental form (zero-valent Hg^0) is insoluble in the scrubber slurry and not removed. Therefore, an additional process of zero-valent mercury conversion is required to complete mercury capture. In some cases, halogens may be added to the gas stream for this purpose. However, it is preferable that additives not be used because of the potential for any residual additive material to cause further gas purification problems.

Mercury in gas streams can occur as three forms: (1) particulate mercury, (2) gaseous oxidized mercury, and (3) gaseous metallic mercury. The mercury in natural gas streams is present predominantly as elemental mercury. However, there is also the possibility that the mercury could be present in other forms: inorganic (such as $HgCl_2$), organic (such as CH_3HgCH_3, $C_2H_5HgC_2H_5$) and organo-ionic (such as $ClHgCH_3$) compounds. The mercury needs to be recovered (removed) for the gas stream because of the value of the metal and, moreover, the potential for the metal to damage aluminum heat exchangers to the point of catastrophic failure. The mercury forms an amalgam with aluminum, resulting in a mechanical failure and gas leakage.

When chlorides are present in the gas stream, such as gas streams from municipal solid waste plants, the oxidized part is mainly mercury chloride ($HgCl_2$). Particulate mercury is removed in good dust separators, while oxidized mercury can be removed in wet and dry scrubbers. The wet scrubber must be designed so that removed oxidized mercury is not reduced to the metallic form and then evaporates from the liquid. If there is hydrogen chloride in the gas stream a prescrubber with low pH is often necessary. Dry scrubbers with additives—activated carbon is the most common additive—can remove both oxidized and metallic mercury and particulate mercury is removed in efficient dust separators. Oxidized mercury, which often is the dominating part of the mercury in the gas stream can be removed in wet and dry scrubbers. The wet scrubber must be designed so that removed oxidized mercury is not reduced to the metallic form and then evaporates from the liquid.

In the United States, there is a rapid deployment of technologies to remove mercury from flue gas. This is typically accomplished by absorption on sorbents or by capture in inert solids as part of the glue gas desulfurization process (Scala et al., 2013). Such scrubbing can lead to meaningful recovery of sulfur for further industrial use (Mokhatab et al., 2006; Speight, 2014a). However, there are other options for removal of mercury from gas streams.

One option is the use of a membrane to remove the mercury (Scholes and Ghosh, 2017) or a combined membrane adsorbent process for mercury removal (Corvini et al., 2002) or the use of a multibed process (Savary and Travers, 2003; Savary, 2004). The adsorbents (created by UOP) can be used for effective mercury removal in existing molecular sieve adsorption units. Since cryogenic plants need to have dry inlet streams, molecular sieve dryers already exist in most plants with NGL recovery. The UOP adsorbents are molecular sieve products that contain silver on the outside surface of the molecular sieve pellet or bead. Mercury from the process fluid (either gas or liquid) amalgamates with the silver, and a mercury-free dry process fluid is obtained (Fig. 7.10) (Corvini et al., 2002). Adding a layer of one of the adsorbents to an existing dryer results in the removal of both the design water load and the mercury without requiring a larger dryer.

Regenerative mercury-removal adsorbents not only dry these streams but also remove mercury to less than 0.01 μg per normal cubic meter. Since the sorption sites for mercury removal are separate from and additive to the dehydration sites, mercury removal is accomplished by replacing a portion of the dehydration grade

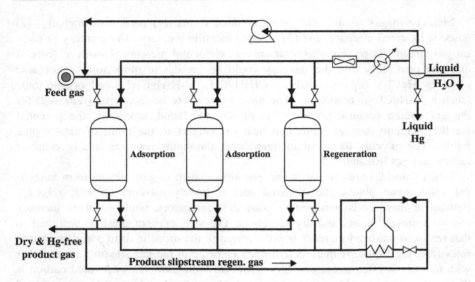

Figure 7.10 UOP mercury removal and recovery process.

molecular sieve with a suitable adsorbent. The dryer bed size does not have to be increased to remove both water and mercury. Moreover, mercury can be completely removed from the adsorbent at conventional dryer regeneration temperatures. Mercury and water are both regenerated from the adsorbents using conventional gas dryer techniques and can be removed from the regeneration gas by condensation when the mercury level in the feed gas is high.

7.10.3 Radioactive residues

During the recovery of natural gas, enhanced concentrations naturally occurring radioactive materials (generally known as NORM) may occur. Isotopes such as radium-226 and radium-228 and the daughter products such as lead-210 may also occur in sludge that accumulates in oilfield pits, tanks and lagoons. Radon gas in the natural gas streams concentrate in gas processing activities. Radon decays to lead-210, then to bismuth-210 as well as polonium-210 and stabilizes with lead-206. Radon decay elements occur as a shiny film on the inner surface of inlet lines, treating units, pumps and valves associated with propylene, ethane and propane processing systems. In some cases, cutting and reaming oilfield pipe, removing solids from tanks and pits, and refurbishing gas processing equipment may expose employees to particles containing increased levels of alpha emitting radionuclides that could pose health risks if inhaled or ingested.

The brine solution (formation wager) contained in reservoirs of oil and gas is pumped to the surface during drilling. The water is separated from the oil and gas into tanks or pits (where it is referred to as produced water) and as the oil and gas in the formation are removed, much of what is pumped to the surface. While uranium and thorium are not soluble in water, the products of the radioactive decay of these elements may dissolve in the brine. The decay products also may remain in solution or settle out to form sludges that accumulate in tanks and pits or form mineral scales inside pipes and on drilling equipment.

The hazards associated with the occurrence of the decay products include inhalation and ingestion routes of entry as well as external exposure where there has been a significant accumulation of scales. Whenever NORM is detected before, during, or after natural gas production, actions should be taken to report this to the US Environmental Protection Agency (or the relevant state or regional authority) who will make recommendations for the next steps.

7.11   Hydrogen sulfide conversion

The disposition of hydrogen sulfide is an issue with many gas processing operations. Burning hydrogen sulfide as a fuel gas component or as a flare gas component is precluded by safety and environmental considerations since one of the combustion products is the highly toxic sulfur dioxide (SO_2), which is also toxic.

As described earlier, hydrogen sulfide is typically removed from gas streams through an *olamine* process after which application of heat regenerates the olamine and forms an acid gas stream (also called the *tail gas* stream). Following from this, the acid gas stream is treated to convert the hydrogen sulfide elemental sulfur and water. The conversion process utilized in most modern refineries is the *Claus process*, or a variant thereof.

7.11.1   Claus process

The disposition of hydrogen sulfide, a toxic gas that originates in crude oils and is also produced in the coking, catalytic cracking, hydrotreating, and hydrocracking, processes, is an issue with many refiners. Burning hydrogen sulfide as a fuel gas component or as a flare gas component is precluded by safety and environmental considerations since one of the combustion products is the highly toxic sulfur dioxide (SO_2), which is also toxic. As described earlier, hydrogen sulfide is typically removed from the refinery light ends gas streams through an olamine process after which application of heat regenerates the olamine and forms an acid gas stream. Following from this, the acid gas stream is treated to convert the hydrogen sulfide elemental sulfur and water. The conversion process utilized in most modern refineries is the Claus process, or a variant thereof.

The Claus process (Fig. 7.11) involves combustion of approximately one-third of the hydrogen sulfide to sulfur dioxide and then reaction of the sulfur dioxide with the remaining hydrogen sulfide in the presence of a fixed bed of activated alumina, cobalt molybdenum catalyst resulting in the formation of elemental sulfur (Maddox, 1974):

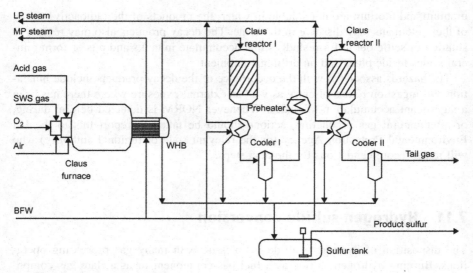

Figure 7.11 The Claus process.

$$2H_2S + 3O_2 \rightarrow 2SO_2 + 2H_2O$$

$$2H_2S + SO_2 \rightarrow 3S + 2H_2O$$

Different process flow configurations are in use to achieve the correct hydrogen sulfide/sulfur dioxide ratio in the conversion reactors.

Overall, conversion of 96%−97% of the hydrogen sulfide to elemental sulfur is achievable in a Claus process. If this is insufficient to meet air quality regulations, a Claus process tail Gas treater is utilized to remove essentially the entire remaining hydrogen sulfide in the tail gas from the Claus unit. The tail gas treater may employ employs a proprietary solution to absorb the hydrogen sulfide followed by conversion to elemental sulfur.

7.11.2 SCOT process

The SCOT (Shell Claus Off-gas Treating) unit is also used to treat *tail gas* and uses a hydrotreating reactor followed by amine scrubbing to recover and recycle sulfur, in the form of hydrogen, to the Claus unit) (Nederland, 2004).

In the process, tail gas (containing hydrogen sulfide and sulfur dioxide) is contacted with hydrogen and reduced in a hydrotreating reactor to form hydrogen sulfide and water. The catalyst is typically cobalt/molybdenum on alumina. The gas is then cooled in a water contractor. The hydrogen sulfide-containing gas enters an amine absorber which is typically in a system segregated from the other refinery amine systems. The purpose of segregation is twofold: (1) the tail gas treater frequently uses a different amine than the rest of the plant and (2) the tail gas is frequently cleaner than the refinery fuel gas (in regard to contaminants) and segregation of the systems reduces maintenance requirements for the SCOT unit. Amines chosen for use in the tail gas system tend to be more selective for hydrogen sulfide and are not affected by the high levels of carbon dioxide in the off-gas.

The hydrotreating reactor converts sulfur dioxide in the off-gas to hydrogen sulfide that is then contacted with a Stretford solution (a mixture of a vanadium salt, anthraquinone disulfonic acid, sodium carbonate, and sodium hydroxide) in a liquid—gas absorber. The hydrogen sulfide reacts stepwise with sodium carbonate and the anthraquinone sulfonic acid to produce elemental sulfur, with vanadium serving as a catalyst. The solution proceeds to a tank where oxygen is added to regenerate the reactants. One or more froth or slurry tanks are used to skim the product sulfur from the solution, which is recirculated to the absorber.

Other tail gas-treating processes include: (1) caustic scrubbing, (2) polyethylene glycol treatment, (3) Selectox process, and (4) sulfite/bisulfite tail gas treating.

References

ASTM, 2017. Annual Book of Standards. ASTM International, West Conshohocken, PA.

Anerousis, J.P., Whitman, S.K., September 1984. An Updated Examination of Gas Sweetening by the Iron Sponge Process. Paper No. SPE 13280. SPE Annual Technical Conference and Exhibition, Houston, TX.

Barbouteau, L., Dalaud, R., 1972. In: Nonhebel, G. (Ed.), Gas Purification Processes for Air Pollution Control. Butterworth and Co, London (Chapter 7).

Bartoo, R.K., 1985. In: Newman, S.A. (Ed.), Acid and Sour Gas Treating Processes. Gulf Publishing, Houston, TX.

Corvini, G., Stiltner, J., Clark, K., 2002. Mercury Removal from Natural Gas and Liquid Streams. Report No. UOP 4022. UOP LLC, Des Plaines, IL. <https://www.uop.com/? document = mercury-removal-from-natural-gas-and-liquid-streams&download = 1>.

Curry, R.N., 1981. Fundamentals of Natural Gas Conditioning. PennWell Publishing Co, Tulsa, OK.

Daiminger, U., Lind, W., 2004. Adsorption-based processes for purifying natural gas. World Refining 14 (7), 32—37.

Deng, L., Hagg, M.B., 2010. Techno-economic evaluation of biogas upgrading process using CO_2 facilitated transport membrane. Int. J. Greenhouse Gas Control 4 (4), 638—646.

Duckworth, G.L., Geddes, J.H., 1965. Natural gas desulfurization by the iron sponge process. Oil Gas J. 63 (37), 94—96.

Foglietta, J.H., 2004. Dew point turboexpander process: a solution for high pressure fields. In: Proceedings. IAPG Gas Conditioning Conference, Neuquen, Argentina (October 18).

Froehlich, P., Clausen, J., 2008. Large scale helium liquefaction and considerations for site services for a plant located in Algeria. In: Advances in Cryogenic Engineering: Transactions of the Cryogenic Engineering Conference (CEC), vol. 53, pp. 985—992.

Fulker, R.D., 1972. In: Nonhebel, G. (Ed.), Gas Purification Processes for Air Pollution Control. Butterworth and Co, London (Chapter 9).

GPSA, 1998. Engineering Data Book, eleventh ed. Gas Processors Suppliers Association, Tulsa, OK.

Gary, J.G., Handwerk, G.E., Kaiser, M.J., 2007. Crude Oil Refining: Technology and Economics, fifth ed. CRC Press, Taylor & Francis Group, Boca Raton, FL.

Geist, J.M., 1985. Refrigeration cycles for the future. Oil Gas J. 83 (5), 56–60.

Grande, C.A., Roussanaly, S., Anantharaman, R., Lindqvist, K., Singh, P., Kemper, J., 2017. CO$_2$ capture in natural gas production by adsorption processes. Energy Procedia 114, 2259–2264.

Hsu, C.S., Robinson, P.R. (Eds.), 2017. Handbook of Crude Oil Technology. Springer International Publishing AG, Cham.

Isalski, W.H., 1989. Separation of Gases. Monograph on Cryogenics No. 5. Oxford University Press, Oxford, pp. 228–233.

Jang, K.-S., Kim, H.-J., Johnson, J.R., Kim, W., Koros, W.J., Jones, C.W., et al., 2011. Modified mesoporous silica gas separation membranes on polymeric hollow fibers. Chem. Mater. 23 (12), 3025–3028.

Jou, F.Y., Otto, F.D., Mather, A.E., 1985. In: Newman, S.A. (Ed.), Acid and Sour Gas Treating Processes. Gulf Publishing Company, Houston, TX (Chapter 10).

Katz, D.K., 1959. Handbook of Natural Gas Engineering.. McGraw-Hill Book Company, New York.

Kindlay, A.J., Parrish, W.R., 2006. Fundamentals of Natural Gas Processing. CRC Press, Taylor & Francis Group, Boca Raton, FL.

Kohl, A.L., Riesenfeld, F.C., 1985. Gas Purification., fourth ed. Gulf Publishing Company, Houston, TX.

Lachet, V., Béhar, E., 2000. Industrial perspective on natural gas hydrates. Oil Gas Sci. Technol. 55, 611–616.

Maddox, R.N., 1974. Gas and Liquid Sweetening, second ed. Campbell Publishing Co, Norman, OK.

Maddox, R.N., 1982. Gas Conditioning and Processing. vol. 4. Gas and Liquid Sweetening. Campbell Publishing Co, Norman, OK.

Maddox, R.N., Bhairi, A., Mains, G.J., Shariat, A., 1985. In: Newman, S.A. (Ed.), Acid and Sour Gas Treating Processes. Gulf Publishing Company, Houston, TX (Chapter 8).

Maple, M.J., Williams, C.D., 2008. Separating nitrogen/methane on zeolite-like molecular sieves. Microporous Mesoporous Mater. 111, 627–631.

Matsui, T., Imamura, S., 2010. Removal of siloxane from digestion gas of sewage sludge. Bioresour. Technol. 101 (1), S29–S32.

Membrane Technology and Research, Inc., 1999. Nitrogen Removal from Natural Gas. Contract Number DE-AC21-95MC32199-02. <http://www.osti.gov/bridge/servlets/purl/780455-PcnOK0/webviewable/780455.pdf>.

Mody, V., Jakhete, R., 1988. Dust Control Handbook. Noyes Data Corp, Park Ridge, NJ.

Mokhatab, S., Towler, B.F., 2007. Nanomaterials hold promise in natural gas industry. Int. J. Nanotechnol. 4 (6), 680–690.

Mokhatab, S., Poe, W.A., Speight, J.G., 2006. Handbook of Natural Gas Transmission and Processing.. Elsevier, Amsterdam.

Nadkarni, R.A.K., 2005. Elemental Analysis of Fuels and Lubricants: Recent Advances and Future Prospects. Publication No. STP1468. ASTM International. West Conshohocken, PA.

Nederland, J., November 2004. Sulphur. University of Calgary, Calgary, AB.

Newman, S.A., 1985. Acid and Sour Gas Treating Processes.. Gulf Publishing, Houston, TX.

Nonhebel, G., 1964. Gas Purification Processes.. George Newnes Ltd, London.

Parkash, S., 2003. Refining Processes Handbook. Gulf Professional Publishing, Elsevier, Amsterdam.

Pitsinigos, V.D., Lygeros, A.I., 1989. Predicting H₂S-MEA equilibria. Hydrocarbon Process. 58 (4), 43−44.

Polasek, J., Bullin, J., 1985. In: Newman, S.A. (Ed.), Acid and Sour Gas Treating Processes.. Gulf Publishing Company, Houston, TX (Chapter 7).

Popat, S.C., Deshusses, M.A., 2008. Biological removal of siloxanes from landfill and digester gases: opportunities and challenges. Environ. Sci. Technol. 42 (22), 8510−8515.

Robeson, L.M., 1991. Correlation of separation factor versus permeability for polymeric membranes. J. Membr. Sci. 62, 165.

Robeson, L.M., 2008. The upper bound revisited. J. Membr. Sci. 320, 390−400.

Rogers, G.S., 1921. Helium-Bearing Natural Gas. USGS Professional Paper No. 121. United States Geological Survey, Reston, VA.

Rojey, A., Jaffret, C., Cornot-Gandolphe, S., Durand, B., Jullian, S., Valais, M., 1997. Natural Gas Production, Processing, Transport. Editions Technip. Institut Français du Petrole, IFP Publications, Paris.

Ryckebosch, E., Drouillon, M., Vervaeren, H., 2011. Techniques for transformation of biogas to biomethane. Biomass Bioenergy 35 (5), 1633−1645.

Sanchez, C., Soler-Illia, G.J.A.A., Ribot, F., Mayer, C.R., Cabuil, V., Lalot, T., 2001. Designed hybrid organic-inorganic nanocomposite from functional nanobuilding blocks. J. Mater. Chem. 13, 3061−3083.

Savary, L., 2004. From purification to liquefaction: gas processing technologies. In: Proceedings. 12th GPA-GCC Technical Conference, Kuwait.

Savary, L., Travers, P., 2003. Axens multibed system − an improved technology for natural gas purification. In: Proceedings. 11th GPA-GCC Technical Conference, Muscat, Oman.

Scala, F., Anacleria, C., Cimino, S., 2013. Characterization of a regenerable sorbent for high temperature elemental mercury capture from flue gas. Fuel 108, 13−18.

Scholes, C.A., Ghosh, U.K., 2017. Review of Membranes for Helium Separation and Purification. Membranes. <https://www.ncbi.nlm.nih.gov/pubmed/28218644> (accessed 16.01.18). <www.mdpi.com/journal/membranes>.

Scholes, C.A., Stevens, D.W., Kentish, S.E., 2012. Membrane gas separation applications in natural gas processing. Fuel 96, 15−28.

Schweigkofler, M., Niessner, R., 2001. Removal of siloxanes in biogases. J. Hazard. Mater. 83 (3), 183−196.

Sharma, S.D., McLennan, K., Dolan, M., Nguyen, T., Chase, D., 2013. Design and performance evaluation of dry cleaning process for syngas. Fuel 108, 42−53.

Soud, H., Takeshita, M., 1994. FGD Handbook. No. IEACR/65. International Energy Agency Coal Research, London.

Speight, J.G., 2000. The Desulfurization of Heavy Oils and Residua, second ed. Marcel Dekker Inc, New York.

Speight, J.G., 2014a. The Chemistry and Technology of Petroleum, fifth ed. CRC Press, Taylor & Francis Group, Boca Raton, FL.

Speight, J.G., 2014b. Oil and Gas Corrosion Prevention. Gulf Professional Publishing, Elsevier, Oxford.

Speight, J.G., 2017. Handbook of Crude Oil Refining. CRC Press, Taylor & Francis Group, Boca Raton, FL.

Speight, J.G., 2018. Handbook of Natural Gas Analysis.. John Wiley & Sons Inc, Hoboken, NJ.

Staton, J.S., Rousseau, R.W., Ferrell, J.K., 1985. In: Newman, S.A. (Ed.), Acid and Sour Gas Treating Processes. Gulf Publishing Company, Houston, TX (Chapter 5).

Ward, E.R., 1972. In: Nonhebel, G. (Ed.), Gas Purification Processes for Air Pollution Control. Butterworth and Co, London (Chapter 8).

Xiong, L., Peng, N., Liu, L., Gong, L., 2017. IOP Conf. Ser.: Mater. Sci. Eng. 171, 012003. <http://iopscience.iop.org/article/10.1088/1757-899X/171/1/012003>.

Yao, P., Boardman, G.D., Li, E.T., 2016. Research progress for removing siloxane from biogas by adsorption. Chem. Ind. Eng. Prog. 35 (2), 604−611.

Zapffe, F., 1963. Iron sponge process removes mercaptans. Oil Gas J. 61 (33), 103−104.

Zhang, L.Q., Shi, L.B., Zhou, Y., 2007. Formation prediction and prevention technology of natural gas hydrate. Nat. Gas Technol. 1 (6), 67−69.

Further reading

Kohl, A.L., Nielsen, R.B., 1997. Gas Purification.. Gulf Publishing Company, Houston, TX.

Gas cleaning processes

8

8.1 Introduction

Gas processing (gas refining) usually involves the use of several integrated unit processes to remove: (1) oil, (2) water, (3) elements such as sulfur, helium, and carbon dioxide, and (4) natural gas liquids (Chapter 7: Process classification). In addition, it is often necessary to install scrubbers and heaters at or near the wellhead that serve primarily to remove sand and other large-particle impurities. The heaters ensure that the temperature of the natural gas does not drop too low and form a hydrate (Chapter 1: History and use) with the water vapor content of the gas stream.

Many chemical processes are available for processing or refining natural gas. However, there are many variables in the choice of process of the choice of refining sequence that dictate the choice of process or processes to be employed. In this choice, several factors must be considered: (1) the types and concentrations of contaminants in the gas, (2) the degree of contaminant removal desired, (3) the selectivity of acid gas removal required, (4) the temperature, pressure, volume, and composition of the gas to be processed, (5) the carbon dioxide—hydrogen sulfide ratio in the gas, and (6) the desirability of sulfur recovery due to process economics or environmental issues.

In addition to hydrogen sulfide and carbon dioxide, gas may contain other contaminants, such as mercaptans (also called *thiols*, R-SH) and carbonyl sulfide (COS). In fact, the variation in the composition of the gases that require cleaning either before being assigned to further use (Chapter 3: Unconventional gas) requires a variation of process types to ensure that specifications are met and the environment is protected. The presence of these impurities may eliminate some of the sweetening processes since some processes remove large amounts of acid gas but not to a sufficiently low concentration. On the other hand, there are those processes that are not designed to remove (or are incapable of removing) large amounts of acid gases. However, these processes also capable of removing the acid gas impurities to very low levels when the acid gases are there in low to medium concentrations in the gas.

Process selectivity indicates the preference with which the process removes one acid gas component relative to (or in preference to) another. For example, some processes remove both hydrogen sulfide and carbon dioxide; other processes are designed to remove hydrogen sulfide only. It is very important to consider the process selectivity for, say, hydrogen sulfide removal compared to carbon dioxide removal that ensures minimal concentrations of these components in the product, thus the need for consideration of the carbon dioxide to hydrogen sulfide in the gas stream.

Natural Gas. DOI: https://doi.org/10.1016/B978-0-12-809570-6.00008-4

To include a description of all of the possible process variations for gas cleaning is beyond the scope of this book. Therefore the focus of this chapter is a selection of the processes that are an integral part within the concept of production of a pipelineable product (methane) for sale to the consumer.

8.2 Glycol processes

Absorption dehydration involves the use of a liquid desiccant to remove water vapor from the gas. Although many liquids possess the ability to absorb water from gas, the liquid that is most desirable to use for commercial dehydration purposes should possess the following properties: (1) high absorption efficiency, (2) relatively easy and economic regeneration, (3) noncorrosive and nontoxic, (4) no operational problems when used in high concentrations, and (5) no interaction with the hydrocarbon portion of the gas, and no contamination by acid gases.

Glycols are the most widely used absorption liquids as they approximate the properties that meet the commercial application criteria. The glycol derivatives, particularly ethylene glycol (EG), diethylene glycol (DEG), triethylene glycol (TEG), and tetraethylene glycol (TREG) are the most appropriate for satisfying these criteria to varying degrees. Water and the glycols show complete mutual solubility in the liquid phase due to hydrogen—oxygen bonds, and their water vapor pressures are very low. A frequently used process is the triethylene glycol process (Fig. 8.1) (Chukwuma and Jacob, 2014). In those situations where inhibition is not feasible or practical, dehydration must be used. Both liquid and solid desiccants may be used, but economics frequently favor liquid desiccant dehydration when it will meet the required dehydration specification.

Liquid desiccant dehydration equipment is simple to operate and maintain. It can easily be automated for unattended operation; e.g., glycol dehydration at a remote production well. Liquid desiccants can be used for sour gases, but additional precautions in the design are needed due to the solubility of the acid gases in the desiccant solution. At very high acid gas content and relatively higher pressures the glycols can also be "soluble" in the gas. Glycol derivatives are typically used for applications where dew point depressions of the order of 15−49°C (59−120°F) are required. DEG, TEG, and TREG are used as liquid desiccants, but TEG is the most common for natural gas dehydration.

In the triethylene glycol process (Fig. 8.1), wet natural gas first enters an inlet separator to remove all liquid hydrocarbon derivatives from the gas stream after which the gas flows to an absorber (contactor) where it is contacted countercurrently and dried by the lean triethylene glycol. The triethylene glycol also absorbs VOCs that vaporize with the water in the reboiler. The dry natural gas existing the absorber passes through a gas/glycol heat exchanger and then into the sales line. The wet or rich triethylene glycol exiting the absorber flows through a coil in the accumulator where it is preheated by hot lean glycol. After the glycol—glycol heat exchanger, the rich glycol enters the stripping column and flows down the packed

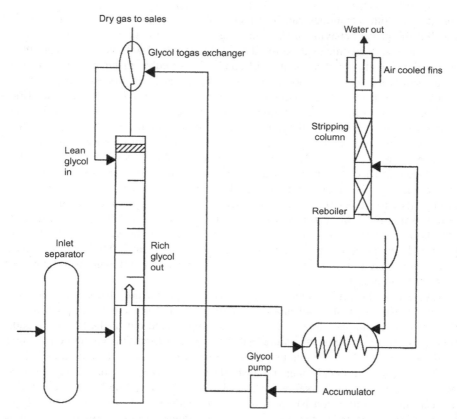

Figure 8.1 The triethylene glycol dehydration system (Manning and Thompson, 1991).

bed section into the reboiler. Steam generated in the reboiler strips absorbed water and VOCs out of the glycol as it rises up the packed bed and the water vapor and desorbed natural gas are vented from the top of the stripper. The hot regenerated lean triethylene glycol flows out of the reboiler into the accumulator (surge tank) where it is cooled via cross exchange with returning rich glycol; it is pumped to a glycol/gas heat exchanger and back to the top of the absorber.

Generally, the contact between a wet gas and glycol can be made in any gas—liquid contact device and is predominantly an absorption/stripping type process, similar to the oil absorption process. The wet gas is dehydrated in the absorber, and the stripping column regenerates the water-free triethylene glycol. The glycol stream should be recharged constantly because some triethylene glycol may react and form higher molecular weight products which should be removed by the filter shown or by distillation of a slip stream.

A distinct difference between the chemical-based (e.g., glycol absorption) and the nonchemical-based (e.g., adsorption) dehydration techniques is that the chemical-based techniques will saturate the gas with chemicals at operational conditions. Both techniques can in principle remove almost all the water from the gas, but the phase behavior of the natural gas leaving the processes will be different. Even though chemicals used for absorption in general will have low vapor pressure and relatively small amounts will condense per cubic meter gas, the effect of condensation has to be considered in design of pipelines and process equipment.

The glycol derivative has a chemical affinity for water and removes water from the gas stream. In this process, a liquid desiccant dehydrator serves to absorb water vapor from the gas stream. Essentially, glycol dehydration involves using a glycol solution, usually either DEG or TEG, which is brought into contact with the wet gas stream in a *contactor*. The glycol solution will absorb water from the wet gas and, once absorbed, the glycol particles become heavier and sink to the bottom of the contactor where they are removed. The natural gas, having been stripped of most of its water content, is then transported out of the dehydrator. The glycol solution, bearing all of the water stripped from the natural gas, is put through a specialized boiler designed to vaporize only the water out of the solution. The boiling point differential between water (100°C, 212°F) and glycol (204°C, 400°F) makes it relatively easy to remove water from the glycol solution, allowing it to be reused in the dehydration process.

As well as absorbing water from the wet gas stream, the glycol solution occasionally carries with it small amounts of methane and other compounds found in the wet gas. In the past, this methane was simply vented out of the boiler. In addition to losing a portion of the natural gas that was extracted, this venting contributes to air pollution and the greenhouse effect. In order to decrease the amount of methane and other compounds that are lost, flash tank separator-condensers work to remove these compounds before the glycol solution reaches the boiler. Essentially, a flash tank separator consists of a device that reduces the pressure of the glycol solution stream, allowing the methane and other hydrocarbon derivatives to vaporize (*flash*). The glycol solution then travels to the boiler, which may also be fitted with air or water-cooled condensers, which serve to capture any remaining organic compounds that may remain in the glycol solution. The regeneration (stripping) of the glycol is limited by temperature; diethylene glycol and triethylene glycol decompose at or before their respective boiling points. Such techniques as stripping of hot triethylene glycol with dry gas (e.g., higher molecular weight hydrocarbon vapors, the *Drizo process*) or vacuum distillation are recommended.

8.3 Olamine processes

Acid gas constituents (hydrogen sulfide and carbon dioxide, CO_2) present in most natural gas streams are a constant reminder of the need for gas processing, specifically acid gas removal technologies. As currently practiced, acid gas removal processes involve the chemical reaction of the acid gases with a solid oxide (such as iron oxide) or selective absorption of the contaminants into a liquid (such as ethanolamine) that is passed countercurrently to the gas. Then the absorbent is stripped of the gas components (regeneration) and recycled to the absorber. The process design will vary and, in practice, may employ multiple absorption columns and multiple regeneration columns. However, depending upon the application, special solutions such as mixtures of amines; amines with physical solvents such as sulfolane and piperazine; and amines that have been partially neutralized with an acid such as phosphoric acid may also be used (Bullin, 2003).

Different amine derivatives can be selected for use, depending on the composition and operating conditions of the feed gas, to meet the product gas specification. Amines are categorized as being primary, secondary, and tertiary, depending upon the degree of substitution of the central nitrogen by organic groups. Primary amines react directly with sulfide, carbon dioxide, and COS.

Examples of primary amines include monoethanolamine (MEA) and the proprietary diglycolamine (DGA) agent. Secondary amines react directly with hydrogen sulfide and carbon dioxide as well as with carbonyl sulfide. The most common secondary amine is diethanolamine (DEA), while diisopropanolamine (DIPA) is another example of a secondary amine that has been used in amine treating systems. Tertiary amines react directly with hydrogen sulfide, react indirectly with carbon dioxide and carbonyl sulfide. The most common examples of tertiary amines are methyldiethanolamine (MDEA) and activated methyldiethanolamine.

In addition, in a refinery other gas streams (process gas) may be added to the natural; gas to be coprocessed and these refinery streams will likely contain mercaptan derivatives (RSH), carbon disulfide (CS_2), or COS. Thus the level of concentration of acid gases in the sour gas is an important consideration for selecting the proper sweetening process. Some processes are applicable for removal of large quantities of acid gas, and other processes have the capacity for removing acid gas constituents to the parts per million (ppm) range. However, whatever the range of nonhydrocarbon constituents in a gas stream, the sweetening process should ensure that the product gas meets pipeline specification or process specifications.

The most commonly used technique for gas processing is to first direct the gas flow through a tower containing an amine (olamine) solution. These processes remove hydrogen sulfide and carbon dioxide from the gas stream by chemical reaction with a material in the solvent solution. The alkanolamine derivatives are the most generally accepted and widely used of the many available solvents for removal of hydrogen sulfide and carbon dioxide from natural gas streams due to their reactivity and availability at low cost. The alkanolamine processes are particularly applicable where acid gas partial pressures are low or low levels of acid gas are considered in the sweet gas. The alkanolamine derivatives are colorless liquids that have a slightly pungent odor and all of the alkanolamine derivatives—with the exception of triethanolamine which decomposes at below its normal boiling point (335°C; 636°F).

Amine derivatives, as used for gas cleaning, absorb sulfur compounds from natural gas and can be reused repeatedly. After desulfurization, the gas flow is directed to the next section, which contains a series of filter tubes. As the velocity of the stream reduces in the unit, primary separation of remaining contaminants occurs due to gravity. Separation of smaller particles occurs as gas flows through the tubes, where they combine into larger particles which flow to the lower section of the unit. Further, as the gas stream continues through the series of tubes, a centrifugal force is generated which further removes any remaining water and small solid particulate matter.

The processes that have been developed to accomplish gas purification vary from a simple once-through wash operation to complex multistep recycling systems (Mokhatab et al., 2006; Speight, 2014). In many cases, the process complexities arise because of the need for recovery of the materials used to remove the contaminants or even recovery of the contaminants in the original, or altered, form (Kohl and Riesenfeld, 1985; Newman, 1985).

The proper selection of the amine can have a major impact on the performance and cost of a sweetening unit. However, many factors must be considered when selecting an amine for a sweetening application (Polasek and Bullin, 1994; Bullin, 2003). Considerations for evaluating an amine type in gas treating systems are numerous. It is important to consider all aspects of the amine chemistry and type since the omission of a single issue may lead to operational issues. While studying each issue, it is important to understand the fundamentals of each amine solution.

MEA and DEA have found the most general application in the sweetening of natural gas streams. Even though a diethanolamine system may not be as efficient as some of the other chemical solvents are, it may be less expensive to install because standard packaged systems are readily available. In addition, it may be less expensive to operate and maintain (Arnold and Stewart, 1999; Jenkins and Haws, 2002).

Monoethanolamine is a stable compound and in the absence of other chemicals suffers no degradation or decomposition at temperatures up to its normal boiling point and readily reacts with hydrogen sulfide and carbon dioxide, thus:

$$2(RNH_2) + H_2S \leftrightarrow (RNH_3)_2S$$

$$(RNH_3)_2S + H_2S \leftrightarrow 2(RNH_3)HS$$

$$2(RNH_2) + CO_2 \leftrightarrow RNHCOONH_3R$$

These reactions are reversible by changing the system temperature. Monoethanolamine also reacts with COS and CS_2 to form heat-stable salts that cannot be regenerated. On the other hand, diethanolamine is a weaker base than monoethanolamine and therefore the diethanolamine system does not typically suffer the same corrosion problems. Diethanolamine reacts with hydrogen sulfide and carbon dioxide in the following manner:

$$2R_2NH + H_2S \leftrightarrow (R_2NH_2)_2S$$

$$(R_2NH_2)_2S + H_2S \leftrightarrow 2(R_2NH_2)HS$$

$$2R_2NH + CO_2 \leftrightarrow R_2NCOONH_2R_2$$

These reactions are reversible. Also, diethanolamine reacts with carbonyl sulfide and with carbon disulfide to form compounds that can be regenerated in the stripping column.

8.3.1 Girdler process

Amine (olamine) washing of natural gas involves chemical reaction of the amine with any acid gases with the liberation of an appreciable amount of heat and it is necessary to compensate for the absorption of heat. Amine derivatives such as ethanolamine (MEA), DEA, triethanolamine, MDEA, DIPA, and DGA have been used in commercial applications (Kohl and Riesenfeld, 1985; Speight, 2014, 2017; Polasek and Bullin, 1994).

Processes that use MDEA became popular with the natural gas industry because of its high selectivity for hydrogen sulfide over carbon dioxide. This high selectivity allows for a reduced solvent circulation rate, as well as a richer hydrogen sulfide feed to the sulfur recovery unit (SRU). The reaction of methyldiethanolamine with hydrogen sulfide is almost instantaneous. However, the reaction of methyldiethanolamine with carbon dioxide is much slower; the reaction rate of methyldiethanolamine with carbon dioxide is slower than that of carbon dioxide with ethanolamine.

Diethanolamine also removes carbonyl sulfide and carbon disulfide partially as its regenerable compound with carbonyl sulfide and carbon disulfide without much solution losses.

One key difference among the various specialty amines is selectivity toward hydrogen sulfide, instead of removing both hydrogen sulfide and carbon dioxide, as generic amines such as monoethanolamine and diethanolamine do, some products readily remove hydrogen sulfide to specifications, but allow controlled amounts of carbon dioxide to slip through. When the methyldiethanolamine process was developed in the mid-1970s, it was principally destined for the sweetening of gases which did not require complete carbon dioxide removal or required the removal of only a controlled part of the carbon dioxide. The selectivity of methyldiethanolamine-based products can lead to more energy savings. For example, allowing carbon dioxide to remain in the treated gas reduces the amount of acid gas in the amine that needs to be regenerated, thus reducing the amount of energy required.

A series of chemical activators used with methyldiethanolamine offers the most cost-effective answer to complete or controlled acid gas removal from sour to very sour natural gases. However, there are limitations of even the most advanced *activated methyldiethanolamine* only based gas treatment technologies in handling very highly acid gas loaded natural or associated oil field gases; especially for bulk acid gas removal when the acid gases are destined for cycling and/or disposal by reinjection. The activated methyldiethanolamine process is probably the most cost-effective solution today to meet the widest range of applications from complete carbon dioxide removal to bulk hydrogen sulfide and/or carbon dioxide removal even for acid gas reinjection projects (Lallemand and Minkkinen, 2001).

The general process flow diagram for an amine sweetening plant varies little, regardless of the aqueous amine solution used as the sweetening agent (Fig. 8.2).

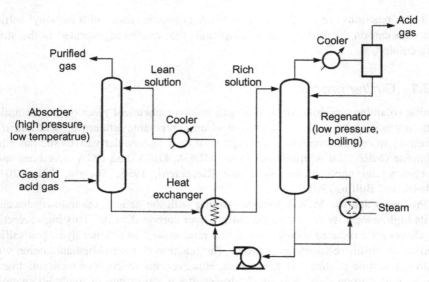

Figure 8.2 The amine (olamine) process for gas sweetening.

The sour gas containing hydrogen sulfide and/or carbon dioxide will nearly always enter the plant through an inlet separator (scrubber) to remove any free liquids and/ or entrained solids. The sour gas then enters the bottom of the absorber column and flows upward through the absorber in intimate countercurrent contact with the aqueous amine solution, where the amine absorbs acid gas constituents from the gas stream. Sweetened gas leaving the top of the absorber passes through an outlet separator and then flows to a dehydration unit (and compression unit, if necessary) before being considered ready for sale.

In many units the rich amine solution is sent from the bottom of the absorber to a flash tank to recover hydrocarbon derivatives that may have dissolved or condensed in the amine solution in the absorber. A small percentage of acid gases will also flash when the pressure is reduced. The heavier hydrocarbon derivatives remain as a liquid, but separate from the aqueous amine, forming a separate liquid layer. Because the hydrocarbon derivatives have a lower density than the aqueous amine, they form the upper liquid layer, and can be skimmed off the top. Therefore a provision should be made to remove these liquid hydrocarbon derivatives. Typically, the flash tanks are designed for 2−3 minutes of retention time for the amine solution while operating half full (Arnold and Stewart, 1999).

The rich solvent is then preheated before entering the top of the stripper column. The amine−amine heat exchanger serves as heat conservation device and lowers total heat requirements for the process. A part of the absorbed acid gases will be flashed from the heated rich solution on the top tray of the stripper, while the remainder of the rich solution flows downward through the stripper in countercurrent contact with vapor generated in the reboiler. The reboiler vapor (primarily steam) strips the acid gases from the rich solution after which the acid gases and the steam leave the top of the stripper and pass overhead through a condenser, where the major portion of the steam is condensed and cooled. The acid gases are separated in the separator and sent to the flare or to processing and the condensed steam is returned to the top of the stripper as reflux.

Lean amine solution from the bottom of the stripper column is pumped through an amine—amine heat exchanger and then through a cooler before being introduced to the top of the absorber column. The amine cooler serves to lower the lean amine temperature to the 100°F range. Higher temperatures of the lean amine solution will result in excessive amine losses through vaporization and also lower acid gas carrying capacity in the solution because of temperature effects.

DEA has a higher boiling temperature than monoethanolamine, requiring other methods of reclaiming such as vacuum distillation in order to prevent thermal degradation of the amine. Moreover, diethanolamine has a slow degradation rate. Consequently, in most cases it is not practical, economical, or necessary to reclaim DEA solutions. Solution purification is maintained by mechanical and carbon filtration, and by caustic or soda ash addition to the system to neutralize the heat-stable amine salts.

Moisture may be removed from hydrocarbon gases at the same time as hydrogen sulfide is removed. Moisture removal is necessary to prevent harm to anhydrous catalysts and to prevent the formation of hydrocarbon hydrates (e.g., $C_3H_8 \cdot 18H_2O$) at low temperatures. A widely used dehydration and desulfurization process is the glycol/amine process, in which the treatment solution is a mixture of ethanolamine and a large amount of glycol. The mixture is circulated through an absorber and a reactivator in the same way as ethanolamine is circulated in the Girbotol process. The glycol absorbs moisture from the hydrocarbon gas passing up the absorber; the ethanolamine absorbs hydrogen sulfide and carbon dioxide. The treated gas leaves the top of the absorber; the spent ethanolamine glycol mixture enters the reactivator tower, where heat drives off the absorbed acid gases and water.

The processes using ethanolamine and potassium phosphate are now widely used. The ethanolamine process, known as the *Girbotol process*, removes acid gases (hydrogen sulfide and carbon dioxide) from liquid hydrocarbon derivatives as well as from natural and from refinery gases. The Girbotol treatment solution is an aqueous solution of ethanolamine, which is an organic alkali that has the reversible property of reacting with hydrogen sulfide under cool conditions and releasing hydrogen sulfide at high temperatures. The ethanolamine solution fills a tower called an absorber through which the sour gas is bubbled. Purified gas leaves the top of the tower, and the ethanolamine solution leaves the bottom of the tower with the absorbed acid gases. The ethanolamine solution enters a reactivator tower where heat drives the acid gases from the solution. Ethanolamine solution, restored to its original condition, leaves the bottom of the reactivator tower to go to the top of the absorber tower, and acid gases are released from the top of the reactivator.

Finally, in an amine absorption process (olamine process), amine absorbers use countercurrent flow through a trayed or packed tower to provide intimate contact between the amine solvent and the sour gas so that the hydrogen sulfide and carbon dioxide molecules can transfer from the gas phase to the solvent liquid phase. In tray columns, a liquid level is maintained on each tray by a weir usually 2–3 in. high. The gas passes up from underneath the trays through openings in the trays such as perforations, bubble caps, or valves, and disperses into bubbles through the liquid, forming a froth. The gas disengages from the froth, travels through a vapor space, providing time for entrained amine solution to fall back down to the liquid on the tray, and passes through the next tray above. On the other hand, in packed column absorption units, the liquid solvent is dispersed in the gas stream, by forming a film over the packing, providing a large surface area for carbon dioxide and hydrogen sulfide transfer from the gas to the liquid solvent. The degree of sweetening achieved is largely dependent on the number of trays or the height of packing available in the absorber.

In most cases a mist eliminator pad is installed near the gas outlet of the absorber (the distance between the top tray and the mist pad is 3–4 ft) to trap entrained solvent, and an outlet knockout drum, similar to the inlet separator for the gas feed, is provided to collect solvent carryover. Some contactors have a water wash consisting of two to five trays at the top of the absorber to minimize vaporization losses of amine, which is often found in low-pressure monoethanolamine systems.

8.3.2 Flexsorb process

The Flexsorb SE process is based on a family of sterically hindered amines in aqueous solutions or other physical solvents. The sterically hindered amines are secondary amines that have a large hydrocarbon group attached to the nitrogen group. The large molecular structure hinders the carbon dioxide approach to the amine. The larger the structure the more difficult it becomes for the carbon dioxide to get close to the amine. They also appear to be unstable to the carbamate form of product and revert easily to the carbonate form found in the tertiary amines. Like tertiary amines, they are capable of a high degree of solvent loading—1 mol/mol, instead of 0.5 mol/mol typical of primary and secondary amines. One version of the solvent, Flexsorb SE Plus, is very selective toward hydrogen sulfide and has been used in several plants for tail gas processing or lean acid gas enrichment.

There are two mixed hindered amine/physical solvent versions of the Flexsorb process. The Hybrid Flexsorb SE process employs a solution of the Flexsorb SE amine, water, and an unspecified physical solvent. The Flexsorb PS solvent consists of a different hindered amine, water, and a physical solvent. Five of these plants are believed to be operating.

8.3.3 Adip process

The Adip process is regenerative amine process removes hydrogen sulfide and carbon dioxide from natural gas, refinery gas and synthesis gas. The process uses aqueous solutions of the secondary amine, diisopropanolamine or the tertiary amine, methyl diethanolamine. Amine concentrations up to 50% w/w can be applied. Hydrogen sulfide can be reduced to low sulfur levels. The process can also be applied to remove H_2S, CO_2, and carbonyl sulfide from liquefied petroleum gas or natural gas liquid to low levels.

The Adip-X process is a regenerative amine process, highly suitable for bulk and deep removal of carbon dioxide from gas streams. The process uses aqueous solutions of the tertiary amine, methyldiethanolamine and an additive.

8.3.4 Purisol process

The Purisol process is used for the removal of acid gases from natural gas, fuel gas, and syngas by physical absorption in N-methyl-pyrrolidone. High concentrations of carbon dioxide are reduced. In the process, a raw gas stream is cooled, and organic sulfur compounds are removed in prewash. Hydrogen sulfide is removed in main absorber by hot regenerated, lean solvent that has been to below ambient temperature. Any traces of the N-methyl-pyrrolidone are removed by backwashing with water.

8.4 Physical solvent processes

Two of the currently most-widely-used physical solvent processes for gas cleaning are *Selexol* and *Rectisol* processes (Kohl and Riesenfeld, 1985; Epps, 1994). The process solvent is a mixture of dimethyl ethers of polyethylene glycol $[CH_3(CH_2CH_2O)_nCH_3]$, where n is between 3 and 9. The solvent is chemically and thermally stable and has a low vapor pressure that limits its losses to the treated gas. The solvent has a high solubility for carbon dioxide, hydrogen sulfide, and carbonyl sulfide. It also has appreciable selectivity for hydrogen sulfide over carbon dioxide.

The principal benefits of physical solvents are: (1) high selectivity for hydrogen sulfide over carbonyl sulfide and carbon dioxide, (2) high loadings at high acid gas partial pressures, (3) solvent stability, and (4) low heat requirements because most of the solvent can be regenerated by a simple pressure letdown. The performance of a physical solvent can be easily predicted. The solubility of a compound in the solvent is directly proportional to its partial pressure in the gas phase, hence, the improvement in the performance of physical solvent processes with increasing gas pressure. Physical solvent processes can be configured to take advantage of their high hydrogen sulfide/carbon dioxide selectivity together with high levels of carbon dioxide recovery.

In the Selexol process, the solvent is composed of a dimethyl ether of polyethylene glycol, which is chemically inert and not subject to degradation. The process also removes carbonyl sulfide, mercaptan derivatives, ammonia, hydrogen cyanide, and metal carbonyl derivatives. A variety of flow schemes permit process optimization and energy reduction and the partial pressure of the acid gas is the key driving force. Typical feedstock conditions range between 300 and 2000 psia, with acid composition (carbon dioxide plus hydrogen sulfide) from 5% v/v to more than 60% v/v by volume. The product specifications achievable depend on the application and can range from ppm up to percent levels of acid gas.

The Selexol process can be configured in various ways, depending on the requirements for the level of hydrogen sulfide/carbon dioxide selectivity, the depth of sulfur removal, the need for bulk carbon dioxide removal, and whether the gas

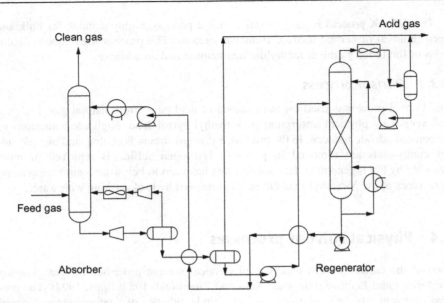

Figure 8.3 The Selexol process.

needs to be dehydrated (Fig. 8.3). The gas stream from the low-pressure flash is combined with the acid gas from the regenerator. This combined gas stream is then sent to a SRU. However, the hydrogen sulfide content could be too low for use in a conventional Claus process.

The Selexol process can, however, be configured to give both a rich acid gas feed to the Claus unit as well as to provide for bulk carbon dioxide removal.

8.4.1 Rectisol process

The Rectisol process is the most widely used physical solvent gas treating process for acid gas removal using an organic solvent at low temperatures. In general, methanol is used for removal of hydrogen sulfide, carbonyl sulfide, and carbon dioxide as well as removal of organic and inorganic impurities. It is possible to produce a clean gas with less than 0.1 ppm sulfur and a carbon dioxide content down to the ppm range. The main advantage over other processes is the use of a cheap, stable, and easily available solvent, a very flexible process and low utilities.

In the process, chilled methanol (methyl alcohol, CH_3OH) is used at a temperature of approximately -40 to $-62°C$ (-40 to $-80°F$). The selectivity (by methanol) for hydrogen sulfide over carbon dioxide at these temperatures is about 6/1, a little lower than that of the Selexol process at its usual operating temperature. However, the solubility of hydrogen sulfide and carbonyl sulfide in methanol, at typical process operating temperatures, is higher than in Selexol and allows for very deep sulfur removal. The high selectivity for hydrogen sulfide over carbon

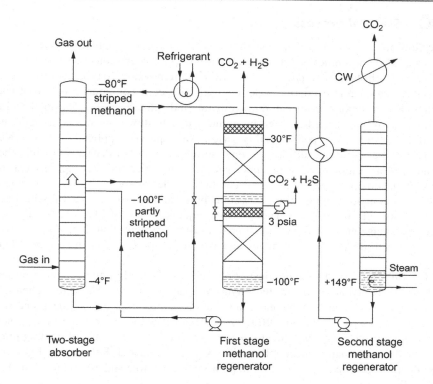

Figure 8.4 The Rectisol process.

dioxide, combined with the ability to remove carbonyl sulfide, is the primary advantage of the process. The need to refrigerate the solvent is the main disadvantage of the process, resulting in high capital and operating costs.

There are many possible process configurations for the Rectisol process, depending on process requirements (see, e.g., Fig. 8.4). Different process layouts are used for selective hydrogen sulfide removal and carbon dioxide, and for nonselective carbon dioxide and hydrogen sulfide.

The process is designed for the bulk removal of carbon dioxide and nearly all of the removal of hydrogen sulfide and carbonyl sulfide take place in the bottom section of the absorber. The methanol solvent contacting the feed gas in the first stage of the absorber is stripped in two stages of flashing via pressure reduction. The regenerated solvent is virtually free of sulfur compounds but contains some carbon dioxide. The acid gas leaving the first stage solvent regenerator is suitable for a Claus process. The second stage of absorption is designed for the removal of the remaining sulfur compounds and carbon dioxide. The solvent from the bottom of the second stage of the absorber is stripped deeply in a steam-heated regenerator and is returned to the top of the absorption column after cooling and refrigeration.

8.4.2 Sulfinol process

The Sulfinol process, developed in the early 1960s, is a combination process that uses a mixture of amines and a physical solvent. The solvent consists of an aqueous amine and sulfolane. The process is used for the removal of hydrogen sulfide, carbonyl sulfide, mercaptan derivatives, other organic sulfur compounds and all or part of the carbon dioxide from natural, synthetic, and refinery gases. The total sulfur compounds in the treated gas can be reduced to ultra-low ppm levels, as required for refinery fuel and pipeline quality gases. An improved application is to selectively remove hydrogen sulfide, carbonyl sulfide, mercaptan derivatives, and other organic sulfur compounds for pipeline specification, while coabsorbing only part of the carbon dioxide. Deep removal of carbon dioxide removal for liquefied natural gas plants is another application, as well as bulk carbon dioxide removal with flash regeneration of the solvent. The process sequence-Sulfinol/Claus/SCOT can be used advantageously with an integrated Sulfinol system that handles selective hydrogen sulfide removal upstream and the SCOT process that treats the Claus offgas.

The Sulfinol mixed solvent consists of a chemical-reacting alkanolamine, water, and physical solvent sulfolane (tetra-hydrothiophene dioxide). The actual chemical formulation is customized for each application. Unlike aqueous amine processes, the process removes carbonyl sulfide, mercaptan derivatives, and other organic sulfur compounds to stringent total sulfur specifications. A wide range of treating pressures and contaminant concentrations can be accommodated. Refinery fuel gas and gas pipeline specifications, such as 40 ppm v/v total sulfur and 100 ppm v/v hydrogen sulfide are readily attained.

The process is in many respects identical to the familiar amine method and its equipment components similar to those found in amine units. The main difference is that while the conventional amine process employs a fairly diluted concentration of amine in water, removing the acid gas by chemical reaction, the Sulfinol system uses a mixture of highly concentrated amine and a physical solvent removing the acid gases by physical and chemical reactions. The concentrations of the amine and the physical solvent vary with the type of feed gas in each application. Common Sulfinol mixtures are in the range of 40% amine (also called DIPA), 40% sulfolane (an organic solvent), and 20% water.

Sulfinol has a good affinity for most of the acid gases and has the ability to release these gases in the regenerator upon pressure reduction and heat application. When operating under suitable conditions, it is capable of removing twice as much acid gas as a 20% MEA solution.

The Sulfinol-D process uses DIPA, while Sulfinol-M uses MDEA. The mixed solvents allow for better solvent loadings at high acid gas partial pressures and higher solubility of carbonyl sulfide and organic sulfur compounds than straight aqueous amines.

The Sulfinol-D process is primarily used in cases where selective removal of hydrogen sulfide is not of primary concern, but where partial removal of organic sulfur compounds (mercaptans, RSH, and carbon disulfide) is desired, typically in natural gas and refinery applications. The Sulfinol-D configuration is also able to remove some carbonyl sulfide via physical solubility in sulfolane and partial hydrolysis to hydrogen sulfide induced by the DIPA. However, complete removal of carbonyl sulfide by Sulfinol-D cannot be guaranteed. Unlike solvents that use other primary and secondary amines (MEA and DEA) Sulfinol-D is claimed not to be degraded by these sulfur compounds.

Sulfinol-M is used when a higher degree of hydrogen sulfide selectivity is needed. Hydrogen sulfide selectivity in Sulfinol-M is controlled by the relative reaction rate of the reaction of hydrogen sulfide with methyldiethanolamine as well as by the physical solubility of hydrogen sulfide and carbon dioxide in the solvent.

8.5 Metal oxide processes

The use of solids for sweetening gas (typically in batch-type process) is based on adsorption of the acid gases on the surface of the solid sweetening agent, or reaction with some component on that surface. The solids processes are usually best applied to gases containing low-to-medium concentrations of hydrogen sulfide or mercaptan derivatives. The solids processes tend to be highly selective and do not normally remove significant quantities of carbon dioxide. Consequently, the regenerated hydrogen sulfide stream from the process is usually high purity and, in addition, pressure has relatively little effect on the adsorptive capacity of a sweetening agent.

8.5.1 Iron sponge process

An example of a hydrogen sulfide scavenger process is the iron sponge process (also called the dry box process), which is the oldest and still the most widely used batch process for sweetening of natural gas and natural gas liquids (Anerousis and Whitman, 1984). The process was implemented during the 19th century and has been in use in Europe and the United States for over 100 years. Hydrogen sulfide scavengers are appropriate for use at the low concentrations of hydrogen sulfide where conventional chemical absorption and physical solvents are not economical. During recent years, hydrogen sulfide scavenger technology has been expanded with many new materials coming on the market and others being discontinued. Overall, the simplicity of the process, low capital costs, and relatively low chemical (iron oxide) cost continue to make the process an ideal solution for hydrogen sulfide removal. In addition, pressure has relatively little effect on the adsorptive capacity of a sweetening agent. The use of iron sponge process for sweetening sour gas is based on adsorption of the acid gases on the surface of the solid sweetening agent.

In the process (Fig. 8.5), the sour gas should pass down through the bed. In the case where continuous regeneration is to be utilized, a small concentration of air is added to the sour gas before it is processed. This air serves to continuously

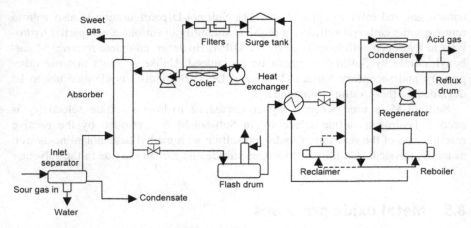

Figure 8.5 Typical iron sponge process flow sheet.

regenerate the iron oxide, which has reacted with hydrogen sulfide, which serves to extend the on-stream life of a given tower but probably serves to decrease the total amount of sulfur that a given weight of bed will remove. The number of vessels containing iron oxide can vary from one to four. In a two-vessel process, one of the vessels would be on stream removing hydrogen sulfide from the sour gas, while the second vessel would either be in the regeneration cycle or having the iron sponge bed replaced.

When periodic regeneration is used a tower is operated until the bed is saturated with sulfur and hydrogen sulfide begins to appear in the sweetened gas stream. At this point the vessel is removed from service and air is circulated through the bed to regenerate the iron oxide. Regardless of the type of regeneration process used, a given iron oxide bed will lose activity gradually and eventually will be replaced. For this reason, the reactor vessels should be designed to minimize difficulties in replacing the iron sponge in the beds. The change out of the beds is hazardous. Exposure to air when dumping a bed can cause a sharp rise in temperature, which can result in spontaneous combustion of the bed. Care must be exercised in opening the tower to the air. The entire bed should be wetted before beginning the change out operation. In some process options, the iron sponge may be operated with continuous regeneration by injecting a small amount of air into the sour gas stream. The air regenerates ferric sulfide, while hydrogen sulfide is removed by ferric oxide. This process is not as effective at regenerating the bed as the batch process and requires a higher pressure air stream (Arnold and Stewart, 1999).

The *sponge* consists of wood shavings impregnated with a hydrated form of iron oxide. The wood shavings serve as a carrier for the active iron oxide powder. Hydrogen sulfide is removed by reacting with iron oxide to form ferric sulfide. The process is usually best applied to gases containing low to medium concentrations (300 ppm) of hydrogen sulfide or mercaptans. This process tends to be highly selective and does not normally remove significant quantities of carbon dioxide. As a result, the hydrogen sulfide stream from the process is usually of high purity. The use of iron sponge process for sweetening sour gas is based on adsorption of the acid gases on the surface of the solid sweetening agent followed by chemical reaction of ferric oxide (Fe_2O_3) with hydrogen sulfide:

$$2Fe_2O_3 + 6H_2S \rightarrow 2Fe_2S_3 + 6H_2O$$

The reaction requires the presence of slightly alkaline water and a temperature below 43°C (110°F) and bed alkalinity should be checked regularly, usually on a daily basis. A pH level on the order of 8–10 should be maintained through the injection of caustic soda with the water. If the gas does not contain sufficient water vapor, water may need to be injected into the inlet gas stream.

The ferric sulfide produced by the reaction of hydrogen sulfide with ferric oxide can be oxidized with air to produce sulfur and regenerate the ferric oxide:

$$2Fe_2S_3 + 3O_2 \rightarrow 2Fe_2O_3 + 6S$$

$$S + O_2 \rightarrow SO_2$$

The regeneration step, i.e., the reaction with oxygen is exothermic and air must be introduced slowly so the heat of reaction can be dissipated. If not and the air is introduced rapidly, there is the potential for the bed to ignite the bed. Some of the elemental sulfur produced in the regeneration step remains in the bed which, after several cycles, the sulfur will form a cake over the ferric oxide thereby decreasing the reactivity of the bed. Typically, after approximately 10 cycles, depending upon the sulfur content of the gas stream, the bed must be removed, and a new bed introduced into the reactor.

Iron oxide reactors have taken on a number of configurations from conventional box vessels to static tower purifiers. The selection depends upon the process application. Static tower purifiers are used in high-pressure applications, since a longer bed depth gives a greater efficiency, and the total pressure drop is a smaller fraction of the available pressure. Static tower purifiers are also fitted with trays to eliminate compaction for lower pressure applications. Conventional boxes consist of large rectangular vessels, either built into the ground, or supported on legs to save footprint. They are composed of several layers, with a typical bed depth of at least 2 ft.

Generally, the iron oxide process is suitable only for small to moderate quantities of hydrogen sulfide. Approximately 90% of the hydrogen sulfide can be removed per bed, but bed clogging by elemental sulfur occurs and the bed must be discarded and the use of several beds in series is not usually economical. Removal of larger amounts of hydrogen sulfide from gas streams requires a continuous process, such as the Ferrox process or the Stretford process. The *Ferrox process* is based on the same chemistry as the iron oxide process except that it is fluid and continuous. The Stretford process employs a solution containing vanadium salts and anthraquinone disulfonic acid (Maddox, 1974).

The natural gas should be wet when passing through an iron sponge bed as drying of the bed will cause the iron sponge to lose its capacity for reactivity. If the gas is not already water-saturated or if the influent stream has a temperature greater than 50°C (approximately 120°F), water with soda ash is sprayed into the top of the contactor to maintain the desired moisture and alkaline conditions during operation.

The advantages of the iron sponge process include: (1) providing complete removal of small to medium concentrations of hydrogen sulfide without removing carbon dioxide, (2) requiring relatively small investment, for small to moderate gas volumes, compared with other processes, (3) being equally effective at any operating pressure, and (4) being used to remove mercaptan derivatives or convert them to disulfides (RSSR). On the other hand, the disadvantages of the iron sponge process are: (1) it is a batch process, requiring duplicate installation or flow interruption of processed gas, (2) it is prone to hydrate formation when operated at higher pressures and at temperatures in the hydrate-forming range, (3) the process will effectually remove ethyl mercaptan that has been added for odorization, and (4) coating of the iron sponge with entrained oil or distillate will require more frequent change out of the sponge bed.

The *Slurrisweet process* uses an iron oxide slurry is similar to those for dry iron oxide processes, except with a higher proportion of the magnetite form of iron oxide (Fe_3O_4). Any foaming and settling problems of the iron oxide particles were solved using a silicon-based defoamer with additives and a dispersant, respectively. Also, corrosion was inhibited by using an epoxy coating on the vessel. Injection of air at 5% mole concentration of hydrogen sulfide extended the batch life and stabilized the spent chemical (Schaack and Chan, 1989a).

Iron oxide suspensions, like iron oxide slurries, rely upon hydrated ferric oxide as the active regenerable agent. However, iron oxide suspensions react in a basic environment with an alkaline compound, followed by the reaction of the hydrosulfide with iron oxide to form iron sulfide (Kohl and Nielsen, 1997). The iron is then regenerated by aeration. Thus:

$$H_2S + Na_2CO_3 \rightarrow NaHS + NaHCO_3$$

$$Fe_2O_3 + 3NaHS + 3NaHCO_3 \rightarrow Fe_2S_3 + 3Na_2CO_3 + 3H_2O$$

Iron oxide suspensions were the precursors to the chelated iron processes.

8.5.2 Adsorption processes

Adsorption (or solid bed) dehydration is the process where a solid desiccant is used for the removal of water vapor from a gas stream. The solid desiccants commonly used for gas dehydration are those that can be regenerated and, consequently, used over several adsorption—desorption cycles. In fact, there are several solid desiccants which possess the physical characteristic to adsorb water from natural gas but the most popular are (1) alumina, (2) silica gel, and (3) silica—alumina gel.

Alumina is a hydrated form of aluminum oxide (Al_2O_3) and is the least expensive adsorbent. It is activated by driving off some of the water associated with it in its hydrated form ($Al_2O_3 \cdot 3H_2O$) by heating. It produces an excellent dew point depression values as low as $-100°F$, but requires much more heat for regeneration. Also, it is alkaline and cannot be used in the presence of acid gases, or acidic chemicals used for well treating. The tendency to adsorb higher molecular weight hydrocarbon derivatives is high, and it is difficult to remove these during regeneration. It has good resistance to liquids, but little resistance to disintegration due to mechanical agitation by the flowing gas.

Silica gel and silica—alumina gel are granular, amorphous solids manufactured by chemical reaction. Gels manufactured from sulfuric acid and sodium silicate reaction are called silica gels and consist almost solely of silicon dioxide (SiO_2). Alumina gels consist primarily of some hydrated form of aluminum trioxide (Al_2O_3). Silica—alumina gels are a combination of silica and alumina gel. Gels can dehydrate gas to as low as 10 ppm and have the greatest ease of regeneration of all desiccants. They adsorb high-molecular weight hydrocarbon derivatives but release them relatively more easily during regeneration. Since these gels are acidic, they can handle sour gases but not alkaline materials such as caustic or ammonia. Although there is no reaction with hydrogen sulfide, sulfur can deposit and block their surface and, therefore, the gels are useful if the content of hydrogen sulfide in the gas stream is less than 5—6% v/v.

Solid desiccants or absorbents are commonly used for dehydrating gases in cryogenic processes. The use of solid adsorbent has been extended to the dehydration of liquid. Solid adsorbents remove water from the hydrocarbon stream and release it to another stream at higher temperatures in a regeneration step. In a dry desiccant bed, the adsorbate components are adsorbed at different rates. A short while after the process has begun, a series of adsorption zones appear. The distance between successive adsorption zone fronts is indicative of the length of the bed involved in the adsorption of a given component. Behind the zone, the component entering the vessel has been removed from the gas; ahead of the zone, the concentration of that component is zero. Note the adsorption sequence: methane and ethane are adsorbed almost instantaneously, followed by the higher molecular weight hydrocarbon derivatives, and finally by water that constitutes the last zone. Almost all the hydrocarbon derivatives are removed after 30—40 minutes after which dehydration begins. Water displaces the hydrocarbon derivatives on the adsorbent surface if enough time is allowed. At the start of dehydration cycle, the bed is saturated with methane as the gas flows through the bed. Then ethane replaces methane, and propane is adsorbed next. Finally, water will replace all the hydrocarbon derivatives. For good dehydration, the bed should be switched to regeneration just before the water content of outlet gas reaches an unacceptable level. The regeneration of the bed consists of circulating hot dehydrated gas to strip the adsorbed water, then circulating cold gas to cool the bed down.

The solid desiccants generally are used in dehydration systems consisting of two or more towers and associated regeneration equipment, such as the use of a two-tower system pressure-swing adsorption system (Grande et al., 2017). One tower is

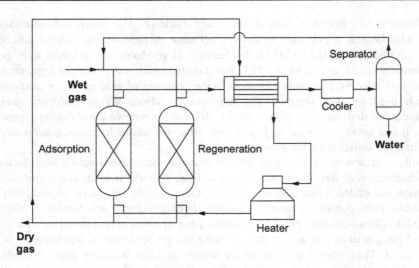

Figure 8.6 The two-tower (pressure-swing) adsorption system for gas dehydration.

on-stream adsorbing water from the gas while the other tower is being regenerated
and cooled (Fig. 8.6). Hot gas is used to drive off the adsorbed water from the desiccant, after which the tower is cooled with an unheated gas stream. The towers are
switched before the on-stream tower becomes water saturated. In this configuration,
part of the dried gas is used for regeneration and cooling and is recycled to the inlet
separator.

Solid desiccant units generally cost more to buy and operate than glycol units
and the use of these units is typically limited to applications such as gases with (1)
a high content of hydrogen sulfide, (2) very low-water dew point requirements, (3)
simultaneous control of water and hydrocarbon dew points, and (4) special cases
such as oxygen-containing gases. In processes where cryogenic temperatures are
encountered, solid desiccant dehydration usually is preferred over conventional
methanol injection to prevent hydrate and ice formation. Solid desiccants are also
often used for the drying and sweetening of natural gas liquids.

8.5.3 Other processes

The so-called dry sorption processes are used to scavenge hydrogen sulfide and
organic sulfur compounds (mercaptans) from gas streams through reactions with a
solid-based media. They are typically nonregenerable processes although some are
partially regenerable, losing activity upon each regeneration cycle. Most dry sorption processes are governed by the reaction of a metal oxide with hydrogen sulfide
to form a metal sulfide compound. For regenerable reactions, the metal sulfide compound can then react with oxygen to produce elemental sulfur and a regenerated
metal oxide. The primary metals used for dry sorption processes are iron and zinc.

Dry sorption processes can be categorized into two subgroups: (1) oxidation to
sulfur and (2) oxidation to oxides of sulfur. Since these processes rely on oxidation,

gas constituents that cannot be oxidized under the process conditions will not be removed (Kohl and Riesenfeld, 1985). This is advantageous when dealing with biogas, since only hydrogen sulfide, mercaptan derivatives and, in some cases, carbon dioxide will be removed, with minor losses of methane due to adsorption. The main product of sulfur oxidation to oxides of sulfur is sulfur dioxide. Because this is a controlled exhaust gas, poses a threat to equipment, due to corrosion and poisoning of fuel cell membranes and requires additional gas processing to achieve air discharge standards, it will not be addressed as a gas processing option.

Zinc oxide has also been used to clean gas tams. For example, at increased temperatures (205−370°C, 400−700°F), zinc oxide has a rapid reaction rate, therefore providing a short mass transfer zone, resulting in a short length of unused bed and improved efficiency. At operating temperatures, the zinc oxide sorbent has a maximum sulfur loading of 0.3−0.4 kilogram of sulfur per kilogram of sorbent. An enhanced form of the zinc oxide process (the Puraspec process) can operate more efficiently at reduced temperatures (38−205°C, 100−400°F), which is due to an increased porosity and decreased density resulting in higher obtainable sulfur loading per pound of media. In the process, fixed beds of chemical absorbents provide effectively total irreversible selective removal of impurities from wet or dry hydrocarbons without feedstock losses. Radial flow reactor designs are available if low system pressure drop is required. The process displays greater efficiency at the higher operating temperatures (205−370°C, 400−700°F) (Carnell, 1986; Spicer and Woodward, 1991; Carnell et al., 1995; Rhodes et al., 1999).

In a slurry process, the zinc oxide slurry reactor consists of a simple vertical bubble contactor with the gas providing sufficient agitation to keep zinc oxide particles in suspension in addition to a dispersant (Kohl and Nielsen, 1997):

$$ZnO + H_2S \rightarrow ZnS + H_2O$$

$$ZnO + H_2S \rightarrow Zn(OH)(HS)$$

Zinc mercaptide ($Zn(OH)(HS)$) is a minor product and the majority of the hydrogen sulfide is converted to zinc sulfide. Zinc mercaptide will form a sludge in the reactor and can contribute to foaming (GPSA, 1998).

Slurry processes were developed as alternatives to iron sponge. Slurries of iron oxide have been used to selectively absorb hydrogen sulfide (Fox, 1981; Kattner et al., 1988; Samuels, 1988). The chemical cost for these processes is higher than that for iron sponge process but this is partially offset by the ease and lower cost with which the contact tower can be cleaned out and recharged. Obtaining approval to dispose of the spent chemicals, even if they are nonhazardous, is time consuming. An example of such a process is the Sulfa-Check process that selectively removes hydrogen sulfide and mercaptan derivatives from natural gas, in the presence of carbon dioxide (Dobbs, 1986). The process uses sodium nitrite ($NaNO_2$).

The *Sulfa-Check process* is used to selectively remove hydrogen sulfide and mercaptan derivatives from natural gas, in the presence of carbon dioxide (Schaack and Chan, 1989b). The original process is accomplished in a one-step single vessel design using an aqueous solution of sodium nitrite buffered to stabilize the pH above 8. Also, there is enough strong base to raise the pH of the fresh material to 12.5. Removal of hydrogen sulfide is not affected under short contact times since the reaction is almost instantaneous. Sodium hydroxide and sodium nitrite are consumables in the processes and cannot be regenerated. The reaction with hydrogen sulfide forms elemental sulfur, ammonia, and caustic soda:

$$NaNO_2 + 3H_2S \leftrightarrow NaOH + NH_3 + 3S + H_2O$$

Other reactions forming the oxides of nitrogen do occur (Burnes and Bhatia, 1985) and carbon dioxide in the gas reacts with the sodium hydroxide to form sodium carbonate and sodium bicarbonate. The spent solution is slurry of fine sulfur particles in a solution of sodium and ammonium salts (Manning and Thompson, 1991).

The *Chemsweet process* is a batch process for the removal of hydrogen sulfide from natural gas (Manning, 1979). The chemicals of choice for the process are a mixture of zinc oxide (ZnO), zinc acetate [(CH3COO)$_2$Zn, ZnAc$_2$], water, and a dispersant to keep the zinc oxide particles in suspension. When one part is mixed with five parts of water the acetate dissolves and provides a controlled source of zinc ions that react instantaneously with the bisulfide and sulfide ions that are formed when hydrogen sulfide dissolves in water. In addition to a chemical reaction to remove the hydrogen sulfide, the zinc oxide also replenishes the zinc acetate. Thus:

Sweetening:

$$ZnAc_2 + H_2S \leftrightarrow ZnS + 2HAc$$

Regeneration:

$$ZnO + 2HAc \leftrightarrow ZnAc_2 + H_2O$$

The overall reaction is:

$$ZnO + H_2S \leftrightarrow ZnS + H_2O$$

The presence of carbon dioxide in the natural gas is of little consequence to the process because the pH of the Chemsweet slurry is low enough to prevent significant absorption of the carbon dioxide, even when the ratio of carbon dioxide to hydrogen sulfide is high (Manning and Thompson, 1991; Kohl and Nielsen, 1997).

The addition of zinc acetate establishes a more controlled reaction. Zinc acetate dissolves to equilibrium to produce zinc ions, which then react with hydrogen sulfide yielding zinc sulfide and acetic acid. The addition of acetic acid causes more zinc to dissociate, thereby controlling the concentration of zinc ions in solution (Kohl and Nielsen, 1997). While a zinc slurry plant can be easy to operate and install, the high cost of the reactant and difficulty of disposing zinc sulfide wastes make it less attractive.

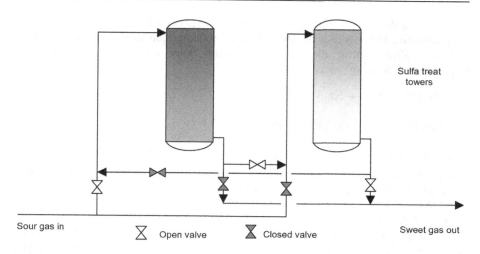

Figure 8.7 The SulfaTreat process.

The *SulfaTreat* process (Fig. 8.7) is also a batch-type process for the selective removal of hydrogen sulfide and mercaptans from natural gas. The process is dry, using no free liquids, and can be used for natural gas applications where a batch process is suitable. The SulfaTreat system is a more recent development using iron oxide on a porous solid material. Unlike the iron sponge process, the SulfaTreat material is nonpyrophoric and has a higher capacity than iron sponge on a volumetric or mass basis.

In the process, the inlet gas stream enters the unit through a small inlet separator to separate out any entrained liquid, salt water, and any liquid hydrocarbon in the gas stream. The sour feed gas then passes to the top of the contactor where the SulfaTreat product (a dry, granular, free-flowing material of uniform porosity and permeability) only reacts with sulfur-containing compounds. This eliminates any side reactions with carbon dioxide, which could reduce its efficiency. As the SulfaTreat bed dries out, the rate of reaction decreases. Optimum efficiency may require measuring water content of the gas and injecting a sufficient amount of water into the influent gas to maintain a water-saturated gas and evidence of free water in the outlet gas.

The chemistry of the SulfaTreat is similar to the chemistry of the iron sponge process:

$$Fe_3O_4 + 4H_2S \rightarrow 3FeS + 3H_2O + S$$

$$Fe_3O_4 + 6H_2S \rightarrow 3FeS_2 + 4H_2O + 2H_2$$

$$Fe_2O_3 + 3H_2S \rightarrow Fe_2S_3 + 3H_2O$$

Another *fixed bed dry sorption process* makes use of a hydroxide media similar to that of caustic scrubbing processes (Sofnolime process) (Kohl and Nielsen,

1997). It is a dry process that claims it has a synergistic mixture of hydroxides in a granular solid. The media is capable of removing hydrogen sulfide, carbon dioxide, carbonyl sulfide, sulfur dioxide, and organic sulfur compounds. The process employs the following reactions:

$$2NaOH + H_2S \rightarrow Na_2S + H_2O$$

$$Ca(OH)_2 + CO_2 \rightarrow CaCO_3H_2O$$

The process removes both hydrogen sulfide and carbon dioxide but gas streams with a high carbon dioxide content, such as biogas, will exhaust the media rapidly. The packed bed is supported between layers of ceramic balls on the top and bottom, with the flow direction in the upward direction.

While not strictly a metal oxide process, *iron chelate* solutions have been developed with greater acceptability due to the nonhazardous classification of the working solution (Kohl and Nielsen, 1997). Two processes are worthy of note here: (1) LO-CAT and (2) SulFerrox. The LO-CAT and SulFerrox processes differ mainly on vessel configurations, iron concentrations, proprietary chelates and additives to optimize the process (Dalrymple and Trofe, 1989). The main advantages of chelated iron solutions are the catalytic nature of the reactant, reduced plant footprint, and ability to reclaim elemental sulfur (Muely and Ruff, 1972).

The technology involves the use of an iron chelate-type catalyst that converts hydrogen sulfide to elemental sulfur by the reduction of ferric ions to ferrous ions. The ferric ions are regenerated by contact with air. Thus:

$$2H_2S + O_2 \rightarrow 2H_2O + 2S$$

$$4Fe^{3+} + 2H_2S \rightarrow 4Fe^{2+} + 2S + 4H^+$$

$$4Fe^{2+} + O_2 + 2H_2O \rightarrow 4Fe^{3+} + 4OH^-$$

When the gas stream contains sufficient oxygen (50% of the hydrogen sulfide concentration), the oxidation and regeneration reactions can take place in the same vessel. Typical systems include three vessels: (1) a precontactor, (2) an absorber column, and (3) an oxidizer. The organic chelating agent stabilizes the ferric ions over a wide range of pH; however, the rate of the oxidation stage is dependent on the pH. Therefore the chelated iron solution is maintained within a pH of 6−10.

In addition, the *BioDeNOx process* is a biological process removes nitrogen oxides from flue gases. An iron chelate selectively absorbs the nitrogen oxides which are reduced to nitrogen with ethanol in the presence of microorganisms. This uses a wet gas scrubber to contact the circulation liquid with flue gas feed and absorb the nitrogen oxides (NO_x). In the sump underneath the scrubber, the absorbed nitrogen oxides are biologically reduced to nitrogen and ethanol is also consumed and, thus, the iron chelate solution is regenerated. The presence of oxygen and acid compounds in the flue gas, such as hydrogen chloride and hydrogen fluoride, oxidizes a part of iron chelate to the ferric (Fe^{3+}) state. Therefore a purge and makeup of iron chelate is necessary to eliminate this oxidized ferric material. To minimize iron chelate consumption, a nanofiltration can be installed and bleed from the unit is passed through the filter and the chelate is recovered.

Another biological-type process is the THIOPAQ process which involves the biological desulfurization of high-pressure natural gas, synthesis gas, fuel gas streams, acid gas from amine regeneration, and treatment of spent caustic. The process is the oxidation of hydrogen sulfide to elemental sulfur using sulfur bacteria (Thiobacilli), which are naturally occurring bacteria that are not genetically modified. In the process, the gas stream is sent to a caustic scrubber in which the hydrogen sulfide reacts to produce sodium sulfide, which is converted to elemental sulfur and caustic by the bacteria when air is supplied in the bioreactor. The sulfur particles are covered with a (bio) macropolymer layer, which maintains the sulfur in a milk like suspension that does not cause fouling or plugging. The sulfur slurry produced can be concentrated to a cake containing 60% w/w dry matter. This cake can be used directly for agricultural purposes, or as feedstock for sulfuric acid manufacturing or, alternately, the biological sulfur slurry can be purified further by melting to high-quality sulfur to meet Claus sulfur specifications (http://www.paqell.com/thiopaq/about-thiopaq-o-and-g/).

8.6 Methanol-based processes

Methanol is probably one of the most versatile solvents in the natural gas processing industry. Historically, methanol was the first commercial organic physical solvent and has been used for hydrate inhibition, dehydration, gas sweetening, and liquids recovery (Kohl and Nielsen, 1997). Most of these applications involve low temperature where methanol's physical properties are advantageous compared with other solvents that exhibit high viscosity problems or even solids formation. Operation at low temperatures tends to suppress methanol's most significant disadvantage, high solvent loss. Furthermore, methanol is relatively inexpensive and easy to produce making the solvent a very attractive alternate for gas processing applications.

Methanol has favorable physical properties relative to other solvents except for vapor pressure. The benefits of the low viscosity of methanol at low temperature are manifested in the pressure drop improvement in the cold box of injection facilities and improved heat transfer. Methanol has a much lower surface tension relative to the other solvents. High surface tension tends to promote foaming problems in contactors. Methanol processes are probably not susceptible to foaming. However, the primary drawback of methanol is the high vapor pressure that is several times greater than that of the glycols or amines. To minimize methanol losses and enhance water and acid gas absorption, the absorber or separator temperatures are usually less than $-20°F$.

The high vapor pressure of methanol may initially appear to be a significant drawback because of high solvent losses. However, the high vapor pressure also has significant advantages. Although often not considered, lack of thorough mixing of the gas and solvent can pose significant problems. Because of the high vapor pressure, methanol is completely mixed in the gas stream before the cold box. Glycols, because they do not completely vaporize, may require special nozzles and nozzle placement in the cold box to prevent freeze-up. Solvent carry-over to other downstream processes may also represent a significant problem. Since methanol is more volatile than glycols, amines, and other physical solvents including lean oil, methanol is usually rejected in the regeneration step of these downstream processes. The stripper concentrates the methanol in the overhead condenser where it can be removed and further purified. Unfortunately, if glycols are carried over to amine units, the glycol becomes concentrated in the solution and potentially starts to degrade and possibly dilute the amine solution.

The use of methanol has been further exploited in the development of the Rectisol process either alone or as toluene—methanol mixtures are used to more selectively remove hydrogen sulfide and slip carbon dioxide to the overhead product (Ranke and Mohr, 1985). Toluene has an additional advantage insofar as carbonyl sulfide is more soluble in toluene than in methanol. The Rectisol process was primarily developed to remove both carbon dioxide and hydrogen sulfide (along with other sulfur-containing species) from gas streams resulting from the partial oxidation of coal, oil, and petroleum residua. The ability of methanol to absorb these unwanted components made it the natural solvent of choice. Unfortunately, at cold temperatures, methanol also has a high affinity for hydrocarbon constituents of the gas streams. For example, propane is more soluble in methanol than in carbon dioxide. There are two versions of the Rectisol process— the two-stage and the once-through. The first step of the two-stage process is desulfurization before shift conversion; the concentrations of hydrogen sulfide and carbon dioxide are about 1% and 5% by volume, respectively. Regeneration of the methanol following the desulfurization of the feed gas produces high sulfur feed for sulfur recovery. The once-through process is only applicable for high pressure partial oxidation products. The once-through process is also applicable when the hydrogen sulfide to carbon dioxide content is unfavorable, in the neighborhood of 1:50 (Esteban et al., 2000).

The physical/chemical combined purification process in which the alkanolamine (mono- or diethanol amine) mixed with methanol is more successful than a single physical solvent used in many of the gas treatment plants. The main advantage of this solvent lies in the good physical absorption of physical solvent component in combination with the chemical reaction of the amine. The combination of a chemically active amine with a low boiling point polar physical solvent such as methanol offers major advantages in the absorption of carbon dioxide and sulfur components as: (1) low sulfur content in the product gas, (2) low content of carbon dioxide in the purified gas, (3) removal by absorption of trace components, such as hydrogen cyanide, carbonyl sulfide, mercaptan derivatives, and higher molecular weight hydrocarbon derivatives, (4) relative low regeneration temperature, since methanol has a boiling point lower than the boiling point of water, and (5) the solvent is noncorrosive so that carbon steel equipment can be used.

Recently, a process using methanol has been developed in which the simultaneous capability to dehydrate, to remove acid gas, and to control hydrocarbon dew point (Rojey and Larue, 1988; Rojey et al., 1990). The IFPEXOL-1 is used for water removal and hydrocarbon dew point control; the IFPEXOL-2 process is used for acid gas removal. The novel concept behind the IFPEXOL-1 process is to use a portion of the water-saturated inlet feed to recover the methanol from the aqueous portion of the low temperature separator. That approach has solved a major problem with methanol injection in large facilities, the methanol recovery via distillation. Beyond that very simple discovery, the cold section of the process is remarkably similar to a basic methanol injection process. Modifications to the process include water washing the hydrocarbon liquid from the low temperature separator to enhance the methanol recovery. The IFPEXOL-2 process for acid gas removal is very similar to an amine type process except for the operating temperatures. The absorber operates below $-20°F$ to minimize methanol losses, and the regenerator operates at about 90 psi. Cooling is required on the regenerator condenser to recover the methanol. This process usually follows the IFPEXOL-1 process so excessive hydrocarbon absorption is not as great a problem (Minkkinen and Jonchere, 1997).

8.7 Alkali washing processes

Alkali washing processes typically fall under the general banner of *scrubber-based processes* (e.g., chemical scrubbers, gas scrubbers) which are a diverse group of gas cleaning processes that can be used to remove some particulate matter and/or gases from industrial exhaust streams (Fig. 8.8). Traditionally, the term *scrubber* is used to refer to devices that use a liquid to wash remove contaminants from gas streams. More recently, the term has also been used to describe systems that inject a dry reagent or a slurry into a contaminated gas stream to wash out acid gases.

The process efficiency for the removal of contaminants is improved by increasing residence time of the gas stream in the scrubber or by increasing the surface area of the scrubber solution by the use of a spray nozzle or a packed tower. A wet scrubbing system has the potential to increase the proportion of water in the treated gas stream.

Wet scrubbers can also be used for heat recovery from hot gases by flue-gas condensation. In this option, water from the scrubber drain is circulated through a cooler to the nozzles at the top of the scrubber. The hot gas enters the scrubber at the bottom and, if the gas temperature is above the dew point of the water, it is initially cooled by evaporation of the water drops. Further cooling causes water vapor to condense thereby adding to the amount of circulating water.

8.7.1 Alkali washing

Caustic scrubbing for hydrogen sulfide removal with caustic scrubbing is only economical when small amounts of hydrogen sulfide are present and suitable means of disposing the spent solution are available (Kohl and Nielsen, 1997). The chemistry

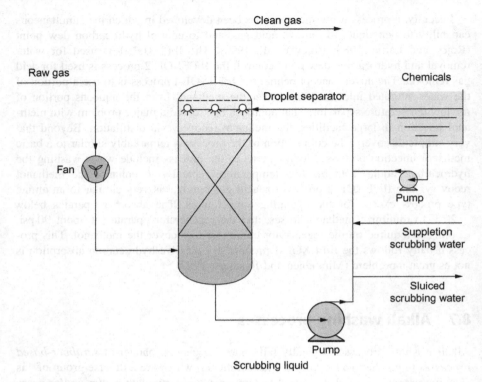

Figure 8.8 A gas scrubbing system.

is simple and to some extent, depends upon the concentration of hydrogen sulfide in the gas stream efficient. Thus:

$$NaOH + H_2S \rightarrow NaHS + H_2O$$

$$2NaOH + H_2S \rightarrow Na_2S + H_2O$$

$$2NaOH + CO_2 \rightarrow Na_2CO_3 + H_2O$$

Carbonate washing is a mild alkali process for emission control by the removal of acid gases (such as carbon dioxide and hydrogen sulfide) from gas streams (Speight, 2014, 2017) and uses the principle that the rate of absorption of carbon dioxide by potassium carbonate increases with temperature. It has been demonstrated that the process works best near the temperature of reversibility of the reactions:

$$K_2CO_3 + CO_2 + H_2O \rightarrow 2KHCO_3$$

$$K_2CO_3 + H_2S \rightarrow KHS + KHCO_3$$

In the Benfield process, acid gases are scrubbed from the feed in an absorber column using potassium carbonate solution with Benfield additives to improve performance and avoid corrosion.

Water washing, in terms of the outcome, is analogous to washing with potassium carbonate (Kohl and Riesenfeld, 1985), and it is also possible to carry out the desorption step by pressure reduction. The absorption is purely physical and there is also a relatively high absorption of hydrocarbon derivatives, which are liberated at the same time as the acid gases.

8.7.2 Hot potassium carbonate process

The hot potassium carbonate process has been utilized successfully for bulk removal of carbon dioxide from a number of gas mixtures. It has been used for sweetening natural gases containing both carbon dioxide and hydrogen sulfide. The process is not suitable for sweetening gas mixtures containing little or no carbon dioxide, as the potassium bisulfide should be very difficult to regenerate if carbon dioxide is not present.

However, there are not only advantages to the use of this process but there are also disadvantages. The advantages are: (1) it is a continuous circulating system employing an inexpensive chemical, (2) it is an isothermal system in that both absorption and desorption of the acid gas are conducted at as nearly a uniform high temperature as can be obtained, thus requiring no heat exchange equipment in the fluid circulating system, and (3) the desorption by stripping is accomplished with a smaller steam rate than required for an amine plant. On the negative side, the disadvantages are: (1) the process will not commercially reduce the hydrogen sulfide content to many pipeline specifications, (2) the process is similar to other acid gas removal processes in that it is also prone to corrosion, and (3) the process is also liable to problems relating to suspended solids and foaming.

The process using *potassium phosphate* is known as phosphate desulfurization, and it is used in the same way as the Girbotol process to remove acid gases from liquid hydrocarbon derivatives as well as from gas streams. The treatment solution is a water solution of tripotassium phosphate (K_3PO_4), which is circulated through an absorber tower and a reactivator tower in much the same way as the ethanolamine is circulated in the Girbotol process; the solution is regenerated thermally.

8.7.3 Other processes

Other processes include the *Alkazid process* (Fig. 8.9), which removes hydrogen sulfide and carbon dioxide using concentrated aqueous solutions of amino acids. The hot potassium carbonate process (Fig. 8.10) decreases the acid content of natural and refinery gas from as much as 50% to as low as 0.5% and operates in a unit similar to that used for amine treating.

The *Giammarco-Vetrocoke* process is used for hydrogen sulfide and/or carbon dioxide removal (Fig. 8.11). In the hydrogen sulfide removal section, the reagent

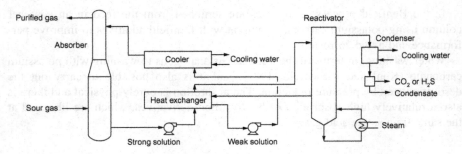

Figure 8.9 The Alkazid process flow diagram.

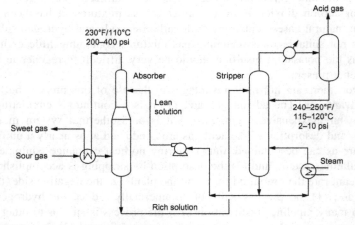

Figure 8.10 The hot potassium carbonate process.

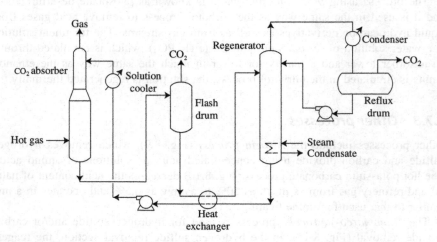

Figure 8.11 The Giammarco-Vetrocoke process.

consists of sodium or potassium carbonates containing a mixture of arsenites and arsenates; the carbon dioxide removal section utilizes hot aqueous alkali carbonate solution activated by arsenic trioxide or selenous acid or tellurous acid.

The *Catacarb process* employs a modified potassium salt solution containing a stable and nontoxic catalyst and corrosion inhibitor. In the process a catalyst is used to activate a carbonate solution in the absorption and desorption of carbon dioxide, thus overcoming the above disadvantage of carbonate scrubbing. Several other catalysts and inhibitors are also used in this process—the choice depends on the composition of the gas stream to be treated. This process is also capable of removing trace amounts of other acid gases such as carbonyl sulfide, carbon disulfide, and mercaptan derivatives.

The *Merox process* (Fig. 8.12) (UOP, 2003) is used to treat end product streams by rendering any mercaptan sulfur compounds inactive. This process can be used for treating liquefied petroleum gas, natural gasoline, and higher molecular weight fractions. The method of treatment is the extraction reaction of the sour feedstock containing mercaptans (RSH) with caustic soda (NaOH) in a single, multistage extraction column using high efficiency trays. The extraction reaction is shown by the following equation:

$$RSH + NaOH \leftrightarrow NaSR + H_2O$$

After extraction, the extracted mercaptans in the form of sodium mercaptide derivatives (NaSR) are catalytically oxidized to water insoluble disulfide derivatives (RSSR):

$$4NaSR + O_2 + sH_2O \rightarrow 2RSSR + 4NaOH$$

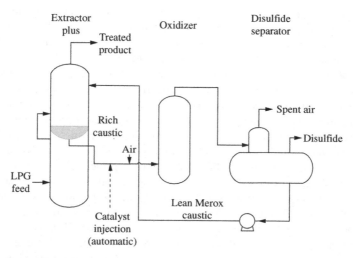

Figure 8.12 The Merox process.

The disulfide oil is decanted and sent to fuel or to further processing in a hydro-treater. The regenerated caustic is then recirculated to the extraction column.

The Merox solution gives a very high degree of removal of mercaptans in a liquid stream. If more complete removal is desired, Merox also provides a fixed-bed catalytic conversion of mercaptans to disulfides. For the treatment of low-boiling feedstocks such as liquefied petroleum gas, no sweetening is required, since the mercaptan derivatives are removed by extraction, while other feedstocks (such as natural gas liquids) containing higher molecular weight mercaptans may require a combination of Merox extraction and sweetening.

The Sulfa-Check process uses sodium nitrite ($NaNO_2$) as the basis for the caustic process. Gas streams with elevated oxygen levels with the Sulfa-Check process will produce some NO_x in the gas stream.

8.8 Membrane processes

Until recently, the use of *membranes* for gas separation has been limited to carbon dioxide removal (Alderton, 1993). Improvements in membrane technology have now made membranes competitive in other applications in the natural gas area. New membrane materials and configurations exhibit superior performance and offer improved stability against contaminants found in natural gas. The new membranes are targeted at three separations: nitrogen, carbon dioxide/hydrogen sulfide, and natural gas liquids (Baker et al., 2002; Baker and Lokhandwala, 2008). The process uses a two-step membrane system design; the methane-selective membranes do not need to be operated at low temperatures, and capital and operating costs are within economically acceptable limits.

The use of membranes has several advantages compared to the other technologies such as absorption and adsorption processes. For instance, the possibility of having a large membrane area in a small module volume due to high packing density allows lower space and weight requirements for membrane units for the same production, and consequently reduced investment. Moreover, the process is more environment-friendly compared to the absorption process due to the nonuse of chemicals. In addition to the previously mentioned advantages, membranes also have the advantage of reduced operating costs, no moving parts, and being suitable for remote locations. Its tolerance of motion makes membrane technology promising for offshore and subsea applications (Jahn et al., 2012; Tierling et al., 2011).

Several carbon dioxide selective membrane systems for carbon dioxide removal from natural gas have been installed onshore or topside on platforms. For small fields with low carbon dioxide content in the gas stream, membrane systems work quite effectively, and the carbon dioxide content in the product can be quite close to the pipeline specification (<2%). However, for bigger fields and high carbon dioxide content in the feed, the membrane treated streams still have a high carbon dioxide content (e.g., >6%) and need further treatment. For small fields membrane technologies are economically competitive compared to absorption, while for large fields absorption is still more favorable. However, due to limitations in weight and size on platforms, membrane processes can be more advantageous for offshore or subsea operations.

The membrane processes for carbon dioxide removal have been more extensively researched compared to hydrogen sulfide removal. Several commercial membrane processes for carbon dioxide removal are available (Jahn et al., 2012). These membranes can tolerate variations in the carbon dioxide concentration in the feed stream within a large range (e.g., 3%–90%). In addition, the membrane systems have the possibility of treating quite a varied feed gas flow (e.g., 3–700 MM ft^3 per day).

New membranes have been developed (Lokhandwala and Jacobs, 2000) for the gas industry. For example, the membranes allow permeation of condensable vapors, such as C_{3+} hydrocarbon derivatives, aromatic derivatives, and water vapor, while rejecting the noncondensable gases, such as methane, ethane, nitrogen, and hydrogen. During the past 15 years, more than 50 systems have been installed in the chemical process industry worldwide. The main applications are nitrogen removal, recovery of natural gas liquids, and dew point control for associated natural gas and fuel gas conditioning for gas turbines and engines (Hall and Lokhandwala, 2004).

In another process (Lokhandwala, 2000), a membrane-based process for upgrading natural gas that contains C_{3+} hydrocarbon derivatives and/or acid gas is described. The conditioned natural gas can be used as fuel for gas-powered equipment, including compressors, in the gas field or the processing plant. Optionally, the process can be used to produce natural gas liquids.

Important factors for the selection of membrane material for carbon dioxide removal are high carbon dioxide permeability and carbon dioxide–methane selectivity, as well as high thermals and mechanical stability (George et al., 2016; Sridhar et al., 2007). The most commonly used commercial membranes for natural gas sweetening are cellulose acetate/triacetate and polyimide membrane. In recent years, a large number of membrane materials for carbon dioxide–methane separation have been developed and tested on a laboratory scale, including both inorganic and polymeric membrane materials.

The conventional polymeric membranes used in natural gas processing involve several challenges, typically including contamination, physical aging, and plasticization (Adewole et al., 2013). The proper pretreatment of natural gas before membrane processes is important to avoid performance decline due to contamination; the different additives used in the oil and gas industry processing are destructive for membranes, and particles and viscous substances in the feed can block membrane pores. The main contaminants in carbon dioxide removal from natural gas are higher molecular weight hydrocarbon derivatives, BTEX (benzene, toluene, ethylene benzene, and xylenes), and hydrogen sulfide. The presence of higher molecular weight hydrocarbon derivatives in the membrane system decreases the membrane permeation ratio as the hydrocarbon derivatives slowly coat the membrane surface (George et al., 2016). The control of the humidity level of the gas is also important for some polymeric membranes, as water vapor causes swelling of the membranes and hence changes their performance (George et al., 2016).

Inorganic membranes have also been investigated for gas separation. Inorganic membranes can have higher selectivity and permeability compared to polymeric membranes. In general, inorganic membranes also have excellent resistance to harsh environments, and at high temperatures and pressures (George et al., 2016; Jusoh et al., 2016).

8.9 Molecular sieve processes

Molecular sieves can be used for removal of sulfur compounds from gas streams. Hydrogen sulfide can be selectively removed to meet 4 ppm v/v specification. The sieve bed can be designed to dehydrate and sweeten simultaneously. In addition, molecular sieve processes can be used for removal of carbon dioxide from gas streams.

Molecular sieves are a crystalline form of alkali metal (calcium or sodium) alumina—silicates, very similar to natural clay minerals. They are highly porous, with a very narrow range of pore sizes, and very high surface area. Manufactured by ion-exchange, molecular sieves are the most expensive adsorbents and possess highly localized polar charges on their surface that act as extremely effective adsorption sites for polar compounds such as water and hydrogen sulfide. Molecular sieves are alkaline and subject to attack by acids, but special acid-resistant sieves are available for very sour gases.

In a molecular sieve, the pore openings in a given structure are all the same size and are determined by the molecular structure of the crystal and the size of molecules present in the crystal. The pores are formed by driving off water of crystallization that is present during the synthesis process. Also, molecular sieves have the large surface area typical of any solid adsorbent and, however, molecular sieves have highly localized polar charges. These localized charges are the reason for the very strong adsorption of polar or polarizable compounds on molecular sieves. This also results in much higher adsorptive capacities for these materials by molecular sieves than by other adsorbents, particularly in the lower concentration ranges.

Since the pore size range is narrow, molecular sieves exhibit selectivity toward adsorbates on the basis of their molecular size and tend not to adsorb bigger molecules such as the higher molecular weight hydrocarbon derivatives. The regeneration temperature is very high. They can produce a water content as low as 1 ppm. Molecular sieves offer a means of simultaneous dehydration and desulfurization and are therefore the best choice for sour gases.

Molecular sieves are highly selective for the removal of hydrogen sulfide (as well as other sulfur compounds) from gas streams and over continuously high absorption efficiency. They are also an effective means of water removal and thus offer a process for the simultaneous dehydration and desulfurization of gas. Gas that has excessively high-water content may require upstream dehydration, however. The molecular sieve process (Fig. 8.13) is similar to the iron oxide process. Regeneration of the bed is achieved by passing heated clean gas over the bed.

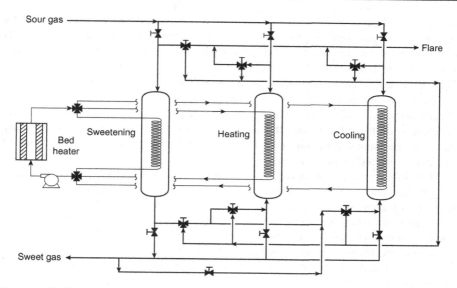

Figure 8.13 The molecular sieve process.

As the temperature of the bed increases, it releases the adsorbed hydrogen sulfide into the regeneration gas stream. The sour effluent regeneration gas is sent to a flare stack, and up to 2% of the gas seated can be lost in the regeneration process. A portion of the natural gas may also be lost by the adsorption of hydrocarbon components by the sieve. In this process, unsaturated hydrocarbon components, such as olefins and aromatics, tend to be strongly adsorbed by the molecular sieves. The molecular sieves are susceptible to poisoning by such chemicals as glycols and require thorough gas cleaning methods before the adsorption step. Alternatively, the sieve can be offered some degree of protection by the use of *guard beds* in which a less expensive catalyst is placed in the gas stream before contact of the gas with the sieve, thereby protecting the catalyst from poisoning. This concept is analogous to the use of guard beds or attrition catalysts in the petroleum industry (Speight, 2014, 2017).

Regeneration of a molecular sieve bed concentrates the hydrogen sulfide into a small regeneration stream which must be treated or sent to disposal. During the regeneration cycle, the hydrogen sulfide will exhibit a peak concentration in the regeneration gas and the peak is approximately 30 times the concentration of the hydrogen sulfide in the inlet stream.

8.10 Sulfur recovery processes

Sulfur is present in natural gas principally as hydrogen sulfide and, in crude oil, as sulfur-containing compounds which are converted to hydrogen sulfide during processing (Speight, 2014, 2017). The hydrogen sulfide, together with some or all of any carbon dioxide present, is removed from the natural gas or refinery gas by means of one gas treating processes. The side stream from acid gas treating units consists mainly of hydrogen sulfide/or carbon dioxide. Carbon dioxide is usually vented to the atmosphere but sometimes is recovered for carbon dioxide floods. Hydrogen sulfide could be routed to an incinerator or flare, which would convert the hydrogen sulfide to sulfur dioxide. The release of hydrogen sulfide to the atmosphere may be limited by environmental regulations. There are many specific restrictions on these limits, and the allowable limits are revised periodically. In any case, environmental regulations severely restrict the amount of hydrogen sulfide that can be vented or flared in the regeneration cycle.

Most sulfur recovery processes use chemical reactions to oxidize hydrogen sulfide and produce elemental sulfur. These processes are generally based either on the reaction of hydrogen sulfide and oxygen or hydrogen sulfide and sulfur dioxide. Both reactions yield water and elemental sulfur. These processes are licensed and involve specialized catalysts and/or solvents. These processes can be used directly on the produced gas stream. Where large flow rates are encountered, it is more common to contact the produced gas stream with a chemical or physical solvent and use a direct conversion process on the acid gas liberated in the regeneration step.

8.10.1 Claus process

Sulfur recovery from sulfur-containing gas streams typically involves application of the famous Claus process using the reaction between hydrogen sulfide and sulfur dioxide (produced in the Claus process furnace from the combustion of hydrogen sulfide with air and/or oxygen) yielding elemental sulfur and water vapor:

$$2H_2S(g) + SO_2(g) \rightarrow (3/n)S_n(g) + 2H_2O(g)$$

Therefore higher conversions for this exothermic, equilibrium-limited reaction call for low temperatures, which lead to low reaction rates and require the use of a catalyst. The catalytic conversion is usually carried out in a multistage fixed-bed adsorptive reactors process to counteract the severe equilibrium limitations at high conversions (Sassi and Gupta, 2008).

Currently, the Claus sulfur recovery process (Fig. 8.14) is the most widely used technology for recovering elemental sulfur from sour gas ((Maddox, 1974). Most of the world's sulfur is produced from the acid gases coming from gas treating. Conventional three-stage Claus plants can approach 98% sulfur recovery efficiency. However, since environmental regulations have become stricter, sulfur recovery plants are required to recover sulfur with over 99.8% efficiency. To meet these stricter regulations, the Claus process underwent various modifications and add-ons.

The add-on modifications to the Claus plant can be considered as a separate operation from the Claus process, in which case it is often called a tail gas treating process. Other sulfur recovery processes can replace the Claus process where it is uneconomic or cannot meet the required specifications. Usually such processes are

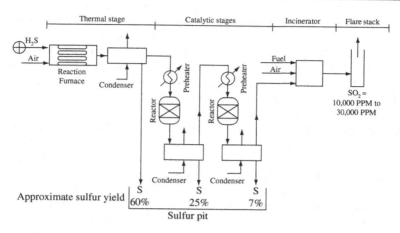

Figure 8.14 The Claus process.

used in small-scale plants, or where the hydrogen sulfide content of the acid gas is too low for a Claus plant or one of its modified versions.

The chemistry of the Claus process involves partial oxidation of hydrogen sulfide to sulfur dioxide and the catalytically promoted reaction of hydrogen sulfide and sulfur dioxide to produce elemental sulfur. The reactions are staged and are as follows:

Thermal stage: $2H_2S + 3O_2 \rightarrow 2SO_2 + 2H_2O$

Thermal and catalytic stage: $SO_2 + 2H_2S \rightarrow 3S + 2H_2O$

In the process, acid gas from the acid gas removal process, along with overhead gases from sour water stripping and a small amount of recycle from the tail gas treating unit (not shown), are burned in the Claus furnace with sufficient air or oxygen to produce an overall gas mixture with the desired 2 to 1 stoichiometric ratio of hydrogen sulfide (H_2S) to sulfur dioxide (SO_2) for conversion to sulfur and water. A substantial amount of sulfur (approximately 67% w/w of the total sulfur recovered) is thermally formed directly in the furnace by the above reactions. As the hot furnace exhaust is cooled in the waste heat boiler, the gaseous sulfur is condensed and removed from the gases. Removal of the sulfur from the right sides of the reactions provides driving force for further conversion in the downstream catalytic reactor stages, which occur at increasingly lower temperatures, also favoring more complete conversion to sulfur. The gases are reheated and enter the first catalytic reactor, where a high conversion (approximately 75% conversion) of the remaining gases takes place, followed by cooling, sulfur condensation, and removal.

The first stage of the process converts hydrogen sulfide to sulfur dioxide and to sulfur by burning the acid-gas stream with air in the reaction furnace. This stage provides sulfur dioxide for the next catalytic phase of the reaction. Multiple catalytic stages are provided to achieve a more complete conversion of the hydrogen sulfide. Each catalytic stage consists of a gas reheater, a reactor, and a condenser. Condensers are provided after each stage to condense the sulfur vapor and separate it from the main stream. Conversion efficiencies of 94%–95% can be attained with two catalytic stages, while up to 97% conversion can be attained with three catalytic stages. The effluent gas is incinerated or sent to another treating unit for *tail-gas treating* before it is exhausted to atmosphere.

The sulfur products are cooled and condensed, generating low pressure steam. Condensed sulfur product is stored in an underground molten sulfur pit, where it is later pumped to truck loading for shipment. Claus tail gas from the last stage sulfur condenser is sent to a tail gas treatment unit to remove unconverted H_2S, SO_2, and COS before disposal. The sulfur recovery depends upon such parameters as (1) feed composition, (2) age of the catalyst, and (3) the number of reactor stages. Typical sulfur recovery efficiencies for Claus plants are 90%–96% for a two-stage plant and 95%–98% for a three-stage plant. Because of equilibrium limitations and other sulfur losses, overall sulfur recovery efficiency in a Claus unit usually does not exceed 98%.

The off-gas leaving a Claus plant is referred to as tail gas and, in the past was burned to convert the unreacted hydrogen sulfide to sulfur dioxide, before discharge to the atmosphere, which has a much higher toxic limit. However, the increasing standards of efficiency required by the pressure from environmental protection has led to the development of a large number of Claus tail gas clean-up units, based on different concepts, in order to remove the last remaining sulfur species (Gall and Gadelle, 2003).

The oxygen-blown Claus process was originally developed to increase capacity at existing conventional Claus plants and to increase flame temperatures of gases having low hydrogen sulfide content. The process has also been used to provide the capacity and operating flexibility for sulfur plants where the feed gas is variable in flow and composition such as often found in refineries.

In the *Selectox process*, the first stage thermal reactor (the furnace) replaced the Claus with a catalytic oxidation step for the conversion of hydrogen sulfide in dilute acid-gas streams to liquid sulfur. The catalytic oxidation reactor can operate at much lower temperatures than the furnace and maintain a more stable flame temperature with gas streams having a low hydrogen sulfide content. The *Recycle Selectox process* is an all-catalytic process in which there are no flames at any point in the process. A special catalyst bed replaces the acid gas burner in a conventional Claus plant and the catalyst occupies the top few inches of the first bed, where it promotes the selective oxidation of hydrogen sulfide to sulfur dioxide. The remainder of the bed is filled with Claus catalyst where the Claus reaction occurs to approximately 80% completion. The highly exothermic nature of these reactions requires that the feed gas be monitored for the concentration of hydrogen sulfide to avoid overheating.

Typically, a Claus sulfur plant can normally achieve high sulfur recovery efficiencies. For lean acid gas streams, the recovery typically ranges from 93% for two stage units (two catalytic reactor beds) up to 96% for three-stage units. For richer acid gas streams, the recovery typically ranges from 95% for two-stage units up to

97% for three-stage units. Since the Claus reaction is an equilibrium reaction, complete removal of hydrogen sulfide and sulfur dioxide is not practical in a conventional Claus plant. In addition, the concentration of contaminants in the acid gas can also limit recovery. For facilities where higher sulfur recovery levels are required, the Claus plant is usually equipped with a tail gas cleanup unit to either extend the Claus reaction or capture the unconverted sulfur compounds for recycle to the Claus plant.

Finally, liquid sulfur flowing from the Claus plant to the sulfur pit contains typically 250–350 ppm v/v of hydrogen sulfide and the sulfur is degassed using an active gas liquid contacting system to release dissolved gas. In this operation, sulfur from the pit is pumped into the degassing tower where it is contacted countercurrently with hot compressed air over a fixed catalyst bed. The degassed sulfur is returned to the product section of the sulfur pit.

The SuperClaus process consists of a thermal stage followed by three or four catalytic reaction stages with sulfur removed between stages by condensers. The first two or three reactors are filled with standard Claus catalyst, while the last reactor is filled with the selective oxidation catalyst. In the thermal stage, the acid gas is burned with a less-than-stoichiometric amount of controlled combustion air so that the tail gas leaving the last Claus reactor typically contains 0.5–0.9 vol. % of hydrogen sulfide. There are two main principles applied in operating the SuperClaus process: (1) operating the Claus plant with excess hydrogen sulfide to suppress the sulfur dioxide content in the Claus tail gas and (2) selective oxidation of the remaining hydrogen sulfide by the process catalyst selectively converts the hydrogen sulfide in the presence of water vapor and excess oxygen to elemental sulfur.

The DynaWave wet gas scrubber, which is a reverse jet scrubber that performs desulfurization in a wet gas environment, can be used as a follow-on to the SuperClaus process and the incinerator. In the process, the scrubbing liquid is injected, through a nonrestrictive jet nozzle, countercurrent to the inlet incinerator flue gas. Liquid, containing caustic reagent, collides with the down-coming gas to create the region of extreme turbulence (the froth zone) with a high rate of mass transfer. Quench, sulfur dioxide removal, and removal of particulate matter occur in this zone. The clean, saturated gas and charged liquid continue through a separation vessel. The saturated gas continues through the vessel to mist removal devices. The liquid descends into the vessel sump for recycle back to the reverse jet nozzle. In the vessel sump, oxidation air is used to convert sodium sulfite (Na_2SO_3) to sodium sulfate (Na_2SO_4).

In the *Clauspol process*, the Claus tail gas is contacted countercurrently with an organic solvent in a low pressure drop packed column. Hydrogen sulfide and sulfur dioxide are absorbed in the solvent and react to form liquid elemental sulfur according to the Claus reaction, which is promoted by an inexpensive dissolved catalyst. The solvent is pumped around the contactor, and the heat of reaction is removed through a heat exchanger to maintain a constant temperature above the sulfur melting point. Due to the limited solubility of sulfur in the solvent, pure liquid sulfur separates from the solvent and is recovered from settling section at the bottom of the contactor. This process allows sulfur recovery up to 99.8% and the recovery level can be customized by adapting the size of the contactor.

Sulfuric acid synthesis is an alternative to sulfur recovery via the Claus process and, in this case, hydrogen sulfide is first burned in a furnace to form sulfur dioxide which is then converted to sulfur trioxide (SO_3), which is then scrubbed with water or a recycled weak sulfuric acid stream to yield 98% sulfuric acid. Thus:

$$2SO_2 + O_2 \rightarrow SO_3$$

$$SO_3 + H_2O \rightarrow H_2SO_4$$

In the process, gas from the final reactor beds enters the absorbing towers, where the produced sulfur trioxide reacts with the excess water in a circulating, strong (98%) sulfuric acid stream, creating additional sulfuric acid. This incrementally raises the concentration of the sulfuric acid so that water is introduced as needed to maintain the sulfuric acid at 98.5% v/v as the final product. The catalytic oxidation of sulfur dioxide to sulfur trioxide is highly exothermic, and the equilibrium becomes increasingly unfavorable for sulfur trioxide formation as temperature increases to approximately 425°C (800°F). For this reason, special catalytic converters (reactors) are designed as multistage reactor bed units with air cooling between each bed for temperature control.

8.10.2 Redox process

Liquid redox sulfur recovery processes are liquid-phase oxidation processes which use a dilute aqueous solution of iron or vanadium to remove hydrogen sulfide selectively by chemical absorption from sour gas streams. These processes can be used on relatively small or dilute hydrogen sulfide stream to recover sulfur from the acid gas stream or, in some cases, they can be used in place of an acid gas removal process. The mildly alkaline lean liquid scrubs the hydrogen sulfide from the inlet gas stream, and the catalyst oxidizes the hydrogen sulfide to elemental sulfur. The reduced catalyst is regenerated by contact with air in the oxidizer(s). Sulfur is removed from the solution by flotation or settling, depending on the process.

The *Selectox process* is based on replacing the Claus first stage thermal reactor (the furnace) with a catalytic oxidation step. A catalytic oxidation reactor can operate at much lower temperatures than the furnace and maintain a more stable flame temperature with lean hydrogen sulfide feeds.

Two versions of the process are offered. The first option is a once-through option, treating acid gases with up to 5% hydrogen sulfide and in which sulfur recovery is on the order of 84−94% w/w for hydrogen sulfide in the feed gas of 2−5% v/v, respectively. In the second option, which is a recycle version that handles hydrogen sulfide concentrations on the order of 5−100% v/v, the gas is recycled from the Selectox reactor condenser to cool the reactor outlet temperature not to exceed 370°C (700°F). The temperature limit is set so that carbon steel can be used for the reactor vessel.

Approximately 80% v/v of the hydrogen sulfide is converted to sulfur in the Selectox reactor. The reactions are of the typical Claus type: first reaction is the conversion of hydrogen sulfide to sulfur dioxide as the predominant reaction, then the hydrogen sulfide reacts with the sulfur dioxide to form elemental sulfur. The rest of the sulfur is recovered in conventional Claus stages. Up to 98% w/w sulfur recovery efficiency can be achieved for a 50% hydrogen sulfide acid gas feed to the recycle version of the Selectox process. Higher sulfur recoveries can only be had with tail gas treating, such as the Beavon Stretford Reactor (BSR)/Selectox process.

In the BSR/Selectox tail gas process, the gases are first hydrogenated to hydrogen sulfide in the BSR, then they proceed to another Selectox reactor stage. Sulfur recoveries up to 99.3% w/w have been reported for BSR/Selectox.

8.10.3   Wet oxidation processes

The wet oxidation processes are based on reduction—oxidation (Redox) chemistry to oxidize the hydrogen sulfide to elemental sulfur in an alkaline solution containing an oxygen carrier. Vanadium and iron are the two oxygen carriers that are used. The best example of a process using the vanadium carrier is the Stretford process. The most prominent examples of the processes using iron as a carrier are the LO-CAT process and the SulFerox process.

The Stretford process using vanadium finds little use now because of the toxic nature of the vanadium solution and iron-based processes are more common.

Both the LO-CAT process and the SulFerox process are essentially the same in principle. The SulFerox process differs from the LO-CAT in that the oxidation and the regeneration steps are carried out in separate vessels and sulfur is recovered from the filters, melted, and sent to sulfur storage. Also, the SulFerox process uses a higher concentration of iron chelates (about 2%−4% by weight vs 0.025%−0.3% by weight for the LO-CAT process). Both processes are capable of up to 99 + % sulfur recovery. However, using the processes for Claus tail gas treating requires hydrolysis of all the sulfur dioxide in the tail gas to hydrogen sulfide because the sulfur dioxide will react with the buffering base potassium hydroxide (KOH) and form potassium sulfate (K_2SO_4), which will consume the buffering solution and quickly saturate it.

8.10.4   Tail gas treating processes

The sulfur content of natural gas and especially crude oil is increasing and with the tightening of the sulfur content in fuels, refiners and gas processors are pushed for additional sulfur recovery capacity. At the same time, environmental regulatory agencies of many countries continue to promulgate more stringent standards for sulfur emissions from oil, gas, and chemical processing facilities. It is necessary to develop and implement reliable and cost-effective technologies to cope with these requirements. In response to this trend, several new technologies are now emerging to comply with the most stringent regulations. The typical sulfur recovery efficiencies for a Claus plant are on the order of 90−96% w/w for a two-stage plant and 95−98% w/w for a three-stage plant. Most environmental agencies require the sulfur recovery efficiency to be in the range of 98.5−99.9 + % w/w and, as a result, there is the need to reduce the sulfur content of the Claus plant tail gas.

Tail gas treating involves the removal of the remaining sulfur compounds from gases after sulfur recovery. Tail gas from a typical Claus process, whether a conventional Claus or one of the extended versions of the process, usually contains small but varying quantities of carbonyl sulfide, carbon disulfide, hydrogen sulfide, and sulfur dioxide as well as sulfur vapor. In addition, there may be hydrogen, carbon monoxide, and carbon dioxide in the tail gas.

In order to remove the rest of the sulfur compounds from the tail gas, all of the sulfur bearing species must first be converted to hydrogen sulfide, which is then absorbed into a solvent and the clean gas vented or recycled for further processing.

8.10.5 Hydrogenation and hydrolysis processes

The reduction of carbonyl sulfide, carbon disulfide, sulfur dioxide, and sulfur vapor in Claus tail gas to hydrogen sulfide is necessary when sulfur recovery of 99.9 + % is required. Usually the sulfur recovery level is set by the allowable emissions of sulfur from the tail gas incinerator. In addition, the reduction of carbonyl sulfide is done on raw synthesis gas when the downstream acid gas removal process is unable to remove carbonyl sulfide to a sufficient extent to meet sulfur emissions regulations from combustion of the cleaned fuel gas. These sulfur compounds are reduced to hydrogen sulfide by hydrogenation or by hydrolysis, at a raised temperature, over a catalytic bed.

In these processes, elemental sulfur and sulfur dioxide are reduced mainly via hydrogenation, while carbonyl sulfide and carbon disulfide are mainly hydrolyzed to hydrogen sulfide. Sulfur and sulfur dioxide are virtually completely converted to hydrogen sulfide when an excess of hydrogen is present.

The hydrogen can be supplied from an outside source, can already be in the Claus tail gas, or obtained by partial oxidation of the fuel gas in a furnace. The tail gas is preheated to the reactor temperature by an in-line burner that combusts fuel gas directly into the tail gas. The same burner can also be used to supply the needed hydrogen by partial combustion of the fuel gas.

When oxygen enrichment is used in the Claus plant, there is often sufficient hydrogen in the tail gas to carry out the reduction without an outside hydrogen source. There is usually sufficient water vapor in the Claus tail gas for the hydrolysis reactions.

The SCOT process (Shell Claus Off-gas Treating process) was developed in the early 1970s and consists of a combination of a catalytic hydrogenation/hydrolysis step and an amine scrubbing unit (Fig. 8.15). In the process, tail gas from the Claus SRU is heated in an in-line burner before entering the hydrogenation reactor, where all sulfur species are converted to hydrogen sulfide (H_2S). Hydrogenation reactor effluent is then cooled by generating low-pressure steam, followed by additional cooling by cooling water exchange. Residual hydrogen sulfide in the cooled tail gas

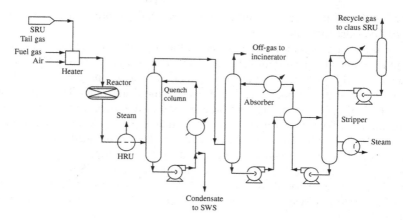

Figure 8.15 The SCOT Process.

is removed with amine in a countercurrent packed absorber. The treated tail gas from the absorber top is incinerated before being vented to the atmosphere. The rich solvent from the amine absorber is pumped to the regenerator after heat exchange against the hot lean solvent from the regenerator. Acid gases are stripped from the solvent in the trayed regenerator via a steam reboiler. The hot lean solvent from the regenerator bottom is pumped back to the absorber after being heat exchanged with rich solvent and cooling water to lower its temperature. Acid gas from the amine regenerator overhead is recycled back to the Claus plant for sulfur recovery. The reactions are:

$$SO_2 + 3H_2 \rightarrow 2H_2O + H_2S$$

$$S_8 + 8H_2 \rightarrow 8H_2S$$

$$COS + H_2O \rightarrow CO_2 + H_2S$$

$$CS_2 + 2H_2O \rightarrow CO_2 + H_2S$$

When carbon monoxide is also present in the tail gas then the following reactions could also take place:

$$SO_2 + 3CO \rightarrow COS + 2CO_2$$

$$S_8 + 8CO \rightarrow 8COS$$

$$H_2S + CO \rightarrow COS + H_2$$

$$H_2O + CO \rightarrow CO_2 + H_2$$

The last reaction (shift reaction) is very rapid, and the presence of CO does not seem to favor the first three reactions.

The Claus tail gas, after being reduced in the reactor, is cooled in a quench column and scrubbed by a Sulfinol solution. The clean tail gas goes to a Claus incinerator and the acid gas rich solution is regenerated in a stripping column. The acid gas off the top of the stripper is recycled back to the Claus plant for further conversion of the hydrogen sulfide. The absorber is operated at near atmospheric pressure and the amine solvent is not highly loaded with acid gases. Because the solution is not highly loaded, unlike high-pressure operation, there is no need for an intermediate flash vessel and the loaded solution goes directly to a stripper.

Early SCOT units used diisopropanolamine in the Sulfinol (Sulfinol-D) solution. Methyldiethanolamine-based Sulfinol (Sulfinol-M) was used later to enhance hydrogen sulfide removal and to allow for selective rejection of carbon dioxide in the absorber.

To achieve the lowest possible hydrogen sulfide content in the treated gas, the Super-SCOT configuration was introduced. In this version, the loaded Sulfinol-M solution is regenerated in two stages. The partially stripped solvent goes to the middle of the absorber, while the fully stripped solvent goes to the top of the absorber. The solvent going to the top of the absorber is cooled below that used in the conventional SCOT process.

The SCOT process can be configured in various ways. For example, it can be integrated with the upstream acid gas removal unit if the same solvent is used in both units. Another configuration has been used to cascade the upstream gas cleanup with the SCOT unit. A hydrogen sulfide-lean acid gas from the upstream gas treating unit is sent to a SCOT process with two absorbers. In the first absorber, the hydrogen sulfide-lean acid gas is enriched, while the second absorber treats the Claus tail gas. A common stripper is used for both SCOT absorbers and, in this latter configuration, different solvents could be used in both the upstream and the SCOT units.

8.11   Process selection

Each of the previous treating processes has advantages relative to the others for certain applications; therefore, in selection of the appropriate process, the following facts should be considered (Morgan, 1994; GPSA, 1998): (1) tail gas clean-up requirements, (2) type and concentration of impurities in the sour gas, (3) specifications for the residue gas, (4) specifications for the acid gas, (5) temperature and pressure at which the sour gas is available and at which the sweet gas must be delivered, (6) volume of gas to be processed, (7) hydrocarbon composition of the gas, and (8) selectivity required for acid gas removal.

Furthermore, technologies in gas processing that have evolved considerably in the last four decades and have moved toward major trends in natural gas processing in the 21st century. These technologies have arisen to allow a much more rigorous optimization of several variables such as: (1) operational flexibility and (2) the types of gas that are offered from processing. Also, increased attention to thermodynamic efficiency and better use of refrigeration and recompression since 1980 have led to development of enhanced recovery processes for natural gas liquids.

Thus the selection of an optimum process will depend on conditions and composition of the inlet gas, cost of fuel and energy, product specifications, and relative product values.

Decisions in selecting a gas treating process can many times be simplified by gas composition and operating conditions. High partial pressures (50 psi) of acid gases enhance the probability of using a physical solvent. The presence of significant quantities of higher molecular weight hydrocarbon derivatives in the feed discourages using physical solvents. Low partial pressures of acid gases and low outlet specifications generally require the use of amines for adequate treating.

In general, the batch and amine processes are used for over 90% of all onshore wellhead applications. Amine processes are preferred when the lower operating cost, the chemical cost for this process is prohibitive, and justifies the higher equipment cost. The key is the sulfur content of the feed gas where below 20-pound sulfur per day, batch processes are more economical, and over 100-pound sulfur per day amine solutions are preferred (Manning and Thompson, 1991).

One important aspect of gas processing that is not typically the process for extraction (separation) of helium from gas stream. At present, there are variety of techniques for the separating and recovering of helium from helium-bearing gas mixtures. Such techniques include (1) membrane technique, (2) pressure swing adsorption technique, and (3) cryogenic technique (Mukhopadhyay, 1980; Froehlich and Clausen, 2008; Lim et al., 2013). Among these techniques, cryogenic processes are the most economical method and have been commonly used to produce helium at high recovery and/or purity from natural gas streams or other gas streams containing low purity helium. However, due to the standard boiling temperatures of methane, nitrogen, hydrogen, and helium, condensation-based distillation or an integrated process involving condensation, distillation, and cryogenics can be used to extract helium from natural gas streams. To meet the high requirements of the product purities, the separation between methane and nitrogen by distillation process leaves the methane and other hydrocarbon derivatives the bottom product while the overhead gas, which is the mixture of nitrogen, helium, and hydrogen, can be further separated by cryogenic condensation process.

References

Adewole, J.K., Ahmad, A.L., Ismail, S., Leo, C.P., 2013. Current challenges in membrane separation of CO_2 from natural gas: a review. Int. J. Greenhouse Gas Control 17, 46−65.

Alderton, P.D., October 27, 1993. Natural gas treating using membranes. In: Proceedings. 2nd GPA Technical Meeting, GCC Chapter, Bahrain.

Anerousis, J.P., Whitman, S.K., 1984. An updated examination of gas sweetening by the iron sponge process. SPE 13280. In: Proceedings. SPE Annual Technical Conference & Exhibition, Houston, Texas.

Arnold, K., Stewart, M., 1999. Surface Production Operations: Vol. 2: Design of Gas-Handling Systems and Facilities, second ed. Gulf Professional Publishing, Houston, TX.

Baker, R.W., Lokhandwala, K.A., 2008. Natural gas processing with membranes: an overview. Ind. Eng. Chem. Res. 47 (7), 2109–2121.

Baker, R.W., Lokhandwala, K.A., Wijmans, J.G., Da Costa, A.R., 2002. Two-Step Process for Nitrogen Removal from Natural Gas. United States Patent 6,425,267.

Bullin, J.A., September 25–27, 2003. Why not optimize your amine sweetening unit? In: Proceedings. GPA Europe Annual Conference, Heidelberg, Germany.

Burnes, E.E., Bhatia, K., 1985. Process for Removing Hydrogen Sulfide from Gas Mixtures. United States Patent, 4, 515, 759.

Carnell, P.J.H., 1986. Gas sweetening with a new fixed bed adsorbent. In: Proceedings. Laurance Reid Gas Conditioning Conference, University of Oklahoma, Norman, Oklahoma.

Carnell, P.J.H., Joslin, K.W., Woodham, P.R., 1995. Oil Gas J., 52.

Chukwuma, N., Jacob, G., 2014. Optimization of triethylene glycol (TEG) dehydration in a natural gas processing plant. Int. J. Res. Eng. Technol. 3 (6), 346–350.

Dalrymple, D.A., Trofe, T.W., 1989. An overview of liquid redox sulfur recovery, Chem. Eng. Prog., March. pp. 43–49.

Dobbs, J.B., March 1986. One step process. In: Proceedings. Laurence Reid Gas Conditioning Conference, Norman, Oklahoma.

Epps, R., February 27–March 2, 1994. Use of Selexol solvent for hydrocarbon dewpoint control and dehydration of natural gas. In: Proceedings. Laurance Reid Gas Conditioning Conference, Norman, Oklahoma.

Esteban, A., Hernandez, V., Lunsford, K., March 2000. Exploit the benefits of methanol. In: Proceedings. 79th GPA Annual Convention, Atlanta, Georgia.

Fox, I., 1981. Process for Scavenging Hydrogen Sulfide from Hydrocarbon Gases. United States Patent, 4, 246, 274.

Froehlich, P., Clausen, J., 2008. Large scale helium liquefaction and considerations for site services for a plant located in Algeria. In: Proceedings. Advances in Cryogenic Engineering, Transactions of the Cryogenic Engineering Conference (CEC), vol. 53, pp. 985–992.

GPSA, 1998. Hydrocarbon treating, Engineering Data Book, eleventh ed. Gas Processors Suppliers Association, Tulsa, OK.

Gall, A.L., Gadelle, D., February, 2003. Technical and commercial evaluation of processes for Claus tail gas treatment. In: Proceedings. GPA Europe Technical Meeting, Paris, France.

George, G., Bhoria, N., Al Hallaq, S., Abdala, A., Mittal, V., 2016. Polymer membranes for acid gas removal from natural gas. Sep. Purif. Technol. 158, 333–356.

Grande, C.A., Roussanaly, S., Anantharaman, R., Lindqvist, K., Singh, P., Kemper, J., 2017. CO_2 capture in natural gas production by adsorption processes. Energy Procedia 114, 2259–2264.

Hall, P., Lokhandwala, K.A., March 2004. Advances in membrane materials provide new gas processing solutions. In: Proceedings. GPA Annual Convention, New Orleans, Louisiana.

Jahn, J., Van Den Bos, P., Van Den Broeke, L.J., 2012. Evaluation of membrane processes for acid gas treatment. In: Proceedings. SPE International Production and Operations Conference & Exhibition. Society of Petroleum Engineers, Richardson, TX, pp. 14–16.

Jenkins, J.L., Haws, R., 2002. Understanding gas treating fundamentals. Pet. Technol. Q. 61–71.

Jusoh, N., Yeong, Y.F., Chew, T.L., Lau, K.K., Shariff, A.M., 2016. Current development and challenges of mixed matrix membranes for CO_2/CH_4 separation. Sep. Purif. Rev. 45, 321–344.

Kattner, J.E., Samuels, A., Wendt, R.P., 1988. J. Pet. Technol. 40 (9), 1237.

Kohl, A.L., Nielsen, R.B., 1997. Gas Purification. Gulf Publishing Company, Houston, TX.

Kohl, A.L., Riesenfeld, F.C., 1985. Gas Purification, fourth ed. Gulf Publishing Company, Houston, TX.

Lallemand, F., Minkkinen, A., May 23, 2001. High sour gas processing in an ever-greener world. In: Proceedings. 9th GPA GCC Chapter Technical Conference, Abu Dhabi.

Lim, W., Choi, K., Moon, I., 2013. Current status and perspectives of liquefied natural gas (LNG) plant design. Ind. Eng. Chem. Res. 52, 3065–3088.

Lokhandwala, K.A., 2000. Fuel Gas Conditioning Process. United States Patent 6,053,965.

Lokhandwala, K.A., Jacobs, M.L., March 2000. New Membrane Application in Gas Processing paper presented at the GPA Annual Convention, Atlanta, GA.

Maddox, R.N., 1974. Gas and Liquid Sweetening, second ed. Campbell Petroleum Series, Norman, OK.

Manning, F.S., Thompson, R.E., 1991. Oil Field Processing of Petroleum. Vol. 1: Natural Gas. Pennwell Publishing Company, Tulsa, OK.

Manning, W.P., 1979. Chemsweet, a new process for sweetening low-value sour gas. Oil Gas J. 77 (42), 122–124.

Minkkinen, A., Jonchere, J.P., May 6, 1997. Methanol simplifies gas processing. In: Proceedings. 5th GPA-GCC Chapter Technical Conference, Bahrain.

Mokhatab, S., Poe, W.A., Speight, J.G., 2006. Handbook of Natural Gas Transmission and Processing. Elsevier, Amsterdam.

Morgan, D.J., November 30, 1994. Selection criteria for gas sweetening. In: Proceedings. GPA Technical Meeting, GCC Chapter, Bahrain.

Muely, W.C., Ruff, C.D., 1972. New method for mercaptan & H_2S removal from gases and liquids. Paper Trade J. 34–36.

Mukhopadhyay, M., 1980. Helium sources and recovery processes with special reference to India. Cryogenics 244–246.

Newman, S.A. (Ed.), 1985. Acid and Sour Gas Treating Processes.. Gulf Publishing Company, Houston, TX.

Polasek, J., Bullin, J.A., 1994. Selecting amines for sweetening units. In: Proceedings. GPA Regional Meeting, Tulsa, Oklahoma.

Ranke, G., Mohr, V.H., 1985. The Rectisol wash new developments in acid gas removal from synthesis gas. In: Newman, S.A. (Ed.), Acid and Sour Gas Treating Processes. Gulf Publishing Company, Houston, TX.

Rhodes, E.F., Openshaw, P.J., Carnell, P.J.H., 1999. Fixed-bed technology purifies rich gas with H_2S, Hg. Oil Gas J. 58.

Rojey, A., Larue, J., 1988. Integrated Process for the Treatment of a Methane-Containing Wet Gas in Order to Remove Water Therefrom. United States Patent 4,775,395.

Rojey, A., Procci, A., Larue, J., 1990. Process and Apparatus for Dehydration, De-acidification, and Separation of Condensate from a Natural Gas. United States Patent 4,979,966.

Samuels, A., 1988. Gas Sweetener Associates. Technical Manual, 3-88, Metairie, Louisiana.

Sassi, M., Gupta, A.K., 2008. Sulfur recovery from acid gas using the Claus process and high temperature air combustion (HiTAC) Technology. Am. J. Environ. Sci. 4 (5), 502–511.

Schaack, J.P., Chan, F., 1989a. Formaldehyde-methanol, metallic-oxide agents head scavenger list. Oil Gas J. 87 (3), 51–55.

Schaack, J.P., Chan, F., 1989b. Caustic-based process remains attractive. Oil Gas J. 87 (5),
    81–82.
Speight, J.G., 2014. The Chemistry and Technology of Petroleum, fifth ed. CRC Press,
    Taylor & Fancies Group, Boca Raton, FL.
Speight, J.G., 2017. Handbook of Petroleum Refining. CRC Press, Taylor & Fancies Group,
    Boca Raton, FL.
Spicer, G.W., Woodward, C., 1991. H_2S control keeps gas from big offshore field on spec.
    Oil Gas J. 76.
Sridhar, S., Smitha, B., Aminabhavi, T.M., 2007. Separation of carbon dioxide from natural
    gas mixtures through polymeric membranes – a review. Sep. Purif. Rev. 36, 113–174.
Tierling, S., Jindal, S., Abascal, R., 2011. Considerations for the use of carbon dioxide
    removal membranes in an offshore environment. In: Proceedings. Offshore Technology
    Conference. OTC, Brazil.
UOP, 2003. Merox Process for Mercaptan Extraction. UOP 4223-3 Process Technology and
    Equipment Manual, UOP LLC, Des Plaines, Illinois.

Further reading

Xiong, L., Peng, N., Gong, L., 2017. Helium extraction and nitrogen removal from LNG
    boil-off gas. In: Proceedings. IOP Conf. Series: Materials Science and Engineering, vol.
    171, Conference 1. IOP Publishing, Philadelphia, Pennsylvania. <http://iopscience.iop.
    org/article/10.1088/1757-899X/171/1/012003>.

Gas condensate

9

9.1 Introduction

The fluids present in a reservoir are the result of a series of thermodynamic changes of pressure and temperature suffered by the original mixture of hydrocarbons over time and during its migration from the container rock to the trap (Chapter 2: Origin and production). In addition, just as crude oil, can vary in composition over the life of a production well (Whitson and Belery, 1994; Speight, 2014a), gas condensate wells can also experience a significant decrease in condensate productivity and condensate composition once the flowing bottomhole pressure drops below the dew point pressure (Wheaton and Zhang, 2000; Fahimpour and Jamiolahmady, 2014). Also, reservoir pressure and temperature increase with depth and their relative relationship will influence the behavior of the light and heavy components that the fluid could contain (Wheaton, 1991). Typically, the content of low-boiling constituents in a mixture of hydrocarbons increases with temperature and depth, which may result in reservoirs near the critical point; gas condensate reservoirs are included in this kind of fluids.

Natural gas reservoirs have also been divided into three groups: (1) a dry gas, (2) a wet gas, and (3) a retrograde-condensate gas (Chapter 1: History and use). A dry-gas reservoir is a reservoir that produces a single composition of gas that is constant in the reservoir, wellbore, and lease-separation equipment throughout the life of a field. Some liquids may be recovered by processing in a gas plant. A wet-gas reservoir is a reservoir that produces a single gas composition to the producing well perforations throughout its life. Gas condensate will form either while flowing to the surface or in lease-separation equipment (Thornton, 1946).

Gas condensate fields are typically a single subsurface accumulation of gaseous hydrocarbon derivatives including low-boiling hydrocarbon derivatives (C_5 to C_8 hydrocarbon derivatives) and, less frequently, constituents with a higher molecular weight (typically C_9 to C_{12} hydrocarbon derivatives). In case of the isothermal reduction of the formation pressure, part of the condensate constituents will condense in the form of gas condensate. Reservoirs with the relative amount of at least 5−10 g of condensate per cubic meter are usually referred to as gas condensate reservoirs. Gas condensate reservoirs can be confined in any formations that from suitable traps and arbitrarily defined in two ways: (1) primary gas condensate reservoirs formed at the depths in excess of 10,000 ft below the surface that are separate from crude oil accumulations and (2) secondary gas condensate reservoirs formed through partial vaporization of the constituents of the crude oil. There are saturated gas condensate fields (the formation pressure is equal to the pressure of initial condensation) and nonsaturated gas condensate fields (the initial condensation pressure is lower than the formation pressure) according to the thermobaric conditions.

Natural Gas. DOI: https://doi.org/10.1016/B978-0-12-809570-6.00009-6

Gas condensate (sometimes referred to as *naphtha*, *low-boiling naphtha*, or *ligroin*) contains significant amounts of C_{5+} components (usually to C_8 or C_{10} components depending upon the source), and they exhibit the phenomenon of *retrograde condensation* at reservoir conditions, in other words, as pressure decreases, increasing amounts of liquid condenses in the reservoir (down to about 2000 psia). This results in a significant loss of in situ condensate reserves that may only be partially recovered by revalorization at lower pressures.

There is also a refinery-produced hydrocarbon product that is also somewhat interchangeable with natural gasoline and middle-of-the-road lease condensates, and that is naphtha, specifically low-boiling naphtha (also called light naphtha). Most of the naphtha is produced from crude oil in the first step of the refining process—distillation. Low-boiling naphtha is composed primarily of pentane hydrocarbon derivatives and hexane hydrocarbon derivatives with minor amounts of higher molecular weight hydrocarbon derivatives. Condensate can be processed through a splitter (effectively a stand-alone, small crude oil distillation tower) that separates the condensate boiling range materials from the lower boiling natural gas liquids (NGLs) and higher boiling hydrocarbon derivatives to yield a clean condensate product that can be used as both a feedstock for refinery upgrading processes and for the production of petrochemicals. Thus the low-boiling naphtha fraction is similar in properties to the typical lease condensate and natural gasoline and is included in this chapter as a means of comparison.

A retrograde-condensate gas reservoir initially contains a single-phase fluid, which changes to two phases (condensate and gas) in the reservoir when the reservoir pressure decreases. Additional condensate forms with changes in pressure and temperature in the tubing and during lease separation. From a reservoir standpoint, dry and wet gas can be treated similarly in terms of producing characteristics, pressure behavior, and recovery potential. Studies of retrograde-condensate gas reservoirs must consider changes in condensate yield as reservoir pressure declines, the potential for decreased well deliverability as liquid saturations increase near the wellbore, and the effects of two-phase flow on wellbore hydraulics.

Thus, by way of review in the context of this chapter, natural gas may be produced from any one of three types of gas wells and these are: (1) crude oil wells and the natural gas that is produced from crude oil wells is called *associated gas*, which is a gas that can exist separate from the crude oil in the underground formation, or dissolved in the crude oil and, also, condensate produced from oil wells is often referred to as *lease condensate*; (2) dry-gas wells, which typically produce only raw natural gas that does not contain any hydrocarbon liquids and such gas is called *nonassociated* gas, also condensate from dry gas is extracted at gas processing plants and is often called *plant condensate*; and (3) condensate wells, which are wells that produce raw natural gas along with NGL, also, such gas is also called *associated* gas and may also be referred to as *wet gas* (Chapter 2: Origin and production and Chapter 4: Composition and properties).

The term *gas condensate* (or *condensate*) is often applied to any liquid composed of low-boiling hydrocarbons produced from a gas well (Tables 9.1 and 9.2). However, the term *condensate reservoir* should be applied only to those reservoirs

Table 9.1 Typical chemical and physical properties of condensate

Property	Comment/value
Appearance	Amber to dark brown
Physical form	Liquid
Odor	Due to the presence of hydrogen sulfide
Vapor pressure	5−15 psia (Reid VP) @ 37.8°C (100°F)
Initial boiling point/range	−29 to 427°C (−20 to 800°F)
Solubility in water	Negligible
Specific gravity (water = 1)	0.6−0.8 @ 15.6°C (60°F)
Bulk density	6.25 lbs/gal
VOC content (% v/v)	50
Evaporation rate (n-butyl acetate = 1)	1
Flash point	−46°C (−51°F)
Lower explosive limits (% v/v in air)	1.1
Upper explosive limits (% v/v in air)	6.0
Autoignition temperature	310°C (590°F)

Table 9.2 Properties of gas condensate from Bibiyana Gas Field, Bangladesh (Sujan et al., 2015)

Property	Value	Method
Composition (by distillation)		
Naphtha (% v/v)	50	
Kerosene (% v/v)	23	
Diesel (% v/v)	24	
Density at 15°C (kg/L)	0.8184	ASTMD1298
API gravity	43	ASTM D287
Kinematic viscosity, cSt, 70°F	1.43	ASTM D445
Kinematic viscosity, 122°F	0.999	ASTM D445
Pour point (°C)	< − 20	ASTM D97
Aniline point (°C)	43	ASTM D611
Flash point (°C)	22	ASTM D93
Sulfur content (% w/w)	0.25	ASTM D129
Carbon residue (% w/w)	1.50	ASTM D189
Ash content (% w/w)	0.0075	ASTM D482

situations in which condensate is formed in the reservoir because of retrograde behavior. Wet-gas reservoirs can always be treated as containing single-phase gas in the reservoir, while retrograde-condensate reservoirs may not. Wet-gas reservoirs generally produce low-boiling liquids with gravities similar to those for retrograde condensates, but with yields less than approximately 20 standard stock tank barrels per million cubic feet (STB/MMscf).

However, the terms used in discussing gas condensate have not been formally accepted as a standard nomenclature and tend to be general descriptions of the condensate rather than chemical descriptions. For example, both distillate and condensate are used synonymously to describe the low-boiling liquid produced during the recovery of natural gas and it has become the standard procedure to adopt somewhat arbitrarily a set of terms, and to explain the meaning of each term as introduced to avoid confusion. Generally, some wells produce, together with large volumes of gas, a water-white or light straw-colored liquid which resembles gasoline or kerosene in appearance. This liquid has been called distillate because of the resemblance to products obtained in the refinery by distilling the volatile components from crude oil. Also, the liquid has been called condensate because it is condensed out of the gas produced by the well. The latter term shall be used throughout this chapter.

Condensate differs substantially from conventional crude oil insofar as (1) the color of crude oil typically varies from dark green to black—typically, condensate is almost colorless; (2) crude oil typically contains some naphtha which is often incorrectly referred to as gasoline—condensate is almost equivalent in boiling range to the low-boiling naphtha fraction of crude oil; (3) crude oil usually contains dark-colored, high molecular weight nonvolatile components—condensate does not contain any dark-colored, high molecular weight nonvolatile components; and (4) the API gravity of crude oil, which is a measure of the weight per unit volume or density, is commonly less than 45 degrees—condensate typically has an API gravity on the order of 50 degrees and higher (Speight, 2014a, 2015). Furthermore, although the difference between retrograde-condensate and wet gas is notable, there is much less distinction between wet gas and dry gas. For both wet gas and dry gas, reservoir engineering calculations are based on a single-phase reservoir gas. The only issue is whether there is a sufficient volume of produced liquid to be considered in such calculations as material balance or wellbore hydraulics. Retrograde systems require more-complex calculations using equation of state that depends upon data produced in an analytical or research laboratory.

When production of a hydrocarbon reservoir commences, analysis plays a major role in determining the types of fluids that are present, along with their main physicochemical characteristics (Speight, 2014a, 2015). Generally, that information is obtained by performing a pressure—volume—temperature (PVT) analysis on a fluid sample of the reservoir as well as other physical analyses which relate to phase behavior and other phenomena, especially the viscosity—temperature relationship (Whitson et al., 1999; Loskutova et al., 2014). Conventional production measurements, such as a drill stem test (DST) are the only parameters that can be measured almost immediately after a well is completed (Breiman and Friedman, 1985; Kleyweg, 1989; Dandekar and Stenby, 1997).

From a reservoir engineering perspective, the issues which must be addressed (by analysis) in a gas condensate reservoir are: (1) the variation of the condensate yield will vary during the life of a reservoir and (2) the manner in which the two-phase gas/oil flow near to the wellbore affects *gas* productivity (Whitson et al., 1999). Both of these issues are strongly related to the PVT properties of the fluid system (though productivity is more affected by relative permeability effects).

It is in the development of gas condensate reservoirs that analytical data reach a high level of importance. Compositional grading in gas condensate reservoirs is important to (1) the design of the pattern of placement of the production wells, (2) estimation of in-place surface volumes, reserves, and (3) prediction of fluid communication vertically (between geologic layers) and horizontally (between fault blocks) (Organick and Golding, 1952; Niemstschik et al., 1993). Prediction of a potential underlying crude oil reserve is often required in discovery wells which are drilled up-structure and encounter only gas which is near-saturated. In such cases, accurate sampling, accurate analysis (though the use of standard test methods) (Speight, 2015, 2018), and PVT modeling are of paramount importance (Pedersen and Fredenslund, 1987; McCain, 1990; Marruffo et al., 2001). A PVT model should describe accurately the key phase, volumetric, and viscosity behavior dictating the key processes affecting rate—time performance and ultimate recovery of gas and oil at the wellhead.

However, a PVT model may not be capable of accurately describing *all* PVT properties with equal accuracy. Models based on equations of state often have difficulty matching retrograde phenomena (compositional variation of gas, and liquid dropout), particularly when the system is near-critical, or only small amounts of condensation occur just below the dew point (tail-like retrograde behavior). Also, the viscosity is a physical property that is often difficult to predict for reservoir condensates, and a measure of the viscosity is not usually available for completion (sometimes referred to as *tuning*) of the viscosity model. Consequently, it is important to determine which PVT properties are most important for the accurate engineering of reservoir and well performance for a given field development. Different fields require different degrees of accuracy for different PVT properties, dependent on field development strategy (depletion vs gas cycling), low or high permeability, saturated or highly undersaturated, geography (offshore or onshore), and the number of wells available for delineation and development.

The PVT properties that are important to the engineering of all gas condensate reservoirs include: (1) the Z-factor, (2) the gas viscosity, (3) the variation of the condensate composition (C_{7+}) with pressure, and (4) the viscosity of the oil as well as any liquid dropout (Lee et al., 1966; Hall and Yarborough, 1973). These properties are particularly important to reservoirs produced by pressure depletion. For gas condensate reservoirs undergoing gas cycling it may also be important to quantify phase behavior (vaporization, condensation, and near-critical miscibility) which develops in gas cycling *below the dew point*.

Thus an important aspect of any estimation of a reservoir fluid(s) is a matter obtaining preliminary values of properties such as: (1) the amount of heptane and higher boiling constituents (C_{7+}), (2) the molecular weight of the original fluid, (3) the maximum retrograde condensation (MRC), and (4) the dew point pressure, P_d (Nemeth and Kennedy, 1967; Breiman and Friedman, 1985; Potsch and Bräuer, 1996; Dandekar, and Stenby, 1997; Elsharkawy, 2001; Marruffo et al., 2001). Most of these properties are an extremely important aspect of the development of a gas condensate reservoir and the early availability of such data will allow the production engineers to carry out reservoir studies that will ensure an efficient exploitation and maximize the final recovery of the liquids present in the reservoir. The only parameter needed to use these correlations is the value of the gas condensate ratio (GCR) of the fluid during the early stage of production (Paredes et al., 2014). Providing the input data is satisfactory, the empirical equations should be valid for *any* gas condensate reservoir, although a range of usability is proposed for a better performance of the correlations.

9.2 Types of condensate

The hydrocarbon products described in the following paragraphs have been used for years as feedstocks (1) for refinery upgrading processes, (2) for gasoline blending, and (3) as feedstocks for petrochemical processes, among other uses. All of these product categories in the condensate family—lease condensates, plant condensates (natural gasoline), and low-boiling naphtha—are, in general terms, composed of the same hydrocarbon constituents. However, it is only the lease condensates that are produced directly from the wellhead without further processing. This distinction turns out to be important when the specifications derived by means of the application of standard analytical test methods are applied to these products.

9.2.1 Gas condensate

By way of recall, natural-gas condensate is a low-density mixture of hydrocarbon liquids as opposed to natural gas liquids that are typically the low-boiling hydrocarbon derivatives (such as ethane, propane, and the butane isomers that are gases at ambient temperature and pressure) present as gaseous components in the unprocessed natural gas produced. Some low-molecular weight hydrocarbon derivatives within the unprocessed natural gas will condense to a liquid state if the temperature is reduced to a temperature that is below the hydrocarbon dew point temperature at a set pressure.

Typically, gas condensate can be produced along with significant volumes of natural gas and is recovered at atmospheric temperatures and pressures from the wellhead gas production. The raw (unrefined) gas condensate is produced from a reservoir, comes out of the ground as mixtures of various hydrocarbon compounds including NGLs, pentane derivatives (C_{5s}), hexane derivatives (C_{6s}), and depending on the condensate, a mixture of higher molecular weight hydrocarbon derivatives in the heptane (C_7) to decane (C_{10}) or even dodecane (C_{12}) carbon-number range.

9.2.2 Lease condensate

Lease condensate is produced at the wellhead as an unprocessed (unrefined) liquid except for stabilization at or near the wellhead. Lease condensates have wide range of API gravity from 45 to 75 degrees. Lease condensates are produced at many sites in the United States, particularly near to the Gulf Coast of the United States from fields such as Eagle Ford and other shale basins.

Lease condensate can be produced along with significant volumes of natural gas and is typically recovered at atmospheric temperatures and pressures from the well-head gas production. These raw condensates come out of the ground as mixtures of various hydrocarbon compounds including NGLs, pentane derivatives (C_{5+}), hexane derivatives (C_{6+}), and depending on the condensate, varying amounts of higher molecular weight hydrocarbons in the C_7, C_8 and even heavier range.

A lease condensate has an API gravity ranging between 45 and 75 degrees. Those with a high API (which is typically clear or translucent in color) contain substantial amounts of NGLs (including ethane, propane, and butane) and not many of the higher molecular weight hydrocarbon derivatives. The lower API gravity condensates with a lower API gravity (on the order of 45 degrees) level look more like crude oil and have much higher concentrations of the heavier higher molecular weight hydrocarbon derivatives (C_7, C_{8+}). In between, a wide range of condensates vary in color.

The higher API gravity condensate can be difficult to handle due to their high vapor pressure and are usually stabilized at the wellhead. In this wellhead process (as compared to the more detailed processes), the condensate is passed through a stabilizer which may be nothing more than a large tank (Fig. 9.1) or a series of tanks (Fig. 9.2) that allow the high vapor pressure components (the NGLs) to vaporize and to be collected for processing as NGLs. That leaves a stabilized condensate that has a lower vapor pressure which is easier to handle, particularly when it must be shipped by truck or rail.

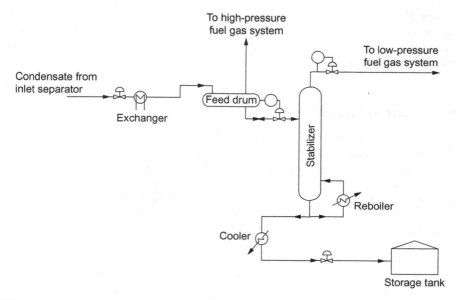

Figure 9.1 The single-tower process for condensate stabilization by fractionation.

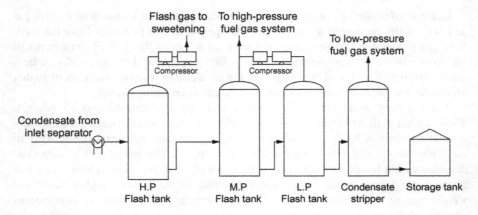

Figure 9.2 Schematic of the condensate stabilization by flash vaporization.

9.2.3 Plant condensate

Plant condensate is a product of plants that process NGLs and is almost equivalent to natural gasoline. Plant condensate is an alternate term that may be applied to natural gasoline, which is composed of hydrocarbon components that are similar to the hydrocarbon constituents in lease condensate, i.e., pentane derivatives (C_{5s}), some hexane derivatives (C_{6s}) and small quantities of higher molecular weight hydrocarbon derivatives. Since it comes from a processing plant, plant condensate is considered to be processed product rather than a naturally occurring product.

Furthermore, since plant condensate (or natural gasoline) is the product of a processing plant, the quality of the natural gasoline (defined by the specifications) is in closer ranges than lease condensates. However, both products can be used interchangeably in some markets, such as crude blending and as a diluent for tar sand bitumen (Speight, 2014a, 2017).

9.2.4 Natural gasoline

Natural gasoline is a mixture of hydrocarbons (extracted from natural gas) that consists mostly of pentane derivatives and higher boiling hydrocarbon. In the beginning of the natural gasoline industry, the only use for natural gasoline was as motor fuel or as a blending agent in the production of motor fuel. Later, the individual components of natural gasoline, namely isobutane derivatives, *n*-butane derivatives, pentane derivatives, and isopentane derivatives, were separated as base stocks for reforming, alkylation, synthetic rubber, and other petrochemicals.

On the other hand, liquefied refinery gases are mixtures of hydrocarbons which are recovered during the refining of crude oil. These materials are composed principally of propane, butane derivatives and are more like, than natural gas, to contain olefin derivatives such as propylene and butylene, and they can be stored and handled at ambient temperatures and at moderate pressures.

9.2.5 Low-boiling naphtha

Low-boiling naphtha (sometimes referred to as *light naphtha*) is another hydrocarbon product that is also somewhat interchangeable with natural gasoline and lease condensates, and is the product designated as naphtha, specifically low-boiling naphtha. Low-boiling naphtha is similar in properties to lease condensate and natural gasoline but it (low-boiling naphtha) is considered to be a refined product since it is produced by refinery distillation or the process of condensate splitting. In fact, most naphtha is produced from crude oil in the first step of the refining process—distillation (Speight, 2014a, 2017). Light naphtha is composed predominantly of pentane and hexane derivatives as well as smaller amounts of higher molecular weight hydrocarbon derivatives.

Another source of naphtha is condensate which can be processed through a stand-alone, small crude oil distillation tower (sometime referred to as a *splitter*). This allows the naphtha to be separated from the more volatile constituents of the condensate and from the higher boiling hydrocarbon derivatives. The naphtha product can be used as both a feedstock for refinery upgrading processes and for petrochemicals production (Sujan et al., 2015).

All of the aforementioned low-boiling hydrocarbon mixtures are in regular use as feedstocks for (1) refinery upgrading processes, (2) gasoline blending, and (3) feedstocks for petrochemical processes, among other uses.

9.3 Production

The operating method for each reservoir should be based upon the characteristics of the gas condensate at reservoir conditions. Other factors such as richness of the gas, size of the reserve, capacities of wells, nature of the reservoir, and mode of occurrence of the gas condensate must be considered. Marketing conditions and other economic factors also are important. Due to the increasing volume of the gas market, and prospects for chemical conversion of gas to liquid fuel by the Fischer—Tropsch process, many operators will be required to make a choice or compromise between complete pressure maintenance and gas sales. This can be done intelligently with known methods of evaluation.

Gas-condensate fields (also called gas-distillate) are today quite commonplace on the Louisiana and Texas Gulf Coast of the United States but are not necessarily confined to that area. The frequency with which they are being discovered and their economic importance in the petroleum industry have increased tremendously with the trend toward deeper drilling. A common understanding of best operating practices by oil operators, land or royalty owners, regulatory bodies, and legislators is necessary if all parties of interest are to cooperate in realizing the maximum potentialities from these fields. During the past decade in particular, much has been learned and written regarding each of the many phases of gas-condensate operations.

There are many different equipment configurations for the processing required to separate natural gas condensate from a raw natural gas. In the example used here

(Fig. 9.3), the raw natural gas feedstock from a gas well or a group of wells is cooled to lower the gas temperature to below the hydrocarbon dew point at the feedstock pressure which condenses a large part of the gas condensate hydrocarbons. The feedstock mixture of gas, liquid condensate, and water is then routed to a high-pressure separator vessel where the water and the raw natural gas are separated and removed. The raw natural gas from the high-pressure separator is sent to the main gas compressor. Then, the gas condensate from the high-pressure separator flows through a throttling control valve to a low-pressure separator. The reduction in pressure across the control valve causes the condensate to undergo a partial vaporization (flash vaporization). The raw natural gas from the low-pressure separator is sent to a booster compressor which raises the gas pressure and sends it through a cooler, and then to the main gas compressor. The main gas compressor raises the pressure of the gases from the high-pressure and low-pressure separators to the pressure that is required for the transportation of the gas by pipeline to the natural gas processing plant.

At the processing plant, the raw natural gas will be dehydrated and acid gases as well as other impurities will be removed from the gas stream after which the ethane, propane, butane isomers, pentane derivative, plus any other high molecular weight hydrocarbon derivatives will also be removed and recovered as valuable by-products (Chapter 7: Process classification). The water removed from both the high- and low-pressure separators may need to be processed to remove hydrogen sulfide (H_2S) before the water can be disposed of underground or reused in some fashion. Some of the raw natural gas may be reinjected into the producing formation to help maintain the reservoir pressure, or for storage pending later installation of a pipeline.

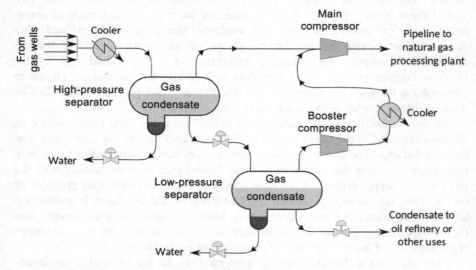

Figure 9.3 Example of a gas condensate separation process.

At the raw natural gas processing plant, the gas is dehydrated, and acid gases and other impurities are removed from the gas (Chapter 6: History of gas processing and Chapter 7: Process classification). Then the ethane, propane butane isomers, and the pentanes and higher molecular weight hydrocarbon derivatives (the C_{5+} fraction) are also removed and recovered as by-products. The water removed from both the high- and low-pressure separators will probably need to be processed to remove hydrogen sulfide before the water can be disposed of or reused in some fashion.

However, the production of condensate can be complicated because of the pressure sensitivity of some condensates. During production, there is a risk of the condensate changing from gas to liquid if the reservoir pressure drops below the dew point pressure. The reservoir pressure can be maintained by fluid injection if gas production is preferable to liquid production.

9.4 Condensate stabilization

Gas condensate is lighter than crude oil, higher molecular weight than NGLs, and, in its natural form, can be dangerous to store and transport. Therefore stabilizing is often required to ensure the condensate meets safety specifications, usually measured by vapor pressure. In addition, condensate is often considered to be an extremely high-quality light crude oil. In comparison to typical crude oil, condensate needs to undergo fewer refining processes and is therefore very economical from the start. Because of the less complex refining process condensate and the potential used of condensate as a blend-stock for various products, it is a resource that is very much in demand.

Whenever gas condensate is produced from a well, a major issue that arises because of the separation of gas condensate and further use of the condensate, is the presence of more volatile constituents such as the hydrocarbon gases: methane, ethane, propane, and butane isomers that are dissolved in the condensate liquids. In some cases, the gas condensate recovered from natural gas may be shipped without further processing but is stabilized often for blending into the crude oil stream and thereby sold as crude oil. In the case of raw gas condensate, there are no specifications for the product other than the process requirements.

Condensate stabilization is the process of increasing the amounts of the intermediate constituents (C_3 to C_5 hydrocarbon derivatives) and higher molecular weight constituents (C_6 and C_{6+} hydrocarbon derivatives) in the condensate. This process is performed primarily to reduce the vapor pressure of the condensate liquids so that a vapor phase is not produced upon flashing the liquid to atmospheric storage tanks. In other word, the scope of this process is to separate the very light hydrocarbon gases, methane and ethane in particular, from the heavier hydrocarbon components (C_3+).

Thus an example of condensate stabilization is the use of a flash procedure to remove the lower boiling hydrocarbon derivatives. For example, after degassing and dewatering in the production separation process, pressurized liquid condensate enters the condensate stabilizer and flows through an exchanger in which hot, stabilized condensate is used to preheat the unstabilized condensate. After preheating, the unstabilized condensate flows to the line heater, where it is heated to stabilization temperature on the order of 95−120°C (205−250°F). The heated, unstabilized condensate is sent to a flash distilled in the condensate separator at approximately 35−45 psig to remove low-density hydrocarbon vapors and any remaining water. The stabilized condensate flows through the plate exchanger for cooling and then off-skid to atmospheric storage. Vapors from the condensate separator are routed through an air-cooled condenser to the NGLs separator, where condensed propane and butane are recovered.

The stabilized liquid condensate typically has a vapor pressure specification, as the product will be injected into a pipeline or transport pressure vessel, which has definite pressure limitations. Condensates may contain a relatively high percentage of intermediate components and can be separated easily from entrained water due to its lower viscosity and greater density difference with water. Thus condensate stabilization should be considered for each gas well production facility and stabilization. Stabilization of condensate streams can be accomplished through either flash vaporization or fractionation.

9.4.1 Flash vaporization

Stabilization by flash vaporization is a simple operation employing only two or three flash tanks. This process is similar to stage separation utilizing the equilibrium principles between vapor and condensate phases. Equilibrium vaporization occurs when the vapor and condensate phases are in equilibrium at the temperature and pressure of separation.

In the process (Figs. 9.1 and9.2), condensate from the inlet separator after passing through the exchanger enters to the high-pressure flash tank, where the pressure is maintained at 600 psia. A decrease in pressure, on the order of 300 psia is obtained, which assists flashing of large amounts of lighter ends, which join the sour vapor stream after recompression. The vapor can either be processed further and put into the sales gas or be recycled into the reservoir and used as gas lift to produce more crude oils. The bottom liquid from the high-pressure tank flows to the middle pressure flash tank operated at 300 psia. Additional methane and ethane are released in this tank. The bottom product is withdrawn again to the low-pressure tank operated at 65 psia. To ensure efficient separation, condensate is degassed in the stripper vessel at the lowest possible pressure prior to storage. This reduces excess flashing of condensate in the storage tank and reduces the inert gas blanket pressure required in it.

9.4.2 Stabilization by fractionation

In this single-tower process (Fig. 9.1), low-boiling constituents such as methane, ethane, propane, and butane isomers are removed and recovered and the residue (that is nonvolatile under the process conditions) remains at the bottom of the column is composed mainly of pentanes and higher molecular weight hydrocarbons. The bottom product is thus a liquid free of all gaseous components able to be stored safely at atmospheric pressure.

The process, in its simplicity, is a derivation of the atmospheric tower for the distillation of crude oil in the refinery and uses the principle of reflux distillation. Thus as the liquid descends in the column, it becomes leaner in lower boiling constituents and richer in higher boiling constituents. At the bottom of the tower some of the liquid is circulated through a reboiler to add heat to the tower. As the gas goes up from tray to tray, more and more of the heavy ends get stripped out of the gas at each tray and the gas becomes richer in the light ends and leaner in the heavy ends.

Overhead gas from the stabilizer, which would seldom meet market specifications for the natural gas market, is then sent to the low-pressure fuel gas system through a back-pressure control valve that maintains the tower pressure to set point. Liquids leaving the bottom of the tower have undergone a series of stage flashes at ever-increasing temperatures, driving off the light components, which exit the top of the tower. These liquids must be cooled to a sufficiently low temperature to keep vapors from flashing to atmosphere in the condensate storage tank.

In some cases, the tower process may operate as a nonreflux tower which is a simpler but less-efficient operation than the reflux tower operation since a condensate stabilization column with reflux will recover more intermediate components from the gas stream.

9.4.3 Condensate storage

Once the condensate has been stabilized, the stabilized product may require storage prior to sales. Condensate is stored between production and shipping operations in condensate storage tanks, which are usually of floating roof type (external and internal). If the condensate does not meet the specifications, the off-specification condensate may be routed to an off-specification condensate storage fixed roof tank (vertical and horizontal) until it is recycled to the condensate stabilization unit by the relevant recycle pump if the latter is available at the plant.

Most likely, breathing loss emissions result from hydrocarbon vapors that are released from the tank by expansion or contraction caused by changes in either temperature or pressure. Working loss represents those emissions that occur due to changes in the liquid level caused by either filling or emptying the tank itself. For floating roof tanks, breathing losses are a result of evaporative losses through rim seals, deck fittings, and deck seam losses. Withdrawal losses occur as the level drops, and thus the floating roof is lowered. Some liquid remains on the inner surface of the tank wall and evaporates when the tank is emptied. For an internal floating roof tank that has a column-supported fixed roof, some liquid also clings to the columns and evaporates. Evaporative loss occurs until the tank is filled and the exposed surfaces are again covered.

9.5 Properties

Briefly and by way of an explanation that relates to the properties of the liquid, *gas condensate* is a mixture of light hydrocarbon liquids obtained by condensation of hydrocarbon vapors: predominately butane, propane, and pentane with some heavier hydrocarbon derivatives and relatively little methane or ethane. *NGLs* are the hydrocarbon liquids that condense during the processing of hydrocarbon gases that are produced from oil or gas reservoir and *natural gasoline* is a mixture of liquid hydrocarbon derivatives extracted from natural gas that is suitable for blending with the various refinery streams that are used to blend to produce the final sales gasoline (Speight, 2014a, 2017).

To complete the explanation, gas condensate is generally recognized as a collection group of hydrocarbons that do not fit easily into mainstream product categories. Other definitions place condensates as liquid hydrocarbons somewhere between crude oil and NGLs (Table 9.3). Technically, all condensates are similar to natural gasoline, the highest boiling of the NGLs. However, the term *condensate* can refer to several products made up of similar hydrocarbon compounds. Along with naphtha, the main uses of condensate fall into the general areas of (1) a blend-stock for gasoline and other liquid fuels and (2) a blend-stock for solvents.

On the other hand, naphtha is produced in the refinery by any one of the several methods that include (1) fractionation of straight-run, cracked, and reforming distillates, or even fractionation of crude petroleum; (2) solvent extraction; (3) hydrogenation of cracked distillates; (4) polymerization of unsaturated compounds (olefins); and (5) alkylation processes. In fact, the naphtha that has a low-to-high boiling range ($0-200°C$, $32-390°F$) is often a combination of product streams from more than one of these processes and may contain constituents (such as olefins and even diolefins) that are not typically found in the natural (nonthermal) gas condensate. On occasion, depending upon availability, gas condensate, NGLs, and natural gasoline are combined with the naphtha streams to complement the composition and volatility requirements of a liquid stream for gasoline manufacture. In addition, the types of uses found for naphtha demands compatibility with the many other materials employed in the formulation of the naphtha, including gas condensate and natural gasoline. Thus the solvent properties of a given fraction must be carefully measured and controlled. For most purposes volatility is important, and, because of the wide use of naphtha in industrial and recovery plants, information on some other fundamental characteristics is required for plant design.

When a condensed liquid (such as gas condensate) is present in gas streams, the analysis is more complicated. In the case of gas condensate, besides bulk analysis, there may be interest on surface-composition (often quite distinct to that of the bulk phase). Compositional analysis in which the components of the mixture are identified, may be achieved by (1) physical means, which is measurement of physical properties, (2) pure chemical means, which is measurement of chemical properties, or more commonly (3) by physicochemical means. Gas analysis is even more may be dangerous and difficult if the composition is a complete unknown. However, when some main constituents are known to occur, the analysis gains accuracy

Table 9.3 Examples in the variation in the definition of gas condensate in different states

State	Definition[a]
Colorado	Hydrocarbons which are in the gaseous state under reservoir conditions and which become liquid when temperature or pressure is reduced; a mixture of pentanes and higher hydrocarbons
Louisiana	Liquid production from noncrude wells as classified by the commissioner of conservation
Montana	The liquid produced by the condensation of a vapor or gas either after it leaves the reservoir or while still in the reservoir. Condensate is often called distillate, drips, or white oil
North Dakota	The liquid hydrocarbons recovered at the surface that result from condensation due to reduced pressure or temperature of petroleum hydrocarbons existing in a gaseous phase in the reservoir
Oklahoma	A liquid hydrocarbon which (1) was produced as a liquid at the surface, (2) existed as gas in the reservoir, and (3) has an API gravity greater than or equal to 50 degrees, unless otherwise proven
South Dakota	Liquid hydrocarbons that were originally in the gaseous phase in the reservoir
Texas	Condensate is liquid produced from a gas well. A gas well is a well that produces more than 100,000 SCF of gas for every barrel of liquids
Wyoming	Lease condensate: liquid hydrocarbons which are separated from other components of the natural gas production stream on the lease or before the inlet to a natural gas processing facility

[a]Since these definitions are subject to change, anyone wishing to proceed further (in a technical or legal sense) with a definition should check the state authority as well as the EIA definition.

(and may be easier) if the known component is removed; this is particularly important in the case of water vapor, which may condense on the instruments, or constituents when the molecular behavior may complicate spectral analyses. A more comprehensive list and description of the properties is available elsewhere (Speight, 2018).

Finally, the term *petroleum solvent* is often used synonymously with *naphtha*. However, naphtha may also be produced from tar sand bitumen, coal tar, and oil shale kerogen by thermal processes as well as by the destructive distillation of wood and from synthesis gas (mixtures of carbon monoxide and hydrogen produced by the gasification of coal and/or biomass or other feedstocks) which is then converted to liquid products by the Fischer–Tropsch process (Davis and Occelli, 2010; Chadeesingh, 2011; Speight, 2011, 2013, 2014a,b). For that reason and to remain within the context of this book, this chapter deals only with the low-boiling naphtha fraction that is produced by the processing of crude oil in refineries (Speight, 2014a, 2015, 2017, 2018; Hsu and Robinson, 2017).

9.5.1 Chemical composition

In terms of composition, natural-gas condensate is a low-density mixture of hydrocarbon liquids that are present as gaseous components in the raw natural gas produced from many natural gas fields. Natural gas condensate is also called condensate, or gas condensate, or sometimes natural gasoline because it contains hydrocarbon derivatives (*n*-hydrocarbon and other hydrocarbon isomers) within the naphtha boiling range (Tables 9.4 and 9.5). Some gas species within the raw natural gas will condense to a liquid state if the temperature is reduced to below the hydrocarbon dew point temperature at a set pressure (Elsharkawy, 2001). Thus condensates are a group of hydrocarbon derivatives that don't fit easily into mainstream product categories and have been defined as liquid hydrocarbon derivatives that lie between crude oil and NGLs. However, the reality is that most condensates differ significantly from crude oils and condensates are not like NGLs but are similar only to natural gasoline, the highest boiling of the NGLs. One the other hand, gas condensate and natural gasoline are often comparable to the lower boiling fraction of naphtha.

More generally, naphtha is an intermediate hydrocarbon liquid stream derived from the crude oil and is usually desulfurized and then catalytically reformed to produce high-octane naphtha before blending into the streams that make up gasoline. Because of the variations in crude oil composition and quality as well as differences in refinery operations, it is difficult (if not impossible) to provide a definitive, single definition of the word naphtha since each refinery produces a site-specific naphtha—often with a unique boiling range (unique initial point and final boiling point) as well as other physical and compositional characteristics. On a chemical basis, (petroleum) naphtha is difficult to define precisely because it can contain varying amounts of the constituents (paraffin derivatives, naphthene derivatives, aromatic derivatives, and olefin derivatives) in different proportions

Table 9.4 General summary of product types and distillation range

Product	Lower carbon limit	Upper carbon limit	Lower boiling point (°C)	Upper boiling point (°C)	Lower boiling point (°F)	Upper boiling point (°F)
Refinery gas	C_1	C_4	−161	−1	−259	31
Liquefied petroleum gas	C_3	C_4	−42	−1	−44	31
Naphtha	C_5	C_{17}	36	302	97	575
Gasoline	C_4	C_{12}	−1	216	31	421
Kerosene/diesel fuel	C_8	C_{18}	126	258	302	575
Aviation turbine fuel	C_8	C_{16}	126	287	302	548
Fuel oil	C_{12}	>C_{20}	216	421	>343	548
Lubricating oil	>C_{20}		>343		>649	
Wax	C_{17}	>C_{20}	302	>343	575	>649
Asphalt	>$C20$		>343		>649	
Coke	>$C50$		>1000		>1832	

Table 9.5 Increase in the number of isomers with carbon number

Carbon atoms	Number of isomers
1	1
2	1
3	1
4	2
5	3
6	5
7	9
8	18
9	35
10	75
15	4347
20	366,319
25	36,797,588
30	4,111,846,763
40	62,491,178,805,831

depending on the source, in addition to the potential isomers of the paraffin derivatives that exist in the naphtha boiling range (the C_5 to C_8 or the C_5 to C_{10} boiling range) (Tables 9.4 and 9.5).

In the refinery, naphtha is an unrefined or refined low-boiling distillate fraction, which is usually boiling below 250°C (480°F), but often with a fairly wide boiling range depending upon the crude oil from which the naphtha is produced as well as the process which generates the naphtha. More specifically, there is a range of special-purpose hydrocarbon fractions that can be described as *naphtha*. For example, the 0−100°C (32−212°F) fraction from the distillation of crude oil is referred to as *light virgin naphtha* and the 100−200°C (212−390°F) fraction is referred to as *heavy virgin naphtha*. The product stream from the fluid catalytic cracker is often split into three fractions: (1) boiling <105°C / <220°F is *light fluid catalytic cracking (FCC) naphtha*, (2) the fraction boiling from 105°C to 160°C (220°F to 320°F) is *intermediate FCC naphtha*, and (3) the fraction boiling from 160°C to 200°C (320°F to 390°F) is referred to as *heavy FCC naphtha* (Occelli, 2010) These boiling ranges can vary from refinery to refinery and even within a refinery when the crude oil feedstock changes or blends of crude oil are used as refinery feedstock.

The so-called *petroleum ether* solvents are specific boiling range naphtha solvents, as is *ligroin*. Thus the term *petroleum solvent* describes a special liquid hydrocarbon fractions obtained from naphtha and used in industrial processes and formulations. These fractions are also referred to as *industrial naphtha*. Other solvents include *white spirit* that is subdivided into *industrial spirit* (distilling between 30°C and 200°C, 86°F and 392°F) and *white spirit* (light oil with a distillation range of 135−200°C, 275−392°F). The special value of naphtha as a solvent lies in its stability and purity.

Regulatory requirements have enhanced the need for better test methods to control manufacturing and the distribution of gasoline. The addition of alcohol and ether as important blending components to gasoline to meet air quality standards has necessitated modifying some existing test methods and the development of new procedures. The desire to reduce manufacturing costs, coupled with the regulatory requirements, have enhanced the application of more cost-effective test methods including rapid screening procedures and wider use of online analyzers. During the early 1950s, instrumental analytical techniques, such as mass spectrometry, infrared, and ultraviolet spectroscopy, were being explored and used for hydrocarbon composition and structural analysis. Beginning with the mid-1950s, publications on gas chromatography began to appear in the literature, and this new technique was soon being used for analyzing a wide variety of hydrocarbon streams. As commercial instrumentation was developed, the application of gas chromatography grew rapidly, with volumes of information being published from its beginning up to the present time. Recently, more rapid spectrometry methods such as infrared and near-infrared and the use of hyphenated analytical techniques, e.g., gas chromatography-mass-spectroscopy (GC-MS), have been applied with success to the characterization of low-boiling distillates.

The proportion of aromatic constituents in hydrocarbon liquids is a key characteristic that will affect a variety of properties of the oil including its boiling range, viscosity, stability, and compatibility of the oil with polymers. In the current context, this is particularly the case for gas condensate and, especially, for natural gasoline where the context of aromatic constituents may influence the compatibility of the gas condensate or the natural gasoline as a blend-stock with other refinery liquids. The aromatic hydrogen and aromatic carbon contents determined by a standard test method can be used to evaluate changes in aromatic contents of hydrocarbon oils due to changes in processing conditions and to develop processing models in which the aromatic content of the hydrocarbon oil is a key processing indicator. Existing methods for estimating the content of the aromatic constituents use physical measurements, such as refractive index, density, and number of average molecular weight or infrared absorbance and often depend on the availability of suitable standards. These procedures do not require standards of known aromatic hydrogen or aromatic carbon contents and are applicable to a wide range of hydrocarbon liquids with the caveat that the hydrocarbon liquids must be soluble in chloroform at ambient temperature.

However, it must be recognized that composition varies with depth in petroleum reservoirs (Speight, 2014a) and component segregation due to gravitational forces is usually given as the physical explanation for the variation in composition. The result of gravitational segregation is that a gas condensate gets richer at greater depths, with increasing C_{7+} mole fraction (and dew point pressure) (Whitson and Belery, 1994). However, not all fields show compositional gradients with depth as predicted by the isothermal model. Some fields show practically no gradient over large depths, while other oil fields have gradients larger than predicted by the isothermal model (Høier and Whitson, 1998). Variation in C_{7+} composition with depth will obviously affect the calculation of initial surface condensate in place, compared with a calculation based on a constant composition.

Sulfur compounds are most commonly removed or converted to a harmless form by chemical treatment with lye, doctor solution, copper chloride, or similar treating agents (Speight, 2014a, 2017). Hydrorefining processes (Speight, 2014a, 2017) are also often used in place of chemical treatment. When used as a solvent, gas condensate and natural gasoline can be blended with naphtha (subject to incompatibility restrictions) (Chapter 6: History of gas processing) (Speight, 2014a) and are selected because of the low content of sulfur-containing constituents. Such blends, also with a low content of aromatic derivatives may have a slight odor, but the aromatic derivatives increase the solvent power of the blend and there is not always the need to remove the aromatic derivatives from the blend (or prior to the blending operation) unless an odor-free product is specified.

9.5.2 Physical properties

The physical properties of gas condensate depend, as should be expected, on the types of hydrocarbon derivatives present in these liquids. In general, the aromatic hydrocarbon derivatives having the highest solvent power and the straight-chain aliphatic compounds the lowest. The solvent properties can be assessed by estimating the amount of the various hydrocarbon types present be made. This method provides an indication of the solvent power of the condensate on the basis that aromatic constituents and naphthene-type constituents provide dissolving ability that paraffinic constituents do not.

When a hydrocarbon reservoir is discovered it is important to know the type of fluids that are present as well as their main physicochemical characteristics, this is normally obtained by performing a PVT analysis to a representative fluid sample of the reservoir. In most cases, having a PVT analysis can take several months, which limits the number and type of reservoir studies that can be carried out during this period (Paredes et al., 2014). The only parameter that can be measured almost immediately after a well is completed, are conventional production measurements. In some cases, this production measurement can be obtained before completing the well by using special testing or measuring equipment such as a DST.

It is important to obtain preliminary values of properties such as: molar percentage of heptane and heavier components (% mole of C_{7+}), molecular weight of the original fluid (MW), MRC, and dew point pressure (P_d). Most of these properties are very important for exploitation of gas condensate reservoir and their early availability will allow engineers to carry out reservoir studies that will ensure an efficient exploitation and maximize the final recovery of the liquids present in the reservoir.

Fluids present in an oil reservoir are the result of a series of thermodynamic changes of pressure and temperature suffered by the original mixture of hydrocarbons over time and during its migration from the container rock to the trap. Reservoir pressure and temperature increase with depth and their relative relationship will influence the behavior of the light and heavy components that the fluid could contain. In general, the content of low-boiling constituents in a mixture of hydrocarbons increases with temperature and depth, which may result in reservoirs near the critical point; gas condensate reservoirs are included in this kind of fluids (Ovalle et al., 2007).

Certain fluid properties are required for studies related to management of gas/condensate reservoirs or prediction of condensate reserves. Often these studies must begin before laboratory data become available, or possibly when laboratory data are not available. Correlations to estimate values of these properties have been developed that are based solely on commonly available field data. These properties are the dew point pressure of the reservoir fluid, changes in the surface yield of condensate as reservoir pressure declines, and changes in the specific gravity of the reservoir gas as reservoir pressure declines (Gold et al., 1989). No correlations based solely on field data have been published for any of these properties.

Reservoir fluid properties are used to characterize the condition of a fluid at a given state. A reliable estimation and description of the properties of hydrocarbon mixtures is fundamental in petroleum and natural engineering analysis and design. Fluid properties are not independent, just as pressure, temperature, and volume are not independent of each other. Equations of state provide the means for the estimation of the PVT relationship, and from them many other thermodynamic properties can be derived. Compositions are usually required for the calculation of the properties of each phase.

The field data required are initial producing GCR from the first-stage separator, initial stock tank liquid gravity in degrees API, specific gravity of the initial reservoir gas, reservoir temperature, and selected values of reservoir pressure. The dew point pressure correlation is based on data of 615 samples of gas condensates with worldwide origins. The other two correlations are based on 851 lines of constant-volume-depletion data from 190 gas-condensate samples, also with worldwide origins.

There are many condensate sources, and each has its own unique gas condensate composition. In general, gas condensate has a specific gravity ranging from 0.5 to 0.8, and is composed of hydrocarbons such as propane, butane, pentane, hexane, and often higher molecular weight hydrocarbons up to decane. Natural gas compounds with more carbon atoms (e.g., pentane, or blends of butane, pentane, and other hydrocarbons with additional carbon atoms) exist as liquids at ambient temperatures. Additionally, condensate may contain other impurities such as: (1) hydrogen sulfide, H_2S; (2) thiols, traditionally also called mercaptans and denoted as RSH, where R is an organic group such as methyl, ethyl, propyl, and the like; (3) carbon dioxide, CO_2; (4) straight-chain alkane derivatives having from 2 to 10 carbon atoms, denoted as C_2 to C_{12}; (5) cyclohexane and perhaps other naphthene derivatives; and (6) aromatic derivatives such as benzene, toluene, xylene isomers, and ethylbenzene (Pedersen et al., 1989).

The primary difficulties in producing condensate reservoirs are as follows: (1) liquid deposition near the wellbore causes a decrease in gas deliverability that can approach 100% in a reservoir with less than 50-md permeability and (2) a large amount of the most valuable hydrocarbon components is left in the reservoir rather than produced. Thus the compositional analyses of gas condensate are used to describe the fluid makeup on a component basis, including calculation of

British thermal unit (energy content) of gases and optimization of separator conditions for liquid yield. In addition, the suitability of the condensate as a blending component in a gasoline plant (Speight, 2014a, 2015, 2017) is an important aspect of determining the compatibility of the condensate with the other components of the blend.

Correlation equations for gas condensates based on readily available field data have been developed. The correlations can be used to predict dew point pressures, decreases in surface condensate yields after reservoir pressure has decreased below dew point pressure, and decreases in reservoir-gas specific gravity at reservoir pressures below dew point pressure. A value of dew point pressure is essential data for any reservoir study. A reasonably accurate estimate of dew point pressure for a specific reservoir fluid is necessary in situations in which laboratory data are not available or before laboratory data are obtained. Laboratory measurements of dew point pressure and other gas properties of 615 gas condensates with worldwide origins were used to develop a dew point pressure correlation based on initial producing GCR, initial stock tank oil gravity, and specific gravity of the original reservoir gas. This is the first proposed dew point pressure correlation that does not require some laboratory-measured quantity.

Estimation of decreases in producing yields after the reservoir pressure drops below the dew point pressure is necessary for accurate prediction of condensate reserves. The reduction in surface yields can be as much as 75% during the primary production of a gas condensate. This reduction must be considered in the prediction of ultimate recovery of a condensate. A surface-yield correlation has been developed that is a function of a selected reservoir pressure, initial stock tank oil gravity, specific gravity of the original reservoir gas, and reservoir temperature. The dataset included laboratory studies of 190 gas-condensate samples. This is the first proposal offered in petroleum literature of a correlation to estimate the decreases in surface yield.

The defining property that distinguishes crude oil and gas condensate black and volatile oils is the content of the equilibrium gases. The volatilized-oil (also called lease condensate or distillate) content of a gas represents its condensable liquid portion. Condensable refers to the portion that condenses or drops out during pressure reduction and ultimately results as stock tank liquid. Condensation may take place within the reservoir as the gas passes through the lease separators. Physically, intermediate-hydrocarbon components, typically the C_2 to C_7, dominate this fraction. Gas condensates and wet gases also contain volatilized oil. Volatilized oil is reported conventionally as part of the crude-oil reserves and production and should not be confused with and is distinctly different from natural-gas liquids. Natural-gas liquids are derived from the gas-processing plant and are, therefore, products of a gas processing plant. The volatilized-oil content of gases is quantified in terms of their volatilized-oil/gas ratio, typically expressed in units of STB/MMscf or stock tank m^3 per standard m^3 of separator gas.

Finally, a comment on the sampling and analysis of gas condensate which invokes similar principles to this used in the sampling of gas (ASTM, 2017; Speight, 2018).

Before a field development starts, the primary goal of sampling is to obtain representative samples (Chapter 5: Recovery, storage, and transportation) of any of the fluids (including gas and gas condensate) found in the reservoir at initial conditions. It may be difficult to obtain a representative sample because of two-phase flow effects near the wellbore. This occurs when a well is produced with a flowing bottomhole pressures below the saturation pressure of the reservoir fluids. It is also commonly thought that nonrepresentative samples of reservoir are the result if gas coning (or liquid coning) occurs during sampling.

Briefly, coning is a production problem in which cap gas or bottom water infiltrates the perforation zone in the near-wellbore area and reduces oil production. Gas coning is distinctly different from, and should not be confused with, free-gas production caused by a naturally expanding gas cap. Likewise, water coning should not be confused with water production caused by a rising water/oil contact from water influx. Coning is a rate-sensitive phenomenon generally associated with high producing rates. Coning is typically a near-wellbore phenomenon and only develops once the pressure forces drawing fluids toward the wellbore overcome the natural buoyancy forces that segregate gas and water from oil.

The most representative in situ samples are usually obtained when the reservoir fluid is single phase at the point of sampling, be it bottomhole or at the surface. Even this condition, however, may not ensure representative sampling. And, samples obtained during gas coning in an oil well can provide accurate in situ representative samples if a proper laboratory procedure is followed (Fevang and Whitson, 1994).

Because reservoir fluid composition can vary horizontally, between fault blocks, and as a function of depth, it is essential that a sample of reservoir fluid is obtained that is representative of the volume being drained by the well during the test. Unfortunately, the concept of a representative sample is typically *a sample that correctly reflects the composition of reservoir fluid at the depth or depths being tested.* If there is any doubt that a sample is not representative (according to the above definition), then it is often the best course of action to not use the sample. If such a sample is used for analysis, the validity of the PVT analysis done on the nonrepresentative sample, and consequently the measured data should not be used when developing the PVT model.

9.5.3 Color

Condensate liquids are generally colorless (*water white*) or near-colorless or may even be light in color (brown, orange, or green) and have an API gravity that is typically on the order of 40 and 60°API (Table 9.6). The yields of these liquids can be as high as 300 STB/MMscf. It has been suggested (McCain, 1990) that when yields are below approximately 20 STB/MMscf, even though phase-behavior considerations may show retrograde behavior, the amount of liquid dropout in the reservoir can be insignificant.

Table 9.6 Examples of the composition of dry gas, wet gas, and gas condensate

Component or property	Dry gas	Wet gas	Condensate
Carbon dioxide, CO_2	0.10	1.41	2.37
Nitrogen, N_2	2.07	0.25	0.31
Methane, C_1	86.12	92.46	73.19
Ethane, C_2	5.91	3.18	7.80
Propane, C_3	3.58	1.01	3.55
iso-Butane, $i\text{-}C_4$	1.72	0.28	0.71
n-Butane, $n\text{-}C_4$		0.24	1.45
iso-Pentane, $i\text{-}C_5$	0.50	0.13	0.64
n-Pentane, $n\text{-}C_5$		0.08	0.68
Hexane derivatives, C_{6s}		0.14	1.09
Heptanes plus, C_{7+}		0.82	8.21

9.5.4 Density

Density (the mass of liquid per unit volume at 15°C or the amount of mass contained in a unit volume of fluid) and the related terms *specific gravity* (the ratio of the mass of a given volume of liquid at 15°C to the mass of an equal volume of pure water at the same temperature), and *relative density* (same as *specific gravity*) are important properties of petroleum products as it is a part of product sales specifications, although it only plays a minor role in studies of product composition. Usually a hydrometer, pycnometer, or digital density meter is used for the determination in all these standards.

Density is the single-most important property of a fluid (Table 9.7) and the density of gas condensate is dependent upon the density and relative amounts of the hydrocarbon constituents. For liquids, density is high, which translates to a very high molecular concentration and short intermolecular distances. For gases, density is low, which translates to low molecular concentrations and large intermolecular distances (Rayes et al., 1992; Piper et al., 1999).

Density is an important parameter for condensate and related liquids and the determination of density (specific gravity) provides a check on the uniformity of the condensate and it permits calculation of the weight per gallon. The temperature at which the determination is carried out and for which the calculations are to be made should also be known and be in accordance with the volatility of the sample. Any such method must be subject to vapor pressure constraints and are used with appropriate precautions to prevent vapor loss during sample handling and density measurement. In addition, some test methods should not be applied if the samples are sufficiently colored and that the absence of air bubbles in the sample cell cannot be established with certainty. The presence of such bubbles can have serious consequences for the reliability of the test data.

The all-important parameter to calculate density is the Z-factor, both for the liquid and vapor phases. Empirical correlations for Z-factor of natural gases were

Table 9.7 Density of various hydrocarbon liquids—possible constituents of gas condensate and natural gasoline

Hydrocarbon (phase)	Formula	Molecular weight	Density
Benzene	C_6H_6	78.114	0.877
Decane	$C_{10}H_{18}$	142.285	0.73
Heptane	C_7H_{16}	100.204	0.684
Hexane	C_6H_{14}	86.177	0.66
Hexene	C_6H_{12}	84.161	0.673
Isopentane	C_5H_{12}	72.15	0.626
Octane	C_8H_{18}	114.231	0.703
Toluene	C_7H_8	92.141	0.867

developed before the advent of digital computers. Although their use is in decline, they can still be used for fast estimates of the Z-factor. These methods are invariably based on some type of corresponding states development. According to the theory of corresponding states, substances at corresponding states will exhibit the same behavior (and hence the same Z-factor) (Standing and Katz, 1942; Hall and Yarborough, 1973).

The specific gravity is the ratio of fluid density to the density of a reference substance, both defined at the same pressure and temperature. These densities are usually defined at standard conditions (14.7 psia and 60°F). For a condensate, oil, or a liquid, the reference substance is water. By definition, the specific gravity of water is unity and, using the API scale, water has an API gravity of 10. Light crude oils have an API greater than or equal to 45 degrees, while the API gravities of gas condensates range between 50 and 70°API.

9.5.5 Dew point pressure

The dew point pressure is the pressure where an incipient liquid phase condenses from a gas phase. Practically, the dew point marks the pressure where (1) reservoir gas phase composition changes and becomes leaner and (2) condensate accumulation starts in the reservoir. These two changes can have a profound effect on reservoir and well performance—or, they may have little impact.

The importance of the *actual dew point pressure* will vary from reservoir to reservoir, but in most situations accurate dew point determination is *not important*. First, in the context of compositional variation with pressure (and associated variation of condensate yield with pressure) accurate determination of the thermodynamic dew point pressure is *not* of particular importance. In fact, knowledge of the specific dew point at all as long as the variation of composition (C_{7+} content) with pressure is well defined "near" the thermodynamic dew point. Second, when the bottomhole flowing pressure drops below the dew point and two phases start flowing near the wellbore, gas relative permeability drops and well *productivity* drops.

Another (less common) need for dew point pressure is when an underlying saturated oil zone *may* exist, and a PVT model is used to predict the existence and location of the gas—oil contact (GOC). In this case, the PVT model dew point should be tuned precisely to an accurately measured dew point pressure. It is not uncommon that a predicted GOC may vary tens of meters per bar of uncertainty in the (PVT model) dew point pressure. Thus an accurate description of the dew point pressure will have an impact on the prediction of initial oil and gas in place, placement of delineation wells, and potential field development strategy. In this situation, accurate dew point measurement and equally accurate modeling of the measured dew point should be given due attention.

9.5.6 Flammability

Gas condensate and natural gasoline are, like naphtha, (1) readily flammable, (2) will evaporate quickly from most surfaces, and (3) must be very carefully contained at all times. Condensate can be ignited by heat, sparks, flames, or other sources of ignition (such as static electricity, pilot lights, mechanical/electrical equipment, and electronic devices such as cell phones). The vapors may travel considerable distances to a source of ignition where they can ignite, flash back, or explode. The condensate vapors are heavier than air and can accumulate in low areas. If container is not properly cooled, it can rupture in the heat of a fire. Hazardous combustion/decomposition products, including hydrogen sulfide, may be released by this material when exposed to heat or fire. If the condensate contains a high percentage of aromatic constituents, it can also be smoky, toxic, and carcinogenic. Some condensate-based fuels have a reduced aromatic content, but many are naturally high or augmented in aromatic derivatives that arise from blends with aromatic naphtha.

The flash point is the lowest temperature at atmospheric pressure (760 mmHg, 101.3 kPa) at which application of a test flame will cause the vapor of a sample to ignite under specified test conditions. The sample is deemed to have reached the flash point when a large flame appears and instantaneously propagates itself over the surface of the sample. The flash point data is used in shipping and safety regulations to define *flammable* and *combustible* materials. Flash point data can also indicate the possible presence of highly volatile and flammable constituents in a relatively nonvolatile or nonflammable material. Since the flash point of gas condensate and the flash point of natural gasoline are low, the test method can also indicate the possible presence of even more highly volatile and flammable constituents in these two liquids.

Flash point for a hydrocarbon or a fuel is the minimum temperature at which vapor pressure of the hydrocarbon is sufficient to produce the vapor needed for spontaneous ignition of the hydrocarbon with the air in the presence of an external source, i.e., spark or flame. From this definition, it is clear that hydrocarbon derivatives with higher vapor pressures (lighter compounds) have lower flash points. Generally, flash point increases with an increase in boiling point. Flash point is an important parameter for safety considerations, especially during storage and transportation of volatile petroleum products (i.e., liquefied petroleum gas, light naphtha, and gasoline) in a high-temperature environment.

The prevalent temperature within and around a storage tank should always be less than the flash point of the fuel to avoid possibility of ignition. Flash point is used as an indication of the fire and explosion potential of a petroleum product. Flash point should not be mistaken with *fire point*, which is defined as the minimum temperature at which the hydrocarbon will continue to burn for at least 5 seconds after being ignited by a flame. For such materials, ignition is dependent upon the thermal and kinetic properties of the decomposition, the mass of the sample, and the heat transfer characteristics of the system. Also, the method can be used, with appropriate modifications, for chemicals that are gaseous at atmospheric temperature and pressure of which gas condensate and natural gasoline are example.

9.5.7 Formation volume factor

The formation volume factor (FVF) relates the volume of oil at stock tank conditions to the volume of oil at elevated pressure and temperature in the reservoir. Values typically range from approximately 1.0 bbl/STB for crude oil systems containing little or no solution gas to nearly 3.0 bbl/STB for highly volatile oils.

The FVF of a natural gas (B_g) relates the volume of 1 lb-mol of gas at reservoir conditions to the volume of the same lb-mol of gas at standard conditions, as follows:

$$B_g = \frac{\text{(Volume of 1 unit of gas at reservoir conditions)}}{\text{(Volume of 1 unit of gas, SCF)}}$$

Those volumes are, evidently, the specific molar volumes of the gas at the given conditions. The FVF of a condensate (B_o) relates the volume of 1 lb-mol of liquid at reservoir conditions to the volume of that liquid once it has gone through the surface separation facility:

$$B_o = \frac{\text{(Volume of 1 unit of liquid at reservoir conditions)}}{\text{(Volume of 1 unit of liquid after separation)}}$$

The FVF can also be seen as the volume of reservoir fluid required to produce 1 barrel of oil in the stock tank.

9.5.8 Solubility

Other methods that are applicable to hydrocarbon liquids typically involve determination of the surface tension from which the solubility parameter is calculated and then provides an indication of dissolving power and compatibility. A similar principal is applied to determine the amount of insoluble material in lubricating oil using *n*-pentane and can be applied to liquid fuels. The insoluble constituents measured can also assist in evaluating the performance characteristics of a liquid fuel in determining the cause of equipment failure and line blockage (Speight, 2014a; Speight and Exall, 2014).

9.5.9 Solvent power

For hydrocarbon liquids, solvent tests are generally performed to ensure the quality of a given product as supplied by the producer to the consumer. In this case the purpose would be to provide data related to the properties of gas condensate and natural gasoline to the refiner and the potential benefits or adverse effect of using gas condensate or natural gasoline as blending stock in the refinery. Many solvent tests are of a somewhat empirical nature such as aniline point and mixed aniline point. The data from these test methods and the need for the test methods are frequently cited in specifications and serve a useful function as control tests. Solvent purity, however, is typically monitored mainly by gas chromatography, with individual nonstandardized tests routinely being used by the associated industry.

9.5.10 Sulfur content

Sulfur-containing components exist in some gas condensate and natural gasoline samples. Individual sulfur components can be speciated (ASTM, 2017) and the method uses a gas chromatographic capillary column coupled with either a sulfur chemiluminescence detector or atomic emission detector. The total sulfur content (especially the content of hydrogen sulfide) is an important test parameter in hydrocarbon liquid that are designated as liquid fuels or are used as blending stock for sales (Kazerooni et al., 2016).

9.5.11 Surface tension

Surface tension and interfacial tension exist when two phases are present, and the phases can be gas/oil, oil/water, or gas/water. The surface tension between gas and crude oil ranges from near zero to approximately 34 dynes/cm and is a function of pressure, temperature, and the composition of each phase. Interfacial tension is the force that holds the surface of a particular phase together and is normally measured in dynes/cm.

More specifically, the surface tension is a measure of the surface free energy of liquids, i.e., the extent of energy stored at the surface of liquids. Although it is also known as interfacial force or interfacial tension, the name *surface tension* is usually used in systems where the liquid is in contact with gas. Qualitatively, it is described as the force acting on the surface of a liquid that tends to minimize the area of its surface, resulting in liquids forming droplets with spherical shape, for instance. Quantitatively, since its dimension is of force over length (lbf/ft in English units), it is expressed as the force (in lbf) required to break a film of 1 ft of length. Equivalently, it may be restated as being the amount of surface energy (in lbf-ft) per square feet.

The interfacial tension is similar to surface tension in that cohesive forces are also involved. However, the main forces involved in interfacial tension are adhesive forces (tension) between the liquid phase of one substance and either a solid, liquid, or gas phase of another substance. The interaction occurs at the surfaces of the substances involved, that is at their interfaces.

The gas—liquid interfacial tension at high pressures is commonly measured by a pendant-drop apparatus (ASTM, 2017). In this technique, a liquid droplet is allowed to hang from the tip of a capillary tube in a high pressure visual cell filled with its equilibrated vapor. The shape of liquid droplet at static conditions, controlled by the balance of gravity and surface forces, is determined and related to the gas—liquid interfacial tension. The pendant-drop method can also be applied to measure the interfacial tension of hydrocarbon-water systems.

The productivity of gas condensate wells can significantly be declined as the condensate bank evolves around the wellbore. Wettability alteration of formation minerals from strongly liquid-wet to intermediate gas-wet conditions using liquid-repellent fluorinated chemicals has shown promising results to mitigate such liquid-blockage issues (Fahimpour and Jamiolahmady, 2014).

9.5.12 Vapor pressure

The vapor pressure or equilibrium vapor pressure is defined as the pressure exerted by a vapor that is in thermodynamic equilibrium with the condensed phase (solid or liquid) at a given temperature in a closed system. The equilibrium vapor pressure is an indication of the evaporation rate of a liquid. A substance with a high vapor pressure at normal temperatures is often referred to as volatile.

The vapor pressure of any substance increases nonlinearly with temperature and the atmospheric boiling point of a liquid (also known as the normal boiling point) is the temperature at which the vapor pressure equals the ambient atmospheric pressure. With any incremental increase in that temperature, the vapor pressure becomes sufficient to overcome atmospheric pressure and lift the liquid to form vapor bubbles inside the bulk of the substance. Bubble formation deeper in the liquid requires a higher pressure, and therefore higher temperature, because the fluid pressure increases above the atmospheric pressure as the depth increases. More pertinent to low-boiling hydrocarbon mixtures such as condensate, the vapor pressure that a single component in the mixture contributes to the total pressure in the system is called the partial pressure.

The vapor pressure of liquids such as gas condensate, natural gasoline, and gasoline is a critical physical test parameter for hydrocarbon liquids. By way of definition, the vapor pressure of a substance is the pressure of a vapor in thermodynamic equilibrium with its condensed phase in a closed system. The Reid vapor pressure (RVP) is the measure of the absolute pressure exerted by a liquid at 37.8°C (100°F), at a vapor to liquid ratio of 4:1. The true vapor pressure (TVP) is the equilibrium vapor pressure of a mixture when the vapor to liquid ratio = 0, e.g., floating roof tanks.

Typically, the lighter condensates (higher API gravity) can be difficult to handle due to their high vapor pressure and are usually stabilized at the wellhead (often referred to as stabilization in the field) insofar as the condensate is run through a stabilizer which may be nothing more than a big tank that allows the high vapor pressure components (the NGLs) to vaporize and to be collected for processing as NGLs. That leaves a stabilized condensate that has a lower vapor pressure which is easier to handle, particularly when it must be shipped by truck or rail.

The primary quality criterion for the vapor pressure of gas condensate is the RVP, which is a common measure of the volatility of condensate, natural gasoline, naphtha, and gasoline and is the absolute vapor pressure exerted by a liquid at 37.8°C (100°F) (Speight, 2015, 2018; ASTM, 2017). The RVP is affected by atmospheric pressure (plant elevation) and maximum ambient temperature. Thus, to store the condensate in floating roof storage tanks, it is very crucial to control the RVP at the desired level (especially in warm seasons). The emissions of gas condensate from a storage tank are normally categorized as occurring from standing storage losses or working losses (sometime referred to as breathing losses, which can be confusing). The term breathing loss refers to those emissions that result without any corresponding change in the liquid level within the storage tank.

The RVP differs slightly from the TVP of a liquid due to small sample vaporization and the presence of water vapor and air in the confined space of the test equipment. More specifically, the RVP is the absolute vapor pressure and the TVP is the partial vapor pressure.

9.5.13 Viscosity

Fluid viscosity is a measure of its internal resistance to flow and, thus, viscosity is an indicator of flow properties of the condensate. The most commonly used unit of viscosity is the centipoise, which is related to other units as follows:

$$1 \text{ cP} = 0.01 \text{ poise} = 0.000672 \text{ lbm/ft-}s = 0.001 \text{ Pa-}s$$

Natural gas viscosity is typically expected to increase both with pressure and temperature (Lee et al., 1966).

9.5.14 Volatility

Distillation, as a means of determining the boiling range (hence the *volatility*) of petroleum and petroleum products has been in use since the beginning of the petroleum industry and is an important aspect of product specifications. Depending on the design of the distillation unit, either one or two naphtha steams may be produced: (1) a single naphtha with an end point of about 205°C (400°F) and similar to straight-run gasoline or (2) this same fraction divided into low-boiling naphtha (light naphtha) and high-boiling naphtha (heavy naphtha). The end point of the light naphtha is varied to suit the subsequent subdivision of the naphtha into narrower boiling fractions and may be of the order of 120°C (250°F). On the other hand, condensate is almost always equivalent to the low-boiling naphtha fraction.

Naphtha (petroleum naphtha), which distills below 240°C (465°F), is a generic term applied to refined, partly refined, or unrefined petroleum products and the liquid products isolated from natural gas streams and is the volatile fraction of the petroleum, which is used as a solvent or as a precursor to gasoline. In fact, not less than 10% v/v of material should distill below 75°C (167°F); not less than 95% v/v of the material should distill below 240°C (465°F) under standard distillation conditions, although there are different grades of naphtha within this extensive boiling range that have different boiling ranges (Hori, 2000; Parkash 2003; Pandey et al., 2004; Gary et al., 2007; Speight, 2014a, 2017; Hsu and Robinson, 2017) and the focus of this chapter is on the low-boiling liquid fraction of gas streams (gas condensate, natural gasoline) which are typically equivalent to the low-boiling fraction of naphtha (boiling range 0−200°C, 32−392°F).

Gas condensate, natural gasoline, and naphtha are often referenced by a boiling range, which is the defined temperature range in which the fraction distills. The ranges are determined by standard methods (ASTM, 2017), it is being especially necessary to use a recognized method since the initial and final boiling points which ensure conformity with volatility requirements and absence of *heavy ends* are affected by the testing procedure. Thus one of the most important physical parameters of gas condensate and natural gasoline is the boiling range distribution. The significance of the boiling range distribution is the indication of volatility that dictates the evaporation rate that is an important property for gas condensate and natural gasoline used in coatings and similar application where the premise is that the gas condensate and natural gasoline evaporates over time.

Since the end use dictates the required composition of condensate blends with natural gasoline, and petroleum-derived naphtha, most blends are available in both high and low solvency categories and the various text methods can have major significance in some applications and lesser significance in others. Hence, the application and significance of tests must be considered in the light of the proposed end use. To emphasize this point, naphtha contains varying amounts of its constituents viz., paraffin derivatives, naphthene derivatives, aromatic derivatives, and olefins in different proportions in addition to potential isomers of paraffin that exist in naphtha boiling range. Naphtha resembles gasoline in terms of boiling range and carbon number, being a precursor to gasoline. Naphtha is used as automotive fuel, engine fuel, and jet-B (naphtha type).

Volatility, solvent properties (dissolving power), purity, and odor determine the suitability of condensate for a particular use. The use of condensate (but specifically naphtha) as an incendiary device in warfare, and as an illuminant dates back to AD 1200. Condensate, like naphtha, may be characterized as lean (high paraffin content) or rich (low paraffin content). The rich naphtha with higher proportion of naphthene content is easier to process in the Platforming unit (Parkash, 2003; Gary et al., 2007; Speight, 2014a, 2017; Hsu and Robinson, 2017).

If spilled or discharged in the environment, condensate represents a threat of the toxicity of the constituents to land and/or to aquatic organisms. A significant spill may cause long-term adverse effects in the aquatic environment. The constituents of naphtha predominantly fall in the C_5 to C_{12} carbon range: alkanes, some cycloalkanes, perhaps aromatic derivatives. On the other hand, naphtha may also contain a preponderance of aromatic constituents (up to 65%), others contain up to 40% alkenes, while all of the others are aliphatic in composition, up to 100%.

The determination of the boiling range distribution of distillates such as gas condensate and natural gasoline by gas chromatography (ASTM, 2017) not only helps identify the constituents but also facilitates on-line controls at the refinery. This test method is designed to measure the entire boiling range of gas condensate and natural gasoline with either high or low RVP (ASTM, 2017). In the method, the sample is injected into a gas chromatographic column that separates hydrocarbon derivatives in boiling point order.

While pure hydrocarbon derivatives such as pentane, hexane, heptane, benzene, toluene, xylene, and ethyl benzene which may be characterized by a fixed boiling point, the constituents of gas condensate and natural gasoline (being mixtures of many hydrocarbon derivatives) are typically less easy to identify. The distillation test does however give a useful indication of their volatility. The data obtained should include the initial and final temperatures of distillation together with sufficient temperature and volume observations to permit a characteristic distillation curve to be drawn.

This information is especially important when a formulation includes other volatile liquids, since the performance of the product is affected by the relative volatility of the constituents. An illustration of the importance of this aspect is found in the use of specifically defined boiling point naphtha in cellulose lacquers, where a mixture with ester, alcohols, and other solvents may be employed. The naphtha does not act as a solvent for the cellulose ester, but is incorporated as a diluent to control the viscosity and flow properties of the mixture. If the solvent evaporates too rapidly blistering of the surface coating may result, while if the solvent evaporates unevenly, leaving behind a higher proportion of the naphtha, precipitation of the cellulose may occur leading to a milky opaqueness known as blushing. Because of the composition, such uses are often prohibited for condensate unless that condensate can be used satisfactorily as a blend without having an adverse influence on the specification of the product.

Although much dependence is placed on the assessment of volatility by distillation methods, some specifications include measurement of drying time by evaporation from a filter paper or dish. Laboratory measurements are expressed as *evaporation rate* either by reference to a pure compound evaporated under similar conditions as the sample under test or by constructing a time weight loss curve under standard conditions. Although the results obtained on the condensate can provide a useful guide, it is preferable, wherever possible, to carry out a performance test on the final product when assessing formulations.

In choosing gas condensate and/or natural gasoline for a particular purpose it is necessary to relate volatility to the fire hazard associated with its use, storage, and transport, and also with the handling of the products arising from the process. This is normally based on the characterization of the gas condensate or the natural gasoline solvent by flash point limits.

9.5.15 Water solubility

Water solubility ranges from very low for the longest-chain alkanes to high solubility for the simplest mono-aromatic constituents. Generally, the aromatic compounds are more soluble than the same-sized alkanes, *iso*-alkanes, and cycloalkanes. This indicates that the components likely to remain in water are the one- and two-ring aromatic derivatives (C_6 to C_{12}). The C_9 to C_{16} alkanes, *iso*-alkanes, and one- and two-ring cycloalkanes are likely to be attracted to sediments based on their low water solubility and moderate to high octanol–water partition coefficient (log K_{ow}) and organic carbon–water partition coefficient (log K_{oc}) values.

References

ASTM, 2017. Annual Book of Standards. ASTM International, West Conshohocken, PA.

Breiman, L., Friedman, J.H., 1985. Estimating optimal transformations for multiple regression and correlation. J. Am. Stat. Assoc. 80 (391), 580–619.

Chadeesingh, R., 2011. The Fischer-Tropsch process. In: Speight, J.G. (Ed.), The Biofuels Handbook. The Royal Society of Chemistry, London, pp. 476–517. , Part 3 (Chapter 5).

Dandekar, A.Y., Stenby, E.H., 1997. Measurement of phase behavior of hydrocarbon mixtures using fiber optical detection techniques. Paper No. SPE38845. Proceedings of the SPE Annual Technical Conference and Exhibition, San Antonio, Texas, 5–8 October, Society of Petroleum Engineers, Richardson, TX.

Davis, B.H., Occelli, M.L., 2010. Advances in Fischer-Tropsch Synthesis, Catalysts, and Catalysis. CRC Press, Taylor & Francis Group, Boca Raton, FL.

Elsharkawy, A.M., 2001. Characterization of the C7 plus fraction and prediction of the dew point pressure for gas condensate reservoirs. Paper No. SPE 68776. Proceedings of the SPE Western Regional Meeting, Bakersfield, California, 26–29 March, Society of Petroleum Engineers, Richardson, TX.

Fahimpour, J., Jamiolahmady, M., 2014. Impact of gas-condensate composition and interfacial tension on oil-repellency strength of wettability modifiers. Energy Fuels 28 (11), 6714–6722.

Fevang, Ø., Whitson, C.H., 1994. Accurate in-situ compositions in petroleum reservoirs. Paper No. SPE 28829. Proceedings of the EUROPEC Petroleum Conference, London, 25–27 October, Society of Petroleum Engineers, Richardson, TX.

Gary, J.G., Handwerk, G.E., Kaiser, M.J., 2007. Petroleum Refining: Technology and Economics, fifth ed. CRC Press, Taylor & Francis Group, Boca Raton, FL.

Gold, D.K., McCain Jr., W.D., Jennings, J.W., 1989. An improved method of the determination of the reservoir gas specific gravity for retrograde gases. J. Pet. Technol. 41 (7), 747–752. Paper No. SPE-17310-PA. Society of Petroleum Engineers, Richardson, TX.

Hall, K.R., Yarborough, L., 1973. A new equation of state for Z-factor calculations. Oil Gas J. 71 (18), 82–92.

Hori, Y., 2000. In: Lucas, A.G. (Ed.), Modern Petroleum Technology. Volume 2: Downstream. John Wiley & Sons Inc, New York (Chapter 2).

Hsu, C.S., Robinson, P.R. (Eds.), 2017. Handbook of Petroleum Technology. Springer International Publishing AG, Cham.

Høier, L., Whitson, C.H., 1998. Miscibility variation in compositional grading reservoirs. Paper No. SPE 49269. Proceedings of the SPE Annual Technical Conference and Exhibition, New Orleans, LO, 27–30 September, Society of Petroleum Engineers, Richardson, TX.

Kazerooni, N.M., Adib, H., Sabet, A., Adhami, M.A., Adib, M., 2016. Toward an intelligent approach for H_2S content and vapor pressure of sour condensate of South Pars Natural Gas Processing Plant. J. Nat. Gas Sci. Eng. 28, 365–371.

Kleyweg, D., 1989. A set of constant PVT correlations for gas condensate systems. Paper No. SPE 19509. Society of Petroleum Engineers, Richardson, TX.

Lee, A., Gonzalez, M., Eakin, B., 1966. The viscosity of natural gases. J. Pet. Technol. 18, 997−1000. SPE Paper No. 1340, Society of Petroleum Engineers, Richardson, TX.

Loskutova, Y.V., Yadrevskaya, N.N., Yudina, N.V., Usheva, N.V., 2014. Study of viscosity-temperature properties of oil and gas-condensate mixtures in critical temperature ranges of phase transitions. Procedia Chem. 10, 343−348.

Marruffo, I., Maita, J., Him, J., Rojas, G., 2001. Statistical forecast models to determine retrograde dew point pressure and the C_{7+} percentage of gas condensate on the basis of production test data from Eastern Venezuelan Reservoirs. Paper No. SPE69393. Proceedings of the SPE Latin American and Caribbean Petroleum Engineering Conference, Buenos Aires, 25−28 March, Society of Petroleum Engineers, Richardson, TX.

McCain Jr., W.D., 1990. The Properties of Petroleum Fluids, second ed. PennWell Books, Tulsa, OK.

Nemeth, L.K., Kennedy, H.T., 1967. A correlation of dew point pressure with fluid composition and temperature. Paper No. SPE-1477-PA. Society of Petroleum Engineers, Richardson, TX.

Niemstschik, G.E., Poettmann, F.H., Thompson, R.S., 1993. Correlation for determining gas condensate composition. Paper No. SPE 26183. Proceedings of the SPE Gas Technology Symposium, Calgary, 28−30 June, Society of Petroleum Engineers, Richardson, TX.

Occelli, M.L., 2010. Advances in Fluid Catalytic Cracking: Testing, Characterization, and Environmental Regulations. CRC Press, Taylor & Francis Group, Boca Raton, FL.

Organick, E.I., Golding, B.H., 1952. Prediction of saturation pressures for condensate-gas and volatile-oil mixtures. Trans. AIME 195, 135−148.

Ovalle, A.P., Lenn, C.P., McCain, W.D., 2007. Tools to manage gas/condensate reservoirs; novel fluid-property correlations on the basis of commonly available field data. Paper No. SPE-112977-PA. SPE Reservoir Evaluation & Engineering Volume. Society of Petroleum Engineers, Richardson, TX.

Pandey, S.C., Ralli, D.K., Saxena, A.K., Alamkhan, W.K., 2004. Physicochemical characterization and application of naphtha. J. Sci. Ind. Res. 63, 276−282.

Paredes, J.E., Perez, R., Perez, L.P., Larez, C.J., 2014. Correlations to estimate key gas condensate properties through field measurement of gas condensate ratio. Paper No. SPE-170601-MS. Proceedings of the SPE Annual Technical Conference and Exhibition, Amsterdam, the Netherlands, 27−29 October, Society of Petroleum Engineers, Richardson, TX.

Parkash, S., 2003. Refining Processes Handbook. Gulf Professional Publishing, Elsevier, Amsterdam.

Pedersen, K.S., Fredenslund, A., 1987. An improved corresponding states model for the prediction of oil and gas viscosities and thermal conductivities. Chem. Eng. Sci. 42, 182−186.

Pedersen, K.S., Thomassen, P., Fredenslund, A., 1989. Characterization of gas condensate mixtures, C_{7+} fraction characterization. In: Chorn, L.G., Mansoori, G.A. (Eds.), Advances in Thermodynamics. Taylor & Francis Publishers, New York.

Piper, L.D., McCain Jr., W.D., Corredor, J.H., 1999. Compressibility factors for naturally occurring petroleum gases. Gas Reservoir Eng. 52, 23−33. SPE Reprint Series Society of Petroleum Engineers, Richardson, TX.

Potsch, K.T., Bräuer, L., 1996. A novel graphical method for determining dew point pressures of gas condensates. Paper No. SPE 36919. Proceedings of the SPE European Petroleum Conference, Milan, Italy, 22−24 October, Society of Petroleum Engineers, Richardson, TX.

Rayes, D.G., Piper, L.D., McCain, W.D. Jr., Poston, S.W., 1992. Two-phase compressibility factors for retrograde gases. Paper No. SPE-20055-PA. Society of Petroleum Engineers, Richardson, TX.

Speight, J.G. (Ed.), 2011. The Biofuels Handbook. Royal Society of Chemistry, London.

Speight, J.G., 2013. The Chemistry and Technology of Coal, third ed. CRC Press, Taylor & Francis Group, Boca Raton, FL.

Speight, J.G., 2014a. The Chemistry and Technology of Petroleum, fifth ed. CRC Press, Taylor & Francis Group, Boca Raton, FL.

Speight, J.G., 2014b. Gasification of Unconventional Feedstocks. Gulf Professional Publishing, Elsevier, Oxford.

Speight, J.G., 2015. Handbook of Petroleum Product Analysis, second ed. John Wiley & Sons Inc, Hoboken, NJ.

Speight, J.G., 2017. Handbook of Petroleum Refining. CRC Press, Taylor & Francis Group, Boca Raton, FL.

Speight, J.G., 2018. Handbook of Natural Gas Analysis. John Wiley & Sons Inc, Hoboken, NJ.

Speight, J.G., Exall, D.I., 2014. Refining Used Lubricating Oils. CRC Press, Taylor & Francis Group, Boca Raton, FL.

Standing, M.B., Katz, D.L., 1942. Density of natural gases. Trans. AIME 146, 140−149.

Sujan, S.M.A., Jamal, M.S., Hossain, M., Khanam, M., Ismail, M., 2015. Analysis of gas condensate and its different fractions of Bibiyana Gas Field to produce valuable products. Bangladesh J. Sci. Ind. Res. 50 (1), 59−64.

Thornton, O.F., 1946. Gas-condensate reservoirs-a review. Paper No API-46-150. Proceedings of the API Drilling and Production Practice, New York, 1 January. API-46-150. <https://www.onepetro.org/conference-paper/API-46-150> (accessed 01.11.17).

Wheaton, R.J., 1991. Treatment of variations of composition with depth in gas-condensate reservoirs. Paper No. SPE 18267. Society of Petroleum Engineers, Richardson, TX.

Wheaton, R.J., Zhang, H.R., 2000. Condensate banking dynamics in gas condensate fields: compositional changes and condensate accumulation around production wells. Paper No. SPE 62930. Proceedings of the SPE Annual Technical Conference and Exhibition, Dallas, Texas, 1−4 October, Society of Petroleum Engineers, Richardson, TX.

Whitson, C.H., Belery, P., 1994. Compositional gradients in petroleum reservoirs. Paper No. SPE 28000. Proceedings of the SPE Centennial Petroleum Engineering Symposium held in Tulsa, Oklahoma, 29−31 August, Society of Petroleum Engineers, Richardson, TX.

Whitson, C.H., Fevang, Ø., Yang, T., 1999. Gas condensate PVT: What's really important and why? Proceedings of the IBC Conference on the Optimization of Gas Condensate Fields. London, UK, 28−29 January, IBC UK Conferences Ltd., London. <http://www.ibc-uk.com> (accessed 20.10.17).

Further reading

DiSanzo, F.P., Giarrocco, V.J., 1988. Analysis of pressurized gasoline-range liquid hydrocarbon samples by capillary column and PIONA analyzer gas chromatography. J. Chromatogr. Sci. 26, 258−401.

Part III

Energy Security and the Environment

Energy security and the environment

10

10.1 Introduction

All fossil fuels—coal, crude oil, and natural gas—release pollutants into the atmosphere when burned to provide energy. However, natural gas—being composed predominantly of methane which combusts to carbon dioxide and water—is considered the most environment-friendly fossil fuel. It is cleaner burning than coal or petroleum because it contains less carbon than some of its fossil fuel cousins. Natural gas once cleaned (Chapter 7: Process classification and Chapter 8: Gas cleaning processes) also contains much less sulfur and nitrogen compounds than, say coal, and when burned natural gas emits less ash (particulate matter) and soot into the air than coal or petroleum fuels. In fact, among the conventional fossil fuels (coal, crude oil, and natural gas), the consumption growth of natural gas outpaced the other fossil fuel types, i.e., coal and petroleum (BP, 2017). This is attributable to stronger demands of natural gas in industrial and residential heating, increased installations of natural gas-based electric power plants, and new discoveries of large natural gas deposits. In the 21st century, the world already experienced several times significant shortages and price hikes of natural gas, mainly due to imbalances between supply and demand.

Natural gas is the third most used energy source in the United States at 23% of the energy requirements, after crude oil and coal. Industry consumers are, by far, the largest consumer of natural gas in manufacturing goods, followed by utilities for electric power generation, residential consumers for heating homes and cooking, and then commercial users mainly for building heating (Speight and Islam, 2016). Industrial use of natural gas contribute to manufacturing a wide variety of goods including plastics, fertilizers, photographic films, inks, synthetic rubber, fibers, detergents, glues, methanol, ethers, insect repellents, and much more. Natural gas is popularly used in electric power generation. Natural gas burns cleaner and more efficiently than coal. It has less emission-related problems than other popular fossil fuels.

However, natural gas has only a limited market share as a transportation fuel, even though it can be used in regular internal combustion engines. This is mainly due to its low energy density per volume, unless natural gas is compressed under very high pressure. In addition, more than 50% of the residential homes in the United States use natural gas as the main heating fuel. It is obvious that any major disruption in natural gas supply would bring out unique but quite grave consequences in the nation's energy management, at least for a short term and for a certain affected region, since natural gas is heavily utilized by both electric power

Natural Gas. DOI: https://doi.org/10.1016/B978-0-12-809570-6.00010-2

generating utilities and residential homes. Furthermore, the regional energy dependence problem has been somewhat mitigated by deregulation of utilities, which altered the business practices of electric utilities and natural gas industry. The deregulation allows natural gas customers to purchase natural gas from suppliers other than the local utility, thus providing choices for consumers and eventually resulting in better economic balance of the energy costs.

Thus natural gas is an extremely important source of energy for reducing pollution and maintaining a clean and healthy environment (Speight, 1993; Speight, 2007, 2014, 2017, 2018). In addition, as an abundant and secure source of energy in the United States because of the domestic reserves (BP, 2017), the use of natural gas also offers a number of environmental benefits over other sources of energy, particularly other fossil fuels. Furthermore, there are environmental concerns with the use of any fuel. As with other fossil fuels, burning natural gas produces carbon dioxide, which is a very important and effective greenhouse gas. Many scientists believe that increasing levels of carbon dioxide and other greenhouse gases in the atmosphere of the Earth are changing the global climate. However, there are issues that have been raised that are related to the means by which the carbon dioxide in ice cores (the leading evidence for increases on carbon dioxide in the atmosphere) have been measured that throw doubt upon the contribution to the carbon dioxide in the atmosphere by human activities (Speight and Islam, 2016).

Thus, as with other fuels, natural gas also affects the environment when it is produced, stored, and transported. Because natural gas is made up mostly of methane (another greenhouse gas), there is the potential for leaks of methane into the atmosphere from wells, storage tanks, and pipelines. In addition, exploring and drilling for natural gas will always have some impact on land and marine habitats but new technologies have greatly reduced the number and size of areas disturbed by drilling (often referred to as *environmental footprints*). Satellites, global positioning systems, remote sensing devices, and 3-D and 4-D seismic technologies make it possible to discover natural gas reserves while drilling fewer wells. Plus, use of horizontal drilling and directional drilling make it possible for a single well to produce gas from much bigger areas of the reservoirs (Speight, 2016).

Once brought from underground (Chapter 4: Composition and properties), the natural gas is refined to remove impurities such as water, other gases, sand, and other compounds (Chapter 6: History of gas processing and Chapter 7: Process classification). Some hydrocarbons (ethane, propane, and the butane isomers) are removed and sold separately, including propane and butane (Chapter 7: Process classification and Chapter 8: Gas cleaning processes). Other impurities, such as hydrogen sulfide are also removed and used to produce sulfur (Chapter 7: Process classification and Chapter 8: Gas cleaning processes), which is then also sold separately. After refining, the clean natural gas (methane) is transmitted through a network of pipelines, thousands of miles of which exist in the United States alone. From these pipelines, natural gas is delivered to the industrial and domestic consumers (Chapter 5: Recovery, storage, and transportation).

The use of fossil fuels for energy contributes to a number of environmental problems. Natural gas, as the cleanest of the fossil fuels, can be used in many ways to help reduce the emissions of pollutants into the atmosphere. Burning natural gas in the place of other fossil fuels emits fewer harmful pollutants into the atmosphere, and an increased reliance on natural gas can potentially reduce the emission of many of these most harmful pollutants (EIA, 2006). Pollutants emitted in the United States, particularly from the combustion of fossil fuels, have led to the development of many pressing environmental problems. Natural gas, emitting fewer harmful chemicals into the atmosphere than other fossil fuels, can help to mitigate some of these environmental issues.

On a unit basis, natural gas emits lower quantities of greenhouse gases and criteria pollutants than other fossil fuels. This occurs in part because natural gas is more easily fully combusted, and in part because processed (cleaned) natural gas contains fewer impurities than any other fossil fuel. For example, coal mined in the United States typically contains 1.6% w/w sulfur (a consumption-weighted national average). Crude oil burned at electric utility power plants ranges from 0.5% to 1.4% sulfur. Diesel fuel has less than 0.05%, while the current national average for motor gasoline is 0.034% sulfur. Comparatively, natural gas when used for power (electricity) generation has less than 0.005% sulfur compounds.

However, the issue of natural gas and any adverse effects on the environment may be due to leaks of methane which can outweigh any carbon dioxide production from combustion from natural gas (and other fossil fuel combustion). In fact, the occurrence of natural gas leaks into the atmosphere from oil and natural gas wells, storage tanks, pipelines, and processing plants are the source of approximately one-third of the total methane emissions and approximately 4% of total greenhouse gas emissions in the United States.

As long as many countries have fossil fuel-based economies, fossil fuel combustion will lead to environmental problems. In addition, the venting or leaking of natural gas into the atmosphere can have a significant effect with respect to greenhouse gases because methane, the principal component of natural gas, is much more effective in trapping these gases than carbon dioxide. The exploration, production, and transmission of natural gas, as well, can have adverse effects on the environment.

This chapter addresses the many environmental aspects related to the use of natural gas, including the environmental impact of natural gas relative to other fossil fuels and some of the potential applications for increased use of natural gas. These issues include: (1) greenhouse gas emissions, (2) smog, air quality, and acid rain, (3) industrial and electric generation emissions, and (4) pollution from the transportation sector—natural gas vehicles.

10.2 Natural gas and energy security

Energy security is the uninterrupted availability of energy sources at an affordable price (IEA, 2018) or, put another way, a particular aspect of energy security is assuring access to a ready supply of energy (US DOE, 2017). Thus energy security has many dimensions: long-term energy security mainly deals with timely invest-ments to supply energy in line with economic developments and sustainable envi-ronmental needs. Short-term energy security focuses on the ability of the energy system to react promptly to sudden changes within the supply-demand balance. Lack of energy security is thus linked to the negative economic and social impacts of either physical unavailability of energy, or prices that are not competitive or are overly volatile (US DOE, 2017).

Historically, energy security for many countries was primarily associated with the supply of crude oil. Thus using the international oil market as the example, prices are allowed to adjust in response to changes in supply and demand, the risk of physical unavailability is limited to extreme events. However, in many instances, petro-politics (some time called geo-politics) does play a role thereby jeopardizing security of crude oil imports (Speight, 2011). Thus supply security concerns are pri-marily related to the economic damage caused by extreme price spikes. The con-cern for physical unavailability of supply is more prevalent in energy markets where transmission systems must be kept in constant balance, such as electricity and, to some extent, natural gas. This is particularly the case in instances where there are capacity constraints or where prices are not able to work as an adjustment mechanism to balance supply and demand in the short term. In fact, analysis of vul-nerability for fossil fuel disruptions, e.g., is based not only on reserves within a country but also on risk factors such as net-import dependence and the political sta-bility of suppliers. Resilience factors include the number of entry points for a coun-try (e.g., ports and pipelines), the level of stocks, and the diversity of suppliers.

10.2.1 Reserves

Reserves are the amount of a resource available for recovery and/or production with the recoverable amount of natural gas being usually tied to the economic aspects of production.

In terms of reserves, natural gas is produced on all continents except Antarctica. The proven reserves of natural gas are of the order of 6588.8 trillion cubic feet (6588.8 Tcf; 1 Tcf = 1×10^{12} cubic feet) of which approximately 374 Tcf exist in the United States and Canada (BP, 2017). It should also be remembered that the total gas resource base (like any fossil fuel or mineral resource base) is dictated by economics. Therefore when resource data are quoted some attention must be given to the cost of recovering those resources. Most important, the economics must also include a cost factor that reflects the willingness to secure total, or a specific degree of, energy independence. The total proved reserves of natural gas are generally taken to be those quantities that geological and engineering information indicates with reasonable certainty can be recovered in the future from known reservoirs under existing economic and operating conditions.

A common misconception about natural gas is that resources are being depleted at an alarming rate and the supplies are quickly running out. In fact, there is a vast amount of natural gas estimated to still be retrieved from a variety of reservoirs, including tight reservoirs and shale reservoirs. However, many proponents of the rapid-depletion theory believe that price spikes indicate that natural gas resources are depleted beyond the point of no return. However, price spikes of any commodity are not always caused by waning resources but can be the outcome of other forces—including foreign economic forces—that are operative in the marketplace.

10.2.2 Energy security

Energy security is the continuous and uninterrupted availability of energy to a specific country or region. The security of energy supply conducts a crucial role in decisions that are related to the formulation of energy policy strategies. The economies of many countries are dependent upon the energy imports (or energy exports) in the notion that their balance of payments is affected by the magnitude of the vulnerability that the countries have in crude oil and natural gas purchases and/or sales (Speight, 2011).

Energy security has been an on-again-off-again political issue in the United States since the first Arab oil embargo in 1973. Since that time, the speeches of various Presidents and the Congress of the United States have continued to call for an end to the dependence on foreign oil and gas by the United States. The congressional rhetoric of energy security and energy independence continues but meaningful suggestions of how to address this issue remain few and far between. Natural gas offers the United States some relief from the geopolitical factors that occur regularly (some would say *every day*) that interfere with crude oil production and supplies.

It is to be hoped that, as difficult as it is because of a variety of factors, reporting data on natural gas reserves in the United States has become less of a political act that statement of crude oil production and reserves in many countries (Speight and Islam, 2016). Nevertheless, the energy literature and numerous statements by officials of oil-and-gas-producing and oil-and-gas-consuming countries indicate that the concept of energy security is elusive. Definitions of energy security range from uninterrupted oil supplies to the physical security of energy facilities to support for biofuels and renewable energy resources. Historically, experts and politicians referred to *security of oil supplies* as *energy security*. Only recently policy makers started to include natural gas supplies in the portfolio of energy definitions.

The past decade has yielded substantial change in the natural gas industry. Specifically, there has been rapid development of technology allowing the recovery of natural gas from shale formations. Since 2000, rapid growth in the production of natural gas from shale formations in North America has dramatically altered the global natural gas market landscape. Indeed, the emergence of shale gas is perhaps the most intriguing development in global energy markets in recent memory.

The security aspects of natural gas are similar, but not identical to those of crude oil. Compared with imports of crude oil, natural gas imports play a smaller role in most importing countries—mainly because it is less costly to transport liquid crude oil and petroleum products than natural gas. Natural gas is transported by pipeline over long distances because of the pressurization costs of transmission; the need to finance the cost of these pipelines encourages long-term contracts that dampen price volatility. In fact, despite some gloom-and-doom prognoses for the depletion of crude oil reserves in the near future, the structure of energy consumption is not expected to significantly change. A marginal reduction of the crude oil influence till 2030 (up to 32%) and equalization of the energy balance of crude oil, natural gas, and coal is expected up to the mid-point of the 21st century.

Moreover, beginning with the Barnett shale in northeast Texas, the application of innovative new techniques involving the use of horizontal drilling with hydraulic fracturing has resulted in the rapid growth in production of natural gas from shale (Speight, 2016). Knowledge of the shale gas resource is not new as geologists have long known about the existence of shale formations, and accessing those resources was long held in the geology community to be an issue of technology and cost. In the past decade, innovations have yielded substantial cost reductions, making shale gas production a commercial reality. In fact, shale gas production in the United States has increased from virtually nothing in 2000 to over 10 billion cubic feet per day (bcfd, 1×10^9 ft^3 per day) in 2010, and it is expected to more than quadruple by 2040, reaching 50% or more of total US natural gas production by the decade starting in 2030.

Natural gas—if not disadvantaged by government policies that protect competing fuels, such as coal—stands to play a very important role in the US energy mix for decades to come. Rising shale gas production has already delivered large beneficial impacts to the United States. Shale gas resources are generally located in close proximity to end-use markets where natural gas is utilized in fuel industry, generate electricity, and heat homes. This offers both security of supply and economic benefits.

The *Energy Independence and Security Act of 2007* (originally named the *Clean Energy Act of 2007*) is an Act of Congress concerning the energy policy of the United States. The stated purpose of the act is "to move the United States toward greater energy independence and energy security, to increase the production of clean renewable fuels, to protect consumers, to increase the efficiency of products, buildings, and vehicles, to promote research on and deploy greenhouse gas capture and storage options, and to improve the energy performance of the Federal Government, and for other purposes.

The bill originally sought to cut subsidies to the petroleum industry in order to promote petroleum independence and different forms of alternative energy. These tax changes were ultimately dropped after opposition in the Senate, and the final bill focused on automobile fuel economy, development of biofuels, and energy efficiency in public buildings and lighting. It was, and still is, felt by many observers that there should have been greater recognition of the role that natural gas can play in energy security. In fact, viewed from the perspective of the energy-importing countries as a whole, diversification in oil supplies has remained constant over the last decade, while diversification in natural gas supplies has steadily increased. Given the increasing importance of natural gas in world energy use, this is an indicator of an increase in overall energy security (Cohen et al., 2011).

However, natural gas is an attractive fuel, and its attraction is growing because of its clean burning characteristics, compared to oil or coal, and because of its price advantage, on an energy equivalent basis, compared to oil. Accordingly, there are predictions of significant future growth in natural gas consumption worldwide and growth in the trade of natural gas by bringing so-called *stranded gas* (including *shale gas*) to market.

Current trends suggest that natural gas will gradually become a global commodity with a single world market, just like oil, adjusted for transportation differences. The outcome of a global gas market is inevitable; once this occurs, the tendency will be toward a world price of natural gas, as with oil today, and the prices of oil and gas each will reach a global equivalence based on energy content (Deutch, 2010).

10.3 Emissions and pollution

Natural gas is an extremely important source of energy for reducing pollution and maintaining a clean and healthy environment. In addition, natural gas offers an abundant and secure source of energy in the United States as well as a number of environmental benefits over other sources of energy, particularly other fossil fuels. In fact, the role that natural gas can play in the future of global energy is inextricably linked to its ability to help address environmental problems. With an increase in the concerns about air quality and climate change looming large, natural gas offers many potential benefits if it is used to displace energy-producing fuels that cause more pollution. The flexibility that natural gas brings to an energy system can also make it a good fit for the rise of variable renewable source of energy.

Natural gas burns more cleanly than other fossil fuels. It has fewer emissions of sulfur, carbon, and nitrogen than coal or oil, and it has almost no ash particles left after burning. Being a clean fuel is one reason that the use of natural gas, especially for electricity generation, has grown so much and is expected to grow even more in the future.

Of course, there are environmental concerns with the use of any fuel. As with other fossil fuels, burning natural gas produces carbon dioxide, which is the most important greenhouse gas. Many scientists believe that increasing levels of carbon dioxide and other greenhouse gases in the Earth's atmosphere are changing the global climate.

Natural gas, although often cited as a relatively clean fuel, is also capable of producing emissions that are detrimental to the environment. While the major constituent of natural gas is methane, there are components such as carbon dioxide (CO), hydrogen sulfide (H_2S), and mercaptans (thiols; R-SH), as well as trace amounts of sundry other emissions. The fact that methane has a foreseen and valuable end-use makes it a desirable product, but in several other situations it is considered a pollutant, having been identified a greenhouse gas.

Also, a characteristic feature of natural gas that contains hydrogen sulfide is the presence of carbon dioxide (generally in the range of 1–4% v/v). In cases where the natural gas does not contain hydrogen sulfide, there may also be a relative lack of carbon dioxide. A sulfur removal process must be very precise, since natural gas contains only a small quantity of sulfur-containing compounds that must be reduced several orders of magnitude. Most consumers of natural gas require less than 4 ppm in the gas.

Some of the nonhydrocarbon constituents of natural gas would not be expected to biologically degrade as these substances do not contain the chemical structure that are amenable to microbial metabolism. For this reason, hydrogen, nitrogen, and carbon dioxide would not be susceptible to biodegradation. Furthermore, carbon dioxide is the final product in the biological mineralization of organic compounds. Although volatilization largely determines environmental distribution of hydrocarbon gases, some constituents have sufficient aqueous solubility that they might be present in aqueous environments at sufficient levels and/or sufficient times to make them potentially available for microbial metabolism. Higher molecular weight hydrocarbon derivatives that are typically found in gas condensate streams (such as the C_5 and C_6 hydrocarbon derivatives) have been shown to be inherently biodegradable in the environment. Not surprisingly, environmental and natural resource-related permits are required for oil and gas operations. A state may handle compliance under a general environmental impact statement. Therefore further environmental analysis may not need to be undertaken on individual oil and gas operations that obtain permits to operate.

In terms of the standard test methods that can be applied, it is of primary importance to recognize that while the major constituent of natural gas is methane, there are components such as carbon dioxide (CO), hydrogen sulfide (H_2S), and mercaptans (thiols; R-SH), as well as trace amounts of sundry other emissions. The fact that methane has a foreseen and valuable end-use makes it a desirable product, but in several other situations it is considered a pollutant, having been identified as one of several greenhouse gases. In addition, the test method may be abbreviated for the analysis of lean natural gases containing negligible amounts of hexanes and higher hydrocarbons, or for the determination of one or more components, as required (Speight, 2018).

Finally, it is necessary to address the issue of particulate matter (including the highly carbonaceous soot) which occurs as a complex emission that is classified as either *suspended particulate matter*, *total suspended particulate matter*, or simply *particulate matter*. Particulate matter is not typically analyzed as a constituent of natural gas but may be a conspicuous constituents of process gas, especially when the process gas is an effluent from a catalytic reactor. Through time, attrition of the catalyst can result in the formation of fine (micro-size) particle that will exist the reactor as part of the gas low. In fact, process gas emissions associated with petroleum refining are more extensive than methane and carbon dioxide and typically include process gases, petrochemical gases, volatile organic compounds (VOCs), carbon monoxide (CO), sulfur oxides (SO_x), nitrogen oxides (NO_x), particulates, ammonia (NH_3), and hydrogen sulfide (H_2S) that must be removed (Chapter 4: Composition and properties). These effluents may be discharged as air emissions and must be treated. Particulate matter emissions are typically determined through spectroscopic analysis of the captured particulate matter that has been collected on a sampling filter (Speight, 2018).

The main sources and pollutants of concern include VOCs) emissions from storage tanks during filling and due to tank breathing, floating roof seals in case of floating roof storage tanks, wastewater treatment units, Fischer–Tropsch (F-T) synthesis units, methanol synthesis units, and product upgrading units. Additional sources of fugitive emissions include nitrogen gas contaminated with methanol vapor from methanol storage facilities; methane (CH_4), carbon monoxide (CO), and hydrogen from synthesis gas (syngas) production units, and F-T or methanol synthesis units.

10.3.1 Greenhouse gas emissions

Global warming, or the *greenhouse effect*, is an environmental issue that deals with the potential for global climate change due to increased levels of atmospheric *greenhouse gases*. These are the gases in the atmosphere that serve to regulate the amount of heat that is kept close to the Earth's surface. It is speculated that an increase in these greenhouse gases will translate into increased temperatures around the globe, which would result in many disastrous environmental effects.

Exhaust gas emissions produced by the combustion of gas or other hydrocarbon fuels in turbines, boilers, compressors, pumps, and other engines for power and heat generation are a significant source of air emissions from natural gas processing facilities. Incineration of oxygenated by-products at gas-to-liquids (GTL) production facilities also generates carbon dioxide and NO_x emissions.

The principle greenhouse gases include water vapor, carbon dioxide, methane, nitrogen oxides, and some manufactures chemicals such as chlorofluorocarbons (CFCs). While most of these gases occur in the atmosphere naturally, levels have been increasing due to the widespread burning of fossil fuels by growing human populations. The reduction of greenhouse gas emissions has become a primary focus of environmental programs in many (but not all) countries around the world.

The majority of greenhouse gas emissions come from carbon dioxide directly attributable to the combustion of fossil fuels. Therefore reducing carbon dioxide emissions can play a huge role in combating the greenhouse effect and global warming. The combustion of natural gas emits almost 30% less carbon dioxide than oil, and just under 45% less carbon dioxide than coal.

One issue that has arisen with respect to natural gas and the greenhouse effect is the fact that methane, the principle component of natural gas, is itself a very potent greenhouse gas. In fact, methane has an ability to trap heat almost 21 times more effectively than carbon dioxide.

Sources of methane emissions include the waste management and operations industry, the agricultural industry, as well as leaks and emissions from the crude oil natural and gas industry itself. It is felt that the reduction in carbon dioxide emissions from increased natural gas use would strongly outweigh the detrimental effects of increased methane emissions. Thus the increased use of natural gas in the place of other, dirtier fossil fuels can serve to lessen the emission of greenhouse gases. Before describing the polluting nature of natural gas, it is worth reviewing the composition of the gas as a means of understanding the nature of the pollutants.

Briefly, natural gas is obtained principally from conventional crude oil and non-associated gas reservoirs, and secondarily from coal beds, tight sandstones, and Devonian shale. Some is also produced from minor sources such as landfills. In the not too distant future, natural gas may also be obtained from natural gas hydrate deposits located beneath the sea floor in deep water on the continental shelves or associated with thick subsurface permafrost zones in the Arctic.

While the primary constituent of natural gas is methane (CH_4), it may contain smaller amounts of other hydrocarbons, such as ethane (C_2H_6) and various isomers of propane (C_3H_8), butane (C_4H_{10}), and pentane (C_5H_{12}), as well as trace amounts of higher boiling hydrocarbons up to octane (C_8H_{18}). Nonhydrocarbon gases, such as carbon dioxide (CO_2), helium (He), hydrogen sulfide (H_2S), nitrogen (N_2), and water vapor (H_2O), may also be present (Chapter 4: Composition and properties). At the pressure and temperature conditions of the source reservoir, natural gas may occur as free gas (bubbles) or be dissolved in either crude oil or brine.

Pipeline-quality natural gas contains at least 80% methane and has a minimum heat content of 870 Btu per standard cubic foot (Chapter 4: Composition and properties). Most pipeline natural gas significantly exceeds both minimum specifications. Since natural gas has by far the lowest energy density of the common hydrocarbon fuels, by volume (not weight) much more of it must be used to provide a given amount of energy. Purified natural gas (specifically methane and *not* the higher boiling constituents) is also much less physically dense, weighing about half as much (55%) as the same volume of dry air at the same pressure. It is consequently buoyant in air, in which it is also combustible at concentrations ranging from 5% to 15% by volume.

10.3.2 Air pollutants

The Earth's atmosphere is a mixture primarily of the gases nitrogen and oxygen, totaling 99%; nearly 1% water; and very small amounts of other gases and substances, some of which are chemically reactive. With the exception of oxygen, nitrogen, water, and the inert gases, all constituents of air may be a source of concern owing either to their potential health effects on humans, animals, and plants, or to their influence on the climate.

The *gaseous pollutants* are carbon monoxide, nitrogen oxides, VOCs, and sulfur dioxide. These are reactive gases that in the presence of sunlight contribute to the formation of ground level ozone, smog, and acid rain. Methane, the principal ingredient in natural gas, is not classed as a VOC because it is not as chemically reactive as the other hydrocarbons, although it is a greenhouse gas.

The nongaseous *particulate matter* consists of metals and substances such as pollen, dust, and larger particles such as soot from wood fires or diesel fuel ignition.

The *greenhouse gases* are water vapor, carbon dioxide, methane, nitrous oxide, and a host of engineered chemicals, such as chlorofluorocarbons. These gases regulate the Earth's temperature and, when the natural balance of the atmosphere is disturbed particularly by an increase or decrease in the greenhouse gases, the Earth's climate could be affected.

The Earth's surface temperature is maintained at a habitable level through the action of certain atmospheric gases known as *greenhouse gases* that help trap the Sun's heat close to the Earth's surface. Most greenhouse gases occur naturally, but concentrations of carbon dioxide and other greenhouse gases in the Earth's atmosphere have been increasing since the Industrial Revolution with the increased combustion of fossil fuels and increased agricultural operations. Of late there has been concern that if this increase continues unabated, the ultimate result could be that more heat would be trapped, adversely affecting Earth's climate.

The major constituent of natural gas, methane, also directly contributes to the greenhouse effect. Its ability to trap heat in the atmosphere is estimated to be 21 times greater than that of carbon dioxide, so although methane emissions amount to only 0.5% of US emissions of carbon dioxide, they account for about 10% of the greenhouse effect of US emissions.

Water vapor is the most common greenhouse gas, at about 1% of the atmosphere by weight, followed by carbon dioxide at 0.04% and then methane, nitrous oxide, and man-made compounds such as the CFCs. Each gas has a different residence time in the atmosphere, from about a decade for carbon dioxide to 120 years for nitrous oxide and up to 50,000 years for some of the chlorofluorocarbons. Water vapor is omnipresent and continually cycles into and out of the atmosphere. In estimating the effect of these greenhouse gases on climate, both the global warming potential (heat-trapping effectiveness relative to carbon dioxide) and the quantity of gas must be considered for each of the greenhouse gases.

10.3.2.1 Emissions during exploration, production, and delivery

When exploration activities for natural gas reservoirs are on land, the activities may disturb vegetation and soil with the vehicles. Drilling a natural gas well on land may require clearing and leveling an area around the well site. Well drilling activities produce air pollution and may disturb people, wildlife, and water resources. Natural gas production can also produce large volumes of contaminated water which requires controlled handling, storage, and treatment so that there is no pollution of the land and other waters.

At sites where natural gas is produced at crude oil wells but is not economical to transport for sale or contains high concentrations of hydrogen sulfide (a toxic gas), it is burned (flared) at well sites. Natural gas flaring will give rise to products carbon dioxide, sulfur dioxide, nitrogen oxides, and many other compounds, depending on the chemical composition of the natural gas and on how well the natural gas burns in the flare. However, flaring is safer than releasing natural gas into the air and results in lower overall greenhouse gas emissions because carbon dioxide is not as strong a greenhouse gas as methane.

The extraction and production of natural gas, as well as other natural gas operations, do have environmental consequences and are subject to numerous laws and regulations. In some areas, development is completely prohibited so as to protect natural habitats, wetlands, and designated wilderness areas. Natural gas production is an industrial activity with the potential for environmental consequences most of which can be identified by application of a variety of standard test methods (Speight, 2014, 2018). Effects can range from water contamination related to drilling and disposal of drilling fluids, air quality degradation from internal combustion engines on drill rigs and trucks, excess dust from equipment transportation, impacts to solitude and night skies from noise and lighting, and safety concerns associated with the large number of vehicles needed to support drilling operations. While the horizontal drilling and hydraulic fracturing practices (Speight, 2016) expected to be used in developing a natural gas resource may have the potential to exert negative environmental effects on the surrounding area, when compared to development of conventional oil and gas resources this development method could result in fewer impacts than conventional vertical wells due to greater flexibility in the location of the well(s).

Hydraulic fracturing (commonly referred to as *fracking*, *fracing*, or *hydrofracking*) of shale, sandstone, and carbonate rock formations is used for production personnel and equipment to gain access to large reserves of natural gas that were previously too expensive to develop significantly reduce the land area that is disturbed to develop natural gas resources (Speight, 2016).

Determining which fracking technique is appropriate for well productivity depends largely on the properties of the reservoir rock from which to extract oil or gas. If the rock is characterized by low permeability, the technique involves drilling a borehole vertically until it reaches a lateral shale rock formation, at which point the drill turns to follow the rock for hundreds or thousands of feet in a horizontal direction (Speight, 2016). In contrast, conventional oil and gas sources are characterized by higher rock permeability, which naturally enables the flow of oil or gas into the wellbore with less intensive hydraulic fracturing techniques than the production of tight gas has required.

The procedure involves pumping liquids under high pressure into a well to fracture the rock, which allows natural gas to escape from the rock. Producing natural gas with this technique has some effects on the environment but, if approached correctly using a multidisciplinary team, the effects on the environment can be minimized. The associated horizontal and directional drilling techniques make it possible to produce more natural gas from a single well than in the past, so fewer wells are necessary to develop a natural gas field. It is necessary however, to advise caution when using hydraulic fracturing.

For example, poor well design construction and maintenance practices—that may be caused by the absence of the presence of a multidisciplinary team—can (or will) increase the chance of a well leak or blowout, so it is extremely important that every aspect of the life cycle of a well—from drilling to completion to well abandonment plugging—is properly executed to reduce any threat to the environment. In addition, the techniques require large amounts of water (and sand or other proppant that keeps the fractures open) which, in some areas of the United States, may affect aquatic habitats and the availability of water for other uses. Also, the hydraulic fracturing fluid, which may contain potentially hazardous chemicals that could be released through spills, leaks, faulty well construction, or other exposure pathways leads to contamination of the surrounding area. The large amounts of wastewater produced at the surface by hydraulic fracturing produces may contain dissolved chemicals and other contaminants that require treatment before disposal or reuse. Because of the quantities of water produced and the complexities inherent in treating some of the wastewater components, the correct treatment and disposal of the wastewater are important.

Any person involved in any aspect of natural exploration, production., processing, and use can (and should) play an important role in the environmental aspects of natural gas production are an important part defining the characteristics of natural gas insofar as it: (1) exist in a gaseous state at room temperature, (2) contains predominantly hydrocarbon compounds with one to four carbon atoms, which are the dominant hazard in the substance, and (3) that after additional fractionation, are sold on the open market in the United States as fungible products. The physical and chemical characteristics require that they be identified through the application of standard test methods that provide reliable data.

The environmental side effects of natural gas production start in what is called the upstream portion of the natural gas industry, beginning with selection of a geologically promising area for possible future natural gas production. An upstream firm will collect all available existing information on the geology and natural gas potential of the proposed area and may decide to conduct new geologic and geophysical studies.

Following analysis of the geologic and geophysical data, permission to drill and produce natural gas from owners of the land and relevant government permitting authorities is necessary. In making leasing and permitting decisions, the potential environmental impacts of future development are often considered. Such considerations include the projected numbers and extent of wells and related facilities, such as pipelines, compressor stations, water disposal facilities, as well as roads and power lines. For example, laying pipelines that transport natural gas from wells usually requires clearing land to bury the pipe. Natural gas wells and pipelines often have engines to run equipment and compressors, which produce air pollutants and noise.

Drilling a gas well involves preparing the well site by constructing a road to it if necessary, clearing the site, and flooring it with wood or gravel. The soil under the road and the site may be so compacted by the heavy equipment used in drilling as to require compaction relief for subsequent farming. In wetland areas, drilling is often accomplished using a barge-mounted rig that is floated to the site after a temporary slot is cut through the levee bordering the nearest navigable stream. However, the primary environmental concern directly associated with drilling is not the surface site but the disposal of drilling waste (spent drilling mud and rock cuttings, etc.).

Drilling of a typical gas well (6000 ft deep) results in the production of about 150,000 pounds of rock cuttings and at least 470 barrels of spent mud. Early industry practice was to dump spent drilling fluid and rock cuttings into pits dug alongside the well and just plow them over after drilling was completed or dump the cuttings directly into the ocean, if the operations are offshore. Currently, the operator must not discharge drilling fluids and solids without permission and it has to be determined whether such waste can be discharged or shipped to a special disposal facility.

Significant volumes of natural gas can also be produced from tight (low permeability) sandstone reservoirs (often referred to arbitrarily as *tight gas*) or shale reservoirs (often referred to as *shale gas*) and coal seams, all of which are unconventional reservoir rocks (Chapter 1: History and use and Chapter 3: Unconventional gas). Through the application of analytical programs (Speight, 2018), a better understanding of the underlying physics of natural gas production from these rocks and the relationship between well completion practices and productivity will be achieved (Speight, 2018). Using the various analytical methods, it will be possible to find ways to maximize recovery of natural gas from these rock formations which, in the mid-to-late 20th century, were considered to be unproductive. Emphasis (when resources of shale gas are to be developed) is typically (and rightly so) often placed on water management during resource development, including the enhancement of water treatment technologies. Resource development based on nothing more than emotional issues rather than on reliable analytical data is to be treated with extreme caution.

Exploration, development, and production activities emit small volumes of air pollutants, mostly from the engines used to power drilling rigs and various support and construction vehicles. As the number of wells increases, such as in the Gulf of Mexico, so do the emissions for exploratory drilling and development drilling, while emissions from supporting activities rise less directly. Offshore development entails some activities not found elsewhere (i.e., platform construction and marine support vessels), but the environmental effects from onshore activities, which include drilling pad and access road construction, especially for development drilling, are many times larger because of the much higher level of activity.

This practice is by no means as common as it was four decades ago and the minor venting and flaring that does occur now is regulated and may happen at several locations: the well gas separator, the lease tank battery gas separator, or a downstream natural gas plant. Whatever the reason for its use, flaring is wasteful economically and harmful environmentally. In many cases, efforts are made (or, should be made) to capture that natural gas, rather than burning it off.

All systems of pipes that transmit any fluid are subject to leaks and, in the case of natural gas, any leak will escape to the atmosphere. Thus throughout the entire process of producing, refining, and distributing natural gas there are losses or fugitive emissions. Production operations account for about 30% of the fugitive emissions, while transmission, storage, and distribution accounts for about 53% of the fugitive emissions.

At onshore and coastal sites, drilling wastes usually cannot be discharged to surface waters and are primarily disposed of by operators on their lease sites. If the drilling fluids are saltwater- or oil-based, they can cause damage to soils and groundwater and on-site disposal is often not permitted, so operators must dispose of such wastes at an off-site disposal facility. The disposal methods used by commercial disposal companies include underground injection, burial in pits or landfills, land spreading, evaporation, incineration, and reuse/recycling. In areas with subsurface salt formations, disposal in man-made salt caverns is an emerging, cost-competitive option. Such disposal poses very low risks to plant and animal life because the formations where the caverns are constructed are very stable and are located beneath any subsurface freshwater supplies. Water-based drilling wastes, if they can be shown to have minimal impact on aquatic life, offshore operators are allowed to discharge such waste into the sea. However, discharging oil-based drilling wastes into the sea is often prohibited and these are generally transported to shore for disposal.

If the drilling fluids are saltwater- or oil-based, they can cause damage to soils and groundwater and on-site disposal is often not permitted, so operators must dispose of such wastes at an off-site disposal facility. The disposal methods include underground injection, burial in pits or landfills, land spreading, evaporation, incineration, and reuse/recycling. In areas with subsurface salt formations, disposal in man-made salt caverns may be permitted; this form of disposal poses very low risks to plant and animal life because the formations where the caverns are constructed are very stable and are located beneath any subsurface freshwater supplies.

The disposal of produced water is often a significant problem for the drilling and production industry. The disposal process varies depending on whether the well is onshore or coproduction of a variable amount of water with the gas is offshore, the local requirements, and the composition of the unavoidable at most locations. Because the water is usually produced fluids, most onshore-produced water is disposed salty, its raw disposal or unintentional spillage on land of by pumping it back into the subsurface through on-site normally interferes with plant growth. In fact, injection water represents the largest volume waste stream generated and its disposal is not practical or economically viable and the produced water is piped or transported to an off-site treatment facility.

In recent years, new drilling technologies such as slim-hole drilling, horizontal drilling, multilateral drilling, coiled tubing drilling, and improved drill bits, have helped to reduce the generated quantity of drilling wastes. Another advanced drilling technology that provides pollution-prevention benefits is the use of synthetic drilling fluids that have a less severe environmental impact and their use results in a much cleaner well bore and less sidewall collapse.

In terms of transportation, there is also the potential for the occurrence of environmental contamination by natural gas.

Natural gas is distributed mainly via pipeline—in the United States more than one million miles of underground pipelines are connected between natural gas fields and major cities. Natural gas can be liquefied by cooling it to $-260°F$ (or, $-162°C$). The liquefied natural gas (LNG) has much more condensed volume which is 615 times lesser in volume compared to the room temperature natural gas and LNG has the fluidity and volume compactness just as other liquid fuels. Therefore LNG is much easier to store or transport and LNG in special tanks can be transported by trucks or ships. In this regard, LNG has some needed characteristics that it can be used as a transportation fuel.

At any point of this transportation train, it must be recognized that there is the potential for leakage and spillage of the gas.

10.3.2.2 Emissions during processing

Gas processing usually poses low environmental risk, primarily because natural gas has a simple and comparatively pure composition. Typical processes performed by a gas plant are separation of the heavier than methane hydrocarbons as liquefied petroleum gases, stabilization of condensate by removal of lighter hydrocarbons from the condensate stream, gas sweetening, and consequent sulfur production and dehydration sufficient to avoid formation of methane hydrates in the downstream pipeline.

Air pollutant emission points at natural gas processing facilities are the glycol dehydration unit reboiler vent, storage tanks and equipment leaks from components handling hydrocarbon streams that contain hazardous air pollutants (HAPs).

Thus potential environmental aspects associated with natural gas processing include the following: (1) air emissions, (2) wastewater, (3) hazardous materials, (4) wastes, and (5) noise. In terms of air emissions, fugitive emissions in natural gas processing facilities are associated with leaks in tubing; valves; connections; flanges; packings; open-ended lines; floating roof storage tank, pump, and compressor seals; gas conveyance systems, pressure relief valves, tanks or open pits/containments, and loading and unloading operations of hydrocarbons.

The identified HAP emission points at natural gas processing plants are the glycol dehydration unit reboiler vent, storage tanks and equipment leaks from components handling hydrocarbon streams that contain HAPs. Other potential HAP emission points are the tail gas streams from amine-treating processes and sulfur recovery units.

Methods vary for removing natural gas contaminants, such as hydrogen sulfide gas, carbon dioxide gas, nitrogen, and water (Chapter 6: History of gas processing and Chapter 7: Process classification). Commonly the hydrogen sulfide is converted to solid sulfur for sale (Chapter 6: History of gas processing and Chapter 7: Process classification). Likewise, the carbons and nitrogen are separated for sale to the extent economically possible but otherwise the gases are vented, while the water is treated before release. Compressor operation at gas plants has a similar impact to that of compressors installed at other locations.

It is sometimes necessary either to vent produced gas into the atmosphere or to flare (burn) it. Worldwide, most venting and flaring occurs when the cost of transporting and marketing gas coproduced from crude oil reservoirs exceeds the netback price received for the gas.

Emissions will result from gas sweetening plants only if the acid waste gas from the amine process is flared or incinerated. Most often, the acid waste gas is used as a feedstock in nearby sulfur recovery or sulfuric acid plants.

When flaring or incineration is practiced, the major pollutant of concern is sulfur dioxide. Most plants employ elevated smokeless flares or tail gas incinerators for complete combustion of all waste gas constituents, including virtually 100% conversion of hydrogen sulfide to sulfur dioxide. Little particulate, smoke, or hydrocarbons result from these devices, and because gas temperatures do not usually exceed 650°C (1200°F), significant quantities of nitrogen oxides are not formed. Some plants still use older, less-efficient waste gas flares. Because these flares usually burn at temperatures lower than necessary for complete combustion, larger emissions of hydrocarbons and particulate, as well as hydrogen sulfide, can occur.

This practice of venting is now by no means as common as it was a few decades ago when oil was the primary valuable product and there was no market for much of the coproduced natural gas. The venting and flaring that does occur now is regulated and may happen at several locations: the well gas separator, the lease tank battery gas separator, or a downstream natural gas plant.

10.3.2.3 Emissions during combustion

In theory, and often but not always in practice, natural gas burns more cleanly than other fossil fuels. It has fewer emissions of sulfur, carbon, and nitrogen than coal or oil, and it has almost no ash particles left after burning. Being a clean fuel is one reason that the use of natural gas, especially for electricity generation, has grown so much and is expected to grow even more in the future.

On a relative basis, natural gas is the cleanest of all the fossil fuels. Composed primarily of methane, the main products of the combustion of natural gas are carbon dioxide and water vapor, the same compounds we exhale when we breathe. Coal and petroleum are composed of much more complex molecules, with a higher carbon ratio and higher nitrogen and sulfur contents. Thus when combusted, coal and oil release higher levels of harmful emissions, including a higher ratio of carbon emissions, NO_x, and sulfur dioxide (SO_2). Coal and fuel oil also release ash particles into the environment, substances that do not burn but instead are carried into the atmosphere and contribute to pollution. The combustion of natural gas, on the other hand, releases very small amounts of sulfur dioxide and nitrogen oxides, virtually no ash or particulate matter, and lower levels of carbon dioxide, carbon monoxide, and other reactive hydrocarbons.

Natural gas is less chemically complex than other fuels, has fewer impurities, and its combustion accordingly results in less pollution. In the simplest case, complete combustive reaction of a molecule of pure methane (CH_4) with two molecules of pure oxygen produces a molecule of carbon dioxide gas, two molecules of water in vapor form, and heat.

$$CH_4 + 2O_2 \rightarrow CO_2 + 2H_2O + \text{Heat}$$

In practice, the combustion process is not always perfect and when the air supply is inadequate, carbon monoxide and particulate matter (soot) are also produced. In fact, since natural gas is never pure methane and small amounts of additional impurities are present, pollutants are also generated during combustion. Thus the combustion of natural gas also produces undesirable compounds but in significantly lower quantities compared to the combustion of coal, petroleum, and petroleum products.

The particulates produced by natural gas combustion are usually less than 1 μm (micrometer) in diameter and are composed of low-molecular weight hydrocarbons that are not fully combusted.

Pollutant emissions from the industrial sector and electric utilities contribute greatly to environmental problems in the United States. The use of natural gas to power both industrial boilers and processes and the generation of electricity can significantly improve the emissions profiles for these two sectors.

Natural gas is becoming an increasingly important fuel in the generation of electricity. As well as providing an efficient, competitively priced fuel for the generation of electricity, the increased use of natural gas allows for the improvement in the emissions profile of the electric generation industry. Power plants in the United States account for 67% of sulfur dioxide emissions, 40% of carbon dioxide emissions, 25% of nitrogen oxide emissions, and 34% of mercury emissions (National Environmental Trust, 2002, "Cleaning up Air Pollution from America's Power Plants"). Coal-fired power plants are the greatest contributors to these types of emissions. In fact, only 3% of sulfur dioxide emissions, 5% of carbon dioxide emissions, 2% of nitrogen oxide emissions, and 1% of mercury emissions come from noncoal-fired power plants.

Essentially, electric generation and industrial applications that require energy, particularly for heating, use the combustion of fossil fuels for that energy. Because of its clean burning nature, the use of natural gas wherever possible, either in conjunction with other fossil fuels, or instead of them, can help to reduce the emission of harmful pollutants.

10.4 Smog and acid rain

Acid rain occurs when the oxides of nitrogen and sulfur that are released to the atmosphere during the combustion of fossil fuels are deposited (as soluble acids) with rainfall, usually at some location remote from the source of the emissions. It is generally believed (the chemical thermodynamics is favorable) that acidic compounds are formed when sulfur dioxide and nitrogen oxide emissions are released from tall industrial stacks. Gases such as sulfur oxides (usually sulfur dioxide, SO_2) as well as the NO_x react with the water in the atmosphere to form acids:

$$SO_2 + H_2O \rightarrow H_2SO_3$$

$$2SO_2 + O_2 \rightarrow 2SO_3$$

$$SO_3 + H_2O \rightarrow H_2SO_4$$

$$2NO + H_2O \rightarrow 2HNO_2$$

$$2NO + O_2 \rightarrow 2NO_2$$

$$NO_2 + H_2O \rightarrow HNO_3$$

Acid rain has a pH less than 5.0 and predominantly consists of sulfuric acid (H_2SO_4) and nitric acid (HNO_3). As a point of reference, in the absence of anthropogenic pollution sources the average pH of rain is −6.0 (slightly acidic; neutral pH = 7.0). In summary, the sulfur dioxide that is produced during a variety of processes will react with oxygen and water in the atmosphere to yield environmentally detrimental sulfuric acid. Similarly, nitrogen oxides will also react to produce nitric acid. Smog and acid rain might be considered to be the end-point of emissions from the use of natural gas.

Particulate emissions also cause the degradation of air quality in the United States. These particulates can include soot, ash, metals, and other airborne particles.

Another acid gas, hydrogen chloride (HCl), although not usually considered to be a major emission, is produced from mineral matter and the brines that often accompany petroleum during production and is gaining increasing recognition as a contributor to acid rain. However, hydrogen chloride may exert severe local effects because it does not need to participate in any further chemical reaction to become an acid. Under atmospheric conditions that favor a buildup of stack emissions in the areas where hydrogen chloride is produced, the amount of hydrochloric acid in rainwater could be quite high.

In addition to hydrogen sulfide and carbon dioxide, natural gas may contain other contaminants, such as mercaptans (R-SH) and carbonyl sulfide (COS). The presence of these impurities may eliminate some of the sweetening processes since some processes remove large amounts of acid gas but not to a sufficiently low concentration. On the other hand, there are those processes that are not designed to remove (or are incapable of removing) large amounts of acid gases. However, these processes are also capable of removing the acid gas impurities to very low levels when the acid gases are there in low to medium concentrations in the gas.

At the global level there is concern that the increased use of any hydrocarbon-based fuels will ultimately raise the temperature of the planet (*global warming*), as carbon dioxide reflects the infrared or thermal emissions from the Earth, preventing them from escaping into space (*greenhouse effect*). If the potential for global warming becomes real will depend upon how emissions into the atmosphere are handled. There is considerable discussion about the merits and demerits of the global warming theory and the discussion is likely to continue for some time. Be that as it may, the atmosphere can only tolerate pollutants up to a limiting value. And that value needs to be determined. In the meantime, efforts must be made to curtail the use of noxious and foreign (nonindigenous) materials into the air.

There are a variety of processes which are designed for sulfur dioxide removal from gas streams (Speight, 2014) but scrubbing process utilizing limestone ($CaCO_3$) or lime [$Ca(OH)_2$] slurries have received more attention than other gas scrubbing processes (Chapter 7: Process classification and Chapter 8: Gas cleaning processes). Most of the gas scrubbing processes are designed to remove sulfur dioxide from the gas streams; some processes show the potential for removal of nitrogen oxide(s).

Natural gas pipelines and storage facilities have a very good safety record. This is very important because when natural gas leaks it can cause explosions. Since raw natural gas has no odor, natural gas companies add a smelly substance to it so that people will know if there is a leak. If you have a natural gas stove, you may have smelled this "rotten egg" smell of natural gas when the pilot light has gone out.

Natural gas has many uses, residentially, commercially, and industrially. Found in reservoirs underneath the Earth, natural gas is commonly associated with oil deposits. Production companies search for evidence of these reservoirs by using sophisticated technology that helps to find the location of the natural gas, and drill wells in the Earth where it is likely to be found.

Once brought from underground, the natural gas is refined to remove impurities like water, other gases, sand, and other compounds. Some hydrocarbons are removed and sold separately, including propane and butane. Other impurities are also removed, like hydrogen sulfide (the refining of which can produce sulfur, which is then also sold separately). After refining, the clean natural gas is transmitted through a network of pipelines, thousands of miles of which exist in the United States alone.

Global warming or *the greenhouse effect* is an environmental issue that deals with the potential for global climate change due to increased levels of atmospheric "greenhouse gases." There are certain gases in our atmosphere that serve to regulate the amount of heat that is kept close to the Earth's surface. Scientists theorize that an increase in these greenhouse gases will translate into increased temperatures around the globe, which would result in many disastrous environmental effects. In fact, the Intergovernmental Panel on Climate Change (IPCC) predicts in its "Third Assessment Report" released in February 2001 that over the next 100 years, global average temperatures will rise by between 2.4°F and 10.4°F.

The principle greenhouse gases include water vapor, carbon dioxide, methane, nitrogen oxides, and some engineered chemicals such as chlorofluorocarbons. While most of these gases occur in the atmosphere naturally, levels have been increasing due to the widespread burning of fossil fuels by growing human populations. The reduction of greenhouse gas emissions has become a primary focus of environmental programs in countries around the world.

One of the principle greenhouse gases is carbon dioxide. Although carbon dioxide does not trap heat as effectively as other greenhouse gases (making it a less potent greenhouse gas), the sheer volume of carbon dioxide emissions into the atmosphere is very high, particularly from the burning of fossil fuels. In fact, according to the EIA in its report "Emissions of Greenhouse Gases in the United States 2000," 81.2% of greenhouse gas emissions in the United States in 2000 came from carbon dioxide directly attributable to the combustion of fossil fuels.

Because carbon dioxide makes up such a high proportion of US greenhouse gas emissions, reducing carbon dioxide emissions can play a huge role in combating the greenhouse effect and global warming. The combustion of natural gas emits almost 30% less carbon dioxide than oil, and just under 45% less carbon dioxide than coal.

One issue that has arisen with respect to natural gas and the greenhouse effect is the fact that methane, the principle component of natural gas, is itself a very potent greenhouse gas. In fact, methane has an ability to trap heat almost 21 times more effectively than carbon dioxide.

According to the Energy Information Administration, although methane emissions account for only 1.1% of total US greenhouse gas emissions, they account for 8.5% of the greenhouse gas emissions based on global warming potential. Sources of methane emissions in the United States include the waste management and operations industry, the agricultural industry, as well as leaks and emissions from the oil and gas industry itself. A major study performed by the Environmental Protection Agency (EPA) and the Gas Research Institute (GRI) in 1997 sought to discover whether the reduction in carbon dioxide emissions from increased natural gas use would be offset by a possible increased level of methane emissions. The study concluded that the reduction in emissions from increased natural gas use strongly outweighs the detrimental effects of increased methane emissions. Thus the increased use of natural gas in the place of other, dirtier fossil fuels can serve to lessen the emission of greenhouse gases in the United States.

Smog and poor air quality is a pressing environmental problem, particularly for large metropolitan cities. Smog, the primary constituent of which is ground level ozone, is formed by a chemical reaction of carbon monoxide, nitrogen oxides, VOCs, and heat from sunlight. As well as creating that familiar smoggy haze commonly found surrounding large cities, particularly in the summertime, smog, and ground level ozone can contribute to respiratory problems ranging from temporary discomfort to long-lasting, permanent lung damage. Pollutants contributing to smog come from a variety of sources, including vehicle emissions, smokestack emissions, paints, and solvents. Because the reaction to create smog requires heat, smog problems are the worst in the summertime.

The use of natural gas does not contribute significantly to smog formation, as it emits low levels of nitrogen oxides, and virtually no particulate matter. For this reason, it can be used to help combat smog formation in those areas where ground level air quality is poor. The main sources of nitrogen oxides are electric utilities, motor vehicles, and industrial plants. Increased natural gas use in the electric generation sector, a shift to cleaner natural gas vehicles, or increased industrial natural gas use, could all serve to combat smog production, especially in urban centers where it is needed the most. Particularly in the summertime, when natural gas demand is lowest and smog problems are the greatest, industrial plants and electric generators could use natural gas to fuel their operations instead of other, more polluting fossil fuels. This would effectively reduce the emissions of smog causing chemicals, and result in clearer, healthier air around urban centers. For instance, a 1995 study by the Coalition for Gas-Based Environmental Solutions found that in the Northeast, smog and ozone-causing emissions could be reduced by 50%−70% through the seasonal switching to natural gas by electric generators and industrial installations.

Particulate emissions also cause the degradation of air quality in the United States. These particulates can include soot, ash, metals, and other airborne particles. A study (Union of Concerned Scientists, 1998, "Cars and Trucks and Air Pollution") showed that the risk of premature death for residents in areas with high airborne particulate matter was 26% greater than for those in areas with low particulate levels. Natural gas emits virtually no particulates into the atmosphere, in fact, emissions of particulates from natural gas combustion are 90% lower than from the combustion of oil, and 99% lower than burning coal.

Acid rain is another environmental problem that affects much of the Eastern United States, damaging crops, forests, wildlife populations, and causing respiratory and other illnesses in humans. Acid rain is formed when sulfur dioxide and nitrogen oxides react with water vapor and other chemicals in the presence of sunlight to form various acidic compounds in the air. The principle source of acid rain causing pollutants, sulfur dioxide and nitrogen oxides, are coal-fired power plants. Since natural gas emits virtually no sulfur dioxide, and up to 80% less nitrogen oxides than the combustion of coal, increased use of natural gas could provide for fewer acid rain causing emissions.

10.5 Natural gas regulations

No matter what one can hear from various pundits, it is advantageous to accept that natural gas is dangerous—when mixed with air, it can form an explosive mixture at specific low concentrations, between a lower explosion limit and an upper explosion limit. However, it can burn as a flame at the point of leakage. In an open, land installation the gas normally disperses quickly. However, in a closed system, simple explosions and boiling liquid, expanding vapor explosions can (will) cause serious damage. The latter is caused if a flame impinges on a vessel containing liquefied gas. Adequate ventilation is required by all safety codes for all enclosed spaces. It is possible to prevent a catastrophe by a chain of preventive measures. Insofar as the condition of the environment is important to floral and faunal life on Earth, any serious disadvantageous disturbance of the environment (pollution) can have serious consequences for continuation of this life.

Furthermore, all carbon-based fuels produce carbon dioxide and water when they are burned. Because of the higher atomic ratio of hydrogen to carbon atom, natural gas produces less carbon dioxide per unit of energy than higher molecular weight hydrocarbon derivatives (such as crude oil constituents) or coal. Also, natural gas impurities can be selectively and relatively easily removed and be completely converted by combustion, resulting in lower emissions of particulate matter. However, any releases of methane and ethane to the atmosphere should be minimized.

The gas industry has been highly regulated for many years mainly as it was regarded as a natural monopoly. In the last 30 years there has been a move away from price regulation and toward liberalization of natural gas markets. These movements have resulted in greater competition in the market and in a dynamic and innovative natural gas industry. Therefore a brief history of the regulation of the natural gas industry is necessary in this introductory chapter.

The regulation and deregulation of the natural gas industry is a 20th century phenomenon in the United States. Since regulation and deregulation are related to use of the commodity, it is appropriate that a brief review of the regulation and deregulation of the natural gas industry be presented in this chapter.

10.5.1 Historical aspects

The regulation of natural gas in the United States dates back to the beginnings of the industry. In the early days of the industry (mid-1800s), natural gas was in limited supply and fuel gas (methane) was manufactured from coal, to be delivered locally, generally within the same municipality in which it was produced. Local governments, seeing the natural monopoly characteristics of the natural gas market at the time, deemed natural gas distribution a business that affected the public interest to a sufficient extent to merit regulation. Because of the distribution network that was needed to deliver natural gas to customers, it was decided that one company with a single distribution network could deliver natural gas more cheaply than two companies with overlying distribution networks and markets. However, economic theory does not predict but *dictates* that a company in a monopoly position, with total control over its market and the absence of any competition will typically take advantage of its position and has incentives to charge overly high prices. The solution, from the point of view of the local governments, was to regulate the rates these natural monopolies charged, and set down regulations that prevented them from abusing their market power.

As the natural gas industry developed, so did the complexity of maintaining regulation. In the early 1900s, natural gas began to be shipped between municipalities. Thus natural gas markets were no longer segmented by municipal boundaries. The first intrastate pipelines began carrying gas from city to city. This new mobility of natural gas meant that local governments could no longer oversee the entire natural gas distribution chain. There was, in essence, a regulatory gap between municipalities. In response to this, state-level governments intervened to regulate the new *intrastate* natural gas market and determine rates that could be charged by gas distributors. This was done by creating public utility commissions and public service commissions to oversee the regulation of natural gas distribution. The first states to do so were New York and Wisconsin, which instituted commissions as early as 1907.

With the advent of technology that allowed the long-distance transportation of natural gas via interstate pipelines, new regulatory issues arose. In the same sense that municipal governments were unable to regulate natural gas distribution that extended beyond their areas of jurisdiction, the state governments were unable to regulate interstate natural gas pipelines.

Between 1911 and 1928, several states attempted to assert regulatory oversight of these interstate pipelines. However, in a series of decisions, the Supreme Court of the United States ruled that such state oversight of interstate pipelines violated interstate commerce because interstate pipeline companies were beyond the regulatory power of state-level government. Without any federal legislation dealing with interstate pipelines, these decisions essentially left interstate pipelines completely unregulated, the second regulatory gap. However, due to concern regarding the monopoly power of interstate pipelines, as well as conglomeration of the industry, the federal government saw fit to step in to fill the regulatory gap created by interstate pipelines.

In 1935, the US Federal Trade Commission stated some concern over the market power that may be exerted by merged electric and gas utilities. By this time, over a quarter of the interstate natural gas pipeline network was owned by only 11 holding companies; companies that also controlled a significant portion of gas production, distribution, and electricity generation. In response to this report, in 1935 Congress passed the Public Utility Holding Company Act to limit the ability of holding companies to gain undue influence over a public utility market. However, the law did not cover the regulation of interstate gas sales.

In the United States, the regulation of natural gas production has traditionally occurred primarily at the state level with most states that produce natural gas and crude oil issuing more rigorous standards that take primacy over federal regulations, as well as additional regulations that control areas not covered at the federal level, such as hydraulic fracturing. Within states, regulation is carried out by a range of agencies.

The specific regulations vary considerably among states, such as different depths for well casing, levels of disclosure on drilling and fracturing fluids, or requirements for water storage. Currently, many states that produce natural gas and crude oil have varying hydraulic fracturing regulations, specifically regulations related to (1) the disclosure of the components of hydraulic fracturing fluids, (2) the proper casing of wells to prevent aquifer contamination, and (3) management of wastewater from flow-back and produced water. The disposal of wastewater by underground injection has emerged as a concern for state regulators due to large interstate flows of wastewater to states with suitable geology for the water disposal as well as the potential for seismic activity near some well sites.

Techniques for treating industrial process wastewater in this sector include source segregation and pretreatment of concentrated wastewater streams. Typical wastewater treatment steps include: grease traps, skimmers, dissolved air floatation, or oil/water separators for separation of oils and floatable solids; filtration for separation of filterable solids; flow and load equalization; sedimentation for suspended solids reduction using clarifiers; biological treatment, typically aerobic treatment, for reduction of soluble organic matter biological oxygen demand (BOD); chemical or biological nutrient removal for reduction in nitrogen and phosphorus; chlorination of effluent when disinfection is required; and dewatering and disposal of residuals in designated hazardous waste landfills. Additional engineering controls may be required for (1) containment and treatment of volatile organics stripped from various unit operations in the wastewater treatment system, (2) advanced metals removal using membrane filtration or other physical/chemical treatment technologies, (3) removal of recalcitrant organics, cyanide, and nonbiodegradable chemical oxygen demand (COD) using activated carbon or advanced chemical oxidation, (4) reduction in effluent toxicity using appropriate technology (such as reverse osmosis, ion exchange, and activated carbon), and (5) containment and neutralization of nuisance odors.

Natural gas processing facilities use and manufacture significant amounts of hazardous materials, including raw materials, intermediate/final products, and byproducts. The handling, storage, and transportation of these materials should to be managed properly to avoid or minimize the environmental impacts from these hazardous materials.

Nonhazardous industrial wastes consist mainly of exhausted molecular sieves from the air separation unit as well as domestic wastes. Other nonhazardous wastes may include office and packaging wastes, construction rubble, and scrap metal.

Hazardous waste should be determined according to the characteristics and source of the waste materials and applicable regulatory classification. In GTL facilities, hazardous wastes may include biosludge; spent catalysts; spent oil, solvents, and filters (e.g., activated carbon filters and oily sludge from oil water separators); used containers and oily rags; mineral spirits; used sweetening; spent amines for CO_2 removal; and laboratory wastes.

Spent catalysts from GTL production are generated from scheduled replacements in natural gas desulfurization reactors, reforming reactors and furnaces, F-T synthesis reactors, and reactors for mild hydrocracking. Spent catalysts may contain zinc, nickel, iron, cobalt, platinum, palladium, and copper, depending on the particular process.

The principal sources of noise in natural gas processing facilities include large rotating machines (e.g., compressors, turbines, pumps, electric motors, air coolers, and fired heaters). During emergency depressurization, high noise levels can be generated due to release of high-pressure gases to flare and/or steam release into the atmosphere.

In terms of the development of unconventional natural gas and crude oil resources, a major environmental concern is the potential contamination of water courses. Apart from water, which makes up 99.5% v/v of the hydraulic fracturing fluid, the fluid also contains chemical additives to improve the process performance. The additives are varied and can include acid, friction reducer, surfactant, gelling agent, and scale inhibitor (Speight, 2016). The composition of the fracturing fluid is tailored to differing geological features and reservoir characteristics to address challenges, including scale build-up, bacterial growth, and proppant transport. Unfortunately, in the past many of the chemical compounds used in the hydraulic fracturing process lacked scientifically based maximum contaminant levels, making it more difficult to quantify their risk to the environment. Moreover, uncertainty about the chemical make-up of fracturing fluids persists because of the limitations on required chemical disclosure (Centner, 2013; Centner and O'Connell, 2014; Maule et al., 2013).

The measures required by state and federal regulatory agencies in the exploration and production of natural gas from deep tight formations have been very effective, e.g., in protecting drinking water aquifers from contamination. In fact, a series of federal laws govern most environmental aspects of natural gas development (Table 10.1). However, federal regulation may not always be the most effective way of assuring the desired level of environmental protection. Therefore most of these federal laws have provisions for granting primacy to the state governments, which have usually developed their own sets of regulations. By statute, the different states may adopt these standards of their own, but they must be at least as protective as the federal principles they replace and, as a result, may be more protective to address local conditions.

Table 10.1 Examples of Federal Laws (*listed alphabetically*) in the United States to monitor hydraulic fracturing projects

Act	Purpose
Clean Air Act	Limits air emissions from engines, gas processing equipment, and other sources associated with drilling and production
Clean Water Act	Regulation of surface discharges of water associated with natural gas and crude oil drilling and production, as well as storm water runoff from production sites
Energy Policy Act	Exempted hydraulic fracturing companies from some regulations; may disclose chemicals through a report submitted to the regulatory authority but in some instances, chemical information may be exempt from disclosure to the public as trade secrets
NEPA	Requires that exploration and production on federal lands be thoroughly analyzed for environmental impacts
NPDES	Requires tracking of any toxic chemicals used in fracturing fluids
Oil Pollution Act	Regulation of ground pollution risks relating to spills of materials or hydrocarbon derivatives into the water table; also regulated under the Hazardous Materials Transport Act
Safe Drinking Water Act	Directs the underground injection of fluids from natural gas and crude oil activities; disclosure of chemical content for underground injections; after 2005, see Energy Policy Act
TSCA	Suggestion that this act be used to regulate the reporting of hydraulic fracturing fluid information

NEPA, National Environmental Policy Act; *NPDES*, National Pollutant Discharge Elimination System; *TSCA*, Toxic Substance Control Act.
NB: The Fracturing Responsibility and Awareness of Chemicals Act (the FRAC Act) was an attempt to define hydraulic fracturing as a federally regulated activity under the Safe Drinking Water Act; no significant moves or passage of this act at the time of writing (https://www.congress.gov/bill/114th-congress/senate-bill/785/text).

State regulation of the environmental practices related to natural gas and crude oil development can more easily address the regional and state-specific character of the activities, compared to a one-size-fits-all management by the federal level. Some of these factors include: geology, hydrology, climate, topography, industry characteristics, development history, state legal structures, population density, and local economics and, thus, the regulation of natural gas and crude oil production is a detailed monitoring of each stage of the development through the many controls at the state level. Each state has the necessary powers to regulate, permit, and enforce all activities—from drilling and fracturing of the well, to production operations, to managing and disposing of wastes, to abandoning and plugging the production well(s). These state powers are a means of assuring that natural gas and crude oil operations do not have an adverse impact on the environment.

Moreover, because of the regulatory make-up of each state—which can vary from state-to-state—different states take different approaches to the regulation and enforcement of resource development, but the laws of each state generally give the state agency responsible for natural gas and crude oil development the discretion to require whatever is necessary to protect the environment, including human health.

In addition, most have a general prohibition against pollution from natural gas and crude oil production. Most of the state requirements are written into rules or regulations but some of the regulations may be added to permits on a case-by-case basis because of (1) environmental review, (2) on-site inspections, (3) commission hearings, and (4) public comments.

Finally, the organization of regulatory agencies within the different states where natural gas and crude oil is produced varies considerably. Some states have several agencies that oversee different aspects (with some inevitable overlap powers) of natural gas and crude oil operations, particularly the requirements that protect the environment. In different states, the various approaches have developed over time to create a structure that best serves the consumers, nonconsumers, and the various industries. The one constant is that each in each state where natural gas and crude oil is produced, there is one agency that has the responsibility for issuing permits for gas and oil development projects. The permitting agencies work with other agencies in the regulatory process and often serve as a central organizing body and a useful source of information about activities related to natural gas and crude production.

10.5.2 Federal regulations

In 1938, the federal government of the United States became involved directly in the regulation of interstate natural gas with the passage of the Natural Gas Act (NGA) which gave the Federal Power Commission (FPC, created in 1920 with the passage of the Federal Water Power Act) jurisdiction over regulation of interstate natural gas sales. The FPC was charged with regulating the rates that were charged for interstate natural gas delivery, as well as limited certification powers.

The rationale for the passage of the NGA was the concern over the heavy concentration of the natural gas industry, and the monopolistic tendencies of interstate pipelines to charge higher than competitive prices due to their market power. While the NGA required that "just and reasonable" rates for pipeline services be enforced, it did not specify any particular regulation of prices of natural gas at the wellhead. The NGA also specified that no new interstate pipeline could be built to deliver natural gas into a market already served by another pipeline. In 1942, these certification powers were extended to cover any new interstate pipelines. This meant that, in order to build an interstate pipeline, companies must first receive the approval of the FPC.

From 1954 to 1960, the FPC attempted to deal with producers and their rates on an individual basis and each producer was treated as an individual public utility, and rates were set based on each producer's cost of service. However, this turned out to be administratively unfeasible, as there were so many different producers and rate cases that a backlog developed. As a result, in 1960, the FPC decided to set rates based on geographic areas in which rates were set for all wells in a particular region. By 1970, rates had been set for only two of the five producing areas. To make matters worse, for most of the areas, prices were essentially frozen at 1959 levels. The problem with determining rates for a particular area based on cost-of-service methodologies was that there existed many wells in each area, with vastly different production costs.

In the 1970s and 1980s, a number of gas shortages and price irregularities indicated that a regulated market was not best for consumers, or the natural gas industry. Into the 1980s and early 1990s, the industry gradually moved toward deregulation, allowing for healthy competition and market-based prices. These moves led to a strengthening of the natural gas market, lowering prices for consumers and allowing for a great deal more natural gas to be discovered.

Since the Energy Policy Act passed in 2005, natural gas and crude oil production in the United States has grown significantly and this rapid growth—along with continued reports of environmental effects—has led to renewed calls for the federal government to provide increased regulation or guidance. This pressure led to the introduction to Congress in 2009 of the Fracturing Responsibility and Awareness of Chemicals Act (the FRAC Act) to define hydraulic fracturing as a federally regulated activity under the Safe Drinking Water Act (Table 10.1). The proposed requires the energy industry to disclose the chemical additives used in the hydraulic fracturing fluid. The Act went did not receive any action, was reintroduced in 2011, and appears to have been held in limbo since that time.

In the absence of new federal regulations, the various states that produce natural gas and crude oil by hydraulic fracturing have continued to use existing natural gas and crude oil and environmental regulations to manage natural gas and crude oil development, as well as introducing individual state regulations for hydraulic fracturing (Table 10.1). In fact, the current regulations are comprised of an overlapping collection of federal, state, and local regulations and permitting systems and these regulations cover different aspects of the development and production of a natural gas and crude oil with the intention that the regulations combine to manage any potential impact on the surrounding environment, including any effects on water management. However, the hydraulic fracturing process that has not previously been regulated under current laws and, therefore, in terms of water management, emissions management, and site activity, the existing regulations are being (must be) reassessed for suitability for application to the hydraulic fracturing process. In the meantime, many states (including Wyoming, Arkansas, and Texas) have already implemented regulations requiring disclosure of the materials used in hydraulic fracturing fluids and the United States Department of the Interior has indicated an interest in requiring similar disclosure for sites on federal lands.

The Bureau of Land Management's (BLM) of the Department of the Interior has proposed draft rules for natural gas and crude oil production on public lands and these proposals would require disclosure of the chemical components used in hydraulic fracturing fluids. The proposed rule requires that an operations plan should be submit to the relevant authority—prior to initiation of the hydraulic fracturing project—that would allow the Bureau of Land Management to evaluate groundwater protection designs based on (1) a review of the local geology, (2) a review of any anticipated surface disturbance, and (3) a review of the proposed management and disposal of project-related fluids. In addition, the Bureau of Land Management would require submittal of the information necessary to confirm wellbore integrity before, during, and after the stimulation operation. Furthermore, before hydraulic fracturing commenced, the company would have to certify that the fluids comply with all applicable Federal, state, and local laws, rules, and regulations. After the conclusion of the hydraulic fracturing stage of the project, a follow-up report would be required in which the actual events that occurred during fracturing activities would be summarized and this report would have to include the specific chemical make-up of the hydraulic fracturing fluid.

On April 17, 2012, the United States EPA released new performance standards and national emissions standards for HAPs in the natural gas and crude oil industries. These rules include the first federal air standards for hydraulically fractured gas wells, along with requirements for other sources of pollution in the natural gas and crude oil industry that currently are not regulated at the federal level. These standards require either flaring or *green completion* on natural gas wells developed prior to January 1, 2015, with only green completions allowed for wells developed on and after that date.

Briefly, *green completion* requires natural gas companies capture the gas at the wellhead immediately after well completion instead of releasing it into the atmosphere or flaring the gas. Thus the green completion systems are systems to reduce methane losses during well completion. After a new well completion or workover, the well bore and formation must be cleaned of debris and fracturing fluid. Conventional methods for doing this include producing the well into an open pit or tank to collect sand, cuttings, and reservoir fluids for disposal. Typically, the natural gas that is produced is vented or flared and the large volume of natural gas that is lost may not only affect regional air quality. When using green completion systems, gas and hydrocarbon liquids are physically separated from other fluids (a form of wellhead gas processing)—there is no venting or flaring of the gas—and delivered directly into equipment that holds or transports the hydrocarbon derivatives for productive use. Furthermore, by using portable equipment to process natural gas and natural gas condensate, the recovered gas can be directed to a pipeline as sales gas. The use of truck-mounted or trailer-mounted portable systems can typically recover more than half of the total gas produced.

Currently, the natural gas industry is regulated to a lesser extent by the Federal Energy Regulatory Commission (FERC). While FERC does not deal exclusively with natural gas issues; it is the primary rule making body with respect to the minimal regulation of the natural gas industry.

Opening up the natural gas industry and the move away from strict regulation has allowed for increased efficiency and technological improvements. Natural gas is now being obtained more efficiently, cheaply, and easily than ever before. However, the search for more natural gas to serve our ever-growing demand requires new techniques and knowledge to obtain it from hard-to-reach places.

Deregulation and the move toward cleaner burning fuels have created an enormous market for natural gas across the country. New technologies are continually developed that allow natural gas to be used in new ways and it is becoming the fuel of choice in the United States and in many countries throughout the world.

References

BP, 2017. Statistical Review of World Energy 2016. British Petroleum, London. <https://www.bp.com/content/dam/bp/en/corporate/pdf/energy-economics/statistical-review-2017/bp-statistical-review-of-world-energy-2017-full-report.pdf>.

Centner, T.J., 2013. Oversight of shale gas production in the United States and the disclosure of toxic substances. Resour. Policy 38, 233–240.

Centner, T.J., O'Connell, L.K., 2014. Unfinished business in the regulation of shale gas production in the United States. Sci. Total Environ. 476–477, 359–367.

Cohen, G., Joutz, F., Loungani, P., 2011. Measuring Energy Security: Trends in the Diversification of Oil and Natural Gas Supplies. IMF Working Paper WP/11/39. International Monetary Fund, Washington, DC.

Deutch, J., 2010. Oil and Gas Energy Security Issues. Resource for the Future. National Energy Policy Institute, Washington, DC.

EIA, February 2006. Annual Energy Outlook 2006 with Projections to 2030. Report DOE/EIA-0383. International Energy Annual. Energy Information Administration, Washington, DC.

IEA, 2018. <https://www.iea.org/topics/energysecurity/> (accessed 21.05.18).

Maule, A.L., Makey, C.M., Benson, E.B., Burrows, I.J., Scammell, M.K., 2013. Disclosure of hydraulic fracturing fluid chemical additives: analysis of regulations. New Solut. 23, 167–187.

Speight, J.G., 1993. Gas Processing: Environmental Aspects and Methods. Butterworth Heinemann, Oxford.

Speight, J.G., 2007. The Chemistry and Technology of Petroleum, fourth ed. CRC-Taylor and Francis Group, Boca Raton, FL.

Speight, J.G., 2011. An Introduction to Petroleum Technology, Economics, and Politics. Scrivener Publishing, Salem, MA.

Speight, J.G., 2014. The Chemistry and Technology of Petroleum, fifth ed. CRC Press, Taylor & Francis Group, Boca Raton, FL.

Speight, J.G., 2016. Handbook of Hydraulic Fracturing. John Wiley & Sons Inc, Hoboken, NJ.

Speight, J.G., 2017. Handbook of Petroleum Refining. CRC Press, Taylor & Francis Group, Boca Raton, FL.

Speight, J.G., 2018. Handbook of Natural Gas Analysis. John Wiley & Sons Inc, Hoboken, NJ.

Speight, J.G., Islam, M.R., 2016. Peak Energy – Myth or Reality. Scrivener Publishing, Beverly, MA.

US DOE, Janauary 2017. Report to Congress. Valuation of Energy Security for the United States. United States Department of Energy, Washington, DC. <https://www.energy.gov/sites/prod/files/2017/01/f34/Valuation%20of%20Energy%20Security%20for%20the%20United%20States%20%28Full%20Report%29_1.pdf>.

Appendix A: Examples of standard test methods for application to fuel gases and condensate

ASTM D1015	Standard Test Method for Freezing Points of High-Purity Hydrocarbons
ASTM D1016	Standard Test Method for Purity of Hydrocarbons from Freezing Points
ASTM D1025	Standard Test Method for Nonvolatile Residue of Polymerization-Grade Butadiene
ASTM D1070	Standard Test Methods for Relative Density of Gaseous Fuels
ASTM D1071	Standard Test Methods for Volumetric Measurement of Gaseous Fuel Samples
ASTM D1072	Standard Test Method for Total Sulfur in Fuel Gases by Combustion and Barium Chloride Titration
ASTM D1142	Standard Test Method for Water Vapor Content of Gaseous Fuels by Measurement of Dew-Point Temperature
ASTM D1217	Standard Test Method for Density and Relative Density (Specific Gravity) of Liquids by Bingham Pycnometer
ASTM D1265	Standard Practice for Sampling Liquefied Petroleum (LP) Gases, Manual Method
ASTM D1267	Standard Test Method for Gage Vapor Pressure of Liquefied Petroleum (LP) Gases (LP-Gas Method)
ASTM D1826	Standard Test Method for Calorific (Heating) Value of Gases in Natural Gas Range by Continuous Recording Calorimeter
ASTM D1835	Standard Specification for Liquefied Petroleum (LP) Gases
ASTM D1837	Standard Test Method for Volatility of Liquefied Petroleum (LP) Gases
ASTM D1838	Standard Test Method for Copper Strip Corrosion by Liquefied Petroleum (LP) Gases
ASTM D1945	Standard Test Method for Analysis of Natural Gas by Gas Chromatography
ASTM D1988	Standard Test Method for Mercaptans in Natural Gas Using Length-of-Stain Detector Tubes
ASTM D2156	Standard Test Method for Smoke Density in Flue Gases from Burning Distillate Fuels
ASTM D2158	Standard Test Method for Residues in Liquefied Petroleum (LP) Gases
ASTM D2420	Standard Test Method for Hydrogen Sulfide in Liquefied Petroleum (LP) Gases (Lead Acetate Method)

ASTM D2421	Standard Practice for Interconversion of Analysis of C5 and Lighter Hydrocarbons to Gas-Volume, Liquid-Volume, or Mass Basis
ASTM D2505	Standard Test Method for Ethylene, Other Hydrocarbons, and Carbon Dioxide in High-Purity Ethylene by Gas Chromatography
ASTM D2593	Standard Test Method for Butadiene Purity and Hydrocarbon Impurities by Gas Chromatography
ASTM D2598	Standard Practice for Calculation of Certain Physical Properties of Liquefied Petroleum (LP) Gases from Compositional Analysis
ASTM D2650	Standard Test Method for Chemical Composition of Gases by Mass Spectrometry
ASTM D2713	Standard Test Method for Dryness of Propane (Valve Freeze Method)
ASTM D3429	Standard Test Method for Solubility of Fixed Gases in Low-Boiling Liquids
ASTM D3588	Standard Practice for Calculating Heat Value, Compressibility Factor, and Relative Density of Gaseous Fuels
ASTM D4051	Standard Practice for Preparation of Low-Pressure Gas Blends
ASTM D4084	Standard Test Method for Analysis of Hydrogen Sulfide in Gaseous Fuels (Lead Acetate Reaction Rate Method)
ASTM D4150	Standard Terminology Relating to Gaseous Fuels
ASTM D4423	Standard Test Method for Determination of Carbonyls in C4 Hydrocarbons
ASTM D4424	Standard Test Method for Butylene Analysis by Gas Chromatography
ASTM D4468	Standard Test Method for Total Sulfur in Gaseous Fuels by Hydrogenolysis and Rateometric Colorimetry
ASTM D4784	Standard Specification for LNG Density Calculation Models
ASTM D4810	Standard Test Method for Hydrogen Sulfide in Natural Gas Using Length-of-Stain Detector Tubes
ASTM D4888	Standard Test Method for Water Vapor in Natural Gas Using Length-of-Stain Detector Tubes
ASTM D4891	Standard Test Method for Heating Value of Gases in Natural Gas and Flare Gases Range by Stoichiometric Combustion
ASTM D5134	Standard Test Method for Detailed Analysis of Petroleum Naphthas through n-Nonane by Capillary Gas Chromatography
ASTM D5504	Standard Test Method for Determination of Sulfur Compounds in Natural Gas and Gaseous Fuels by Gas Chromatography and Chemiluminescence
ASTM D5954	Standard Test Method for Mercury Sampling and Measurement in Natural Gas by Atomic Absorption Spectroscopy
ASTM D6849	Standard Practice for Storage and Use of Liquefied Petroleum Gases (LPG) in Sample Cylinders for LPG Test Methods
ASTM D7551	Standard Test Method for Determination of Total Volatile Sulfur in Gaseous Hydrocarbons and Liquefied Petroleum Gases and Natural Gas by Ultraviolet Fluorescence
ASTM D7607	Standard Test Method for Analysis of Oxygen in Gaseous Fuels (Electrochemical Sensor Method)

Conversion factors

1 General

Crude oil
    $1 \text{ m}^3 = 6.29$ barrels
    1 barrel $= 0.159 \text{ m}^3$
    1 tonne $= 7.49$ barrels
Natural gas
    $1 \text{ m}^3 = 35.3 \text{ ft}^3$
    $1 \text{ ft}^3 = 0.028 \text{ m}^3$
Numbers
    Million $= 1 \times 10^6$
    Billion $= 1 \times 10^9$
    Trillion $= 1 \times 10^{12}$

2 Concentration conversions

1 part per million (1 ppm) = 1 microgram per liter (1 μg/L)
1 microgram per liter (1 μg/L) = 1 milligram per kilogram (1 mg/kg)
1 microgram per liter (μg/L) $\times 6.243 \times 10^8$ = 1 lb per cubic foot (1 lb/ft³)
1 microgram per liter (1 μg/L) $\times 10^{-3}$ = 1 milligram per liter (1 mg/L)
1 milligram per liter (1 mg/L) $\times 6.243 \times 10^5$ = 1 pound per cubic foot (1 lb/ft³)
I gram mole per cubic meter (1 g mol/m³) $\times 6.243 \times 10^5$ = 1 pound per cubic foot (1 lb/ft³)
10,000 ppm = 1% w/w
1 ppm hydrocarbon in soil $\times 0.002$ = 1 lb of hydrocarbons per ton of contaminated soil

3 Temperature conversions

°F = (°C $\times 1.8$) + 32
°C = (°F − 32)/1.8
(°F − 32) $\times 0.555$ = °C
Absolute zero = −273.15°C
Absolute zero = −459.67°F

4 Area

1 square centimeter (1 cm^2) = 0.1550 square inches
1 square meter (1 m^2) = 1.1960 square yards
1 hectare = 2.4711 acres
1 square kilometer (1 km^2) = 0.3861 square miles
1 square inch (1 $in.^2$) = 6.4516 square centimeters
1 square foot (1 ft^2) = 0.0929 square meters
1 square yard (1 yd^2) = 0.8361 square meters
1 acre = 4046.9 square meters
1 square mile (1 mi^2) = 2.59 square kilometers

5 Nutrient conversion factor

1 pound phosphorus × 2.3 (1 lb P × 2.3) = 1 pound phosphorous pentoxide (1 lb P_2O_5)
1 pound potassium × 1.2 (1 lb K × 1.2) = 1 pound potassium oxide (1 lb K_2O)

6 Sludge conversions

1700 lbs wet sludge = 1 yd^3 wet sludge
1 yd^3 sludge = wet tons/0.85
Wet tons sludge × 240 = gallons sludge
1 wet ton sludge × % dry solids/100 = 1 dry ton of sludge

7 Weight conversion

1 ounce (1 oz) = 28.3495 grams (28.3495 g)
1 pound (1 lb) = 0.454 kilogram
1 pound (1 lb) = 454 grams (454 g)
1 kilogram (1 kg) = 2.20462 pounds (2.20462 lb)
1 stone (English, 1 st) = 14 pounds (14 lb)
1 ton (US; 1 short ton) = 2000 lbs
1 ton (English; 1 long ton) = 2240 lbs
1 metric ton = 2204.62262 pounds
1 tonne = 2204.62262 pounds

8 Other approximations

14.7 pounds per square inch (14.7 psi)—1 atmosphere (1 atm)
1 kiloPascal (kPa) × 9.8692 × 10^{-3} = 14.7 pounds per square inch (14.7 psi)
1 yd^3 = 27 ft^3

1 US gallon of water = 8.34 lbs
1 imperial gallon of water−10 lbs
1 yd^3 = 0.765 m^3
1 acre-inch of liquid = 27,150 gallons = 3.630 ft^3
1 ft depth in 1 acre (in situ) = 1613 × (20%−25% excavation factor) = ∼2000 yd^3
1 yd^3 (clayey soils-excavated) = 1.1−1.2 tons (US)
1 yd^3 (sandy soils-excavated) = 1.2−1.3 tons (US)

Glossary

Abandoned well A well not in use because it was a dry hole originally, or because it has ceased to produce. Statutes and regulations in many states require the plugging of abandoned wells to prevent the seepage of oil, gas, or water from one stratum to another.

Abiogenic gas Gas formed by inorganic means.

Abiotic Produced from nonorganism materials.

Absolute pressure Gauge pressure plus barometric pressure; the absolute pressure can be zero only in a perfect vacuum.

Absolute zero The zero point on the absolute temperature scale. It is equal to $-273.16°C$, or 0 K (Kelvin), or $-459.69°F$, or $0°R$ (Rankine).

Absorbed gas Natural gas that has been dissolved into the rock and requires hydraulic fracturing to be released.

Absorbent A material which, due to an affinity for certain substances, extracts one or more such substances from a liquid or gaseous medium with which it contacts, and which changes physically, or both, during the process.

Absorption The process by which the gas is distributed throughout an absorbent (liquid); depends only on physical solubility and may include chemical reactions in the liquid phase (*chemisorption*).

Absorption plant A device or unit that removes hydrocarbon compounds from natural gas, especially casinghead gas; in the plant the gas is run through absorption oil which absorbs the liquid constituents, which are then recovered by distillation.

Abyssal Of or relating to the bottom waters of the ocean.

Accumulate To amass or collect; when oil and gas migrate into porous formations, the quantity collected is called an accumulation.

Accumulation Pressure increase over the maximum allowable working pressure of the vessel during discharge through the pressure relief valve (expressed as a percent of that pressure) is called accumulation.

Accuracy The closeness of the data of an analytical method to the true value.

Acid deposition (acid rain) Occurs when sulfur dioxide (SO_2) and, to a lesser extent, NO_x emissions are transformed in the atmosphere and return to the Earth as dry deposition or in rain, fog, or snow.

Acid gas Impurities in a gas stream usually consisting of carbon dioxide (CO_2), hydrogen sulfide (H2S), carbonyl sulfide (COS), thiols/mercaptans (RSH), and sulfur dioxide (SO_2) —the most common in natural gas are carbon dioxide, hydrogen sulfide, and carbonyl sulfide; see also Sour gas.

Acid gas loading The amount of acid gas, on a molar or volumetric basis, which will be picked up by a solvent.

Acid rain Abnormally acidic rainfall, most often containing dilute concentrations of sulfuric acid or nitric acid.

Adsorption The process by which the gas is concentrated on the surface of a solid or liquid to remove impurities; carbon is a common adsorbing medium which can be regenerated upon *desorption*.

Aerobic bacteria Bacteria which can grow in the presence of oxygen.

Air gun A chamber filled with compressed air, often used offshore in seismic exploration. As the gun is trailed behind a boat, air is released, making a low-frequency popping noise, which penetrates the subsurface rock layers and is reflected by the layers. Sensitive hydrophones receive the reflections and transmit them to recording equipment on the boat.

Alkazid process A process for removal of hydrogen sulfide and carbon dioxide from natural gas using concentrated aqueous solutions of amino acids.

Alluvial fan A large, sloping sedimentary deposit at the mouth of a canyon, laid down by intermittently flowing water, especially in arid climates, and composed of gravel and sand. The deposit tends to be coarse and unworked, with angular, poorly sorted grains in thin, overlapping sheets. A line of fans may eventually coalesce into an apron that grows broader and higher as the slopes above are eroded away.

Amphibole Any group of common rock-forming silicate minerals.

Anaerobic bacteria Bacteria which can grow in the absence of oxygen.

Angle of deflection In directional drilling, the angle at which a well diverts from vertical; usually expressed in degrees, with vertical being 0 degree.

Angle of dip The angle at which a formation dips downward from the horizontal.

Analytical batch Consists of samples which are analyzed together with the same method sequence and the same lots of reagents and with the manipulations common to each sample within the same time period or in continuous sequential time periods.

Anticlinal trap A hydrocarbon trap in which petroleum accumulates in the top of an anticline; see Anticline, Syncline.

Anticline An area of the Earth's crust where folding has made a dome-like shape in the once flat rock layers. Anticlines often provide an environment where natural gas can become trapped beneath the surface of the Earth and extracted; see also Traps, Faults, Permeability, and Porosity.

Antifoam A substance, usually a silicone or long-chain alcohol, added to the treating system to reduce the tendency to foam.

Aquifer An underground porous, permeable rock formation that acts as a natural water reservoir.

Aquifer storage field A subsurface facility for storing natural gas consisting of water-bearing sands topped by an impermeable caprock.

Associated gas Natural gas that over-lies and contacts crude oil in a reservoir. Where reservoir conditions are such that the production of associated gas does not substantially affect the recovery of crude oil in the reservoir, such gas may also be reclassified as nonassociated gas by a regulatory agency. Also called associated free gas; see *gas cap*.

Associated natural gas Gas that occurs as free gas in a petroleum reservoir; see Dissolved natural gas, Nonassociated natural gas, and Gas cap.

Atmospheric discharge The release of vapors and gases from pressure-relieving and depressurizing devices to the atmosphere.

Back pressure Pressure on the discharge side of safety-relief valves is back pressure; the pressure that exists at the outlet of a pressure relief device as a result of pressure in the discharge system.

Balancing item Represents differences between the sum of the components of natural gas supply and the sum of the components of natural gas disposition.

Barrel (bbl) A measure of volume for petroleum products. One barrel is the equivalent of 35 imperial gallons or 42 US gallons or 0.15899 m^3 (9702 in.^{3}). One cubic meter equals 6.2897 barrels.

Base gas The quantity of gas needed to maintain adequate reservoir pressures and deliverability rates throughout the withdrawal season; base gas usually is not withdrawn and remains in the reservoir; all gas native to a depleted reservoir is included in the base gas volume.

Base load requirements (base load storage) Gas that is used to meet seasonal demand increases and the facilities are capable of holding enough natural gas to satisfy long-term seasonal demand requirements.

Basement rock The impervious geological stratum that underlays the reservoir rock and retains gas or oil in a reservoir.

Basin A local depression in the Earth's crust in which sediments can accumulate to form thick sequences of sedimentary rock.

Bcf (billion cubic feet) Gas measurement approximately equal to one trillion (1,000,000,000,000) Btu's; see also Mcf, Tcf, and Quad.

Bed A specific layer of Earth or rock that presents a contrast to other layers of different material lying above, below, or adjacent to it.

Bedrock Solid rock just beneath the soil.

Benthic Relating to the seabed.

Biogenic coal bed methane Methane formed in coal seams by naturally occurring bacteria that are associated with meteoric water recharge at outcrop or subcrop; see Coal bed methane.

Biomass Organic nonfossil material of biological origin constituting a renewable energy source.

Biotic Produced by living organisms.

Bit The cutting or boring element used in drilling oil and gas wells. The bit consists of a cutting element and a circulating element. The cutting element is steel teeth, tungsten carbide buttons, industrial diamonds, or polycrystalline diamonds. These teeth, buttons, or diamonds penetrate and gouge or scrape the formation to remove it. The circulating element permits the passage of drilling fluid and utilizes the hydraulic force of the fluid stream to improve drilling rates. In rotary drilling, several drill collars are joined to the bottom end of the drill pipe column, and the bit is attached to the end of the drill collars. Drill collars provide weight on the bit to keep it in firm contact with the bottom of the hole. Most bits used in rotary drilling are roller cone bits, but diamond bits are also used extensively.

Bitumen A hydrocarbonaceous substance of dark to black color consisting almost entirely of carbon and hydrogen with very little oxygen, nitrogen, or sulfur; bitumen occurs naturally in tar sand (oil sand) formations. This is also the term used in some countries for *asphalt*.

Black shale A thinly bedded shale that is rich in carbon, sulfide, and organic material, formed by anaerobic (lacking oxygen) decay of organic matter; black shale formations occur in thin beds in many areas at various depths and are of interest both historically and economically.

Blowdown The difference between set pressure and reseating pressure of a safety valve expressed in percent of the set pressure or in psi, bar, or kPa.

Blowout An uncontrolled flow of gas, oil, or other well fluids into the atmosphere. A blowout, or gusher, occurs when formation pressure exceeds the pressure applied to it by the column of drilling fluid. A kick warns of an impending blowout; see *kick*.

Blowout preventer (BOP) One of several valves installed at the wellhead to prevent the escape of pressure either in the annular space between the casing and the drill pipe or in open hole (i.e., hole with no drill pipe) during drilling or completion operations. Blowout preventers on land rigs are located beneath the rig at the land's surface; on jackup or platform rigs, at the water's surface; and on floating offshore rigs, on the seafloor.

Boiling point (boiling temperature) The temperature at which the vapor pressure of the substance is equal to atmospheric pressure.

Black shale A thinly bedded shale that is rich in carbon, sulfide, and organic material, formed by anaerobic (lacking oxygen) decay of organic matter; black shales occur in thin beds in many areas at various depths and are of interest both historically and economically.

Blowdown The difference between set pressure and reseating pressure of a safety valve expressed in percent of the set pressure in psi or kPa.

Bottom simulating reflector (BSR) A seismic reflection at the sediment to clathrate stability zone interface caused by the different density between normal sediments and sediments laced with clathrates.

Bottom-supported offshore drilling rig A type of mobile offshore drilling unit that has a part of its structure in contact with the seafloor when it is on site.

Brine An aqueous solution of salts that occur with gas and crude oil; seawater and saltwater are also known as brine.

Bright spot A seismic phenomenon that shows up on a seismic, or record, section as a sound reflection that is much stronger than usual. A bright spot sometimes directly indicates natural gas in a trap.

Btu (British Thermal Unit) A unit of measurement for energy; the amount of heat that is necessary to raise the temperature of one pound of water by $1°F$; see also Bcf, Tcf, and Quad.

Bubble point pressure At a given temperature, the pressure when crude oil releases a bubble of gas from solution.

Built-up back pressure The pressure in the discharge header which develops as a result of flow after that the safety relief valve opens.

Bundled service Gas sales service and transportation service packaged together in a single transaction in which the pipeline, on behalf of the utility, buys gas from producers and then delivers it to the utility.

Calibration The process of adjusting the instrument read-out so that it corresponds to the actual concentration value or a reference standard; involves checking the instrument with a known concentration of a gas or vapor to see that the instrument gives the proper response; calibration results in calibration factors or functions establishing the relationship between the analyzer response and the actual gas concentration introduced to the analyzer; an important element of quality assurance in emission control.

Cap gas Natural gas trapped in the upper part of a reservoir and remaining separate from any crude oil, salt water, or other liquids in the well.

Cap rock (caprock) A disk-like plate of anhydrite, gypsum, limestone, or sulfur overlying most salt domes in the Gulf Coast region; impermeable rock overlying an oil or gas reservoir that tends to prevent migration of oil or gas out of the reservoir.

Carbonaceous material A material which contains carbon as well as some hydrogen and other noncarbon and nonhydrogen elements such as nitrogen, oxygen, and sulfur.

Carbonate rock A rock consisting primarily of a carbonate mineral such as calcite or dolomite, the chief minerals in limestone and dolostone, respectively.

Carbonate washing A chemical conversion processes in which acid contaminants in natural gas are converted to compounds that are not objectionable or that can be removed from the stream with greater ease than the original constituents.

Carbon capture and storage A combination of a number of existing technologies, with the potential to play a major role in the management and reduction of global carbon dioxide (CO_2) levels; the process allows for carbon dioxide emissions released during energy production to be captured and stored underground.

Carbon dioxide fracturing The use of gaseous carbon dioxide to fracture a formation.

Casing Steel pipe placed in an oil or gas well to prevent the wall of the hole from caving in, to prevent movement of fluids from one formation to another and to aid in well control; also used to protect the surrounding Earth and rock layers from being contaminated by petroleum, or the drilling fluids.

Casinghead gas (casing head gas) Natural gas produced with oil in oil wells and is usually the flash gas from the oil reservoir. Casinghead gas collects in the annular space between the well tubing and casing of an oil well. The weight of the casinghead gas contributes to reducing the bottom hole pressure and lowering well production—*specifically by holding a back-pressure against the formation*; reducing the gas pressure on the well casing (annulus) can increase oil production. The goal is to maintain casinghead pressure as close to zero as possible. In areas or the country where it is allowed, these systems are often configured to pull a vacuum.

Catalytic oxidation A chemical conversion process that is used predominantly for destruction of volatile organic compounds and carbon monoxide.

Cathodic protection The method of preventing corrosion in metal structures that involves using electric voltage to slow or prevent corrosion; used in natural gas pipelines to resist corrosion over an extended period of time.

Cenozoic era The time period from 65 million years ago until the present. It is marked by rapid evolution of mammals and birds, flowering plants, grasses, and shrubs, and little change in invertebrates.

CFCs (Chlorofluorocarbons) Gaseous compounds used for cooling; release into the atmosphere has produces ozone depletion.

Chadacryst A crystal enclosed in another crystal.

Chemisorption See Absorption.

Christmas tree The series of pipes and valves that sits on top of a producing gas well; used in place of a pump to extract the gas from the well.

City gate A location at which custody of gas passes from a gas pipeline company to a local distributor.

Clastic rock A sedimentary rock composed of fragments of preexisting rocks. The principal distinction among clastic rocks is grain size; conglomerates, sandstones, and shale are clastic rocks.

Claus process A sulfur recovery process recovering elemental sulfur from sour gas; a major producer of sulfur.

Clean Air Act Amendments of 1990 Legislation to improve the quality of the atmosphere and curb acid rain promotes the use of cleaner fuels in vehicles and stationary sources.

Clinopyroxene A subgroup name for monoclinic pyroxene group of minerals.

Coal A carbonaceous, rocklike material that forms from the remains of plants that were subjected to biochemical processes, intense pressure, and high temperatures. It is used as fuel.

Coalbed methane (coal bed methane) Methane from coal seams; released or produced from the seams when the water pressure within the seam is reduced by pumping from either vertical or inclined to horizontal surface holes; see also Biogenic coal bed methane and Thermogenic coal bed methane.

Coalescer A mechanical process vessel with wettable, high-surface area packing on which liquid droplets consolidate for gravity separation from a second phase (e.g., gas or immiscible liquid).

Coal gas A generic term for gaseous mixture (mainly hydrogen, methane, and carbon monoxide) made from coal by the destructive distillation (i.e., heating in the absence of air) of bituminous coal; also synonymous with *blue gas, producer gas, water gas, town gas, fuel gas manufactured gas*, and *syngas (synthetic natural gas*, SNG).

Coke oven gas The mixture of permanent gases produced by the carbonization of coal in a coke oven at temperatures in excess of 1000°C (1832°F).

Composition The make-up of a gaseous stream.

Compressed natural gas (CNG) Natural gas compressed to a pressure at or above 2900–3600 psi and stored in high-pressure containers; used as a fuel for natural gas-powered vehicles.

Compressibility factor The ratio of the molar volume V_m of the gas to the molar volume V_m^o of an ideal gas at the same pressure and temperature. Thus:

$$Z = V_m / V_m^o$$

The value of Z provides information on the dominant types of intermolecular forces acting in a gas. Thus, when $Z = 1$, there are no intermolecular forces, ideal gas behavior; when $Z < 1$, the attractive forces dominate, gas occupies a smaller volume than an ideal gas; when $Z > 1$, the repulsive forces dominate, gas occupies a larger volume than an ideal gas. All gases approach $Z = 1$ at very low pressures, when the spacing between particles is typically large.

Compression Reduction in volume of natural gas is compressed during transportation and storage.

Concentration The amount of a substance, expressed as mass, volume, or number of particles in a unit volume of a solid, liquid, or gaseous substance.

Concrete gravity rigid platform rig A rigid offshore drilling platform built of steel-reinforced concrete and used to drill development wells. The platform is floated to the drilling site in a vertical position. At the site, one or more tall caissons that serve as the foundation of the platform are flooded so that the platform comes to rest on bottom. Because of the enormous weight of the platform, the force of gravity alone keeps it in place.

Condensate A hydrocarbon liquid stream that consists of varying proportions of butane, propane, pentane, and heavier fractions, with little or no methane or ethane; separated from natural gas; higher molecular weight hydrocarbons that exist in the reservoir as constituents of natural gas, but which are recovered as liquids in separators, field facilities or gas-processing plants; see Gas condensate.

Condensate (lease condensate) Low-boiling liquid hydrocarbons recovered from lease separators or field facilities at associated and nonassociated natural gas wells; mostly pentane derivative and higher molecular weight hydrocarbon derivatives that enter the crude oil stream after production.

Condensate reservoir A reservoir in which both condensate and gas exist in one homogeneous phase. When fluid is drawn from such a reservoir and the pressure decreases below the critical level, a liquid phase (condensate) appears.

Condensate Separator A unit for the removal of condensate from the gas stream at the wellhead through the use of mechanical separators. In most instances, the gas flow into the separator comes directly from the wellhead, since the gas—oil separation process is not needed. Extracted condensate is routed to on-site storage tanks.

Condensate well A well that produces raw natural gas along with low-boiling hydrocarbon liquids; the gas is also *nonassociated* gas and is often referred to as *wet gas*.

Consumption Natural gas consumed within the country, including imports but excluding amounts reinjected, flared, and lost in shrinkage.

Contact In geology, any sharp or well-defined boundary between two different bodies of rock; a bedding plane or unconformity that separates formations. In a petroleum reservoir, a horizontal boundary where different types of fluids meet and mix slightly; e.g., a gas—oil or oil—water contact; also called an interface.

Contaminant removal Removal of contaminants which includes the elimination of hydrogen sulfide, carbon dioxide, water vapor, helium, and oxygen. The most commonly used technique is to first direct the flow through a tower containing an olamine solution which absorbs sulfur compounds. After desulfurization, the gas flow is directed to the next section, which contains a series of filter tubes. As the velocity of the stream reduces in the unit, primary separation of remaining contaminants occurs due to gravity.

Continuous accumulations Petroleum that occurs in extensive reservoirs and is not necessarily related to conventional structural or stratigraphic traps. These accumulations of oil and/or gas lack well-defined down-dip petroleum/water contacts and thus are not localized by the buoyancy of oil or natural gas in water.

Conventional gas Natural gas that is extracted from underground reservoirs using traditional exploration and production methods.

Conventional gas Liquid separator. A vertical or horizontal separator in which gas and liquid are separated by means of gravity settling with or without a mist eliminating device.

Conventional mud A drilling fluid containing essentially clay and water; no special or expensive chemicals or conditioners are added.

Core A cylindrical sample taken from a formation for geological analysis. Usually a conventional core barrel is substituted for the bit and procures a sample as it penetrates the formation. *V*: to obtain a solid, cylindrical formation sample for analysis.

Core analysis Laboratory analysis of a core sample to determine porosity, permeability, lithology, fluid content, angle of dip, geological age, and probable productivity of the formation.

Coring The process of cutting a vertical, cylindrical sample of the formations encountered as an oil well is drilled. The purpose of coring is to obtain rock samples, or cores, in such a manner that the rock retains the same properties that it had before it was removed from the formation.

Cretaceous Of or relating to the geologic period from about 135 million to 65 million years ago at the end of the Mesozoic era, or to the rocks formed during this period, including the extensive chalk deposits for which it was named.

Cricondenbar pressure The maximum pressure at which two phases can coexist.

Cricondentherm temperature The maximum temperature at which two phases can coexist.

Critical point The pressure and temperature of a reservoir fluid where the bubble point pressure curve meets the retrograde dew point pressure curve representing a unique state where all properties of the bubble point oil are identical to the dew point gas.

Cross section A geological or geophysical profile of a vertical section of the Earth.

Crude oil Unrefined liquid petroleum which ranges in gravity from 9 to 55° API and in color from yellow to black, and may have a paraffin, asphalt, or mixed base. If a crude oil, or crude, contains a sizable amount of sulfur or sulfur compounds, it is called a sour crude; if it has little or no sulfur, it is called a sweet crude. In addition, crude oils may be referred to as heavy or light according to API gravity, the lighter oils having the higher gravities.

Crust The outer layer of the Earth, varying in thickness from 5 to 30 miles (10 to 50 km). It is composed chiefly of oxygen, silicon, and aluminum.

Cryogenic process A process involving low temperatures.

CSST (Corrugated Stainless-Steel Tubing) Flexible piping used to install gas service in residential and commercial areas.

Cubic foot (ft³) The volume of a cube, all edges of which measure 1 foot. Natural gas in the United States is usually measured in cubic feet, with the most common standard cubic foot being measured at 60°F and 14.65 pounds per square inch absolute, although base conditions vary from state to state.

Cubic meter (m³) A unit of volume measurement in the SI metric system, replacing the previous standard unit known as the barrel, which was equivalent to 35 imperial gallons or 42 US gallons. The cubic meter equals approximately 62,898 barrels.

Cumulative production Volumes of oil and natural gas liquids that have been produced.

Cutting A piece of rock or dirt that is brought to the surface of a drilling site as debris from the bottom of well; often used to obtain data for logging.

DEA Diethanolamine.

Dead crude oil Crude oil in the reservoir with minimal or no dissolved associated gas; often difficult to produce as there is little energy to drive it.

Decline rate The rate at which the production rate of a well decreases.

Deep water In offshore operations, water depths greater than normal for the time and current technology.

Degradation products Impurities in a treating solution which are formed both reversible and irreversible side reactions.

Delineation well A well drilled in an existing field to determine, or delineate, the extent of the reservoir.

Dehydration Removal of water from gas streams that is accomplished by several methods, such as the use of ethylene glycol (glycol injection) systems as an absorption mechanism to remove water and other solids from the gas stream. Alternatively, adsorption dehydration may be used, utilizing dry-bed dehydrators towers, which contain desiccants such as silica gel and activated alumina, to perform the extraction.

Dekatherm (dth) A unit of energy used primarily to measure natural gas and was developed in about 1972 by the Texas Eastern Transmission Corporation—a natural gas pipeline company; the dekatherm is equal to 10 therms or 1,000,000 British thermal units (MMBtu) or, using the SI system, 1.055 gigajoules (GJ) and is also approximately equal to one thousand cubic feet (Mcf, Mft³) of natural gas or exactly one thousand cubic feet of natural gas with a heating value of 1000 Btu/ft³.

Delivered gas The physical transfer of natural, synthetic, and/or supplemental gas from facilities operated by the responding company to facilities operated by others or to consumers.

Delivery point (receipt point) The point where natural gas is transferred from one party to another. The city gate is the delivery point for a pipeline or transportation company because this is where the gas is transferred to the LDC.

Density The mass of a substance contained in a unit volume (mass divided by volume).

Deplete To exhaust a supply; an oil and/or a gas reservoir is depleted when most or all economically recoverable hydrocarbons have been produced.

Depleted reservoirs Reservoirs that have already been tapped of all their recoverable natural gas.

Depleted storage field A subsurface natural geological reservoir, usually a depleted gas or oil field, used for storing natural gas.

Depth The distance to which a well is drilled, stipulated in a drilling contract as contract depth. Total depth is the depth after drilling is finished.

Derrick A large load-bearing structure, usually of bolted construction. In drilling, the standard derrick has four legs standing at the corners of the substructure and reaching to the crown block; an assembly of heavy beams used to elevate the derrick and provide space to install equipment such as blowout preventers and casingheads; the standard derrick must be assembled piece by piece, it has largely been replaced by the mast, which can be lowered and raised without disassembly.

Desorption See Adsorption.

Detection limit The lowest amount of analyte in a sample which can be detected but not necessarily quantitated as an exact value; often called the *limit of detection (LOD)* which is the lowest concentration level that can be determined statistically different from a blank at a specified level of confidence; determined from the analysis of sample blanks; see Method detection limit.

Development (of a gas or oil field) All operations associated with the construction of facilities to enable the production of oil and gas.

Development drilling Drilling that occurs after the initial discovery of hydrocarbons in a reservoir. Usually, several wells are required to adequately develop a reservoir.

Development well A well drilled in proven territory in a field to complete a pattern of production; an exploitation well.

Deviation Departure of the wellbore from the vertical, measured by the horizontal distance from the rotary table to the target. The amount of deviation is a function of the drift angle and hole depth. The term is sometimes used to indicate the angle from which a bit has deviated from the vertical during drilling.

Dew point temperature The temperature at which a vapor begins to condense and deposit as a liquid; the hydrocarbon dew point is the temperature (at a given pressure) at which the hydrocarbon constituents of any hydrocarbon-rich gas mixture, such as natural gas, will start to condense out of the gaseous phase; the maximum temperature at which such condensation takes place is called the *cricondentherm*.

DGA Diglycolamine.

Diamond bit A drill bit that has small industrial diamonds embedded in its cutting surface. Cutting is performed by the rotation of the very hard diamonds over the rock surface.

Dip The angle at which it lies in relation to a flat line at the surface; often helps the geologist to locate possible traps (reservoirs).

DIPA Diisopropanolamine.

Directional drilling Intentional deviation of a wellbore from the vertical. Although wellbores are normally drilled vertically, it is sometimes necessary or advantageous to drill at an angle from the vertical; controlled directional drilling makes it possible to reach subsurface areas laterally remote from the point where the bit enters the Earth. It often involves the use of deflection tools.

Discovery well The first oil or gas well drilled in a new field that reveals the presence of a hydrocarbon-bearing reservoir. Subsequent wells are development wells.

Dissolved gas Natural gas that is in solution with crude oil in the reservoir.

Dissolved natural gas Gas that occurs in solution in the petroleum in a reservoir; see Associated natural gas.

Dissolved water Water in solution, in oil, at a defined temperature and pressure.

Drill To bore a hole in the Earth, usually to find and remove subsurface formation fluids such as oil and gas.

Drilling fluid Circulating fluid, one function of which is to lift cuttings out of the wellbore and to the surface. It also serves to cool the bit and to counteract downhole formation pressure. Although a mixture of barite, clay, water, and other chemical additives is the most common drilling fluid, wells can also be drilled by using air, gas, water, or oil-base mud as the drilling mud. Also called circulating fluid, drilling mud; see Mud.

Drilling mud Specially compounded liquid circulated through the wellbore during rotary drilling operations; see Drilling fluid, Mud.

Drillship (Drill ship) A self-propelled floating offshore drilling unit that is a ship constructed to permit a well to be drilled from it.

Dry A hole is dry when the reservoir it penetrates is not capable of producing hydrocarbons in commercial amounts.

Dry gas Gas whose water content has been reduced by a dehydration process; gas containing few or no hydrocarbons commercially recoverable as liquid product; also called lean gas.

Dry gas well A well that typically produces only raw natural gas that does not contain any hydrocarbon liquids; the gas is called *nonassociated* gas.

Dry hole Any well that does not produce oil or gas in commercial quantities. A dry hole may flow water, gas, or even oil, but not in amounts large enough to justify production.

Dry natural gas Natural gas which remains after: (1) the liquefiable hydrocarbon portion has been removed from the gas stream and (2) any volumes of nonhydrocarbon gases have been removed where they occur in sufficient quantity to render the gas unmarketable; also known as consumer-grade natural gas.

Ecology Science of the relationships between organisms and their environment.

Eh/redox A measure of the degree of oxygenation of a sediment.

Endowment The sum of cumulative production, remaining reserves, mean undiscovered recoverable volumes, and mean additions to reserves by field growth.

Environment The sum of the physical, chemical, and biological factors that surround an organism; the water, air, and land and the interrelationship that exists among and between water, air, and land and all living things.

EPACT (Energy Policy Act of 1992) Comprehensive energy legislation designed to expand natural gas use by allowing wholesale electric transmission access and providing incentives to developers of clean fuel vehicles.

EPIC (Engineering, Procurement, Installation, Commissioning) An EPIC or "turnkey" contract integrates the responsibility going from the conception to the final acceptance of one or more elements of a production system. It can be awarded for all, or part, of a field development.

Erosion The process by which material (such as rock or soil) is worn away or removed (as by wind or water).

Essexite A dark gray or black holocrystalline plutonic igneous rock.

Ethanol Produced through fermentation of agricultural raw materials (biomass), ethanol is used for various applications: drinks, pharmaceuticals, cosmetics, solvents, chemicals and more and more often in fuels, either in the form of an additive to gasoline (ETBE: Ethyl Tertiary-Butyl Ether) or blended directly with hydrocarbon-based gasoline.

Estimated additional amount in place The volume additional to the proved amount in place that is of foreseeable economic interest. Speculative amounts are not included.

Estimated additional reserves recoverable The volume within the estimated additional amount in place which geological and engineering information indicates with reasonable certainty might be recovered in the future.

Estuary A coastal indentation or bay into which a river empties and where freshwater mixes with seawater.

Evaporation ponds Artificial ponds with very large surface areas that are designed to allow the efficient evaporation of water through exposure to sunlight and ambient surface temperatures.

Exploration The search for reservoirs of oil and gas, including aerial and geophysical surveys, geological studies, core testing and drilling of wildcats.

Exploration well A well drilled either in search of an as-yet-undiscovered pool of oil or gas (a wildcat well) or to extend greatly the limits of a known pool. It involves a relatively high degree of risk. Exploratory wells may be classified as (1) wildcat, drilled in an unproven area; (2) field extension or step-out, drilled in an unproven area to extend the proved limits of a field; or (3) deep test, drilled within a field area but to unproven deeper zones.

Exploratory well A well drilled either in search of a new and as yet undiscovered accumulation of oil or gas, or in an attempt to significantly extend the limits of a known reservoir.

Extraction loss The reduction in volume of natural gas due to the removal of natural gas liquid constituents such as ethane, propane, and butane at natural gas processing plant.

Extraction plant A plant in which products, such as propane, butane, oil, ethane, or natural gasoline, which are initially components of the gas stream, are extracted or removed for sale.

Fabric filter Collectors in which dust is removed from the gas stream by passing the dust-laden gas through a fabric of some type; commonly termed "bag filters" or "baghouses."

Fault When part of the Earth's crust fractures due to forces exerted on it by movement of plates on the Earth's crust; of interest because they often form traps that are natural gas reservoirs; a break in the Earth's crust along which rocks on one side have been displaced (upward, downward, or laterally) relative to those on the other side.

Fault plane A surface along which faulting has occurred.

Fault trap A subsurface hydrocarbon trap created by faulting, in which an impermeable rock layer has moved opposite the reservoir bed or where impermeable gouge has sealed the fault and stopped fluid migration.

Feldspathoid Low silica igneous minerals that would have formed feldspars if only more silica (SiO_2) was present in the original magma.

Field A geographical area in which a number of oil or gas wells produce from a continuous reservoir. A field may refer to surface area only or to underground productive formations as well. A single field may have several separate reservoirs at varying depths; a contiguous area consisting of a single reservoir or multiple reservoirs of petroleum, all grouped on, or related to, a single geologic structural and/or stratigraphic feature.

Fire point The temperature to which gas must be heated under prescribed conditions of the method to burn continuously when the mixture of vapor and air is ignited by a specified flame.

Fischer–Tropsch process The catalytic process by which synthesis gas (syngas; mixtures of carbon monoxide and hydrogen) is converted to hydrocarbon products.

Fixed platform A structure made of steel or concrete, firmly fixed to the bottom of the body of water in which it rests.

Flare A tall stack equipped with burners used as a safety device at wellheads, refining facilities, gas processing plants, and chemical plants; used for the combustion and disposal of combustible gases.

Flare gas Gas or vapor that is flared.

Flash point The temperature to which gas must be heated under specified conditions to give of sufficient vapor to form a mixture with air that can be ignited momentarily by a specified flame; dependant on the composition of the gas and the presence of other hydrocarbon constituents.

Flash tank A vessel used to separate the gas evolved from liquid flashed from a higher pressure to a lower pressure.

Flexible flow line Flexible pipe laid on the seabed for the transportation of production or injection fluids. It is generally an infield line, linking a subsea structure to another structure or to a production facility. Its length ranges from a few hundred meters to several kilometers.

Flexible riser Riser constructed with flexible pipe; see Riser.

Flexsorb process A process that uses sterically hindered amines (olamines) in aqueous solutions or other physical solvents; the molecular structure hinders the carbon dioxide approach to the amine and preferentially removes hydrogen sulfide from the gas stream.

Floaters Floating production units including floating platforms, and floating, production, storage and offloading units (FPSOs).

Floating offshore drilling rig A type of mobile offshore drilling unit that floats and is not in contact with the seafloor (except with anchors) when it is in the drilling mode; floating units include barge rigs, drill ships, and semisubmersibles.

Floating production and system off-loader A floating offshore oil production vessel that has facilities for producing, treating, and storing oil from several producing wells and which puts (offloads) the treated oil into a tanker ship for transport to refineries on land; some floating, production, storage, and offloading units are also capable of drilling, in case they are termed floating production, drilling, and system off-loaders (FPDSOs).

Flow rate The rate that expresses the volume of fluid or gas passing through a given surface per unit of time (e.g., cubic feet per minute).

Flowing well A well that produces oil or gas by its own reservoir pressure rather than by use of artificial means (such as pumps).

Fold A flexure of rock strata (e.g., an arch or a trough) produced by horizontal compression of the Earth's crust; see Anticline and Syncline.

Formation A bed or deposit composed throughout of substantially the same kind of rock; often a lithological unit; either a certain layer of the Earth's crust, or a certain area of a layer; often refers to the area of rock where a reservoir is located; each formation is given a name, frequently as a result of the study of the formation outcrop at the surface and sometimes based on fossils found in the formation.

Formation volume factor (FVF) The ratio of a phase volume (water, oil, gas, or gas plus oil) at reservoir conditions, relative to the volume of a surface phase (water, oil, or gas) at standard conditions resulting when the reservoir material is brought to the surface; denoted mathematically as B_w (bbl/STB), B_0 (bbl/STB), B_g (ft^3/SCF), and B_t (bbl/STB).

Fossil The remains or impressions of a plant or animal of past geological ages that have been preserved in or as rock.

FPSO unit (Floating, Production, Storage, and Offloading unit) A converted or custom-built ship-shaped floater, employed to process oil and gas and for temporary storage of the oil prior to trans-shipment.

FPU (Floating Production Unit) A ship-shaped floater or a semisubmersible used to process and export oil and gas.

Fractionation The process of separating the various natural gas liquids present in the remaining gas stream by using the varying boiling points of the individual hydrocarbons in the gas stream.

Fracturing A method used by producers to extract more natural gas from a well by opening up rock formations using hydraulic or explosive force.

Free water Water produced with oil. It usually settles out within 5 minutes when the well fluids become stationary in a settling space within a vessel.

FSHR (Free Standing Hybrid Riser) A deepwater riser configuration (see Riser) consisting of a vertical rigid pipe section between the seabed and a submerged buoy and a catenary flexible pipe jumper between the submerged buoy and the floater.

Fuel cell technology The chemical interaction of natural gas and certain other metals, such as platinum, gold, and other electrolytes to produce electricity.

Future petroleum The sum of the remaining reserves, mean reserve growth, and the mean of the undiscovered volume. Cumulative production does not contribute to the future petroleum. The terms future oil, future liquid volume, or future endowment are sometimes used as variations of future petroleum to reflect those resources that are yet to be produced.

Gabbro A large group of dark, often coarse-grained, mafic intrusive igneous rocks chemically equivalent to plutonic basalt; formed when molten magma is trapped beneath the surface of the Earth and slowly cools into a holocrystalline mass.

Gas A compressible fluid that completely fills any container in which it is confined. Technically, a gas will not condense when it is compressed and cooled, because a gas can exist only above the critical temperature for its particular composition. Below the critical temperature, this form of matter is known as a vapor, because liquid can exist, and condensation can occur; the terms "gas" and "vapor" are often used interchangeable. The latter, however, should be used for those streams in which condensation can occur and that originate from, or are in equilibrium with, a liquid phase; see Vapor.

Gas cap The gas trapped between the liquid petroleum and the impervious cap rock of the petroleum reservoir; see Associated gas and Reservoir.

Gas cap A free-gas phase overlying an oil zone and occurring within the same producing formation as the oil.

Gas-cap drive Drive energy supplied naturally (as a reservoir is produced) by the expansion of the gas cap. In such a drive, the gas cap expands to force oil into the well and to the surface.

Gas condensate A liquid condensed from natural gas that contain a significant amount of C_{5+} components which exhibits the phenomenon of retrograde condensation at reservoir conditions, in other words, as pressure decreases, increasing amounts of liquid condenses in the reservoir (down to about 2000 psia), which results in a significant loss of in situ condensate reserves that may only be partially recovered by revalorization at lower pressures; see Condensate and Gas condensate reservoir.

Gas condensate reservoir A reservoir that exhibits producing gas–oil ratios from 2500 to 50,000 SCF/STB (400 to 10 STB/MMSCF); see Condensate, Gas condensate, and Gas cycling project.

Gas conditioning The removal of objectionable constituents and addition of desirable constituents.

Gas cycling project A project designed to avoid liquid loss from retrograde condensation can usually be justified for fluids with liquid content higher than about 50–100 STB/MMSCF; offshore, the minimum liquid content to justify cycling is about 100 STB/MMSCF; see Condensate, Gas condensate, and Gas condensate reservoir.

Gas deliverability The deliverability of the gas from the storage facility (also referred to as the deliverability rate, withdrawal rate, or withdrawal capacity) is often expressed as a measure of the amount of gas that can be delivered (withdrawn) from a storage facility on a daily basis.

Gas detection analyzer A device used to detect and measure any gas in the drilling mud as it is circulated to the surface.

Gas holder A gas-tight receptacle or container in which gas is stored for future use (1) at approximately constant pressure (low pressure containers) in which case the volume of the container changes and (2) in containers of constant volume (usually high-pressure containers) in which case the quantity of gas molecules stored varies with the pressure.

Gas hydrates Solid, crystalline, wax-like substances composed of water, methane, and usually a small amount of other gases, with the gases being trapped in the interstices of a water–ice lattice; formed beneath permafrost and on the ocean floor under conditions of moderately high pressure and at temperatures near the freezing point of water.

Gas, liquefied petroleum (LPG) A gas containing certain specific hydrocarbons which are gaseous under normal atmospheric conditions but can be liquefied under moderate pressure at normal temperatures. Propane and butane are the principal examples.

Gas, manufactured A gas obtained by destructive distillation of coal, or by the thermal decomposition of oil, or by the reaction of steam passing through a bed of heated coal or coke, or catalyst beds. Examples are coal gases, coke oven gases, producer gas, blast furnace gas, blue (water) gas, and carbureted water gas. Btu content varies widely.

Gas–liquid separator A vertical or horizontal separators in which gas and liquid are separated by means of gravity settling with or without a mist eliminating device.

Gas–oil separator Commonly a closed cylindrical shell, horizontally mounted with inlets at one end, an outlet at the top for removal of gas, and an outlet at the bottom for removal of oil.

Gasoline A volatile, flammable liquid hydrocarbon refined from crude oils and used universally as a fuel for internal combustion, spark-ignition engines.

Gas pipeline A transmission system for natural gas or other gaseous material. The total system comprises pipes and compressors needed to maintain the flowing pressure of the system.

Gasoline plant A plant in which hydrocarbon components common to the gasoline fractions are removed from "wet" natural gas, leaving a "drier" gas.

Gas processing The preparation of gas for consumer use by removal of the nonmethane constituents; synonymous with gas refining; the separation of constituents from natural gas for the purpose of making salable products and also for treating the residue gas to meet required specifications.

Gas refining See Gas processing.

Gas reservoir A geological formation containing a single gaseous phase. When produced, the surface equipment may or may not contain condensed liquid, depending on the temperature, pressure, and composition of the single reservoir phase.

Gas sand A stratum of sand or porous sandstone from which natural gas is obtained.

Gas-to-liquids (GTL) Transformation of natural gas into liquid fuel (Fischer–Tropsch technology).

Gas well A well completed for production of natural gas from one or more gas zones or reservoirs.

Geochemistry Study of the relative and absolute abundances of the elements of the Earth and the physical and chemical processes that have produced their observed distributions.

Geological survey The exploration for natural gas that involves a geological examination of the surface structure of the Earth to determine the areas where there is a high probability that a reservoir exists.

Geologist A scientist who gathers and interprets data pertaining to the rocks of the Earth's crust.

Geology The science of the physical history of the Earth and its life, especially as recorded in the rocks of the crust.

Geophone Equipment used to detect the reflection of seismic waves during a seismic survey.

Geophysical exploration Measurement of the physical properties of the Earth to locate subsurface formations that may contain commercial accumulations of oil, gas, or other minerals; to obtain information for the design of surface structures, or to make other practical applications. The properties most often studied in the oil industry are seismic characteristics, magnetism, and gravity.

Geophysics The physics of the Earth, including meteorology, hydrology, oceanography, seismology, volcanology, magnetism, and radioactivity.

Geothermal Pertaining to heat within the Earth.

Giammarco-Vetrocoke process A process for hydrogen sulfide and/or carbon dioxide removal from natural gas.

Giant field Recoverable reserves >500 million barrels of oil equivalents.

Girdler process Amine (olamine) washing of natural gas to remove acid gases.

GK6 technology This technology improves the ethylene production of the furnaces by 35% compared to the original capacity. This increase in capacity is achieved by replacing the existing coil. Moreover, this technology allows the furnace to operate on a range of feedstocks from naphtha to heavy oils, with high selectivity and long on-stream time.

Global warming An environmental issue that deals with the potential for global climate change due to increased levels of atmospheric greenhouse gases; see Interglacial period.

Graben A block of the Earth's crust that has slid downward between two faults. Compare *horst*.

Gravel island A man-made construction of gravel used as a platform to support drilling rigs and oil and gas production equipment.

Gravity The attraction exerted by the Earth's mass on objects at its surface; the weight of a body.

Gravity survey An exploration method in which an instrument that measures the intensity of the Earth's gravity is passed over the surface or through the water. In places where the instrument detects stronger- or weaker-than-normal gravity forces, a geologic structure containing hydrocarbons may exist.

Greenhouse effect See Global warming.

Greenhouse gases See Global warming.

Groundwater Water that seeps through soil and fills pores of underground rock formations; the source of water in springs and wells.

Guard bed A bed (usually alumina) which serves as a protector of a more expensive bed (e.g., molecular sieve); serves by the act of attrition and may be referred to as an attrition catalyst.

Gusher An oil well that has come in with such great pressure that the oil jets out of the well like a geyser. In reality, a gusher is a blowout and is extremely wasteful of reservoir fluids and drive energy. In the early days of the oil industry, gushers were common, and many times were the only indication that a large reservoir of oil and gas had been struck; see *blowout*.

HCR (Hybrid Catenary Riser) Riser configuration comprised of two flexible sections of flexible pipe (at the top and the bottom) and a rigid section in the middle.

Heating value The number of British thermal units per cubic foot (Btu/ft^3) of natural gas as determined from tests of fuel samples.

Heat of combustion (energy content) The amount of energy that is obtained from burning natural gas; measured in British thermal units (Btu).

HCFC's (Hydrochlorofluorocarbons) Gaseous compounds that meet current environmental standards for minimizing stratospheric ozone depletion.

High flash stocks Liquids having a closed cup flash point of 55°C (131°F) or over (such as heavy fuel oil, lubricating oils, and transformer oils); does not include any stock that may be stored at temperatures above or within 8°C (14.4°F) of the flash point.

Hole In drilling operations, the wellbore or borehole.

Holocrystalline A rock that is completely crystalline.

Horizontal drilling The method which allows producers to extend horizontal shafts into areas that could not otherwise be reached; especially useful in offshore drilling; categorized as short (extending only 20–40 ft from the vertical), medium (300–700 ft from the vertical) or long (1000–4500 ft from vertical) radius.

Horsehead (balanced conventional beam, sucker rod) pump A common type of cable rod lifting equipment for recovery of oil and gas; so-called because of the shape of the counter weight at the end of the beam.

Horst A block of the Earth's crust that has been raised (relatively) between two faults; compare Graben.

Hybrid riser A riser configuration combining both flexible and rigid pipe technologies; see Riser.

Hydraulic fracturing A process through which small fractures are made in impermeable rock by a pressurized combination of water, sand, and chemical additives; the small fractures are held open by a proppant (such as grains of sand) thereby allowing the natural gas to flow out of the rock and into the wellbore.

Hydrocarbon An organic compound composed of hydrogen and carbon only; the density, boiling point, and freezing point increase as the molecular weight increases; the smallest molecules of hydrocarbons are gaseous; the largest are solids. Petroleum is a mixture of many different hydrocarbons.

Hydrocarbon, saturated A chemical compound of carbon and hydrogen in which all the valence bonds of the carbon atoms are taken up with hydrogen atoms.

Hydrocarbon, unsaturated A chemical compound of carbon and hydrogen in which not all the valence bonds of the carbon atoms are taken up with hydrogen atoms.

Hydrogen sulfide (H_2S) A poisonous, corrosive compound consisting of two atoms of hydrogen and one of sulfur, gaseous in its natural state; found in manufactured gas or process gas made from coal or crude oil containing sulfur and must be removed; also found to some extent in some natural gas.

Ice scour The abrasion of material in contact with moving ice in a sea, ocean, or other body of water.

Ideal gas A gas in which all collisions between atoms or molecules are perfectly elastic and in which there are no intermolecular attractive forces.

IFPEXOL process A methanol-based process for water removal and hydrocarbon dew point control; also used for acid gas removal.

Igneous rock A rock mass formed by the solidification of magma within the Earth's crust or on its surface. It is classified by chemical composition and grain size. Granite is an igneous rock.

Ignition, automatic A means which provides for automatic lighting of gas at the burner when the gas valve controlling flow is turned on and will affect relighting if the flame on the burner has been extinguished by means other than closing the gas burner valve.

Ignition, continuous Ignition by an energy source which is continuously maintained through the time the burner is in service, whether the main burner is firing or not.

Ignition, intermittent Ignition by an energy source which is continuously maintained through the time the burner is firing.

Ignition, interrupted Ignition by an energy source which is automatically energized each time the main burner is fired and subsequently is automatically shut off during the firing cycle.

Ignition, manual Ignition by an energy source which is manually energized and where the fuel to the pilot is lighted automatically when the ignition system is energized.

Ignition temperature The temperature at which a substance, such as gas, will ignite and continue burning with adequate air supply.

Ilmenite A weakly magnetic titanium-iron oxide mineral which is iron-black or steel-gray.

Impermeable Preventing the passage of fluid. A formation may be porous yet impermeable if there is an absence of connecting passages between the voids within it; see Permeability.

Impure natural gas Natural gas as delivered from the well and before processing (refining).

Independent producer A nonintegrated company which receives nearly all of its revenues from production at the wellhead; by the IRS definition, a firm is an Independent if the refining capacity is less than 50,000 barrels per day in any given day or their retail sales are less than $5 million for the year.

Indirect vaporizer A vaporizer in which heat furnished by steam, hot water, the ground, surrounding air, or other heating medium is applied to a vaporizing chamber or to tubing, pipe coils, or other heat exchange surface containing the liquid LP-Gas to be vaporized; the heating of the medium used being at a point remote from the vaporizer.

Interglacial period A geological interval of warmer global average temperature lasting thousands of years that separates consecutive glacial periods within an ice age; the current Holocene interglacial began at the end of the Pleistocene, approximately 11,700 years ago; alternatively known as interglacial, interglaciation.

Intermediate precision The within-laboratory variations, such as different days, different analysts, and different equipment; see Precision.

Interruptible service contracts Contracts that allow a distributing party to temporarily suspend delivery of gas to a buyer in order to meet the demands of customers who purchased firm service.

Interstice A pore space in a reservoir rock.

IPB (Integrated Production Bundle) A patented flexible riser assembly combining multiple functions of production and gas lift, which incorporates both active heating and passive insulation. Used for severe flow assurance requirements.

IRM (Inspection, Repair, and Maintenance) Routine inspection and servicing of offshore installations and subsea infrastructures.

Iron oxide process (iron sponge process, dry box method) A process in which the gas is passed through a bed of wood chips impregnated with iron oxide to scavenge hydrogen sulfide and organic sulfur compounds (mercaptans) from natural gas streams.

Jackup drilling rig A mobile bottom-supported offshore drilling structure with columnar or open-truss legs that support the deck and hull. When positioned over the drilling site, the bottoms of the legs penetrate the seafloor. A jackup rig is towed or propelled to a location with its legs up. Once the legs are firmly positioned on the bottom, the deck and hull height are adjusted and leveled. Also called self-elevating drilling unit.

J-Lay A vertical lay system for rigid pipes.

Joule–Thomson effect The cooling which occurs when a compressed gas is allowed to expand in such a way that no external work is done. The effect is approximately 7°F per 100 psi for natural gas.

Joule–Thomson expansion The throttling effect produced when expanding a gas or vapor from a high pressure to a lower pressure with a corresponding drop in temperature.

Jumper Short pipe (flexible or rigid) sometimes used to connect a flow line to a subsea structure or two subsea structures located close to one another.

Kelly The heavy square or hexagonal steel pipe which goes through the rotary table and turns the drill string (also called grief stem).

Kitchen The underground deposit of organic debris that is eventually converted to petroleum and natural gas.

Knock-out pot A separator used for a bulk separation of gas and liquid, particularly when the liquid volume fraction is high.

Landfill gas Gas produced during the in situ maturation of landfill materials which can include municipal waste and industrial waste.

Landman A person in the petroleum industry or natural gas industry who negotiates with landowners for oil and gas leases, options, minerals, and royalties and with producers for joint operations relative to production in a field; also called a leaseman.

Lava Magma that reaches the surface of the Earth.

Lean gas Natural gas in which methane is the major constituent.

Lease condensate Low-boiling (low molecular weight) liquid hydrocarbons recovered from lease separators or field facilities at associated and nonassociated natural gas wells. Mostly pentanes and heavier hydrocarbons. Normally enters the crude oil stream after production.

Lease fuel Natural gas used in well, field, and lease operations, such as gas used in drilling operations, heaters, dehydrators, and field compressors.

Lease operations Any well, lease, or field operations related to the exploration for or production of natural gas prior to delivery for processing or transportation out of the field gas used in lease operations includes usage such as for drilling operations, heaters, dehydrator units, and field compressors used for gas lift.

Lease separation facility A facility installed at the surface for the purpose of (1) separating gases from produced crude oil and water at the temperature and pressure conditions set by the separator and/or (2) separating gases from that portion of the produced natural gas stream that liquefies at the temperature and pressure conditions set by the separator.

Lease separator A facility installed at the surface for the purpose of separating the full well stream volume into two or three parts at the temperature and pressure conditions set by the separator; for gas wells, these parts include produced natural gas, lease condensate, and water.

Limestone A sedimentary rock rich in calcium carbonate that sometimes serves as a reservoir rock for petroleum.

Limit of quantitation (LOQ) The level above which quantitative results may be determined with acceptable accuracy and precision.

Linearity The ability of the method to elicit results that are directly proportional to analyte concentration within a given range.

Liquefied natural gas The liquid form of natural gas; natural gas (primarily methane) that has been liquefied by reducing its temperature to −260°F at atmospheric pressure.

Liquefied petroleum gas (LPG) The term applied to certain specific hydrocarbons and their mixtures, which exist in the gaseous state under atmospheric ambient conditions but can be converted to the liquid state under conditions of moderate pressure at ambient temperature.

Liquids, natural gas Those liquid hydrocarbon mixtures which are gaseous at reservoir temperatures and pressures but are recoverable by condensation or absorption; gas condensate, natural gasoline, and liquefied petroleum gases are often included in this category.

Lithology The study of rocks; important for exploration and drilling crews to understand lithology as it relates to the production of gas and oil.

LNG (Liquefied Natural Gas) Natural gas, liquefied through the reduction of its temperature to $-162°C$ ($-260°F$), thus reducing its volume by 600 times, allowing its transport by LNG tanker.

Local Distribution Company A retail gas distribution company that delivers natural gas to end-users.

LO-CAT process A wet oxidation process.

Log A systematic recording of data, such as a driller's log, mud log, electrical well log, or radioactivity log. Many different logs are run in wells to discern various characteristics of downhole formation.

Logging Lowering of different types of measuring instruments into the wellbore and gathering and recording data on *porosity*, *permeability*, and types of fluids present near the current well after which the data are used to construct subsurface maps of a region to aid in further exploration.

Lower explosive limit The lower percent by volume of the gas vapor in air at which the gas will explode or inflame; see also Upper explosive limit.

Lower flammability limit The minimum concentration by volume of a combustible substance that is capable of propagating a flame under specified conditions.

Low-flash stocks Liquids having a closed cup flash point under $55°C$ ($131°F$) such as gasoline, kerosene, jet fuels, some heating oils, diesel fuels, and any other stock that may be stored at temperatures above or within $8°C$ ($14.4°F$) of the flash point.

Macrofauna Benthic invertebrates that live on or in the sediment and that are retained on a mesh with an aperture of 0.5 mm.

Magma The hot fluid matter within the Earth's crust that is capable of intrusion or extrusion and that produces igneous rock when cooled.

Magnetic survey An exploration method in which an instrument that measures the intensity of the natural magnetic forces existing in the Earth's subsurface is passed over the surface or through the water. The instrumentation detects deviations in magnetic forces, and such deviations may indicate the existence of underground formations that favor the entrapment of hydrocarbons.

Magnetometer A device to measure small changes in the Earth's magnetic field at the surface, which indicates what kind of rock formations might be present underground.

Major field Recoverable reserves >100 million barrels of oil equivalents.

Manufactured gas Gas obtained by destructive distillation of coal or by the thermal decomposition of oil, or by the reaction of steam passing through a bed of heated coal or coke; examples are coal gases, coke oven gases, producer gas, blast furnace gas, blue (water) gas, carbureted water gas; the Btu content varies widely.

Marsh gas See Natural gas.

Mass concentration The concentration expressed in terms of mass of substance per unit volume (g substance/m^3 gas volume).

Mcf (thousand cubic feet) One thousand cubic feet; a unit of measure that is more commonly used in the low volume sectors of the gas industry, such as stripper well production; see also Btu, Bcf, Tcf, Quad.

MDEA Methyldiethanolamine.

MEA Ethanolamine (monoethanolamine).

Median A statistical measure of the midmost value, such that half the values in a set are greater and half are less than the median.

Meiofauna Benthic invertebrates that live in the sediment and that are retained on a mesh with an aperture of 0.062 mm.

MER See Most efficient recovery rate.

Mercaptan Typically, a hydrocarbon group (usually a methane, ethane, or propane) with a hydrosulfur group ($-SH$) substituted on a terminal carbon atom.

Methane A light, gaseous, colorless, and naturally odorless flammable paraffinic hydrocarbon, CH_4, that has a boiling point of $-25°F$ (and is the chief component of natural gas and an important basic hydrocarbon for petrochemical manufacture); commonly (often incorrectly) known as natural gas; burns efficiently without many by-products.

Methane separation Cryogenic processing and absorption methods are some of the ways to separate methane from natural gas liquids. The cryogenic method is better at extraction of the lighter liquids, such as ethane, than is the alternative absorption method and consists of lowering the temperature of the gas stream to around $-120°F$. The absorption method, on the other hand, uses a "lean" absorbing oil to separate the methane from the natural gas liquids. While the gas stream is passed through an absorption tower, the absorption oil soaks up a large amount of the natural gas liquids. The "enriched" absorption oil, now containing NGLs, exits the tower at the bottom. The enriched oil is fed into distillers where the blend is heated to above the boiling point of the natural gas liquids, while the oil remains fluid. The oil is recycled while the natural gas liquids are cooled and directed to a fractionator tower. Another absorption method that is often used is the refrigerated oil absorption method where the lean oil is chilled rather than heated, a feature that enhances recovery rates somewhat.

Methanogens Methane-producing microorganisms.

Method detection limit (MDL) The minimum concentration of a substance than can be measured and reported with 99% confidence that the analyte concentration is greater than zero; determined from analysis of a sample in a given matrix containing the analyte; see Detection limit.

MFO (Mixed function oxygenases) An enzyme system, located within the cells, which can assist in the metabolism and excretion of contaminants.

Microgabbro Gabbro with finer grain crystals (<1 mm).

Migration The movement of oil, gas, or water through porous and permeable rock.

Mineral rights The rights of ownership, conveyed by deed, of gas, oil, and other minerals beneath the surface of the Earth.

Mist extractor A device installed in the top of scrubbers, separators, tray, or packed vessels, etc. to remove liquid droplets entrained in a flowing gas stream.

Mixed gas A fuel gas in which natural or liquefied petroleum gas is mixed with manufactured gas to give a product of better utility and higher heat content or Btu value.

MMBtu A thermal unit of energy equal to 1,000,000 Btu, i.e., the equivalent of 1000 cubic feet of gas having a heating content of 1,000 Btu per cubic foot, as provided by contract measurement terms.

MMcf A million cubic feet.

MMSCFD Million standard cubic feet per day.

Mobile offshore drilling unit A drilling rig that is used exclusively to drill offshore exploration and development wells and that floats upon the surface of the water when being used.

MODU (*plural*: **MODUs**) Mobile offshore drilling unit often used in conjunction with semi-submersibles and floating, production, storage, and offloading units, which do not have drilling rigs.

Moonpool Opening in a vessel or a platform deck through which drilling, subsea pipe-laying, or construction is conducted. Several vessels are equipped with a moonpool allowing the use of the VLS for flexible pipe and Reel-Lay or J-Lay systems.

Most efficient recovery rate (MER) The rate at which the greatest amount of natural gas may be extracted without harming the formation itself.

MSCC (millisecond catalytic cracker) A catalytic cracking unit of FCC type (FCC stands for *fluid catalytic cracking*).

Mud The liquid circulated through the wellbore during rotary drilling and workover operations. In addition to its function of bringing cuttings to the surface, drilling mud cools and lubricates the bit and the drill stem, protects against blowouts by holding back subsurface pressures, and deposits a mud cake on the wall of the borehole to prevent loss of fluids to the formation; originally a suspension of Earth solids (especially clay) in water, the mud used in modern drilling operations is a more complex, three-phase mixture of liquids, reactive solids, and inert solids—the liquid phase may be freshwater, diesel oil, or crude oil and may contain one or more conditioners; see Drilling fluid.

Mud line The sea bed, unless otherwise specified by the driller.

Multiple completions The result of drilling several different depths from a single well to increase the rate of production or the amount of recoverable gas.

Naphtha An arbitrarily defined crude oil fraction containing primarily aliphatic (linear) hydrocarbons with boiling points ranging from 30°C to 250°C (86°F to 482°F); principal uses are for (1) gasoline blend stock, (2) solvents, (3) paint thinners, (4) and as a feedstock for the production of organic chemicals, and (v) as a feedstock for the production of synthetic natural gas.

Native gas Gas in place at the time that a reservoir was converted to use as an underground storage reservoir in contrast to injected gas volumes.

Natural gas Also called *marsh gas*, *swamp gas*, and *landfill gas*; a gaseous fossil fuel that is found in oil fields and natural gas fields, and in coal beds a mixture of hydrocarbon and small quantities of nonhydrocarbons that exists either in the gaseous phase or is in solution in crude oil in natural underground reservoirs, and which is gaseous at atmospheric conditions of pressure and temperature.

Natural Gas Act Passed in 1938 and give the Federal Power Commission (now the Federal Energy Regulatory Commission or FERC) jurisdiction over companies engaged in interstate sale or transportation of natural gas.

Natural gas co-firing The injection of natural gas with pulverized coal or oil into the primary combustion zone of a boiler.

Natural gas hydrates Solid, crystalline, wax-like substances composed of water, methane, and usually a small amount of other gases, with the gases being trapped in the interstices of a water−ice lattice; formed beneath permafrost and on the ocean floor under conditions of moderately high pressure and at temperatures near the freezing point of water.

Natural gas liquids (NGLs) A hydrocarbon liquid stream *gas condensate*; higher molecular weight hydrocarbons that exist in the reservoir as constituents of natural gas, but which are recovered as liquids in separators, field facilities, or gas-processing plants.

Natural gasoline A term used in the gas processing industry to refer to a mixture of liquid hydrocarbons (mostly pentane, including *iso*-pentane, and higher molecular weight hydrocarbons) extracted from natural gas.

Natural gas plant liquids Hydrocarbons in natural gas that are separated as liquids at natural gas processing plants, fractionating and cycling plants, and in some instances, field facilities; the products obtained include liquefied petroleum gases (ethane, propane, and butanes), pentanes plus, and *iso*-pentane.

Natural Gas Policy Act of 1978 One of the first efforts to deregulate the gas industry and to determine the price of natural gas as dictated by market forces, rather than regulation.

Natural gas processing plant A facility designed to recover natural gas liquids from a stream of natural gas that may or may not have passed through lease separators and/or field separation facilities; the facility controls the quality of the natural gas to be marketed.

Natural gas resource base An estimate of the amount of natural gas available, based on the combination of proved reserves, and those additional volumes that have not yet been discovered, but are estimated to be "discoverable" given current technology and economics.

Natural gas shrinkage, natural gas The reduction in volume of wet natural gas due to the extraction of some of its constituents, such as hydrocarbon products, hydrogen sulfide, carbon dioxide, nitrogen, helium, and water vapor.

Natural gas vehicle (NGV) A car, bus, or truck that is powered by a natural gas, either in compressed or liquefied form, rather than the traditional gasoline or diesel fuel.

Nepheline magnetite A sodium potassium aluminosilicate and magnetite—one of the two common naturally occurring iron oxides (chemical formula Fe_3O_4) and a member of the spinel group of minerals.

NES (National Energy Strategy) A 1991 federal proposal that focused on national security, conservation, and regulatory reform, with options that encourage natural gas use.

Nitrogen extraction Once the hydrogen sulfide and carbon dioxide are processed to acceptable levels, the stream is routed to a nitrogen rejection unit, it is routed through a series of passes through a column and a brazed aluminum plate fin heat exchanger. Helium, if any, can be extracted from the gas stream through membrane diffusion in a pressure swing adsorption (PSA) unit.

Nonassociated natural gas Sometimes called *gas well gas*; gas produced from geological formations that typically do not contain much, if any, crude oil, or higher boiling hydrocarbons (*gas liquids*) than methane; can contain nonhydrocarbon gases such as carbon dioxide and hydrogen sulfide.

Noncombustible Material incapable of igniting or supporting combustion.

Nonhydrocarbon gases Typical nonhydrocarbon gases that may be present in reservoir natural gas, such as carbon dioxide, helium, hydrogen sulfide, and nitrogen.

Nonporous Containing no interstices; having no pores and therefore unable to hold fluids.

Nonstandard method A method that is not taken from authoritative and validated sources; includes methods from scientific journals and unpublished laboratory-developed methods.

Norite A coarse-grained basic igneous rock dominated by essential calcic plagioclase and orthopyroxene. Norites also can contain up to 50% clinopyroxene and can be considered orthopyroxene dominated gabbro.

NO_x (nitrogen oxides) Produced during combustion; precursors to acid deposition (acid rain).

NPC (National Petroleum Council) An advisory body of appointed members whose purpose is to advise the Secretary of Energy.

Off peak period The time during a day, week, month, or year when gas use on a particular system is not at its maximum.

Offshore The geographic area that lies seaward of the coastline. In general, the term "coastline" means the line of ordinary low water along that portion of the coast that is in direct contact with the open sea or the line marking the seaward limit of inland waters.

Offshore drilling Drilling for oil or gas in an ocean, gulf, or sea. A drilling unit for offshore operations may be a mobile floating vessel with a ship or barge hull, a semisubmersible or submersible base, a self-propelled or towed structure with jacking legs (jackup drilling rig), or a permanent structure used as a production platform when drilling is completed. In general, wildcat wells are drilled from mobile floating vessels or from jackups, while development wells are drilled from platforms or jackups.

Offshore oil and gas installation Subsea or surface (platform) oil and gas drilling facilities.

Offshore production platform An immobile offshore structure from which wells are produced.

Offshore rig Any of various types of drilling structures designed for use in drilling wells in oceans, seas, bays, gulfs, and so forth. Offshore rigs include platforms, jackup drilling rigs, semisubmersible drilling rigs, submersible drilling rigs, and drill ships.

Oikocrysts Small, randomly orientated, crystals are enclosed within larger crystals of another mineral; the term is most commonly applied to igneous rock textures. The smaller enclosed crystals are known as chadacrysts, while the larger crystals are known as oikocrysts.

Oil A simple or complex liquid mixture of hydrocarbons that can be refined to yield gasoline, kerosene, diesel fuel, and various other products.

Oil field (oilfield) A field producing oil and gas is termed an oil field when the petroleum contained within has a gas—oil ratio (GOR) of less than 20,000 cubic feet per barrel. If the gas-oil ratio >20,000 cubic feet per barrel, the field is called a gas field.

Oil in place Crude oil that is estimated to exist in a reservoir but that has not been produced.

Oil patch (Slang) the oilfield.

Oil pool A loose term for an underground reservoir where oil occurs. Oil is actually found in the pores of rocks, not in a pool.

Oil seep A surface location where oil appears, the oil having permeated its subsurface boundaries and accumulated in small pools or rivulets. Also called oil spring.

Oil shale A shale containing hydrocarbons that cannot be recovered by an ordinary oil well but that can be extracted by mining and processing.

Oil slick A film of oil floating on water; considered a pollutant.

Oil spill *n* A quantity of oil that has leaked or fallen onto the ground or onto the surface of a body of water.

Oil well A well from which oil is obtained.

Oil zone A formation or horizon of a well from which oil may be produced. The oil zone is usually immediately under the gas zone and on top of the water zone if all three fluids are present and segregated.

Olamine process A process that used an amine derivative (an olamine) to remove acid gas from natural gas streams.

Olamines Compounds such as ethanolamine (monoethanolamine, MEA), diethanolamine (DEA), triethanolamine (TEA), methyldiethanolamine (MDEA), diisopropanolamine (DIPA), and diglycolamine (DGA) that are widely used in gas processing.

Olefins Family of molecules including in particular ethylene and propylene, which constitutes the raw material allowing for the manufacture of many plastics.

Olivine A name for a series between two end members, fayalite and forsterite; fayalite is the iron rich member (Fe_2SiO_4), whereas forsterite is the magnesium-rich member (Mg_2SiO_4).

Onshore oil and gas installation Onshore oil and gas exploration/production.

Order 636 The Federal Energy Regulatory Commission's 1992 order that required pipelines to unbundle their transportation, sales, and storage services.

Organic compounds Chemical compounds that contain carbon atoms, either in straight chains or in rings, and hydrogen atoms. They may also contain oxygen, nitrogen, or other atoms.

Organic rock Rock materials produced by plant or animal life (coal, petroleum, limestone, and so on).

Orthopyroxene An essential constituent of various types of igneous rocks and metamorphic rocks.

Outcrop Part of a formation exposed at the Earth's surface. *V*: to appear on the Earth's surface (as a rock).

PAH Polycyclic aromatic hydrocarbons, which contain more than one fused benzene ring; see PNA.

Paraffin hydrocarbons Saturated hydrocarbon compounds with the general formula C_nH_{2n+2} containing only single bonds and carbon and hydrogen only; sometimes referred to as alkanes or natural gas liquids.

Part concentration The concentration expressed as number of particles of substance per a certain number of particles.

Peak load requirements (peak load storage) The design to have high-deliverability for short periods of time during which the natural gas can be withdrawn from storage quickly as the need arises.

Peak shaving Use of natural gas from storage to supplement the normal amounts delivered to customers during peak-use periods.

Peak use period The period of time when gas use on a particular system is at its maximum.

Peat An organic material that forms by the partial decomposition and disintegration of vegetation in tropical swamps and other wet, humid areas. It is believed to be the precursor of coal.

Pegmatite A holocrystalline, intrusive igneous rock composed of interlocking phaneritic crystals usually larger than 2.5 cm in size; such rocks are referred to as *pegmatitic*.

Permeability A measure of the ease that a fluid can pass through a section of rock through the connecting pore spaces of rock or cement—the unit of measurement is the millidarcy; fluid conductivity of a porous medium; the ability of a fluid to flow within the interconnected pore network of a porous medium.

Permeable rock A porous rock formation in which the individual pore spaces are connected, allowing fluids to flow through the formation.

Petroleum Crude oil; a flammable naturally occurring liquid that may vary from almost colorless to black, occurs in many places in the upper strata of the Earth; a complex mixture of hydrocarbons with small amounts of other substances, and is prepared for use as naphtha, gasoline, and other products by various refining processes.

Petroleum geology The study of oil- and gas-bearing rock formations. It deals with the origin, occurrence, movement, and accumulation of hydrocarbon fuels.

Petroleum reservoir A rock formation that holds oil and gas.

Petroleum rock Sandstone, limestone, dolomite, fractured shale, and other porous rock formations where accumulations of oil and gas may be found.

Petroleum window The conditions of temperature and pressure under which petroleum will form; also called oil window.

pH A measure of the acidity or alkalinity of a solution.

PIGs Robotic agents used to inspect pipeline interior walls for corrosion and defects, measure pipeline interior diameters, remove accumulated debris and for other specialty tasks.

Pip (pipe-in-pipe) Steel pipes assembly consisting of a standard production pipe surrounded by a so-called carrier pipe. The gap between the carrier and production pipes is filled with an insulation material. As the insulation is protected from the external pressure by the carrier pipe, a high thermal performance material can be used.

Pipeline (natural gas) A continuous pipe conduit, complete with such equipment as valves, compressor stations, communications systems, and meters for transporting natural and/or supplemental gas from one point to another, usually from a point in or beyond the producing field or processing plant to another pipeline or to points of utilization. Also refers to a company operating such facilities.

Plagioclase A large group of dark, often phaneritic (coarse-grained), mafic intrusive igneous rocks, chemically equivalent to plutonic basalt; formed when molten magma is trapped beneath the surface of the Earth and slowly cools into a holocrystalline mass.

Plate tectonics Movement of great crustal plates of the Earth on slow currents in the *plastic* mantle.

Platform The structure that supports production and drilling operations. The types of offshore platforms can be either floating or fixed, depending on the location, water depth, climate, and the facility's size.

Play The extent of a petroleum-bearing formation; the activities associated with petroleum development in an area.

Plutonic rock An intrusive igneous rock that is crystallized from magma slowly cooling below the surface of the Earth.

PNA Polynuclear aromatic hydrocarbons, which contain more than one fused benzene ring; see PAH.

Polypropylene Due to its exceptional shock resistance properties, polypropylene is a plastic material used in a wide range of industries including automobile parts, household goods, fibers, and films.

Pool A reservoir or group of reservoirs. The term is a misnomer in that hydrocarbons seldom exist in pools, but, rather, in the pores of rock. *V*: to combine small or irregular tracts into a unit large enough to meet state spacing regulations for drilling.

Porosity The condition of being porous (such as a rock formation); also, the ratio of the volume of empty space to the volume of solid rock in a formation, indicating how much fluid rock can hold; the spaces between grains of sediment in sedimentary rock.

Potential The maximum volume of oil or gas that a well is capable of producing, calculated from well test data.

Precision The agreement between a set of replicate measurements without assumption of knowledge of the true value; see Intermediate precision, Repeatability, and Reproducibility.

Pressure, absolute (psia) Pressure in excess of a perfect vacuum; absolute pressure is obtained by algebraically adding gauge pressure to atmosphere pressure. Pressures reported in "atmospheres" are understood to be absolute. Absolute pressure must be used in equations of state and in all gas-law calculations. Gauge pressures below atmospheric pressure are called "vacuum."

Producer The company generally involved in exploration, drilling, and refining of natural gas.

Producer gas A mixture of flammable gases (principally carbon monoxide and hydrogen) and nonflammable gases (mainly nitrogen and carbon dioxide) made by the partial combustion of carbonaceous substances, usually coal, in an atmosphere of air and steam; producer gas has a lower heating value than other gaseous fuels, but it can be manufactured with relatively simple equipment; it is used mainly as a fuel in large industrial furnaces.

Producing sand A rock stratum that contains recoverable oil or gas.

Producing zone The interval of rock actually producing oil or gas.

Production rate The rate of production of oil and/or gas from a well; usually given in barrels per day (bbls/day) for oil or standard cubic feet (scft^3/day) for gas.

Propping agents Sand, glass beads, epoxy, or silica sand that serve to prop open the newly widened fissures in the formation.

Proved reserves of crude oil According to API standard definitions, proved reserves of crude oil as of December 31 of any given year are the estimated quantities of all liquids statistically defined as crude oil that geological and engineering data demonstrate with reasonable certainty to be recoverable in future years from known reservoirs under existing economic and operating conditions.

Proved resources Part of the resource base that includes the working inventory of natural gas; volumes that have already been discovered and are readily available for production and delivery.

Proved amount in place The volume originally occurring in known natural reservoirs which has been carefully measured and assessed as exploitable under present and expected local economic conditions with existing available technology.

Proved recoverable reserves The volume within the proved amount in place that can be recovered in the future under present and expected local economic conditions with existing available technology.

PSI (pounds per square inch) Pressure measured with respect to that of the atmosphere.

PUHCA (Public Utility Holding Company Act of 1935) Amended by EPACT, to allow power generation by independent power producers (IPPs) without restrictions on corporate structure.

Pyroxene A group of rock-forming inosilicate minerals found in many igneous and metamorphic rocks; variable composition, among which calcium-, magnesium-, and iron-rich varieties predominate.

Quad An abbreviation for a quadrillion (1,000,000,000,000,000) Btu; roughly equivalent to one trillion (1,000,000,000,000) cubic feet, or 1 Tcf; see also Bcf, Mcf, and Tcf.

Quartzgabbro A coarse-grained igneous rock dominated by plagioclase (50% v/v), orthopyroxene (15%), clinopyroxene (25%), and quartz (<5%) with minor biotite and accessory apatite.

Quenching Cooling of a hot vapor by mixing it with another fluid or by partially vaporizing another liquid.

Range The interval between the upper and lower concentration of analyte in sample for which it has been demonstrated that the analytical procedure has an acceptable level of accuracy, precision, and linearity.

Raw natural gas Impure natural gas as delivered from the well and before processing (refining).

Recovery rate The rate at which natural gas can be removed from a reservoir.

Rectisol process A process that used a physical (nonreactive) solvent for gas cleaning.

Redox process A sulfur recovery process that involves liquid-phase oxidation; uses a dilute aqueous solution of iron or vanadium to remove hydrogen sulfide selectively by chemical absorption from sour gas streams; can be used on relatively small or dilute hydrogen sulfide stream to recover sulfur from the acid gas stream or, in some cases, they can be used in place of an acid gas removal process.

Reeled pipe An installation method based on the onshore assembly of long sections of rigid steel pipeline, approximately 0.5-mile-long, which are welded together as they are spooled onto a vessel-mounted reel for transit and subsequent cost-effective unreeling onto the seabed. Minimum welding is done at sea.

Refinery gas Noncondensable gas collected in petroleum refineries.

Reid vapor pressure (RVP) The pressure of the vapor in equilibrium with liquid at 37.8°C (100°F).

Remaining reserves Recoverable volumes of crude oil and natural gas liquids that were originally present and have not yet been produced.

Repeatability The precision under the same operating conditions over a short period of time; see Precision.

Reproducibility The precision between laboratories; see Precision.

Reserve growth (field growth) The increases of estimated petroleum volume that commonly occur as oil and gas fields are developed and produced.

Reserves The amount of a resource available for recovery and/or production; the recoverable amount is usually tied to economic aspects of production.

Reservoir A subsurface, porous, permeable, or naturally fractured rock body in which oil or gas are stored. Most reservoir rocks are limestone, dolomites, sandstones, or a combination of these. The four basic types of hydrocarbon reservoirs are oil, volatile oil, dry gas, and gas condensate. An oil reservoir generally contains three fluids—gas, oil, and water—with oil the dominant product. In the typical oil reservoir, these fluids become vertically segregated because of their different densities. Gas, the lightest, occupies the upper part of the reservoir rocks; water, the lower part; and oil, the intermediate section. In addition to its occurrence as a cap or in solution, gas may accumulate independently of the oil; if so, the reservoir is called as gas reservoir. Associated with the gas, in most instances, are salt water and some oil. Volatile oil reservoirs are exceptional in that during early production they are mostly productive of light oil plus gas, but, as depletion occurs, production can become almost totally completely gas. Volatile oils are usually good candidates for pressure maintenance, which can result in increased reserves. In the typical dry gas reservoir natural gas exists only as a gas and production is only gas plus freshwater that condenses from the flow stream reservoir. In a gas condensate reservoir, the hydrocarbons may exist as a gas, but, when brought to the surface, some of the heavier hydrocarbons condense and become a liquid.

Reservoir rock A permeable rock that may contain oil or gas in appreciable quantity and through which petroleum may migrate.

Residue gas Natural gas from which the higher molecular weight hydrocarbons have been extracted; mostly methane.

Resource A concentration of naturally occurring solid, liquid, or gaseous hydrocarbons in or on the crust of the Earth, some of which is currently or potentially economically extractable.

Retrograde dew point pressure At a given temperature, the pressure when a gas condenses, and the pressure drops below the dew point.

Rich gas A gaseous stream is traditionally very rich in natural gas liquids (NGLs); see Natural gas liquids.

Reburning An effective and economic means of reducing NO_x emissions from all types of industrial and electric utility boilers.

Reproducibility (of an analytical method) A measure of the repeatability of the data using an analytical method.

Reserve additions Volumes of the resource base that are continuously moved from the resource category to the proved resources category.

Reservoir A geological formation that retains or *traps* the gas.

Reservoir characterization Determination of the physical properties of a reservoir (such as porosity, permeability, and fluid saturation) and changes in the distribution of these properties throughout the reservoir.

Reservoir energy The underground pressure in a reservoir that will push the petroleum and natural gas up the wellbore to the surface.

Reservoir water The water found in petroleum reservoirs is usually a *brine* consisting mostly of sodium chloride (NaCl) in quantities from 10 to 350 ppt (‰); seawater has about 35 ppt. Other compounds (electrolytes) found in reservoir brines include calcium (Ca), magnesium (Mg), sulfate (SO_4), bicarbonate (HCO_3), iodide (I), and bromide (Br); sometimes referred as *brine* or *connate water*.

Resources Concentrations of naturally occurring liquid or gaseous hydrocarbons in the Earth's crust, some part of which are currently or potentially economically extractable.

Rig The drilling equipment used to drill the well that can either be installed on a platform or a MODU.

Riser A pipe or assembly of pipes used to transfer produced fluids from the seabed to the surface facilities or to transfer injection fluids, control fluids or lift gas from the surface facilities and the seabed.

Risk analysis The activity of assigning probabilities to the expected outcomes of drilling venture.

Rod pumping The use of a lifting (pumping) method to recover oil from a reservoir.

Rotary bit The cutting tool attached to the lower end of the drill pipe of a rotary drilling rig; the bit does the actual drilling of the hole through the formation.

Rotary drilling A method for drilling wells using a cutting bit attached to a revolving drill pipe.

ROV (remotely operated vehicle) An unmanned subsea vehicle remotely controlled from a vessel or an offshore platform. It is equipped with manipulator arms that enable it to perform simple operations.

R/P (reserves/production) ratio Calculated by dividing proved recoverable reserves by production (gross less reinjected) in a given year.

RSCR (reeled steel catenary riser) Installation of the SCR by the reel-lay method which, compared to conventional installation solutions, allows most of the welding to be performed onshore in a controlled environment, thereby reducing offshore welding which brings many benefits, particularly for fatigue sensitive components of the pipeline.

Ruggedness or robustness A measure of an analytical procedure's capacity to remain unaffected by small, but deliberate variations in method parameters and provides an indication of its reliability during normal usage.

Salt caverns Caverns formed out of existing salt deposits.

Sandstone A sedimentary rock composed of individual mineral grains of rock fragments between 0.06 and 2 mm (0.002 and 0.079 in.) in diameter and cemented together by silica, calcite, iron oxide, and so forth. Sandstone is commonly porous and permeable and therefore a likely type of rock in which to find a petroleum reservoir.

Satellite well Usually a single well drilled offshore by a mobile offshore drilling unit to produce hydrocarbons from the outer fringes of a reservoir.

Saturated condition A condition where an oil and gas are in thermodynamic equilibrium, i.e., the chemical force exerted by each component in the oil phase is equal to the chemical force exerted by the same component in the gas phase, thereby eliminating mass transfer of components from one phase to the other; see Undersaturated condition.

Saturation pressure An oil at its bubble point pressure or a gas at its dew point pressure.

SCADA (supervisory control and data acquisition) Remote controlled equipment used by pipelines and LDCs to operate their gas systems.

SCOT (Shell Claus Off-gas Treating) process A tail gas treating process.

SCR (steel catenary riser) A deepwater steel riser (see Riser) suspended in a single catenary from a platform (typically a floater) and connected horizontally on the seabed.

Scrubber A unit that has been designed for the removal of contaminates from a gas stream.

Seafloor The bottom of the ocean; the seabed.

Sedimentary rock A rock composed of materials that were transported to their present position by wind or water. Sandstone, shale, and limestone are sedimentary rocks.

Seep The surface appearance of oil or gas that results naturally when a reservoir rock becomes exposed to the surface, thus allowing oil or gas to flow out of fissures in the rock.

Seismic Of or relating to an earthquake or earth vibration, including those artificially induced.

Seismic data Detailed information obtained from earth vibration produced naturally or artificially (as in geophysical prospecting).

Seismic method A method of geophysical prospecting using the generation, reflection, refraction detection, and analysis of sound waves in the Earth.

Seismic survey An exploration method in which strong low-frequency sound waves are generated on the surface or in the water to find subsurface rock structures that may contain hydrocarbons. The sound waves travel through the layers of the Earth's crust; however, at formation boundaries some of the waves are reflected back to the surface where sensitive detectors pick them up. Reflections from shallow formations arrive at the surface sooner than reflections from deep formations, and since the reflections are recorded, a record of the depth and configuration of the various formations can be generated. Interpretation of the record can reveal possible hydrocarbon-bearing formations.

Seismic wave The record of an Earth tremor by a seismograph.

Seismograph An instrument used to detect and record earthquakes, is able to pick up and record the vibrations of the Earth that occur during an earthquake; when seismology is applied to the search for natural gas, seismic waves, emitted from a source, are sent into the Earth and the seismic waves interact differently with the underground formation (underground layers), each with its own properties.

Seismology The study of the movement of energy, in the form of seismic waves, through the Earth's crust.

Selexol process A process that used a physical (nonreactive) solvent for gas cleaning.

Semisubmersible drilling rig A floating offshore drilling unit that has pontoons and columns that, when flooded, cause the unit to submerge to a predetermined depth. Living quarters, storage space, and so forth are assembled on the deck. Semisubmersible rigs are self-propelled or towed to a drilling site and anchored or dynamically positioned over the site, or both. In shallow water, some semisubmersibles can be ballasted to rest on the seabed. Semi-submersibles are more stable than drill ships and ship-shaped barges and are used extensively to drill wildcat wells in rough waters such as the North Sea. Two types of semisubmersible rigs are the bottle-type and the column-stabilized.

Separator tank Tanks are usually located at the well site to separate oil, gas, and water before sending each off to be processed at different locations.

Sewage gas A gas produced by the fermentation of sewage sludge low in heating value due to dilution with carbon dioxide, and nitrogen.

Shale A fine-grained, sedimentary rock composed of clay minerals and tiny fragments (silt-sized particles) of other materials.

Shale basin An underground rock formation that serves both as a natural gas generator and a natural gas reservoir.

Shale gas Gas that occurs in low permeability shale; see Unconventional gas.

Shale oil A liquid produced by the thermal decomposition of the kerogen component of oil shale.

Shallow gas Natural gas deposit located near enough to the surface that a conductor or surface hole will penetrate the gas-bearing formations. Shallow gas is potentially dangerous because, if encountered while drilling, the well usually cannot be shut in to control it. Instead, the flow of gas must be diverted.

Shift converter A reactor in which carbon monoxide and water are catalytically converted to hydrogen and carbon dioxide.

Show The appearance of oil or gas in cuttings, samples, or cores from a drilling well.

Shrinkage, natural gas The reduction in volume of wet natural gas due to the extraction of some of its constituents, such as hydrocarbon products, hydrogen sulfide, carbon dioxide, nitrogen, helium, and water vapor.

Slickwater Slickwater fracturing is a method or system of hydraulic fracturing that involves adding chemicals to water to reduce friction and increase the fluid flow; slickwater increases the speed at which the pressurized fluid can be pumped into the wellbore.

Slurry process A process that uses a slurry of iron oxide to selectively absorb hydrogen sulfide.

SO$_2$ (sulfur dioxide) A precursor to acid deposition (acid rain); produced when sulfur is combusted to sulfur dioxide.

Solution gas−oil ratio (GOR) The amount of surface gas that can be dissolved in a stock tank oil when brought to a specific pressure and temperature; denoted mathematically as R_s (SCF/STB).

Solution oil−gas ratio (OGR) The amount of surface condensate that can be vaporized in a surface gas at a specific pressure and temperature; sometimes referred to as liquid content. Denoted mathematically as r_s (STB/MMSCF).

Sour gas Natural gas that contains hydrogen sulfide and/or other acid gases.

SPAR Deep draft surface piercing cylinder type of floater, particularly well adapted to deepwater, which accommodates drilling, top tensioned risers, and dry completions.

Specifications A feedstock specification or product specification is the data that give adequate control of feedstock behavior in a refinery or product quality; the specifications are derived from the set of tests and data limits applicable (in the context of this book) to the natural gas or to a finished product in order to ensure that every batch is of satisfactory and consistent quality at release for sales; the specifications should include all critical parameters in which variations would be likely to affect the safety and in-service use of the product.

Specific gravity (API) A common measure of oil specific gravity, defined by $\gamma_{API} = (141.5/\gamma_o) - 131.5$, with units in °API (degrees API).

Specific gravity (gas) The ratio of density of any gas at standard conditions (14.7 psia and 60°F) to the density of air at standard conditions; based on the ideal gas law ($pV = nRT$), gas gravity is also equal to the gas molecular weight divided by air molecular weight ($M_{air} = 28.97$); denoted mathematically as γ_g (where air $= 1$)

Specific gravity (liquid) The ratio of density of any liquid measured at standard conditions (usually 14.7 psia and 60°F) to the density of pure water at the same standard conditions; denoted mathematically as γ_o (where water $= 1$).

Specificity The ability to assess unequivocally the analyte in the presence of components which may be expected to be present.

Specific volume The volume of a unit weight of a substance at specific temperature and pressure conditions.

Specific weight Weight per unit volume of a substance.

Spinel group A class of minerals of general formulation $A^{2+}B^{3+}_2O^{2-}_4$ which crystallize in the cubic (isometric) crystal system, with the oxide anions arranged in a cubic close-packed lattice and the cations A and B occupying some or all of the octahedral and tetrahedral sites in the lattice; a and b can be divalent, trivalent, or quadrivalent cations, including magnesium, zinc, iron, manganese, aluminum, chromium, titanium, and silicon.

Spool Short length pipe connecting a subsea pipeline and a riser, or a pipe and a subsea structure.

Spot market A method of contract purchasing whereby commitments by the buyer and seller are of a short duration at a single volume price.

Spot purchase Gas purchased on the spot market, which involves short-term contracts for specified amount of gas, at a one-time purchase price.

Spud To begin operations on a well.

Spud cans Cylindrically shaped steel shoes with pointed ends.

SRB Sulfate reducing bacteria.

Standard condition A temperature of 15°C (59°F) and a pressure of one atmosphere (14.7 psi, 101.325 kPa) which also is known as Standard Temperature and Pressure (STP).

Standard method A method that is traceable to a recognized, validated method.

Storage measures Several volumetric measures are used to quantify the fundamental characteristics of an underground storage facility and the gas contained within it. For some of these measures, it is important to distinguish between the characteristic of a facility, such as it's the storage *capacity*, and the characteristic of the natural gas within the facility such as the actual *inventory level*.

Stratigraphic test A borehole drilled primarily to gather information on rock types and sequence.

Stratigraphic trap A petroleum trap that occurs when the top of the reservoir bed is terminated by other beds or by a change of porosity or permeability within the reservoir itself; see Structural trap.

Stretford process A wet oxidation process.

Stripper Wells Natural gas wells that produce less than 60,000 cubic feet of gas per day.

Structural trap A petroleum trap that is formed because of deformation (such as folding or faulting) of the reservoir formation. Compare *stratigraphic trap*.

Structure A geological formation of interest to drillers. For example, if a particular well is on the edge of a structure, the wellbore has penetrated the reservoir (structure) near its periphery.

Subduction zone A deep trench formed in the ocean floor along the line of convergence of oceanic crust with other oceanic or continental crust when one plate (always oceanic) dives beneath the other. The plate that descends into the hot mantle is partially melted. Magma rises through fissures in the heavier, nonliquid (unmelted) crust above, creating a line of plutons and volcanoes that eventually form an island arc parallel to the trench.

Subsea technology All products and services required to install and operate production installations on the seabed.

SulfaTreat process A batch-type process for the selective removal of hydrogen sulfide and mercaptans from natural gas; the process is dry, using no free liquids, and can be used for natural gas applications where a batch process is suitable.

Sulfinol process A combination process that uses a mixture of amines and a physical solvent (an aqueous amine and sulfolane).

SulFerox process A wet oxidation process.

Super compressibility factor A factor used to account for the following effect: Boyle's law for gases states that the specific weight of a gas is directly proportional to the absolute pressure, the temperature remaining constant. All gases deviate from this law by varying amounts, and within the range of conditions ordinarily encountered in the natural gas industry, the actual specific weight under the higher pressure is usually greater than the theoretical. The factor used to reflect this deviation from the ideal gas law in gas measurement with an orifice meter is called the *super compressibility factor, Fpv.* The factor is used to calculate actual volumes from volumes at standard temperatures and pressures from actual volumes. The factor is of increasing importance at high pressures and low temperatures.

Supergiant field Recoverable reserves >5 billion barrels of oil equivalents.

Supplemental gaseous fuels supplies Synthetic natural gas, propane-air, coke oven gas, refinery gas, biomass gas, air injected for Btu stabilization, and manufactured gas commingled and distributed with natural gas.

SURF Subsea Umbilicals Risers Flowlines.

Swamp gas See Natural gas.

Sweet crude oil Oil containing little or no sulfur, especially little or no hydrogen sulfide.

Sweetening process A process for the removal of hydrogen sulfide and other sulfur compounds from natural gas.

Sweet gas Natural gas that contains very little, if any, hydrogen sulfide.

Syncline A trough-shaped configuration of folded rock layers; see Anticline.

Synthesis gas (syngas) Mixtures of carbon monoxide and hydrogen.

Synthetic natural gas (SNG) A manufactured product, chemically similar in most respects to natural gas, resulting from the conversion or reforming of hydrocarbons that may easily be substituted for or interchanged with pipeline-quality natural gas; also referred to as substitute natural gas.

Tail gas The residue gas left after the completion of a treating process designed to remove certain liquids or liquefiable hydrocarbons.

Tail gas treating Removal of the remaining sulfur compounds from gases remaining after sulfur recovery.

Tar sand A sandstone that contains chiefly heavy, tarlike hydrocarbons. Tar sands are difficult to produce by ordinary methods; thus, it is costly to obtain usable hydrocarbons from them.

Tcf (trillion cubic feet) Gas measurement approximately equal to one quadrillion (1,000, 000,000,000,000) Btu's; see also Bcf, Mcf, and Quad.

TEA Triethanolamine.

Tectonic Of or relating to the deformation of the Earth's crust, the forces involved in or producing such deformation, and the resulting rock forms.

Temperature, dew point The temperature at which a vapor begins to condense and deposit as a liquid.

Terminal velocity (drop-out velocity) The velocity at which a particle or droplet will fall under the action of gravity, when drag force just balance gravitational force and the particle (or droplet) continues to fall at constant velocity.

Teta wire Wire with a specific, patented, T-shape used in flexible pipe to resist the radial effect of the internal pressure. Used for high pressure and harsh environments.

Therm A unit of heating value equivalent to 100,000 British thermal units (Btu).

Thermal decomposition The breakdown of a compound or substance by temperature into simple substances or into constituent elements.

Thermal maturity The amount of heat, in relative terms, to which a rock has been subjected; a thermally immature rock has not been subjected to enough heat to begin the process of converting organic material into oil and/or natural gas; a thermally over-mature rock has been subjected to enough heat to convert organic material to graphite. However, these are the two extremes, and there are many intermediate stages of thermal maturity.

Thermogenic Generated or formed by heat, especially via physiological processes.

Thermogenic coal bed methane Methane formed in coal seams by the action of increasing temperature; see Coal bed methane.

Thermogenic gas Gas formed by pressure effects and temperature effects on organic debris.

Three dimensional (3-D) Seismic Survey Allows producers to see into the crust of the Earth to find promising formations for retrieval of gas.

Threshold limit value (TLV) The amount of a contaminant to which a person can have repeated exposure for an 8-hour day without adverse effects.

Tie-back Connection of a satellite subsea development to an existing infrastructure.

Tight formation A petroleum- or water-bearing formation of relatively low porosity and permeability.

Tight gas Natural gas found trapped in impermeable rock and nonporous sandstone or limestone formations, typically at depths greater than 10,000 ft below the surface.

Tight oil Oil produced from petroleum-bearing formations with low permeability such as the Eagle Ford, the Bakken, and other formations that must be hydraulically fractured to produce oil at commercial rates.

TLP (tension leg platform) A floating production unit anchored to the seabed by taut vertical cables, which considerably restrict its heave motion, making it possible to have the wellheads on the platform.

Total organic carbon (TOC) The concentration of material derived from decaying vegetation, bacterial growth, and metabolic activities of living organisms or chemicals in the source rocks.

Town gas A generic term referring to manufactured gas produced for sale to consumers and municipalities.

Trillion cubic feet A volume measurement of natural gas. Approximately equivalent to one Quad; see also Btu's, Bcf, and Mcf.

Trap A body of permeable oil-bearing rock and/or gas-bearing rock surrounded or overlain by an impermeable barrier that prevents oil from escaping; the types of traps are structural, stratigraphic, or a combination of these.

Ultimate analysis The determination of the elements contained in a compound, i.e., carbon, hydrogen, oxygen, nitrogen, sulfur, and other components.

Ulvospinel A mineral from the spinel group of minerals being iron and titanium oxide of the formula: Fe_2TiO_4 (ortho-titanate iron).

Umbilicals An assembly of hydraulic hoses which can also include electrical cables or optic fibers, used to control subsea structures from a platform or a vessel.

Unassociated gas Natural gas unaccompanied by crude oil when produced; also called non-associated gas or gas well gas.

Unbundled services Unbundling, or separating, pipeline transmission, sales and storage services, along with guaranteeing space on the pipelines for all gas shippers.

Unconventional gas Gas that occurs in tight sandstones, siltstones, sandy carbonates, limestone, dolomite, and chalk; natural gas that cannot be economically produced using current technology; see also Shale gas.

Underground gas storage The use of subsurface facilities for storing gas that has been transferred from its original location for the primary purpose of load balancing; usually natural geological reservoirs, such as depleted oil or gas fields or water-bearing sands on the top by and impermeable cap rock.

Undersaturated condition A condition when an oil or a gas is in a single phase but not at its saturation point (bubble point or dew point), i.e., the mixture is at a pressure greater than its saturation pressure; see Saturated condition.

Undiscovered resources Resources inferred from geologic information and theory to exist outside of known oil and gas fields.

Unsaturated compound Any compound having a double bond or triple bond between two adjacent carbon atoms.

Upper explosive limit The higher percent by volume of the gas vapor in air at which the gas will explode or inflame; see Lower explosive limit,

Upper flammability limit The maximum concentration by volume of a combustible substance that is capable of continued propagation of a flame under the specified conditions.

Vacuum A pressure less than atmospheric pressure, measured either from the base of zero pressure or from the base of atmospheric pressure.

Validation The process of establishing the performance characteristics and limitations of a method and the identification of the influences which may change these characteristics and to what extent.

Vapor When the gas phase of a substance is present under conditions when the substance would normally be a solid or liquid (e.g., below the boiling point of the substance), this is referred to as the vapor phase; see Gas.

Vapor density The density of any gas compared to the density of air with the density of air equal to unity.

Vaporizer A device other than a container which receives gas in liquid form and adds sufficient heat to convert the liquid to a gaseous state.

Vent stack The elevated vertical termination of a disposal system that discharges vapors into the atmosphere without combustion or conversion of the relieved fluid.

Verification The confirmation by examination and provision of objective evidence that specified requirements have been fulfilled.

Viscosity The measure of a fluid's thickness, or how well it flows.

VLS (vertical lay system) A Technip proprietary technology for installation of flexible pipes.

Volatile matter Matter which is readily vaporizable at a relatively low temperature.

Volume, specific The volume of a unit weight of a substance at specific temperature and pressure conditions.

Volume concentration The concentration expressed in terms of gaseous volume of substance per unit volume (cm^3 substance/m^3 gas volume).

Wash The removal of impurities from a gas or vapor by passing the gas through water or other liquid which retains or dissolves the impurity.

Washer A shell with internal baffler or packing, so arranged that gas to be cleaned passes up through the baffles countercurrent to the flow of scrubbing liquid down through the washer. The baffler or packing causes intimate contact and mixing of the gas with the liquid stream.

Washer-cooler A washer in the form of a tall tower in which the washing liquid is sprayed in at top is collected in the bottom of the tower and then is cooled and recycled through the tower. Serves a dual purpose of washing the gas free of impurities and also cooling the gas.

Water-producing interval The portion of an oil or gas reservoir from which water or mainly water is produced.

Weathering The breakdown of large rock masses into smaller pieces by physical and chemical climatological processes; the evaporation of liquid by exposing it to the conditions of atmospheric temperatures and pressure.

Weight, specific Weight per unit volume of a substance.

Well The hole made by the drilling bit, which can be open, cased, or both. Also called borehole, hole, or wellbore.

Wellbore (well bore) The channel created by the drill bit.

Well casing A series of metal tubes installed in the freshly drilled hole; serves to strengthen the sides of the well hole, ensure that no oil or natural gas seeps out of the well hole as it is brought to the surface, and to keep other fluids or gases from seeping into the formation through the well.

Well completion The process for completion of a well to allow for the flow of petroleum or natural gas out of the formation and up to the surface; includes strengthening the well hole with casing, evaluating the pressure and temperature of the formation, and then installing the proper equipment to ensure an efficient flow of natural gas out of the well.

Well control The methods used to control a kick and prevent a well from blowing out. Such techniques include, but are not limited to, keeping the borehole completely filled with drilling mud of the proper weight or density during all operations, exercising reasonable care when tripping pipe out of the hole to prevent swabbing, and keeping careful track of the amount of mud put into the hole to replace the volume of pipe removed from the hole during a grip.

Wellhead (well head) The pieces of equipment mounted at the opening of the well to regulate and monitor the extraction of hydrocarbons from the underground formation; prevents leaking of oil or natural gas out of the well and prevents blowouts due to high pressure formations.

Well intervention (well work) Any operation carried out on a crude oil or natural gas well during or at the end of its productive life, which alters the state of the well and/or well geometry, provides well diagnostics, or manages the production of the well.

Well logging A method used for recording rock and fluid properties to find gas and oil containing zones in subterranean formations; the recording of information about subsurface geologic formations, including records kept by the driller and records of mud and cutting analyses, core analysis, drill stem tests, and electric, acoustic, and radioactivity procedures.

Well servicing Intervention in subsea production wells carried out from a floating rig or a dynamically positioned vessel.

Wet gas Natural gas that contains considerable amounts of higher molecular weight hydrocarbons other than methane.

Wet oxidation process Based on reduction—oxidation (Redox) chemistry to oxidize the hydrogen sulfide to elemental sulfur in an alkaline solution containing an oxygen carrier; vanadium and iron are the two oxygen carriers that are used.

Wildcat A well drilled in an area where no oil or gas production exists.

Wireline A cable technology used by operators of crude oil and natural gas wells to lower equipment or measuring devices into the well for the purposes of well intervention, reservoir evaluation, and pipe recovery.

Wobbe Index (Wobbe Number) A number which indicates interchangeability of fuel gases and is obtained by dividing the heating value of a gas by the square root of its specific gravity.

Working gas The quantity of natural gas in the reservoir (reported in thousand cubic feet at standard temperature and pressure) that is in addition to the cushion or base gas; may or may not be completely withdrawn during any particular withdrawal season; conditions permitting, the total working capacity could be used more than once during any season; the volume of gas in the reservoir above the level of base gas and is, simply, the natural gas is that is available for withdrawal and sales.

Working gas capacity The total gas storage capacity minus base gas.

Yield point The stress at which a material exceeds its elastic limit; below this stress, the material will recover its original size on removal of the stress; above this stress, it will not recover its original size on removal of the stress.

Zero gas Gas at atmospheric pressure.

Zone A geographical area. A geological zone, however, means an interval of strata of the geologic column that has distinguishing characteristics from surrounding strata; also, a space or group of spaces within a building with heating and/or cooling requirements sufficiently similar so that comfort conditions can be maintained by a single controlling device.

Index

Note: Page numbers followed by "*f*" and "*t*" refer to figures and tables, respectively.

B

Base gas, 156–158

Beavon Stretford Reactor (BSR), 317

Benzene, toluene, ethylene benzene, and
     xylenes (BTEX), 309

Benzene (C_6H_6), 15, 102, 196

BioDeNOx process, 300–301

Biogas, 59, 72–74, 73t, 219

Biogenic gas. *See* Biogas

Biogenic methane, 27

Biomass, 72
     gasification, 183

Black water, 119

Blending process, 115

BLM. *See* Bureau of Land Management
     (BLM)

Blowout preventer, 48

Blue gas, 194

Blue water gas process (BWG process),
     197

Blushing, 355

Boiling point, 126
     of natural gas hydrocarbons, 126f

Boltzmann constant (k), 117

Breathing losses, 353

Bright spots, 31–32

British thermal units (Btu), 132

British thermal units (MMBtu), 157

BSR. *See* Beavon Stretford Reactor (BSR)

BTEX. *See* Benzene, toluene, ethylene
     benzene, and xylenes (BTEX)

Btu. *See* British thermal units (Btu)

BTX, 102

Bubble point pressure, 44t

Bullet perforators, 52

Bureau of Land Management (BLM),
     388–389

Burning spring, 8

Burning velocity. *See* Rate of flame
     propagation

Butadiene, 93, 112–113

Butane (C_4H_{10}), 15, 102, 108, 111, 165,
     336–337, 370
     splitter or deisobutanizer, 254

Butylene concentrates, 92–93

Butylene isomers, 172

BWG process. *See* Blue water gas process
     (BWG process)

C

C_{5+} (condensate) fraction, 222

C_8 to C_{12} hydrocarbon derivatives, 40–41

Calcium carbonate, 209

Calcium sulfite ($CaSO_3$), 209

Capillary action, 114–115

Carbon
     isotopic composition, 108
     number of natural gas hydrocarbons, 126f

Carbon dioxide (CO_2), 13, 28, 38, 42,
     99–103, 109, 119, 150, 153–154,
     168–169, 189–190, 203, 225, 362,
     367–368, 370
     replacement, 66

Carbon disulfide (CS_2), 201, 203, 281

Carbon monoxide (CO), 59, 77, 91, 99–100,
     196–197, 203, 368–369

Carbonate reservoirs, 41
     porosity, 38

Carbonate washing processes, 262–263,
     304–305

Carbonization, 192

Carbonyl sulfide (COS), 102, 174–175, 201,
     220, 223–225, 277, 379

Carbureted water gas process, 86, 197

Casing head, 53–54

Casinghead gas, 16, 104

Catacarb process, 307

Catalytic cracking processes, 88–89

Catalytic oxidation, 266

Catalytic reformer tail gases, 89

Caustic scrubbing, 303–304

Caustic soda, 210

CBM. *See* Coal bed methane (CBM)

Centipoise (cP), 141, 353

CFCs. *See* Chlorofluorocarbons (CFCs)

Chemical
     composition, 339–343
          increase in number of isomers with
               carbon number, 341t
          product types and distillation range, 340t
     compounds, 65
     conversion processes, 262
     inhibition, 64, 66
     inhibitor injections, 66
     processes, 277
     properties of natural gas, 109–113
     solvent processes, 242, 259

Printed in the United States
By Bookmasters